Optimizing Plant Health under Abiotic-Stress Environments

Abiotic stresses negatively affect crop yields, and crop health is essential to ensuring global food security. There exists an urgent need to develop more abiotic stress-resistant and -tolerant crop varieties to ensure food and nutritional security. *Optimizing Plant Health under Abiotic-Stress Environments: Ensuring Food and Nutritional Security* presents a comprehensive review of important mineral nutrients in plants and effective management strategies under abiotic stress environments. The optimization of significant traits for abiotic stress tolerance creates effective breeding programs contributing to the development of next-generation crop varieties.

FEATURES:

- Discusses plant breeding and using biotechnological approaches to address the issue of food and nutritional security
- Presents information on how mineral nutrient homeostasis influences plant growth, nutrient enhancement, and crop yield under abiotic stresses
- Contains information on biofortification, QTL mapping, Genome-Wide Association Studies (GWAS), and genetic analysis for improving the nutritional value of plants
- Covers agronomical, breeding, physiological, molecular, and biotechnological approaches to promoting yields and enhanced nutrient availability for crop improvement

Optimizing Plant Health under Abiotic-Stress Environments: Ensuring Food and Nutritional Security is ideal for researchers and scientists looking for guidance on ways to improve agricultural productivity and useful for plant breeders, agriculturalists, and students.

Optimizing Plant Health under Abiotic-Stress Environments

Ensuring Food and Nutritional Security

Edited by
M. Iqbal R. Khan, Faroza Nazir, and Kadambot H. M. Siddique

CRC Press
Taylor & Francis Group
Boca Raton London New York

CRC Press is an imprint of the
Taylor & Francis Group, an **informa** business

Designed cover image: Shutterstock

First edition published 2026
by CRC Press
2385 NW Executive Center Drive, Suite 320, Boca Raton FL 33431

and by CRC Press
4 Park Square, Milton Park, Abingdon, Oxon, OX14 4RN

CRC Press is an imprint of Taylor & Francis Group, LLC

ISBN: 978-1-032-61955-2 (hbk)
ISBN: 978-1-032-63037-3 (pbk)
ISBN: 978-1-032-63038-0 (ebk)

DOI: 10.1201/9781032630380

Typeset in Times
by Apex CoVantage, LLC

Contents

About the Editors

M. Iqbal R. Khan is Assistant Professor of Botany at Jamia Hamdard, New Delhi, India. Dr. Khan's current research interests are elucidating physiological and molecular mechanisms associated with abiotic stress tolerance. Working on the metabolism of plants under different abiotic stresses, Dr. Khan has found a significant role of phytohormones in the regulation of plant growth and development and has suggested that phytohormones play an important in controlling stress responses and interact in coordination with each other for defense signal networking to fine tune tolerance mechanisms. He is also exploring the regulatory role of signalling molecules and their impact on nutrient homeostasis and source-sink relationship under abiotic stresses. Dr. Khan has published more than 100 articles and has edited nine books.

He has been recognized as a Clarivate Highly Cited Researcher in 2022 and 2023. He is also a member of the National Academy of Sciences, India (NASI), and a Young Associate of the Indian National Science Academy (INSA). He has received the R. H. Dastur Gold Medal from the Indian Society for Plant Physiology (ISPP), the Young Scientist Platinum Jubilee Award, and the Young Scientist Award from ISPP.

Faroza Nazir is currently working as Post-Doctoral Fellow (DBT-RA) in Jamia Hamdard, New Delhi, India. Dr. Nazir's research focuses on abiotic stress tolerance mechanisms to decipher the nutritional and hormonal signaling pathways for developing resistance in plants. She has found a significant role of phytohormones and signalling molecules in the regulation of plant growth and development, and abiotic stress tolerance, during her PhD tenure. Besides these, she has authored several international publications in peer-reviewed journals and is manuscript reviewer for several plant biology journals.

Kadambot H. M. Siddique is Hackett Professor of Agriculture Chair and Director of the UWA Institute of Agriculture. Professor Siddique is recognized internationally as a leader in crop science and agriculture. He has more than 30 years' experience in agricultural research, research training, and management in Australia and overseas. He has developed a national and international reputation in agricultural science, especially in the fields of crop physiology, production agronomy, farming systems, genetic resources, breeding research in cereal, grain, and pasture legumes, and oilseed crops. Professor Siddique's publications are considered key papers in the aforementioned fields and are widely cited. These publications include numerous highly cited papers on adaptation, physiology, and genetics of crops to dryland environments. Professor Siddique has conducted research on adaptation of crops to water deficits and the phenological, morphological, physiological, biochemical, and genetic traits that enable crops to cope with various abiotic stresses. As a result of Professor Siddique's personal research and with others with whom he collaborates, Australia has become one of the major grain legume exporting nations in the world. His pioneering research on chickpea has contributed enormously to the Australian chickpea industry, which is currently valued at more than $300 million per annum.

Contributors

Mohd Ali
Department of Zoology
Guru Nanak Dev University
Amritsar, Punjab, India

Ghazala Ambrin
Department of Botany
Aligarh Muslim University
Aligarh, India
and
Molecular Biology and Genetics Unit
Jawaharlal Nehru Centre for Advanced Scientific Research
Bangalore, India

Mohd Anas
Department of Botany
Aligarh Muslim University
Aligarh, India

Rizwan A. Ansari
Department of Plant Protection
Aligarh Muslim University
Aligarh, (U.P.), India

Palak Bakshi
Department of Botanical and Environmental Sciences
Guru Nanak Dev University
Amritsar, Punjab, India
and
Department of Botany
Baba Ghulam Shah Badshah University
Rajouri, Jammu and Kashmir, India

Renu Bhardwaj
Department of Botanical and Environmental Sciences
Guru Nanak Dev University
Amritsar, Punjab, India

Tamanna Bhardwaj
Department of Botanical and Environmental Sciences
Guru Nanak Dev University
Amritsar, Punjab, India

Urooj Hassan Bhat
Department of Botany
Aligarh Muslim University
Aligarh, India

Aditi Bisht
Department of Biosciences
Chandigarh University
Chandigarh, India

Rohit Chhabra
Department of Botany and Environmental Study
DAV University
Jalandhar, India

Shreya Chhabra
Department of Botany
Jamia Hamdard
New Delhi, India

Aman Sobia Chishti
Department of Botany
Aligarh Muslim University
Aligarh, India

Shalini Dhiman
Department of Botanical and Environmental Sciences
Guru Nanak Dev University
Amritsar, Punjab, India

Rashke Eram
Department of Biotechnology
Vinoba Bhave University
Hazaribagh, India
and
Indian Institute of Agricultural Biotechnology
Ranchi, India

Qazi Fariduddin
Department of Botany
Aligarh Muslim University
Aligarh, India

Noor Fatima
Department of Botany
Aligarh Muslim University
Aligarh, (U.P.), India

Hyda Haroon
Department of Botany
Jamia Hamdard
New Delhi, India

Mudassara Hasan
Department of Botany
Aligarh Muslim University
Aligarh, Uttar Pradesh, India

Bisma Hilal
Department of Botany
Aligarh Muslim University
Aligarh, India

Nitika Kapoor
Department of Botanical and Environmental Sciences
Guru Nanak Dev University
Amritsar, Punjab, India

Pravneet Kaur
Department of Botany
Jamia Hamdard
New Delhi, India

M. Masroor A. Khan
Department of Botany
Aligarh Muslim University
Aligarh, India

Sayeda Khatoon
Department of Botany
Jamia Hamdard
New Delhi, India

Deepak Kumar
Department of Zoology
Guru Nanak Dev University
Amritsar, Punjab, India

Sarika Kumari
Department of Botany
Jamia Hamdard
New Delhi, India

Isha Madaan
Department of Botany
Punjabi University
Patiala, Punjab, India
and
Government College of Education
Jalandhar, Punjab, India

Moksh Mahajan
Department of Botany
Jamia Hamdard
New Delhi, India

Chirag Maheshwari
Division of Biochemistry
ICAR-Indian Institute of Agricultural Research
New Delhi, India

Irshad Mahmood
Department of Botany
Aligarh Muslim University
Aligarh, (U.P.), India

Firoz Mohammad
Department of Botany
Aligarh Muslim University
Aligarh, Uttar Pradesh, India

Moin Uddin
Department of Botany
Women's College, Aligarh Muslim University
Aligarh, India

Fauzia Naushin
Department of Botany
Women's College, Aligarh Muslim University
Aligarh, Uttar Pradesh, India

Puja Ohri
Department of Zoology
Guru Nanak Dev University
Amritsar, Punjab, India

Muskan Parveen
Department of Botany
Aligarh Muslim University
Aligarh, Uttar Pradesh, India

Sneha Prakash
Department of Botany
Jamia Hamdard
New Delhi, India

Prakriti
Department of Zoology
Guru Nanak Dev University
Amritsar, Punjab, India

Rose Rizvi
Department of Botany
Aligarh Muslim University
Aligarh, Uttar Pradesh, India

Rudra Narayan Sahoo
Department of Botany
Jamia Hamdard
New Delhi, India

Sajad Hussain Shah
Department of Botany
Aligarh Muslim University
Aligarh, Uttar Pradesh, India

Indu Sharma
Department of Life Sciences
University Institute of Sciences, Sant Baba Bhag Singh University
Jalandhar, Punjab, India

Pooja Sharma
Department of Biotechnology
DAV College
Amritsar, Punjab, India

Roohi Sharma
Department of Zoology
Guru Nanak Dev University
Amritsar, Punjab, India

Rajesh Kumar Singh
Biotechnology Division
CSIR-Institute of Himalayan Bioresource Technology
Palampur, Himachal Pradesh, India

Sangram Singh
Department of Botany
Aligarh Muslim University
Aligarh, India

Sarika Singh
Department of Botany
Aligarh Muslim University
Aligarh, India

Geetika Sirhindi
Department of Botany
Punjabi University
Patiala, Punjab, India

Raman Tikoria
Department of Zoology
Lovely Professional University
Phagwara, Punjab, India

Shahid Umar
Department of Botany
Jamia Hamdard
New Delhi, India

Vikram
Department of Zoology
Guru Nanak Dev University
Amritsar, Punjab, India

1 Plant Responses to Heavy Metals
Mechanisms and Adaptations

Mudassara Hasan and Fauzia Naushin

1.1 INTRODUCTION

The principal abiotic stressors pose a danger to global food security because they can negatively impact staple food production, biomass output, and survival by up to 70% (Kaur et al. 2008; Thakur et al. 2010). Increased accumulation of heavy metals (HMs) is one consequence of human disruption of natural biogeochemical cycles, which is very problematic for ecologic, nutritional, and environmental reasons (Nagajyoti et al. 2010; Ali et al. 2013; Riyazuddin et al. 2021; Nazir et al. 2023b). With an atomic weight above 20 and a density over 5 g/cm^3, HMs are a group of constant, not biodegradable inorganic chemical substances that can pollute soil, reservoirs, irrigating or drinking water supplies, and the atmosphere. They can also have deleterious and mutation-causing effects on humans, animals, and plants (Rascio and Navari-Izzo 2011; Wuana and Okieimen 2011; Jahan et al. 2023). Soils include two different types of metals: those that correspond to essential micronutrients (Fe, Mn, Zn, Cu, Mg, Mo, and Ni) for proper growth of plants, and those that are considered non-essential elements (Cd, Sb, Cr, Pb, As, Co, Ag, Se, and Hg) with unclear physiological and biological significance (Rascio and Navari-Izzo 2011; Tangahu et al. 2013; Zhou et al. 2014; Khan et al. 2021a; Nazir et al. 2023a). Both the subterranean and upper surface of plants could take in HMs (Patra et al. 2004). At detrimental concentrations, HMs disrupt metabolic processes and hinder typical plant activity in many ways, including the substitution or alteration of the essential elements of protein structure brought by the bonding of HMs with sulfhydryl groups (Hall 2002; Thao et al. 2015; Asgher et al. 2018; Khan et al. 2021b), impeding functional groups of significant cellular molecules (Hossain et al. 2012), substituting or obstructing the operation of necessary metals in biological molecules such as various pigments or different enzymes (Ali et al. 2013), and negatively impacting the vital functions of plants like photoactivity, respiration, and enzymatic activities (Hossain et al. 2012). In contrast, greater concentrations of HMs are associated with an increased development of reactive oxygen species (ROS), including superoxide free radicals (O_2–), hydroxyl free radicals (OH–), free oxygen radicals (O_2^*), and hydrogen peroxide (H_2O_2). When combined with toxic compounds like methylglyoxal (MG), these ROS may exacerbate oxidative stress in plant cells by upsetting the balance between prooxidant and antioxidant homeostasis (Hossain et al. 2012; Sytar et al. 2013), suggesting that a number of degenerative conditions, including oxidative DNA damage, ion leakage, oxidation of proteins and lipids, redox imbalance, and breakdown of the cell's framework, are the cause of this condition. This condition eventually contributes towards programmed cell death (PCD) pathways (Rascio and Navari 2011; Sharma et al. 2012). When challenged with the difficult circumstances imposed by an elevated level of HMs, plants implement a wide range of both internal and external protective measures for endurance or elimination. Understanding the main genetic mechanisms for better selective breeding is extremely challenging since resistance to this issue is multigenic and quantifiable in nature (Collins et al. 2008). To gain more insight into the perception and transmission of stress signals by the associated molecular regulatory networks, the molecular pathways behind abiotic stress tolerance in plants are being uncovered using a variety of

DOI: 10.1201/9781032630380-1

large-scale sequencing and functional genomics methods (Heidarvand and Amiri 2010; Sanchez et al. 2011). For the purpose of developing transgenic techniques and intelligent breeding to provide plants stress tolerance, it is essential to understand how environmental stimuli are received by plants and trigger adaptive responses. Ultimately, creating or selecting cultivars of high-yielding, stress-tolerant food crops will require combining understanding of the physiological, biochemical, and gene regulatory frameworks.

1.2 PLANT DEFENCE MECHANISMS AGAINST HEAVY METAL STRESS

According to DalCorso et al. (2010), plant adaptation strategies necessitate the integration of numerous intricate physiological and biochemical processes, involving modifications to the systemic gene expression system. Plants employ many strategies to offset the harmful effects of metals or semi-metals. When plants are able to restrict metal absorption, they can "avoid" metal stress, or when they can tolerate high internal metal levels, they can "tolerate" it. Avoidance entails lowering the quantity of metal entering the cell through absorption to cell walls, decreased intake, enhanced outflow, or extracellular precipitation.

Another type of situation involves the chelation of metals within the cell through the use of metallothioneins, phytochelatins, amino acids, natural acids, a substance called glutathione (GSH), vacuole compartmentalization, and over-expression of the glyoxalase and antioxidant defence systems to counteract the harmful effects of ROS and MG (Hall 2002; Yang et al. 2005; Hossain et al. 2010; Yadav 2010; Seth et al. 2012). As previously stated, plants have a complex network of defensive mechanisms that work together to prevent or tolerate HM poisoning (Figure 1.1).

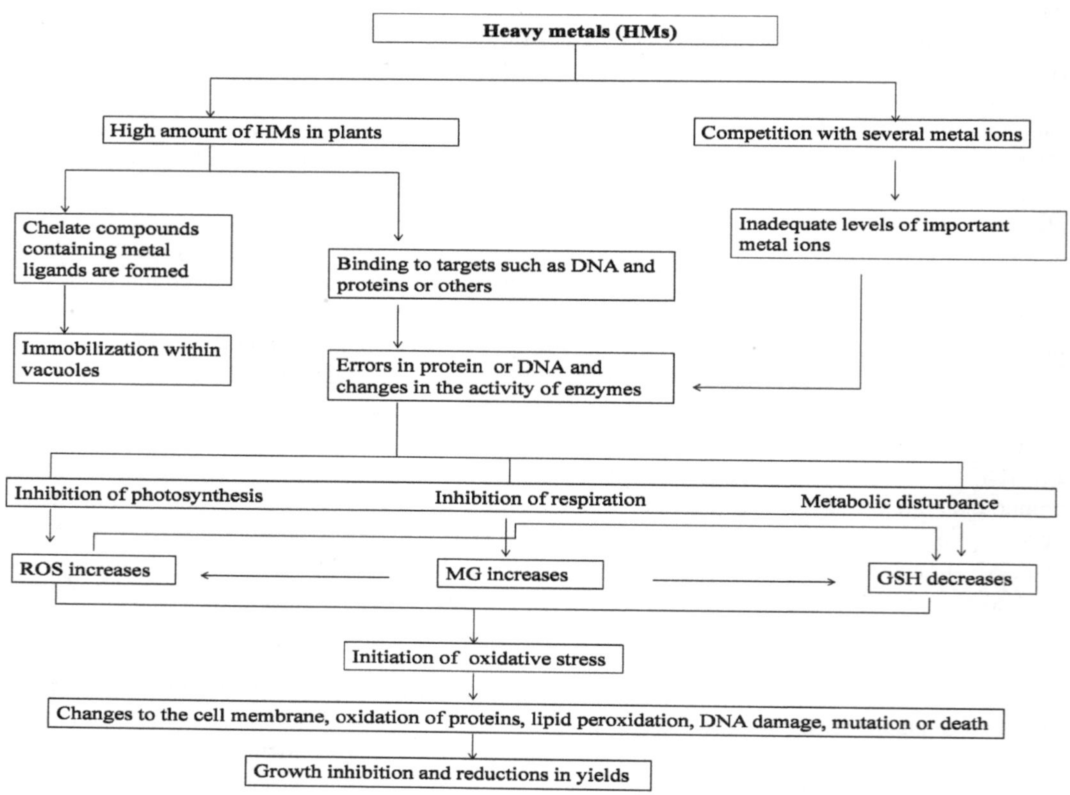

FIGURE 1.1 Potential biochemical and molecular pathways of ROS generation and disruption caused to plant growth due to heavy metals.

Physical obstacles serve as the first line of defence in plants against HM stress. When plants encounter HM stress, certain physical features, such as physically active tissues including trichomes, cell walls, and mycorrhizal symbiosis, as well as thick cuticles, may function as barriers (Hall 2002; Wong et al. 2004; Harada et al. 2010). Trichomes, for example, have the ability to either store HM for detoxifying reasons or produce other secondary metabolites to counteract their harmful effects (Lee et al. 2002; Hauser 2014). The cell wall, composed of cellulose, hemicellulose, pectin, and protein, plays a crucial role. Functional groups like carboxyl, hydroxyl, amino, and aldehyde groups that are present on its surface can bind with metallic ions to prevent heavy metals from passing through membranes. During a lead stress condition, the amount of lead in the radish lateral root cell ranged from 71.08% to 80.40% of the total amount in the lateral root. Following the extraction of hemicellulose from the root walls of rice, tomato, pepper, leaf lettuce, and common cabbage, the amount of zinc accumulated in the shoot rose while the amount considerably decreased in the root cytoderm. Furthermore, the cytoderm of rice roots contains proteins high in cysteines that bind to lead (Yu et al. 2019). When exogenous nitrogen oxides are administered, the quantity of hemicellulose and pectin in the root cytoderm increases, improving the resistance of rice and Kandelias to cadmium (Zhu et al. 2013).

On the other hand, plants activate a number of biological defence manoeuvres to neutralize and mitigate the negative effects of HMs after they pass through biophysical and subcellular barriers and metal ions infiltrate cells and tissues. The fundamental processes of HM build-up, tolerance, and adaptive defences against metal stress are still being studied using physiological, biochemical, and molecular methods. Some of the mechanisms developed by plants include immobilization, membrane marginalization, constraints of intake and transfer, the formation of particular HM carriers, chelation and retention of metals by specific ligands (phytochelatins and metallothioneins), activation of strategies that counteract the impact of reactive oxygen species and methylglyoxal, including the increased expression of antioxidant and glyoxalase systems, stimulation of proline biosynthesis, polyamines, stress proteins, and signalling chemicals such as nitric oxide and salicylic acid (Cobbett 2000; Hossain et al. 2009; Hossain and Fujita 2009; Sharma and Dietz 2009; DalCorso et al. 2010; Hossain et al. 2012). Depending on the plant species, amount of metal tolerance (Solanki and Dhankhar 2011), stages of development, and metal nature, any of these processes and elements may be exploited or up-regulated.

1.2.1 Limitations on the Metal Uptake and Transportation

1.2.1.1 Marginalization of Heavy Metals from the Plants

Plants absorb heavy metals by trapping HM ions with their roots, which then enter the roots and move to the shoot. Depending on the type of HM, metal ions penetrate the plant through either the apoplast (extracellular) or the symplast (intracellular) once they become bio-available to the plant. According to Bubb and Lester (1991), the majority of HM ions enter plant cells through specialized or universal HM ion carriers or channels in an energy-dependent mechanism. Minimizing excess HMs from entering the plant is one method of reducing the damaging effects of HMs (avoidance technique). A plant might do this in one of two ways: by precipitating or by sequestering HMs in the root area. By raising the pH of the root zone or by releasing anions like phosphate, plants can precipitate HMs. Furthermore, it has been demonstrated that in response to Cd stress, the roots of maize (*Zea mays*) and sorghum (*Sorghum bicolor*) exude citrate and malate, respectively (Pinto et al. 2008). These results lend credence to the concept that root exudates' ability to bind HMs may be a key mechanism for stabilizing HMs close to the root, decreasing their toxicity and making them unavailable to the plant. According to a study conducted by Zhu et al. (2011), the oxalate has been observed to be released by the tomato (*Lycopersicon esculentum*). The root apex helps keep Cd from penetrating tomato roots, which increases the cultivar's tolerance to Cd. Their findings imply that the resistant crop possibly used precipitation to exclude the HM as a defence strategy. When administered with Al, it has been discovered that wheat (*Triticum aestivum*) cultivars resistant to Al generate more malic acid than sensitive cultivars (Delhaize 1993; Huang et al. 1996). Moreover, it has been suggested that rice (*Oryza sativa*) particularly uses the oxidants and oxygen (O_2) produced in the rhizosphere to produce iron plaques on the outermost layer of the root, which remove dangerous metals from plants (Liu et al. 2008). According to Liu et al. (2008), iron plaque can hinder rice plants from absorbing Cd by adsorbing and sequestering it on the root surface. While it is reasonable to keep extra HMs out of the plant, certain hyper-accumulator and tolerant plants actually take in more HMs than sensitive plants do (Rascio and Navari-Izzo 2011). According to their findings, the tolerant cultivar could have used the precipitation method to exclude the HM as a defence strategy. When exposed to Al, wheat (*Triticum aestivum*) genotypes have been demonstrated to produce more malic acid than genotypes that are sensitive to Al. Furthermore, it has been proposed that the production of iron plaques on the roots, which are caused by O_2 and other oxidants released in the rhizosphere, is a method by which harmful HMs are excluded from plants, especially in rice (Liu et al. 2008). It was observed that iron plaque can stop rice crops from Cd uptake by adsorbing and sequestering it onto the root surface. Although in order to keep extra HMs out of the plant, certain hyper-accumulator and tolerant plants actually absorb more HMs than do sensitive plants (Rascio and Navari-Izzo 2011), indicating that additional mechanisms must also be in existence.

1.2.1.2 Heavy Metals Restriction Inside Cells

An essential adaptation approach for HM endurance in plants is the cellular elimination of HMs. The apoplastic area contains a significant portion of the HMs in plant roots. An exclusion mechanism was demonstrated, for example, when a sensitive wheat cultivar had more symplastic Al than the tolerant cultivar at equal external Al concentrations (Tice et al. 1992). HM transporter proteins may have a role in preventing hazardous metal ions from moving from the symplastic to the apoplastic area.

1.2.1.3 Heavy Metal Complexation at the Junction of the Cell Wall and Plasma Membrane

Due to the substantial HM build-up at the cell wall–plasma membrane interface, it is believed that HM tolerance may develop there. The cell wall and plasma membrane of Italian ryegrass (*Lolium multiflorum*) limited 60% of the copper present in the roots (Iwasaki et al. 1990). The cation exchange capacity (CEC) of plants is mostly determined by the exchange sites found in their cell walls (Horst and Marschner 1978). Sensitive wheat cultivars have significantly lower cell wall CECs than tolerant cultivars have, while tolerant wheat cultivars use a high CEC to compound HMs at the cell wall and prevent admission into the cell (Masion and Bertsch 1997). However, how the cell wall influences HM tolerance is yet unknown.

1.2.1.4 Movement of Heavy Metals

By controlling the movement and transfer of HMs inside their organs or cells, plants can reduce the negative consequences of excessive HMs. Apart from hyper-accumulators, the most prevalent examples of this are the much greater concentrations of HMs in plant root systems compared to shoots (Qureshi et al. 1985; Wheeler and Power 1995; Zheng et al. 1998), and decreased HM translocation to the shoots is a typical endurance mechanism. Compared to susceptible genotypes, tolerable genotypes have a lower HM concentration in the shoot because this probably works to remove metals from sensitive processing in the shoots. Plants showed a resistance response by evenly dispersing HMs once they reached a plant tissue or organ. The distribution of Mn in the leaves of tolerant cowpea (*Vigna unguiculata*) genotypes is comparatively consistent. However, sensitive genotypes deposit Mn in confined regions of the leaf, which are characterized by dark brown patches brought on by precipitated manganese oxide (Horst 1983).

1.2.2 Plant Cell HM Complexation and Compartmentalization

1.2.2.1 Vacuolar Partitioning or Intracellular Sequestration

A plant employs a number of defence mechanisms to combat the toxicity of an HM once they have entered the cell. One such tactic entails removing the HM from the cell by transferring it outside of it or retaining it inside the vacuole, hence keeping it out of the cytosol or other sensitive metabolic areas (Clemens 2001; DalCorso et al. 2008, 2010). As a result, the central vacuole appears to be an ideal location to store HMs that have gathered in excessive amounts. The majority of solutes are really taken up by two vacuolar proton pumps, the vacuolar proton-ATPase (V-ATPase) and the vacuolar proton pyrophosphatase (V-Ppase). Grasses have the ability to readily transport Zn into vacuoles, and more tolerant clones can do this at higher doses of external Zn than sensitive clones can (Brookes et al. 1981). Either carriers or channels may facilitate the uptake. Now, a variety of gene families have been discovered via the use of potent genetic and molecular approaches that are likely engaged in the process of HM ion assimilation into cells, HM retention in vacuoles, HM reintegration from vacuoles, and HM loading and unloading from xylem. Some HM transporter proteins include the ATP-binding cassette (ABC) transporters, P-type metal ATPases, the multidrug resistance-associated proteins (MRP), the iron-regulated transporter (IRT)-like protein ZIP family, the iron-regulated transporter (ZRT) family proteins, the copper transporter (COPT) family proteins, the cation diffusion facilitator (CDF) family of proteins, and the Ca^{2+}: cation antiporter (CAX) family (Lee et al. 2005; Chiang et al. 2006). The continuous matrix found in plant cell walls can hold varying metal concentrations, act as a cation exchanger, and perhaps partially reject heavy metals (Rauser 1999). Barley (*Hordeum vulgare*) leaf exposure to increased external zinc led to an increase in cellular zinc, but this was entirely offset by an increase in vacuolar zinc, which the cytoplasm perfectly regulated (Brune et al. 1994). Although less often utilized in the research of HM accumulation, vacuole separation or compartmental flow analysis may nevertheless offer useful information. To identify the position of Cd and potential Cd-binding peptides, protoplasts and vacuoles were extracted from the leaves of tobacco (*Nicotiana rustica* var *Pavonii*) seedlings subjected to Cd (Vogeli-Lange and Wagner 1990). Nearly every Cd-binding peptide and Cd found in protoplasts could be found in pure vacuoles. The vacuole is most likely where HM is sequestered and detoxified in *Nicotiana tabacum*, according to computer modelling studies with Zn (Wang et al. 1992) and Cd (Wang et al. 1991). It seems pretty clear from the evidence that HM sequestration occurs in plant vacuoles. Most of these proteins' transport roles, specificities, and molecular sites in plants are yet unclear, and there are still a large number of plant HM transporters that have not yet been discovered at the molecular level.

1.2.2.2 Metal Complex Generation through Phytochelatin Induction

One of the most important mechanisms for HM removal and adaptation in plants under metal stress may be the chelation of HMs in the cytosol by high-affinity ligands. Plants generate two different types of peptide metal binding ligands: phytochelatins (PCs) and metallothioneins (MTs). The latest improvements in our knowledge of several facets of PC production and function are mostly the result of molecular and genetics techniques employed for representing organisms. PCs belong to a group of polypeptides rich in Cys that have the general structural formula (γ-Glu-Cys)*n*-X. Depending on the organism, X can be any of the following: Gly, γ-Ala, Ser, Gln, or Glu, and *n* can range from 2 to 11, but the most prevalent versions contain 2–4 peptides (Yang et al. 2005). The enzyme γ-glutamyl cysteinyl dipeptidyl transpeptidase (PC synthase), which is constitutively produced, is activated by the metal to catalyze the conversion of GSH into PCs (Zenk 1996). Of a number of PC formation-promoting HMs such as Cd, Hg, Ag, Cu, Ni, Au, Pb, As, and Zn, Cd is clearly the best promoter (Grill et al. 1987, 1989). According to Di Toppi et al. (1999), the thiol group (-SH) of Cys allows PCs to mix with Cd ions. PC-Cd aggregates are then gathered in the vacuole by ABC transporters, which limit the amount of free Cd^{2+} that may travel throughout the cytoplasm. Moreover, plants are unable to metabolize or eliminate Cd.

Instead, they efficiently store Cd inside vacuoles via the method of forming Cd-GSH and Cd-PCs complexes (Di Toppi et al. 1999; DalCorso et al. 2008) as well as to move Cd through xylem and phloem vessels across a considerable distance (Mendoza-Cózatl et al. 2008). Further evidence for the occurrence of PCs under HM stress came from the finding that PC synthesis and HM build-up had a high correlation in both roots and shoots. Water hyacinth (*Eichhornia crassipes*) roots were found to take up Cd^{2+}, which was shown to be present as Cd-binding complexes in an experiment involving water hyacinth and Cd stress, but no such complexes were seen in the absence of Cd^{2+} (Fujita 1985). This suggests that these Cd^{2+}-binding complexes develop in response to Cd^{2+} and are necessary for the accumulation of Cd in water hyacinth. With a molecular weight of 4000 , the later Cd-binding complex was discovered to consist of two cadystins: $(\gamma\text{-Glu-Cys})_3$-Gly for cadystin A and $(\gamma\text{-Glu-Cys})_2$-Gly for cadystin B, as well as inorganic sulfur, which is the same as the fission yeast Cd-BPI (Fujita and Kawanishi 1986). Iglesia-Turiño et al. (2006) discovered that Hg uptake was closely linked with PC concentration in *Brassica napus* cultivated under a Hg concentration gradient (0-1,000 M). PCs also have a role in Zn^{2+} and Cu^{+}/Cu^{2+} homeostasis by acting as a transitory storage form for the ions (Grill et al. 1988; Thumann et al. 1991). Arsenate anion was shown to induce PCs in research involving peptide-inducing metal ions (Grill et al. 1987), implying a novel way of PC synthase activation. Nouairi et al. (2009) discovered that cadmium (Cd) treatment increased PC production in *Brassica juncea* leaves by up to 50 M (94.66 mol g^{-1} FW) compared to the control (10.57 mol g^{-1} FW). The PC level in *B. napus* leaves, however, did not significantly differ from the control at any Cd dose after a 15-day treatment. Additionally, research on the expression of the PCS gene in garlic (*Allium sativum*) plants exposed to heavy metals (Zhang et al. 2005) and an investigation conducted in vitro using the marine alga (*Dunalie latertiolecta*; Tsuji et al. 2002) suggested a role for PCs in HM detoxifying and oxidative stress reduction. The HM-detoxification process is more than only chelation of HM ions. After HM ions activate PC synthase and HM chelation by the produced PCs, the HM ion complex is carried to the vacuole where it forms a complex with sulphides or organic acid to remain there (Rauser 1999). It was additionally demonstrated that PCs possess some function in HM distribution (Salt and Rauser 1995), indicating that their cleansing abilities are secondary or a component of an additional complicated system. Although it has been convincingly proven that PCs are important for the removal and accumulation of heavy metals in higher plants, nevertheless, the species- or HM-specific specificity of hyper-accumulation cannot be explained by the formation of HM complexes (Baker et al. 2020). As a result, additional study is needed to establish the precise function PCs play in the HM-tolerance process at the cellular level.

1.2.2.3 Complexing by Metallothioneins

Another class of small, low-molecular-weight, cysteine-rich cytoplasmic metal-binding proteins or polypeptides that are found in a wide variety of eukaryotic species, including fungi, invertebrates, mammals, and plants, are called MTs, and they were first isolated by Margoshes and Vallee (1957) from horse kidney, in addition to a few prokaryotes (Macovei et al. 2010; Du et al. 2012). Unlike PCs, which are the consequence of enzymatic peptide synthesis, MTs are the outcome of the translation of mRNA (Verkleij et al. 2003). While PCs in plants may primarily deal with the process of detoxification of Cd, MTs seem to have preference for a broader spectrum of metals, including Cu, Zn, Cd, and As. Since plant MTs exhibit such a wide range of molecular traits and structural properties (Leszczyszyn et al. 2013), more and more diverse roles are probably played by them in plants than in any other living organism. MTs display various features and functioning depending on the organisms in which they are found. Using cellular sequestration, intracellular metal ion homeostasis, and metal transport modification, these ligands help plants reduce the toxicity of HM (Kohler et al. 2004). MTs are known to be active participants in a range of cellular-related functions. Apart from their role in HM detoxification, they also serve as ROS scavengers (Wong et al. 2004), maintain redox levels (Macovei et al. 2010), repair the plasma membrane (Shruti and Dubey 2010), promote cell development, and repair damaged DNA

(Grennan 2011). Other than HMs, there are several additional intrinsic and extrinsic stimuli that can stimulate the formation and activity of MTs. These include osmotic stress, harsh weather, food deprivation, hormone release, naturally occurring tissue senescence brought on by darkness, traumas, and viral infections (Du et al. 2012).

According to the positioning of Cys residues, plants have several kinds of MT that are often grouped into four separate subgroups (Huang and Wang 2009). They exhibit organ and developmental stage specificity, with type 1 MTs being largely expressed in roots, type 2 MTs being primarily expressed in shoots, type 3 MTs being created in leaves and during fruit ripening, and type 4 MTs being extensively dispersed in the growing seeds (Kohler et al. 2004). According to Freisinger (2011), sequence variability in plant MT is rather high. Each of the four MT subgroups (MT1 through MT4) is further distinguished into distinct isoforms. Freisinger classified sugarcane MT2 into three subclasses – MT2–1, MT2–2, and MT2–3 – which is similar to Memon et al.'s (2001) classification of MT4 in *Arabidopsis* as MT4a and MT4b. It appears that each of the four kinds of MTs and their isoforms found in plants can bind to HMs and function as metal chelators, but growing evidence indicates that, on one side, plant MTs treat different types of metals differently and, on the other side, they can vary within a plant species or across species in terms of their functioning, metal-binding and metal-affinity properties, and tissue location. According to Grennan (2011), there is a substantial possibility that metallothionein (MT) isoforms from types 1, 2 (1a, 2a, and 2b), and 3 are engaged in copper chelation in *Arabidopsis*, whereas MT isoforms from type 4 (4a and 4b) function as zinc linkers. In certain mutants of *Arabidopsis*, as demonstrated by Garcıa-Hernández et al. (1998), MT1 may be more crucial for copper detoxification in leaf veins than in leaf mesophyll. It was shown that MT3's important role in grain-filling and mature seeds of barley is to maintain Zn and Cu homeostasis, while MT4 was engaged in Zn retention (Hegelund et al. 2012). According to studies conducted on soybean, MT1, MT2, and MT3 were more apt to participate in the elimination of harmful levels of Cd, whereas MT4 had Zn-binding properties (Pagani et al. 2012). It is possible that different metallothionein types and their isoforms perform both unique and complementary roles in regulation and HM elimination (Du et al. 2012). However, it suggests that variations in plant genetic structure, complicated diversity in the metal-binding parts of plant MTs (Grennan 2011), and varying sequences and activity of isoforms may be a rationale for the noted pattern. Further investigation is still needed to discover the potential causes for these different and favoured responses of plant MTs towards HMs.

The use of DNA recombinant techniques to engineer plant MTs and express them in plants, yeasts, and bacteria that lack some of these proteins can expand our understanding of MTs and their capabilities while also offering novel opportunities for phytoremediation or bioremediation techniques. MTs may enhance the capacity of transgenic plants to reduce the generation of ROS and strengthen the cellular antioxidant defence mechanism particularly when it involves detoxifying high amounts of metals. According to findings by Xia et al. (2012), the expression of *Elsholtziahai chowensis* metallothionein type 1 (*EhMT1*) in tobacco crops elevated the transgenic tobacco's tolerance to the toxic effects of copper as well as reduced hydrogen peroxide production and increased peroxidase (POD) activity in roots, which improved the capacity of the crops to withstand oxidative stress. *TaMT3*, a metallothionein type 3 from *Tamarix androssowii*, was engineered into tobacco, and Zhou et al. (2014) showed that although this enhanced resistance to Cd stress by a substantial improvement in SOD performance, which increased the capacity of ROS cleaning-up in transgenic plants, it also resulted in a decline in POD activity. Further research is necessary since it appears that different components of the antioxidant system of transgenic plants are affected differently by the produced metallothionein. Plants modified with illicitly produced MTs have been demonstrated to have increased tolerance for metal poisoning. According to Zhigang et al. (2006), BjMT2, a metallothionein type 2 from *B juncea*, boosted copper and Cd tolerance at the seedling stage in *Arabidopsis thaliana* but sharply decreased root growth when there was no presence of HMs. These patterns could imply that MTs may function non-specifically and have diverse effects on organ development in the host plant when expressed ectopically in transgenic plants.

1.2.2.4 Phosphate Derivatives, Amino Acids, and Organic Acids Can Chelate Metals

Internal tolerance and external exclusion are the two types of HM tolerance and detoxifying processes in plants. In the external process of detoxification, organic acids discharged by root systems may combine with HM ions to create a stable HM-ligand complex that alters the flow and availability of the ions, preventing them from entering plants or building in critical areas of roots. Organic acids may bond with HM in the cytosol during the internal HM detoxification process, changing the ions' toxicity into a harmless form (Clemens 2001; Hall 2002; Yang et al. 2005). Plants create a variety of ligands like Al, Cd, Cu, Ni, Co, and Zn. A few examples of carboxylic and amino acids that can bind to HMs and are crucial for resistance and elimination are citrate, malate, oxalate, histidine (His), nicotianamine (NA), and phosphate derivatives (phytate) (Clemens 2001; Hall 2002; Yang et al. 2005; Sharma and Dietz 2009; Singh and Chauhan 2011). Extensive HM ion neutralization is a property of citrate, which is particularly important for Fe (Clemens 2001). For other HMs, such as Cd, Ni, Co, and Zn, citrate also serves as a strong ligand. Studies on many Ni hyper-accumulators have shown that, even before the insertion of PCs into plants, Ni is mostly associated with citrate and the quantity of citrate produced is closely correlated with the accumulated Ni (Lee et al. 1977, 1978). Since then, carboxylic acids have been found to be associated with heavy metals in a significant number of studies. Miyasaka et al. (1991) discovered a connection between citric acid release in snapbeans (*Phaseolus vulgaris* L.) and Al tolerance. Citric acid has been shown to be linked with zinc accumulation and tolerance, and it is considered an essential ligand at low levels of Cd (Godbold et al. 1984; Wagner 1993). When Co ions were applied to the cells of the non-accumulators *Raufolia serpentina* and *Silene cucubalus* as well as the Co hyper-accumulator *Crotalaria cobalticola*, the citrate levels increased (Oven et al. 2002). This further indicates that citrate plays a part in the interaction of HM ions. Also, removal of HMs such as Al and oxalate root secretion has also been reported. In reaction to Al stress, buckwheat (*Fagopyrum esculentum* Moench.) releases oxalic acid from the roots and builds up nontoxic-Al-oxalate in the leaves as a result; detoxification takes place both internally and externally. According to Zheng et al. (1998), buckwheat is more tolerant to Al than Atlas 66, an Al-tolerant wheat cultivar. According to research conducted on wheat and maize, the root-based secretion of malate is facilitated by alkaline-activated anion channels in the plasma membrane. In plants tolerant to zinc, malate has been proposed as an alternative to cytosolic Zn chelators (Mathys 1977). Upon the removal of Al from the roots, malate secretion rapidly decreases to a level comparable to non-Al conditions, indicating processes that are both sensitive to Al and capable of secreting malate. The variances in Al tolerances are due to the movement of malate out of the apical root cells through an Al-activated malate-permeable channel (Zhang et al. 2001). Salazar et al. (1997) found that when seedlings were subjected to 50 μM Al for 24 hours, compared to Al-sensitive genotypes, Al-tolerant genotypes released around three to five times more succinate and nearly ten times more malate. Li et al. (2000) discovered that the Al-induced release of organic acids in rye (*Secale cereale*) resulted from a change in the metabolism of organic acids, while the increased release of malate in wheat appears to be solely due to the activation of an anion channel. Histidine and nicotianamine are involved in the chelation of HM ions inside plant cells and in the xylem sap (Rai 2002). *Thlaspica erulescens*, a Zn hyper-accumulator related to this process, has roots containing potential Zn-His complexes, as noted by Salt et al. (1999). Additionally, it was observed that free His increased up to 36-fold in response to Ni stress in the xylem exudates of *Alyssum lesbiacum*, a Ni hyper-accumulator (Krämer et al. 1996).

When a harmful Ni level was present in the growing media, transgenic plants that overexpressed the *Salmonella typhimurium* ATP phosphoribosyl transferase gene (*StHisG*) accumulated around two times as much histidine as wild-type plants and generated more biomass than ten times as much (Wycisk et al. 2004). This explains the functional association between moderate Ni tolerance and histidine. Another type of free non-proteinogenic amino acid serving as a chelator HMs in plants is nicotianamine. NA is mobile within the plant and has been detected in the sap cells of leaves, roots, and phloem. It has the capacity to bind HM ions such as Fe, Zn, Cu, and Ni. The enzyme NA synthase converts three molecules of S-adenosyl-methionine (SAM) into NA in a single step (Higuchi et al. 1999). NA has been shown to influence Fe homeostasis in solanaceous plants (Ling et al. 1999). In addition to its role in long-distance HM transport, NA is hypothesized to play a role in regulating HM transfer within plant cells (Pich et al. 2001; Takahashi et al. 2003). Weber et al. (2004) identified the genes that are more active in the roots of *Arabidopsis halerii* under controlled conditions than those of *A. thaliana* using *Arabidopsis* Gene Chips. The two genes in *A. halerii* roots that express themselves most highly encode NA synthase and a putative Zn^{2+} absorption mechanism. Additionally, NA synthase and NA levels were greater in the roots of *A. halerii.* All of these results suggest the active roles of NA plays in plant Zn homeostasis and NA synthase in *A. halerii* Zn hyper-accumulation. Phytate has been suggested as a possible connecting agent for HMs. HM deposits and phytotate commonly coexist in plants, including *Zea mays* vacuoles that are tolerant to Al (Vázquez et al. 1999). Zinc-phytate particles in *Deschampsia caespitose* roots were observed to be proportionally more common in tolerant ecotypes than in sensitive ecotypes (Van Steveninck et al. 1987). However, it is unknown whether HM phytate complexes result in a stress or tolerance response. As stated earlier, phosphate, organic acids, and amino acids can all be regarded as by-products of HM detoxification. Further research is required to quantify the organic acids produced as toxicity vs. tolerance response across a range of plant species using various HMs.

1.2.3 Hyper-Accumulating Mechanisms

Plants that exhibit tolerance to HMs often act as excluders, preventing the transfer of metals from the roots to the shoots. Alternatively, they may retain the absorbed metals in the root cells, detoxify them through cytoplasmic chelation, or sequester them in vacuoles. However, a remarkable group of plants known as hyper-accumulators has been known to efficiently and rapidly transport HMs from the xylem to the shoots. According to Salt et al. (1995), transpiration likely facilitates this translocation in the xylem. The majority of the accumulated HMs migrates to the shoots, allowing hyper-accumulators to accumulate high concentrations of HMs even when present in minimal concentrations in the soil (Krämer et al. 2007). HM absorption in hyper-accumulators is especially high at the root membrane level. This might be due to the high and constitutive expression of an HM transporter in the plasma membrane, as demonstrated by *Thlaspica erulescens*, a Zn and Cd hyper-accumulator (Pence et al. 2000; Lombi et al. 2001). According to Lasat et al. (2000), *T. caerulescens* exhibits lower HM build-up in root vacuoles, improved root to shoot translocation, increased absorption into leaf cells, and stronger HM tolerance. According to Pilon-Smits and Pilon (2002), extremely effective intracellular partitioning and chelation may contribute to the high HM tolerance. Different chelators, such as organic acid chelators (Salt et al. 1995) and nicotianamine (Von Wiren et al. 1999), may be implicated in the translocation of HM cations across the xylem. Once in the xylem, HM ions are transported from the xylem apoplast to the shoot symplast by HM transporters located in the shoot cell membrane.

1.2.4 Heat Shock Proteins

Heat shock proteins (HSPs) play the role of chaperones that assist in the normal folding and packaging of proteins, but they may also play a role in the defence and repair proteins during stress. According to studies done by Ireland et al. (2004); Neumann et al. (1994, 1995); Sarry et al. (2006); and Zhen et al. (2007), a number of transition HMs (Zn, Cu, Cd, Hg, Al, Cr) can induce HSPs. In reaction to Cd stress, significant build-up of a large HSP (HSP70) was observed (Sarry et al. 2006). Additionally, Ireland et al. (2004) reported that in response to Cd stress, freshwater plant species and marine macroalgae species increased their HSP70 protein accumulation. Furthermore, it was shown that when cultivated in Curich soil, the roots of *Armeria maritima* plants expressed a small HSP (sHSP; Neumann et al. 1995). Likewise, HSP25 accumulated more in soybean (*Glycine max* (L.) Merr) in response to Al stress (Zhen et al. 2007). Before HM stress, brief heat stress is administered to establish Cd tolerance. The oxidative damage brought on by HM exposure may result in permanent protein denaturation, however the molecular chaperones activated during HM stress may serve to channel this destruction. However, more information is needed about the potential part HSPs play in HM tolerance.

1.2.5 Chemical Modification

Metal-modifying enzymes can be involved in the absorption of heavy metals into organic molecules, such in the conversion of selenium to dimethylselenide, or in altering the oxidation state of metals, such as when toxic Cr(VI) is reduced to nontoxic Cr(III) (De Souza et al. 2000). According to Lytle et al. (1998), water hyacinth *Eichhornia crassipes*, growing in nutritive culture with Cr(VI), accumulated nontoxic Cr(III) in the tissues of their roots and shoots. This indicates that *E. crassipes* detoxified Cr(VI) during root absorption and moved part of the refined Cr to leaf tissues. In dicots, a reductase at the root cell membrane lowers Fe (and maybe Cu) before absorption (Pilon-Smits and Pilon 2002; Robinson et al. 1999). Plants' ability to reduce HMs in their natural environment may be a beneficial detoxifying strategy for phytoremediation.

1.2.6 Transcription Factor (TF) Modification

Metal response element binding transcription factor 1, or metal responsive transcription factor 1 (MTF-1), is a transcription factor that plays a major role in the cellular response and tolerance to heavy metal stress by stimulating the activation of the genes responsible for HM absorption, transport, and detoxification. Different plant species have already shown evidence of TFs connected to the stress tolerance and response in HM (Fusco et al. 2005; van de Mortel et al. 2008). Importantly, Cd-responsive TFs and other stress-related TFs use the same signal transduction route (Singh et al. 2002). TFs from many families, such as ethylene-responsive factor (ERF), basic leucine zipper (bZIP), WRKY, and myeloblastosis protein (MYB), significantly affect the expression of certain stress-related genes in response to Cd stress (Yang et al. 2005).

1.2.7 Glyoxalase Pathway and Antioxidant Defence Activation

Oxidative damage and antioxidative defence systems, which include the enzymes involved in the synthesis and elimination of free radical-scavenging molecules such ascorbate (AsA) and GSH, are often brought on by HM stress (Noctor and Foyer 1998; Sharma and Dietz 2009; Hossain et al. 2010). Furthermore, it was shown that MG, a cytotoxic substance, increased in response to a variety of abiotic stressors, such as HM stress (Singla-Pareek et al. 2003; Hossain et al. 2009). It is primarily detoxified by plants' glyoxalase system and enhanced tolerance to oxidative stress.

1.2.8 Salicylic Acid Generation and Heavy Metal Resistance

Salicylic acid (SA) is a naturally occurring molecule that plays a critical role in regulating a range of physiological and biochemical processes that enable plants to withstand biotic and abiotic stresses (Shi and Zhu 2008; Zhou et al. 2009; Mahajan et al. 2023). In *Thlaspi*, a Ni hyper-accumulator, chronically high SA indicates GSH-mediated Ni tolerance, as observed by Freeman et al. (2004). As GSH has been shown to guard against Ni-induced lipid peroxidation in *Thlaspi hyper-accumulators* and *Arabidopsis*, pre-treatment with SA increases GSH accumulation and provides protection against Ni toxicity (Freeman et al. 2004). Exogenous SA treatment substantially enhanced HM tolerance in plants, according to a large number of studies (Metwally et al. 2003; Shi and Zhu 2008; Zhou et al. 2009). However, SA's protective properties mostly include controlling ROS and antioxidants, inducing gene expression, and adsorbing and distributing substances (Shi and Zhu 2008; Zhou et al. 2009).

TABLE 1.1
Heavy Metal Resistant Plants and Their Characteristics Linked to GSH and the Enzymes That Metabolize Them in the Elimination of ROS and MG

Plant species	Mechanism(s)	Defence components-linked to GSH	References
Arabis paniculata	Zn/Cd hyper-accumulator	High GSH amount	Zeng et al. (2009)
Phragmites australis	Cd accumulator	High GPX activity and AsA and GSH amount	Iannelli et al. (2002)
Sedum alfredii	Cd/Zn hyper-accumulator	High GSH amount and DHAR activity	Jin et al. (2008)
	Pb accumulator		Gupta et al. (2010)
Thlaspica erulescens	Cd tolerant	High GSH amount	Boominathan and Doran (2003)
	Cd accumulator		
Pisum sativum	Cd tolerant	High GSH amount	Metwally et al. (2003)
Arabidopsis thaliana	Ni tolerant	High GSH amount	Freeman et al. (2004)
Pteris vittata	As hyper-accumulator	High GSH amount	Singh et al. (2006)
Nicotiana tabacum	Zinc tolerant Gly-I and Gly-II inserted transgenic plants	High Gly-I and Gly-II activities and high GSH amount	Singla-Pareek et al. (2003)

1.2.9 Synthesis of Proline

Proline is indeed a versatile amino acid with various roles in plant physiology. It serves as a compatible solute, helping plants cope with osmotic stress by regulating water balance. Additionally, proline acts as a signalling molecule in stress responses, contributes to cell wall structure, and plays a role in antioxidative defence mechanisms. Its involvement in stem elongation, embryo development, and the transition from vegetative to reproductive growth highlights its significance in plant development and adaptation to environmental challenges. In response to stresses brought on by a variety of abiotic and biotic factors, larger plants generally create larger levels of proline in response to conditions such excessive salt, dehydration, increased sun ultraviolet (UV) radiation, heavy metals, and oxidative stress (Szabados and Savouré 2010). Proline performs a variety of functions when it comes to helping plants deal with stress, including adaptation, recuperation, and signalling (Fidalgo et al. 2013). There have been several ways through which proline enhances plants tolerance to HM toxicity.

According to Clemens (2006), water balance inequalities, which develop as a consequence of metal overload, are what causes proline to be activated in plants and are not the direct outcome of HM stress. Proline serves as an osmoregulator or osmoprotectant in this aspect. According to Mourato et al. (2012) and Tripathi and Gaur (2004), Proline's HM stress-induced ROS scavenging is mostly accomplished by quenching singlet oxygen and detoxifying hydroxyl radicals. When HMs are present in plants, proline is associated with elevated antioxidant enzyme activity, their defence, maintenance of the cellular redox balance (Mourato et al. 2012), chlorophyll rebuilding, and intracellular pH regulation (Rastgoo et al. 2011).

A study of the literature indicates that proline is mostly concentration-dependent and organ- and metal-specific when it comes to how plants react to HM. It is shown that proline augmentation in cells and HM concentration are linearly correlated in the hyper-accumulator *Cynaras colymus* L. (artichoke; Karimi et al. 2012). Gohari et al. (2012) showed that although proline accumulation in aerial sections was less noticeable than in root, proline concentration in the root of rapeseed (*Brassica napus* L.) surged when the plant was subjected to increasing exposure of Pb^{2+} (100 to 400 M).

The same organ-specific accumulation of proline, with roots containing more proline than shoots, has been observed in numerous studies with a variety of crops, including *Brassica juncea* under Pb and Cd stress (John et al. 2009), *Solanum nigrum* under Cu stress (Fidalgo et al. 2013), wheat under Cd stress, and lemongrass (*Cymbopogon flexuosus* Stapf) exposed to Hg and Cd stress (Handique and Handique 2009).

The distribution pattern of proline in plants under metal stress can vary depending on various factors such as the type of metal, its concentration, and the specific plant species. In some cases, when plants are exposed to high doses of metals like Cd, proline accumulation may be higher in the roots compared to the leaves, as observed in hybrid poplar. However, under different conditions, such as lower metal concentrations or with other metals, proline may accumulate more in the shoots than in the roots. This variability underscores the complex and dynamic nature of plant responses to metal stress. In addition to the type of plant species and the preferred accumulation of metals within plant sections, it appears that temperature, exposure length, and heavy metal concentration are significant experimental settings and factors that affect how proline is enhanced in plants. Nevertheless, rather than permitting metals to reach aboveground sections, the quick induction of proline in roots and formation of proline-metal complexes may offer a better and more efficient strategy to neutralize metal toxicity. Theriappan et al. (2011) used cauliflower seedlings (*Brassica oleracea var. botrytis*) and three HMs (Cd, Hg, and Zn) to observe concentration-dependent accumulation of proline, where an increase in toxicity (up to 1000 M) almost doubled the synthesis of proline. They also demonstrated that Hg was proline's strongest initiator. It was discovered that Cd, Pb, and As were, correspondingly, better evokers of proline in a different investigation utilizing sal seedlings (*Shorea robusta*; Pant et al. 2011). Rastgoo et al. (2011) discovered that Cd treatment resulted in the greatest proline accumulation when they studied the effects of equimolar concentrations (50 μM and *m*100 M) of HMs (Cd, Co, Pb, and Ag) in Gouan (*Aeluropus littoralis*). The study by Zengin and Kirbag (2007) highlights the differential effect of heavy metals on proline accumulation in sunflower seedlings. They found that the extent of proline accumulation varied depending on the type and concentration of the heavy metal. Specifically, they observed that proline content was significantly increased in the order of Hg > Cd > Cu > Pb. This suggests that different heavy metals may elicit varying degrees of stress response in plants, leading to differential accumulation of proline. Factors such as the toxicity threshold of each metal, its concentration, and the specific plant species being studied likely contribute to these differences in proline accumulation patterns. According to Rastgoo et al. (2011), gouan (*Aeluropus littoralis*) was exposed to equal concentrations of heavy metals of 50 μM and 100 *m*M. The application of Cd to the plant produced the largest accumulation of proline. Proline content was highly raised in sunflower seedlings exposed to varying levels of HMs in the following order: Pb < Cu < Cd < Hg, as demonstrated by Zengin and Kirbag (2007). The findings of this study suggest that a variety of factors, such as the concentration and specificity of the HMs, their toxicity threshold, and the kind of plant used in the investigations, may affect a trace metal's ability to accumulate proline. According to Ruscitti et al. (2011), raising the concentration of HM increases the quantity of cell proline to a certain point, after which the accumulation of proline is suppressed when the metal concentration rises over a particular limit. One effective strategy to reduce the toxicity of metals and activate defence mechanisms in plants to offset the toxic effects of HMs is to spray proline on the aerial parts of plants grown under HM stress. As per the findings of Hayat et al. (2013), the application of proline externally to chickpeas under Cd stress elevated the activity of antioxidant enzymes, carbonic anhydrase, and photosynthetic traits, ultimately leading to the plant's increased tolerance to Cd. Furthermore, they observed that the plant produced significantly more naturally derived proline when it received proline from exogenous sources, which improved the plant's ability to withstand the metal stress. Furthermore, it was demonstrated that proline supplemented with nutrients enhanced the development of plants under stress more effectively than proline alone. Hayat et al. (2012) report that exogenous proline may create interactions with Cu, Cd, and Zn to reverse the decrease of nitrate reductase caused by metal toxicity. Proline pre-treatment is said to reduce ROS levels, thereby reducing Hg^{2+} phytotoxicity in rice (Wang et al. 2009). The function of exogenous proline in HM detoxification has to be taken into account more, especially when comparing different enriched extracts that include proline.

1.2.10 Polyamines

All living things include polyamines (PAs), which are organic cations with a low molecular weight. Putrescine, spermidine, and spermine are the three PAs that are most common in plants. They have been proposed as a family of plant growth regulators and second messengers, according to Hussain et al. (2011). Many plant expansion and growth processes are influenced by them. Through transcriptional and translational controls, PAs are essential for appropriate developmental processes (Kusano et al. 2008), even though they participate in a variety of biotic and abiotic stress responses, such as HM stress in plants, which accounts for their physiological importance (Lin and Kao 1999). They have the ability to scavenge ROS and free radicals (Ha et al. 1998), which enables them to act as potent antioxidants under a variety of stress conditions (Groppa and Benavides 2008). When faced with environmental challenges like low temperatures, heavy metals, and inadequate nourishment, proline and PAs are part of the "general adaptation syndrome" (GAS; Groppa and Benavides 2008; Groppa et al. 2008). From several plant species, a number of genes encoding the enzymes necessary for PA metabolism have been described and cloned. Engineered plants that overexpress the biosynthetic genes for PAs are more tolerant to HMs (Franchin et al. 2007) and other environmental stressors (Prabhavathi and Rajam 2007).

1.2.11 Nitric Oxide Generation and Tolerance to Heavy Metals

A popular biologically active signalling molecule, nitric oxide (NO) regulates the quantity and toxicity of ROS and hormones in plants. It also causes transcriptional changes leading to the possibility of the recognition of genes that regulate a variety of functional processes, including vital metabolism, transport, signal transduction, generation of ROS, defence, and cell death (Lamattina et al. 2003; del Rıo et al. 2004). While NO has antioxidant qualities of its own, it also shields plants from oxidative stress by controlling overall cellular redox homeostasis mechanisms, encouraging the conversion of $O_2^{\bullet-}$ to H_2O_2 and O_2, and increasing the activities of enzymes that eliminate H_2O_2 (Lamattina et al. 2003; Shi et al. 2007; Zheng et al. 2009; Siddiqui et al. 2011). Apart from its direct capacity to eliminate ROS and regulate lipid peroxidation by blocking LOX, NO can also protect cells from oxidative stressors by stimulating the synthesis of GSH. The Cd hyper-accumulator *Solanum nigrum* has attracted a lot of interest lately. According to Xu et al. (2011), NO levels in *S. nigrum* roots increased quickly in response to Cd stress, and Cd-induced NO can reduce oxidative damage and maintain intracellular antioxidant capacity. These results suggest that NO accumulation brought on by Cd is advantageous for plants' defence against Cd stress. Additionally, exogenous NO administration changed the way ROS-scavenging enzymes functioned, activated H^+-ATPase and H^+-PPase in the tonoplast or plasma membrane, and significantly reduced the growth inhibition in tomato plants caused by $CuCl_2$. According to these findings, exogenous NO may successfully cause tomato seedlings to modify their physiological and biochemical defences against Cu toxicity while also maintaining their basic metabolic function and normal development under HM stress. Exogenous NO also improves Cd tolerance in the roots of *Madicago truncatula* seedlings by increasing the synthesis of proline and total GSH and reducing oxidative damage (Zhao et al. 2020). According to Laspina et al. (2005), NO considerably reduces the oxidative damage that Cd exposure causes to sunflower (*Helianthus annuus* L.) leaf tissue. The aversion of hindering growth and destruction of chlorophyll (Chl), the recovery of GSH levels and catalase (CAT) activity, and the enhancement of AsA concentration and ascorbate peroxidase (APX) activity were the main factors leading to this finding. Plant HM tolerance is extremely complicated and includes interconnected physiological, biochemical, and metabolic plant processes, as the previous explanations have made perfectly evident.

1.3 PLANT CELL GENERATION OF REACTIVE OXYGEN SPECIES

During several processes of metabolism in organelles of cells, higher plants create ROS. Yet, the level of generation is markedly increased through HM stress. Organelles with high rates of electron flow or highly oxidizing metabolic activity, such as peroxisomes, mitochondria, and chloroplasts, are primarily responsible for the formation of ROS in plant cells (Apel and Hirt 2004; Mhamdi et al. 2010). Specifically, when photosynthetic carbon fixation is slowed down, which is frequently the case when abiotic stressors are present, the Chl is the primary generator of ROS with the potential to create substantial levels of superoxide (O_2•−) by means of Mehler reaction and H_2O_2 (Takahashi and Murata 2008). According to general consensus, HM (Cd) affects PS II's water oxidizing system by substituting Ca^{2+} and Mn^{2+} ions in the reaction centre; this inhibits PS II reaction and causes the electron transport in Chl to become decoupled (Baszyński et al. 1980; Faller et al. 2005; Krantev et al. 2008). According to Siedlecka and BaszyńAski (1993), Cd has been shown to block electron transport on the reducing side of PS I. HM has detrimental effects that are also seen during the carboxylation stage of photosynthesis. The primary targets of HM's impact are the vital CO_2 fixation enzymes phosphoenol pyruvate carboxylase (PEPC) and ribulose 1,5-bisphosphate carboxylase (Rubisco). The activity is lowered and its structure is altered by substituting Cd^{2+} ions for Mg^{2+} ions, which are essential cofactors in carboxylation processes. Additionally, Cd^{2+} ions may cause Rubisco activity to shift toward oxygenation reactions, and the glycolate produced is transported from the Chl to peroxisomes where it undergoes oxidation by glycolate oxidase (GO) to produce H_2O_2 (Takahashi and Murata 2008). Xanthine oxidase (XO) and GO, flavin oxidase (FO), and β-oxidation are peroxisomal enzymes that produce O_2•− and H_2O_2, respectively (Corpas et al. 2001; del Río et al. 2006; Mhamdi et al. 2010). HM stress increased peroxisomal mobility and was associated with a rise in ROS (Rodríguez-Serrano et al. 2009; Sharma and Dietz 2009). In reaction to exposure to Cd, leaf peroxisomes rapidly produce H_2O_2 (Romero-Puertas et al. 2004). It has been proposed that plant mitochondria play a significant role in the production of ROS brought on by abiotic stress. A typical supply of ubiquinone (Q) is reduced by a number of dehydrogenase complexes that make up the mitochondrial electron transport chain. According to Heyno et al. (2008), HM-induced loss of functionalities and mitochondrial respiratory chain pathway susceptibility have both been documented. Free HM cations, whether redox active or inactive, may have a substantial role in the development and maintenance of oxidative stress (Romero-Puertas et al. 2004). The primary sources of O_2•− and H_2O_2 generating apoplastic enzymes are cell wall-associated peroxidases and plasma membrane-bound NADPH oxidases (Mhamdi et al. 2010). NADPH oxidase-dependent ROS initiation was observed in response to Ni stress in wheat (Hao et al. 2006), Pb stress in *Vicia faba* (Pourrut et al. 2008), and Cd and Cu stress in *Arabidopsis* (Remans et al. 2010). In addition to these metabolic ROS sources, the Fenton reaction and the Haber-Weiss reaction also allow for the formation of •OH from H_2O_2 and O_2•−, respectively, in the presence of redox-active HMs like Cr, Co, Cu, Fe, and V (Halliwell 2006). Extremely reactive •OH in the cell severely deteriorates membranes and other macromolecules, including proteins, DNA, lipids, and pigments used in photosynthetic processes (Foyer and Noctor 2005; Gechev et al. 2006). In most cases, antioxidants that are already present in the area quickly scavenge ROS near the source of their generation. However, H_2O_2 can leak in the cytosol and diffuse to neighbouring compartments when the regional antioxidant capacity is unable to handle ROS generation. The activation of stress-responsive pathways, proteins, and genes has been shown to be regulated by ROS (particularly H_2O_2), in addition to their negative effects on cells (Hossain et al. 2012; Sharma and Dietz 2009). Given that H_2O_2 is promptly created in response to HM stress, it is likely a crucial molecule that initiates signal transmission and HM tolerance in plants (Seth et al. 2012).

1.4 MG AND ROS SYSTEMS FOR SCAVENGING AND DETOXIFICATION IN PLANTS

The plant antioxidant system plays a crucial role in regulating ROS levels within plant cells, thereby protecting against oxidative damage. This system comprises various enzymes, including superoxide dismutase (SOD), ascorbate peroxidase (APX), monodehydroascorbate reductase (MDHAR), dehydroascorbate reductase (DHAR), and glutathione reductase (GR), which act as antioxidants. Ascorbic acid and glutathione (GSH), both of which are cofactors of enzymes in the antioxidant and glyoxalase pathways, are also important components of the antioxidant system. They not only serve as cofactors but also directly quench ROS and influence gene expression associated with both biotic and abiotic stress responses, thereby promoting plant defence and survival. Maintaining an appropriate concentration of ROS, particularly hydrogen peroxide (H_2O_2), is crucial for enhancing plant growth and strengthening tolerance to environmental stresses. This regulation of ROS levels can affect gene expression and redox signalling pathways, ultimately contributing to plant resilience against various stressors (Neill et al. 2002). Similar to this, the glyoxalase system, which is found in the cytoplasm of cells and cellular organelles, notably mitochondria, is a vital part and primary mechanism of cellular metabolism of MG in living systems. Comparably, the glyoxalase system is a crucial component and the main process of cellular metabolism of MG in living systems. Studies have examined the role of the glyoxalase pathway in the ability to withstand abiotic stress (Hossain and Fujita 2009). The two enzymes that make up the widely distributed glyoxalase pathway are glyoxalase I (EC 4.4.1.5, lactoylglutathionelyase) and glyoxalase II (EC 3.1.2.6, hydroxyacylglutathione hydrolase). Using GSH as a cofactor in a two-step process, these enzymes work in cooperation to transform MG and other 2-oxoaldehydes to their 2-hydroxyacids (Thornalley 1990). Hemithioacetal is produced by the spontaneous interaction of GSH and MG, and Gly I then transforms it into S-D-lactoylglutathione (SLG). The second step involves Gly II catalyzing the hydrolysis of SLG to D-lactate while GSH is recycled back into the system. As a result, MG detoxification is highly reliant on the presence of cellular GSH. Lack of GSH prevents hemithioacetal from being produced, which causes MG to build up. The glyoxalase system catalyzes irreversible processes. This shunt pathway's presence and extensive variation attest to its basic significance in biological processes. Recent studies have shed light on the role of the glyoxalase system in stress tolerance and its interaction with oxidative defence mechanisms in plants. Molecular cloning of associated genes has deepened our understanding of how this system operates biologically and has provided a framework for investigating its physiological role in higher plants' defences against various abiotic stressors, including heavy metals. Studies by researchers such as Hossain et al. (2009, 2010, 2012) and Saxena et al. (2005) have been instrumental in elucidating these mechanisms.

1.5 INTERACTIONS BETWEEN PLANT MICROBES AND HEAVY METAL TOLERANCE

Fungi play crucial roles in nearly all ecosystems and possess significant potential for bioremediation owing to their extensive genetic and physiological diversity (Lanfranco et al. 2002; Ouziad et al. 2005; Hossain et al. 2010). Among these, arbuscular mycorrhizal (AM) fungi, which form widespread root symbioses, are particularly valuable. Many vascular plant species establish mutualistic relationships with symbiotic mycorrhizal fungi, such as AM fungi, under various environmental conditions. These fungi benefit from the photosynthetic products provided by plants, which enhance their mineral nutrition. Additionally, they can confer increased resilience to specific stresses and toxins on the host plants (Johansson et al. 2004; Leung et al. 2007). Plant-fungal mutualisms may also serve as early warning systems, alerting symbiotic plants to impending stress and allowing them to activate defence mechanisms and mitigate negative impacts before non-symbiotic plants (Ruscitti et al. 2011; Hasan and Naushin 2022; Hasan et al. 2023, 2024).

While the majority of descriptions regarding the symbiosis of mycorrhiza between plant roots and HMs fall beneath the category of bioremediation approaches, it is impossible to ignore the beneficial characteristics of this relationship, which support plants' general defence mechanisms against external stressors like HMs. The other main strategies used by mycorrhizal fungi to lessen the effects of HM stress on plants include establishing a barrier by collecting metals into cortical cells (Turnau 1998), attaching metals to cell walls or mycelium and maintaining them in their vacuoles or other organelles (Hall 2002), releasing glutathione and heat-shock protein (Hildebrandt et al. 2007), separating or chelating metals in the soil matrix through the production of glycoprotein, and forming phosphate metallic complexes within the hyphae (de Andrade and da Silveira 2008; Aloui et al. 2011; Garg and Chandel 2011). The various tactics used by AM fungi to combat the toxicity of HMs imply that different mycorrhizal fungus species may operate selectively or adopt a restorative role that is appropriate for the prevailing situations in the rhizosphere or plant. The immobilization of HMs in soil was demonstrated to drastically inhibit the translocation of Cd, Ni, and Zn from soil to various areas of the plant in ryegrass (*Lolium perenne* L.), which has a symbiotic interaction with AM (Takács et al. 2001). The retention of metals in the mycorrhizosphere area is mostly attributed to changes in soil pH brought on by the actions of AM fungi (Bano and Ashfaq 2013). According to Huang et al. (2005), the AM fungus (*Glomus mosseae*) reduced the amount of excess Zn, Cu, and Pb available to maize growing in HM-contaminated soil by binding the metals to organic matter or absorbing them into its organs, hence reducing the potential that the host plant would absorb the metals. ROS production causes the antioxidant defence mechanisms of plants exposed to high levels of HMs to be depleted. AM fungal association can minimize or prevent ROS generation and increase the activity of detoxifying enzymes in plants. According to Abad and Khara (2007), wheat plants colonized by several AM fungus species and exposed to hazardous levels of Cd exhibited significantly higher levels of the antioxidants APX and GPX in their roots and shoots than non-AM wheat. They also noticed that *Glomus veruciforme* and *Glomuse tunicatum* were the two fungus species that were most successful in activating the protective proteins. Under Cd- and Pb-contaminated soils, mycorrhizal pigeon pea (*Cajanus cajan* L.) plants show reduced lipid peroxidation and electrolyte leakage as well as improved SOD, CAT, and POX activity (Garg and Aggarwal 2012). According to research by Farshian et al. (2007), lettuce plants (*Lactuca sativa* L.) exposed to $ZnSO_4$ and infected with AM fungus (*Glomuse tunicatum*) exhibited greater cell protein content and, consequently, more antioxidant enzymes than plants that were not inoculated. Sugar and chlorophyll levels decreased in both AM-inoculated and non-AM-inoculated plants. On the other hand, higher levels of chlorophyll were seen in maize plants treated with the AM fungus (*Glomus intraradices*) under Cr stress, as per the research conducted by Rahmaty and Khara (2011). It implies that either plant or metal specificity can account for this difference in the results. The effects of the AM fungus on the production and up-regulation of other metabolites, such PCs, MTs, and proline, in plants under metal stress are not well understood. Two types of mycorrhizal fungus were injected into maize plants that were stressed by Cu and Cd in an experiment (*Acaulo sporalaevis* and *Glomus mosseae*), Abdelmoneim et al. (2014) found that proline build-up was lower in AM-inoculated maize than in non-AM-inoculated maize, and that other variables were related to the effectiveness of the fungal and plant partnership in reducing the negative effects of HM.

Ruscitti et al. (2011) observed that when mycorrhizal inoculation (*Glomus mosseae* and *Glomus intraradices*) was combined with Cr-stressed pepper plants (*Capsicum annuum* L.), the result was a rise in leaf proline content but a reduction in root proline concentration. According to Amanifar et al. (2014), mycorrhizal tomato (*Lycopersicon esculentum* L.) plants treated with Pb and infected with *Glomus* species showed no change in shoot proline concentration, despite a large rise in root proline content. They proposed a number of possible explanations, including HM type, plant species, and the unusual pattern of proline induction seen in metal-stressed plants infected with AM fungus. Moreover, it seems that the developmental stage, inoculation approach, and duration of being subjected to toxic metals may all have an impact on how proline forms in the presence of AM. More study is needed to determine if AM fungus and the protective compounds that metal-stressed plants produce in large quantities interact antagonistically or synergistically. Major crops including rice, wheat, and maize have been shown to be hosts for AM fungi (Bhalerao 2013). This suggests that appropriate fungus species should be found and cultivated in order to increase plant resistance to HM toxicity. The improved HM tolerance of mycorrhizal plants may be mostly attributed to efficient fungal HM-tolerance mechanism(s), due to the up-regulation of ROS metabolic genes. It has been discovered that plant associations with mycorrhizal fungus improve plant fitness in response to HM stress by employing GSH-dependent detoxifying pathways.

Further research endeavours might furnish insights into the precise regulatory functions executed by the AMF fungus concerning the GSH-dependent HM tolerance of plants or other abiotic stressors. It will thus be easier to choose the best AM isolates and plant fungal combinations for bioremediation and soil protection if the molecular processes governing HM tolerance in this symbiotic system are well understood (Turnau et al. 2001; Weyens et al. 2011). Therefore, more investigation is required to identify the precise processes behind AMF-induced HM tolerance in connection to plant GSH metabolism.

1.6 CONCLUSIONS AND FUTURE PROSPECTS

Indeed, a plant's ability to mitigate the toxicity of heavy metals underscores its remarkable capacity to adapt and survive in diverse environments. To effectively address the challenges posed by heavy metal contamination, it is crucial to deepen our understanding of the precise mechanisms underlying the activities of biological molecules involved in detoxification processes and their interactions with different plant species. Over the past decade, significant progress has been made in unravelling the complex systems that govern agricultural plants' responses to abiotic stressors, including heavy metal exposure. This progress has been driven by advancements in molecular biology, genomics, and bioinformatics, allowing researchers to elucidate key pathways and molecular players involved in plant stress responses. By continuing to explore these intricate mechanisms and their variations across plant species, researchers can develop more targeted and effective strategies for enhancing plant resilience to heavy metal contamination. Such efforts hold promise for improving crop productivity, safeguarding food security, and mitigating environmental pollution caused by heavy metals. Plants often become resistant to heavy metals by avoiding chelation, cytoplasmic complexation, extracellular complexation, and absorption. Plants must be able to mobilize their defence systems in order to do this, which include encouraging the production of antioxidant enzymes and preventing or repairing oxidative stress-related damage. In the future, improving agricultural production in the face of abiotic stressors like HMs will require precise control over plant responses. This necessitates a thorough understanding of the physiological and molecular processes underlying stress resistance in plants. By unravelling the intricacies of these mechanisms, researchers can develop targeted strategies to enhance stress tolerance in crop plants, ultimately leading to improved agricultural resilience and productivity. Continued investigation into how plants resist heavy metals will be crucial for developing sustainable agricultural practices and safeguarding global food security in the face of environmental challenges.

1.7 AUTHORSHIP CONTRIBUTION

Conceptualization, M.H.; Writing – Original draft preparation, M.H.; Writing – Review and editing, M.H. and F.N. All authors have read and agreed to the published version of the book chapter.

1.8 DECLARATION OF COMPETING INTEREST

The authors declare that they have no known competing financial interests or personal relationships that could have appeared to influence the work reported in this book chapter.

REFERENCES

Abad, A. K., and Jalil Khara. "Effect of cadmium toxicity on the level of lipid peroxidation and antioxidative enzymes activity in wheat plants colonized by arbuscular mycorrhizal fungi." *Pakistan Journal of Biological Sciences: PJBS* 10, no. 14 (2007): 2413–2417.

Abdelmoneim, T. S., T. A. Moussa, O. A. Almaghrabi, and Ismail Abdelbagi. "Investigation the effect of arbuscular mycorrhizal fungi on the tolerance of maize plant to heavy metals stress." *Life Science Journal* 11, no. 4 (2014): 255–263.

Ali, Hazrat, Ezzat Khan, and Muhammad Anwar Sajad. "Phytoremediation of heavy metals – concepts and applications." *Chemosphere* 91, no. 7 (2013): 869–881.

Aloui, Achref, Ghislaine Recorbet, Franck Robert, Benoît Schoefs, Martine Bertrand, Céline Henry, Vivienne Gianinazzi-Pearson, Eliane Dumas-Gaudot, and Samira Aschi-Smiti. "Arbuscular mycorrhizal symbiosis elicits shoot proteome changes that are modified during cadmium stress alleviation in *Medicago truncatula*." *BMC Plant Biology* 11, no. 1 (2011): 1–17.

Amanifar, S., N. Aliasgharzad, M. Toorchi, and M. Zarei. "Lead phytotoxicity on some plant growth parameters and proline accumulation in mycorrhizal tomato (*Lycopersicon esculentum* L.)." *International Journal of Biosciences (IJB)* 4, no. 10 (2014): 80–88.

Apel, Klaus, and Heribert Hirt. "Reactive oxygen species: metabolism, oxidative stress, and signal transduction." *Annual Review of Plant Biology* 55 (2004): 373–399.

Asgher, Mohd, M. Iqbal R. Khan, Naser A. Anjum, Susheel Verma, Dhiraj Vyas, Tasir S. Per, Asim Masood, and Nafees A. Khan. "Ethylene and polyamines in counteracting heavy metal phytotoxicity: a crosstalk perspective." *Journal of Plant Growth Regulation* 37 (2018): 1050–1065.

Baker, Alan J. M., S. P. McGrath, Roger D. Reeves, and J. A. C. Smith. "Metal hyperaccumulator plants: a review of the ecology and physiology of a biological resource for phytoremediation of metal-polluted soils." In *Phytoremediation of contaminated soil and water*, pp. 85–107. 2020.

Bano, Syeda Asma, and Darima Ashfaq. "Role of mycorrhiza to reduce heavy metal stress." *Natural Science* 2013 (2013).

Baszyńki, Tadeusz, Lucyna Wajda, Maria Krol, Danuta Wolińska, Zbigniew Krupa, and Anna Tukendorf. "Photosynthetic activities of cadmium-treated tomato plants." *Physiologia Plantarum* 48, no. 3 (1980): 365–370.

Bhalerao, Satish A. "Arbuscular mycorrhizal fungi: a potential biotechnological tool for phytoremediation of heavy metal contaminated soils." *International Journal of Science and Nature* 4, no. 1 (2013): 1–15.

Boominathan, Rengasamy, and Pauline M. Doran. "Cadmium tolerance and antioxidative defenses in hairy roots of the cadmium hyperaccumulator, Thlaspi caerulescens." *Biotechnology and Bioengineering* 83, no. 2 (2003): 158–167.

Brookes, A., J. C. Collins, and D. A. Thurman. "The mechanism of zinc tolerance in grasses." *Journal of Plant Nutrition* 3, nos. 1–4 (1981): 695–705.

Brune, A., W. Urbach, and K.-J. Dietz. "Compartmentation and transport of zinc in barley primary leaves as basic mechanisms involved in zinc tolerance." *Plant, Cell & Environment* 17, no. 2 (1994): 153–162.

Bubb, J. M., and J. N. Lester. "The impact of heavy metals on lowland rivers and the implications for man and the environment." *Science of the Total Environment* 100 (1991): 207–233.

Chiang, Huai-Chih, Jing-Chi Lo, and Kuo-Chen Yeh. "Genes associated with heavy metal tolerance and accumulation in Zn/Cd hyperaccumulator *Arabidopsis halleri*: a genomic survey with cDNA microarray." *Environmental Science & Technology* 40, no. 21 (2006): 6792–6798.

Clemens, Stephan. "Molecular mechanisms of plant metal tolerance and homeostasis." *Planta* 212 (2001): 475–486.

Clemens, Stephan. "Toxic metal accumulation, responses to exposure and mechanisms of tolerance in plants." *Biochimie* 88, no. 11 (2006): 1707–1719.

Cobbett, Christopher S. "Phytochelatins and their roles in heavy metal detoxification." *Plant Physiology* 123, no. 3 (2000): 825–832.

Collins, Nicholas C., François Tardieu, and Roberto Tuberosa. "Quantitative trait loci and crop performance under abiotic stress: where do we stand?" *Plant Physiology* 147, no. 2 (2008): 469–486.

Corpas, Francisco J., Juan B. Barroso, and Luis A. del Rı́o. "Peroxisomes as a source of reactive oxygen species and nitric oxide signal molecules in plant cells." *Trends in Plant Science* 6, no. 4 (2001): 145–150.

DalCorso, Giovanni, Silvia Farinati, and Antonella Furini. "Regulatory networks of cadmium stress in plants." *Plant Signalling & Behavior* 5, no. 6 (2010): 663–667.

DalCorso, Giovanni, Silvia Farinati, Silvia Maistri, and Antonella Furini. "How plants cope with cadmium: staking all on metabolism and gene expression." *Journal of Integrative Plant Biology* 50, no. 10 (2008): 1268–1280.

de Andrade, S. A. L., and Adriana P. D. da Silveira. "Mycorrhiza influence on maize development under Cd stress and P supply." *Brazilian Journal of Plant Physiology* 20 (2008): 39–50.

De Souza, M. P., E. A. H. Pilon-Smits, and N. Terry. "The physiology and biochemistry of selenium volatilization by plants." In *Phytoremediation of toxic metals: using plants to clean-up the environment*, pp. 171–190. New York: John Wiley & Sons, Inc., 2000.

del Rıo, Luis A., F. Javier Corpas, and Juan B. Barroso. "Nitric oxide and nitric oxide synthase activity in plants." *Phytochemistry* 65, no. 7 (2004): 783–792.

del Río, Luis A., Luisa M. Sandalio, Francisco J. Corpas, José M. Palma, and Juan B. Barroso. "Reactive oxygen species and reactive nitrogen species in peroxisomes. Production, scavenging, and role in cell signaling." *Plant Physiology* 141, no. 2 (2006): 330–335.

Delhaize, Emmanuel, Peter R. Ryan, and Peter J. Randall. "Aluminum tolerance in wheat (*Triticum aestivum* L.) (II. Aluminum-stimulated excretion of malic acid from root apices)." *Plant Physiology* 103, no. 3 (1993): 695–702.

Di Toppi, L. Sanità, and R. Gabbrielli. "Response to cadmium in higher plants." *Environmental and Experimental Botany* 41, no. 2 (1999): 105–130.

Du, Jia, Jing-Li Yang, and Cheng-Hao Li. "Advances in metallotionein studies in forest trees." *Plant Omics* 5, no. 1 (2012): 46–51.

Faller, Peter, Katharina Kienzler, and Anja Krieger-Liszkay. "Mechanism of Cd2+ toxicity: Cd2+ inhibits photoactivation of Photosystem II by competitive binding to the essential Ca2+ site." *Biochimica et BiophysicaActa (BBA)-Bioenergetics* 1706, nos. 1–2 (2005): 158–164.

Farshian, Shadi, Jalil Khara, and Parviz Malekzadeh. "Influence of arbuscular mycorrhizal fungus (*Glomus etunicatum*) with lettuce plants under zinc toxicity in nutrient solution." *Pakistan Journal of Biological Sciences: PJBS* 10, no. 14 (2007): 2363–2367.

Fidalgo, Fernanda, Manuel Azenha, António F. Silva, Alexandra de Sousa, Ana Santiago, Pedro Ferraz, and Jorge Teixeira. "Copper-induced stress in *Solanum nigrum* L. and antioxidant defense system responses." *Food and Energy Security* 2, no. 1 (2013): 70–80.

Foyer, Christine H., and Graham Noctor. "Redox homeostasis and antioxidant signaling: a metabolic interface between stress perception and physiological responses." *The Plant Cell* 17, no. 7 (2005): 1866–1875.

Franchin, Cinzia, Tiziana Fossati, Elena Pasquini, Guido Lingua, Stefano Castiglione, Patrizia Torrigiani, and Stefania Biondi. "High concentrations of zinc and copper induce differential polyamine responses in micropropagated white poplar (*Populus alba*)." *Physiologia Plantarum* 130, no. 1 (2007): 77–90.

Freeman, John L., Michael W. Persans, Ken Nieman, Carrie Albrecht, Wendy Peer, Ingrid J. Pickering, and David E. Salt. "Increased glutathione biosynthesis plays a role in nickel tolerance in Thlaspi nickel hyperaccumulators." *The Plant Cell* 16, no. 8 (2004): 2176–2191.

Freisinger, Eva. "Structural features specific to plant metallothioneins." *JBIC Journal of Biological Inorganic Chemistry* 16 (2011): 1035–1045.

Fujita, Masayuki. "The presence of two Cd-binding components in the roots of water hyacinth cultivated in a Cd2+-containing medium." *Plant and Cell Physiology* 26, no. 2 (1985): 295–300.

Fujita, Masayuki, and Takahiro Kawanishi. "Purification and characterization of a Cd-binding complex from the root tissue of water hyacinth cultivated in a Cd2+-containing medium." *Plant and Cell Physiology* 27, no. 7 (1986): 1317–1325.

Fusco, Nicola, Lorenza Micheletto, Giovanni Dal Corso, Lorena Borgato, and Antonella Furini. "Identification of cadmium-regulated genes by cDNA-AFLP in the heavy metal accumulator *Brassica juncea* L." *Journal of Experimental Botany* 56, no. 421 (2005): 3017–3027.

Garcıa-Hernández, Margarita, Angus Murphy, and Lincoln Taiz. "Metallothioneins 1 and 2 have distinct but overlapping expression patterns in Arabidopsis." *Plant Physiology* 118, no. 2 (1998): 387–397.

Garg, Neera, and Nalini Aggarwal. "Effect of mycorrhizal inoculations on heavy metal uptake and stress alleviation of *Cajanus cajan* (L.) Millsp. genotypes grown in cadmium and lead contaminated soils." *Plant Growth Regulation* 66 (2012): 9–26.

Garg, Neera, and Shikha Chandel. "Arbuscular mycorrhizal networks: process and functions." *Sustainable Agriculture* 2 (2011): 907–930.

Gechev, Tsanko S., Frank Van Breusegem, Julie M. Stone, Iliya Denev, and Christophe Laloi. "Reactive oxygen species as signals that modulate plant stress responses and programmed cell death." *Bioessays* 28, no. 11 (2006): 1091–1101.

Godbold, D. L., W. J. Horst, J. C. Collins, D. A. Thurman, and H. Marschner. "Accumulation of zinc and organic acids in roots of zinc tolerant and non-tolerant ecotypes of Deschampsiacaespitosa." *Journal of Plant Physiology* 116, no. 1 (1984): 59–69.

Gohari, M., A. R. Habib-Zadeh, and M. Khayat. "Assessing the intensity of tolerance to lead and its effect on amount of protein and proline in root and aerial parts of two varieties of rape seed (*Brassica napus* L.)." *Journal of Basic and Applied Scientific Research* 2, no. 1 (2012): 935–938.

Grennan, Aleel K. "Metallothioneins, a diverse protein family." *Plant Physiology* 155, no. 4 (2011): 1750–1751.

Grill, Erwin, Susanne Löffler, Ernst-L. Winnacker, and Meinhart H. Zenk. "Phytochelatins, the heavy-metal-binding peptides of plants, are synthesized from glutathione by a specific γ-glutamylcysteine dipeptidyl transpeptidase (phytochelatin synthase)." *Proceedings of the National Academy of Sciences* 86, no. 18 (1989): 6838–6842.

Grill, Erwin, Josef Thumann, Ernst-Ludwig Winnacker, and Meinhart H. Zenk. "Induction of heavy-metal binding phytochelatins by inoculation of cell cultures in standard media." *Plant Cell Reports* 7, no. 6 (1988): 375–378.

Grill, Erwin, Ernst-L. Winnacker, and Meinhart H. Zenk. "Phytochelatins, a class of heavy-metal-binding peptides from plants, are functionally analogous to metallothioneins." *Proceedings of the National Academy of Sciences* 84, no. 2 (1987): 439–443.

Groppa, María Daniela, and María Patricia Benavides. "Polyamines and abiotic stress: recent advances." *Amino Acids* 34 (2008): 35–45.

Groppa, María D., Myriam S. Zawoznik, María L. Tomaro, and María P. Benavides. "Inhibition of root growth and polyamine metabolism in sunflower (*Helianthus annuus*) seedlings under cadmium and copper stress." *Biological Trace Element Research* 126 (2008): 246–256.

Gupta, D. K., H. G. Huang, X. E. Yang, B. H. N. Razafindrabe, and M. Inouhe. "The detoxification of lead in Sedum alfredii H. is not related to phytochelatins but the glutathione." *Journal of Hazardous Materials* 177, no. 1–3 (2010): 437–444.

Ha, Hyo Chol, Nilantha S. Sirisoma, Periannan Kuppusamy, Jay L. Zweier, Patrick M. Woster, and Robert A. Casero Jr. "The natural polyamine spermine functions directly as a free radical scavenger." *Proceedings of the National Academy of Sciences* 95, no. 19 (1998): 11140–11145.

Hall, J. L. "Cellular mechanisms for heavy metal detoxification and tolerance." *Journal of Experimental Botany* 53, no. 366 (2002): 1–11.

Halliwell, Barry. "Reactive species and antioxidants. Redox biology is a fundamental theme of aerobic life." *Plant Physiology* 141, no. 2 (2006): 312–322.

Handique, Gautam, K., and A. K. Handique. "Proline accumulation in lemongrass (*Cymbopogon flexuosus* Stapf.) due to heavy metal stress." *Journal of Environmental Biology* 30, no. 2 (2009): 299–302.

Hao, Fushun, Xuechen Wang, and Jia Chen. "Involvement of plasma-membrane NADPH oxidase in nickel-induced oxidative stress in roots of wheat seedlings." *Plant Science* 170, no. 1 (2006): 151–158.

Harada, Emiko, Ji-Ah Kim, Andreas J. Meyer, Rüdiger Hell, Stephan Clemens, and Yong-Eui Choi. "Expression profiling of tobacco leaf trichomes identifies genes for biotic and abiotic stresses." *Plant and Cell Physiology* 51, no. 10 (2010): 1627–1637.

Hasan, Mudassara, and Fauzia Naushin. "Potentials of co-inoculation of microbial organisms and sewage sludge on growth of a pulse crop and microbial population." *Research Journal of Agricultural Sciences* 13, no. 4 (2022): 1046–1050.

Hasan, Mudassara, Fauzia Naushin, Mohd Irfan Naikoo, and Fareed Ahmad Khan. "Co-applications of rhizobium and arbuscular mycorrhizal fungi synergistically alleviate heavy metal uptake and toxicity in lentil plants (*Lens culinaris* Medik.) grown in sewage sludge amended soil." *South African Journal of Botany* 165 (2024): 349–366.

Hasan, Mudassara, Fauzia Naushin, Hina Shaher, and D. J. Bagyaraj. "Influence of sewage sludge, rhizobium and arbuscular mycorrhizal fungi on nutrient uptake, growth, photosynthetic and biochemical attributes in *Cicer arietinum* L." *Brazilian Journal of Botany* 46, no. 4 (2023): 1161–1176.

Hauser, Marie-Theres. "Molecular basis of natural variation and environmental control of trichome patterning." *Frontiers in Plant Science* 5 (2014): 320.

Hayat, Shamsul, Quaiser Hayat, Mohammed Naseer Alyemeni, and Aquil Ahmad. "Proline enhances antioxidative enzyme activity, photosynthesis and yield of *Cicer arietinum* L. exposed to cadmium stress." *Acta Botanica Croatica* 72, no. 2 (2013): 323–335.

Hayat, Shamsul, Qaiser Hayat, Mohammed Nasser Alyemeni, Arif Shafi Wani, John Pichtel, and Aqil Ahmad. "Role of proline under changing environments: a review." *Plant Signaling & Behavior* 7, no. 11 (2012): 1456–1466.

Hegelund, Josefine Nymark, Michaela Schiller, Thomas Kichey, Thomas Hesselhøj Hansen, Pai Pedas, Søren Husted, and Jan Kofod Schjoerring. "Barley metallothioneins: MT3 and MT4 are localized in the grain aleurone layer and show differential zinc binding." *Plant Physiology* 159, no. 3 (2012): 1125–1137.

Heidarvand, Leila, and Reza Maali Amiri. "What happens in plant molecular responses to cold stress?" *Acta Physiologiae Plantarum* 32 (2010): 419–431.

Heyno, Eiri, Cornelia Klose, and Anja Krieger-Liszkay. "Origin of cadmium-induced reactive oxygen species production: mitochondrial electron transfer versus plasma membrane NADPH oxidase." *New Phytologist* 179, no. 3 (2008): 687–699.

Higuchi, Kyoko, Kazuya Suzuki, Hiromi Nakanishi, Hirotaka Yamaguchi, Naoko-Kishi Nishizawa, and Satoshi Mori. "Cloning of nicotianamine synthase genes, novel genes involved in the biosynthesis of phytosiderophores." *Plant Physiology* 119, no. 2 (1999): 471–480.

Hildebrandt, Ulrich, Marjana Regvar, and Hermann Bothe. "Arbuscular mycorrhiza and heavy metal tolerance." *Phytochemistry* 68, no. 1 (2007): 139–146.

Horst, W. J. "Factors responsible for genotypic manganese tolerance in cowpea (*Vigna unguiculata*)." *Plant and Soil* 72 (1983): 213–218.

Horst, Walter, J., and H. Marschner. "Effect of excessive manganese supply on uptake and translocation of calcium in bean plants (*Phaseolus vulgaris* L.)." *ZeitschriftfürPflanzenphysiologie* 87, no. 2 (1978): 137–148.

Hossain, Mohammad Anwar, and Masayuki Fujita. "Purification of glyoxalase I from onion bulbs and molecular cloning of its cDNA." *Bioscience, Biotechnology, and Biochemistry* 73, no. 9 (2009): 2007–2013.

Hossain, Mohammad Anwar, Mirza Hasanuzzaman, and Masayuki Fujita. "Up-regulation of antioxidant and glyoxalase systems by exogenous glycinebetaine and proline in mung bean confer tolerance to cadmium stress." *Physiology and Molecular Biology of Plants* 16 (2010): 259–272.

Hossain, Mohammad Anwar, Mohammad Zakir Hossain, and Masayuki Fujita. "Stress-induced changes of methylglyoxal level and glyoxalase I activity in pumpkin seedlings and cDNA cloning of glyoxalase I gene." *Australian Journal of Crop Science* 3, no. 2 (2009): 53.

Hossain, Mohammad Anwar, Pukclai Piyatida, Jaime A. Teixeira da Silva, and Masayuki Fujita. "Molecular mechanism of heavy metal toxicity and tolerance in plants: central role of glutathione in detoxification of reactive oxygen species and methylglyoxal and in heavy metal chelation." *Journal of Botany* 2012 (2012).

Huang, Guo-Yong, and You-Shao Wang. "Expression analysis of type 2 metallothionein gene in mangrove species (*Bruguiera gymnorrhiza*) under heavy metal stress." *Chemosphere* 77, no. 7 (2009): 1026–1029.

Huang, Jianwei W., Didier M. Pellet, Lisa A. Papernik, and Leon V. Kochian. "Aluminum interactions with voltage-dependent calcium transport in plasma membrane vesicles isolated from roots of aluminum-sensitive and-resistant wheat cultivars." *Plant Physiology* 110, no. 2 (1996): 561–569.

Huang, Yi, Shu Tao, and You-jian Chen. "The role of arbuscular mycorrhiza on change of heavy metal speciation in rhizosphere of maize in wastewater irrigated agriculture soil." *Journal of Environmental Sciences* 17, no. 2 (2005): 276–280.

Hussain, Syed Sarfraz, Muhammad Ali, Maqbool Ahmad, and Kadambot H. M. Siddique. "Polyamines: natural and engineered abiotic and biotic stress tolerance in plants." *Biotechnology Advances* 29, no. 3 (2011): 300–311.

Iannelli, Maria Adelaide, Fabrizio Pietrini, Lucia Fiore, Luca Petrilli, and Angelo Massacci. "Antioxidant response to cadmium in Phragmites australis plants." *Plant Physiology and Biochemistry* 40, no. 11 (2002): 977–982.

Iglesia-Turiño, Santiago, Anna Febrero, Olga Jauregui, Cristina Caldelas, Jose Luis Araus, and Jordi Bort. "Detection and quantification of unbound phytochelatin 2 in plant extracts of Brassica napus grown with different levels of mercury." *Plant Physiology* 142, no. 2 (2006): 742–749.

Ireland, H. Elyse, Steve J. Harding, Graham A. Bonwick, Michael Jones, Christopher J. Smith, and John H. H. Williams. "Evaluation of heat shock protein 70 as a biomarker of environmental stress in *Fucus serratus* and *Lemna minor*." *Biomarkers* 9, no. 2 (2004): 139–155.

Iwasaki, Kōzō, Katsutoshi Sakurai, and Eiichi Takahashi. "Copper binding by the root cell walls of Italian ryegrass and red clover." *Soil Science and Plant Nutrition* 36, no. 3 (1990): 431–439.

Jahan, Urfi, Uzma Kafeel, Mohd Irfan Naikoo, Mohd Kaifiyan, Mudassara Hasan, and Fareed Ahmad Khan. "Trophic transfer, bioaccumulation, and detoxification of lead and zinc via sewage sludge applied soil-barley-aphid-ladybird food chain." *Water, Air, & Soil Pollution* 234, no. 8 (2023): 508.

Jin, Xiaofen, Xiaoe Yang, Ejazul Islam, Dan Liu, and Qaisar Mahmood. "Effects of cadmium on ultrastructure and antioxidative defense system in hyperaccumulator and non-hyperaccumulator ecotypes of Sedum alfredii Hance." *Journal of Hazardous Materials* 156, no. 1–3 (2008): 387–397.

Johansson, Jonas F., Leslie R. Paul, and Roger D. Finlay. "Microbial interactions in the mycorrhizosphere and their significance for sustainable agriculture." *FEMS Microbiology Ecology* 48, no. 1 (2004): 1–13.

John, Riffat, Parvaiz Ahmad, K. Gadgil, and Satyawati Sharma. "Heavy metal toxicity: effect on plant growth, biochemical parameters and metal accumulation by *Brassica juncea* L." *International Journal of Plant Production* 3, no. 3 (2009): 66–75.

Karimi, L. Naser, Masumeh Khanahmadi, and Bahareh Moradi. "Accumulation and phytotoxicity of lead in *Cynara scolymus*." *Indian Journal of Science and Technology* (2012): 3634–3641.

Kaur, G., S. Kumar, H. Nayyar, and H. D. Upadhyaya. "Cold stress injury during the pod-filling phase in chickpea (*Cicer arietinum* L.): effects on quantitative and qualitative components of seeds." *Journal of Agronomy and Crop Science* 194, no. 6 (2008): 457–464.

Khan, M. Iqbal R., Priyanka Chopra, Himanshu Chhillar, Mohammad Abass Ahanger, Sofi Javed Hussain, and Chirag Maheshwari. "Regulatory hubs and strategies for improving heavy metal tolerance in plants: chemical messengers, omics and genetic engineering." *Plant Physiology and Biochemistry* 164 (2021a): 260–278.

Khan, M. Iqbal R., Badar Jahan, Mohamed F. AlAjmi, Md Tabish Rehman, Noushina Iqbal, Mohammad Irfan, Zebus Sehar, and Nafees A. Khan. "Crosstalk of plant growth regulators protects photosynthetic performance from arsenic damage by modulating defense systems in rice." *Ecotoxicology and Environmental Safety* 222 (2021b): 112535.

Kohler, Annegret, Damien Blaudez, Michel Chalot, and Francis Martin. "Cloning and expression of multiple metallothioneins from hybrid poplar." *New Phytologist* (2004): 83–93.

Krämer, Ute, Janet D. Cotter-Howells, John M. Charnock, Alan J. M. Baker, and J. Andrew C. Smith. "Free histidine as a metal chelator in plants that accumulate nickel." *Nature* 379, no. 6566 (1996): 635–638.

Krämer, Ute, Ina N. Talke, and Marc Hanikenne. "Transition metal transport." *FEBS Letters* 581, no. 12 (2007): 2263–2272.

Krantev, Alexander, Rusina Yordanova, Tibor Janda, Gabriella Szalai, and Losanka Popova. "Treatment with salicylic acid decreases the effect of cadmium on photosynthesis in maize plants." *Journal of Plant Physiology* 165, no. 9 (2008): 920–931.

Kusano, T., T. Berberich, C. Tateda, and Y. Takahashi. "Polyamines: essential factors for growth and survival." *Planta* 228 (2008): 367–381.

Lamattina, Lorenzo, Carlos García-Mata, Magdalena Graziano, and Gabriela Pagnussat. "Nitric oxide: the versatility of an extensive signal molecule." *Annual Review of Plant Biology* 54, no. 1 (2003): 109–136.

Lanfranco, Luisa, Angelo Bolchi, Emanuele Cesale Ros, Simone Ottonello, and Paola Bonfante. "Differential expression of a metallothionein gene during the presymbiotic versus the symbiotic phase of an arbuscular mycorrhizal fungus." *Plant Physiology* 130, no. 1 (2002): 58–67.

Lasat, Mitch M., Nicole S. Pence, David F. Garvin, Stephen D. Ebbs, and Leon V. Kochian. "Molecular physiology of zinc transport in the Zn hyperaccumulator *Thlaspi caerulescens*." *Journal of Experimental Botany* 51, no. 342 (2000): 71–79.

Laspina, N. V., M. D. Groppa, M. L. Tomaro, and M. P. Benavides. "Nitric oxide protects sunflower leaves against Cd-induced oxidative stress." *Plant Science* 169, no. 2 (2005): 323–330.

Lee, Julian, Roger D. Reeves, Robert R. Brooks, and Tanguy Jaffré. "Isolation and identification of a citrato-complex of nickel from nickel-accumulating plants." *Phytochemistry* 16, no. 10 (1977): 1503–1505.

Lee, Julian, Roger D. Reeves, Robert R. Brooks, and Tanguy Jaffré. "The relation between nickel and citric acid in some nickel-accumulating plants." *Phytochemistry* 17, no. 6 (1978): 1033–1035.

Lee, Miyoung, Kiyoul Lee, Joohyun Lee, Eun Woon Noh, and Youngsook Lee. "AtPDR12 contributes to lead resistance in Arabidopsis." *Plant Physiology* 138, no. 2 (2005): 827–836.

Lee, Sangman, Jae Sun Moon, Leslie L. Domier, and Schuyler S. Korban. "Molecular characterization of phytochelatin synthase expression in transgenic *Arabidopsis*." *Plant Physiology and Biochemistry* 40, no. 9 (2002): 727–733.

Leszczyszyn, Oksana I., Hasan T. Imam, and Claudia A. Blindauer. "Diversity and distribution of plant metallothioneins: a review of structure, properties and functions." *Metallomics* 5, no. 9 (2013): 1146–1169.

Leung, H. M., Z. H. Ye, and Ming Hung Wong. "Survival strategies of plants associated with arbuscular mycorrhizal fungi on toxic mine tailings." *Chemosphere* 66, no. 5 (2007): 905–915.

Li, Xiao Feng, Jian Feng Ma, and Hideaki Matsumoto. "Pattern of aluminum-induced secretion of organic acids differs between rye and wheat." *Plant Physiology* 123, no. 4 (2000): 1537–1544.

Lin, Chuan Chi, and Ching Huei Kao. "Excess copper induces an accumulation of putrescine in rice leaves." *Botanical Bulletin of Academia Sinica* 40 (1999).

Ling, Hong-Qing, Gudrun Koch, Helmut Bäumlein, and Martin W. Ganal. "Map-based cloning of chloronerva, a gene involved in iron uptake of higher plants encoding nicotianamine synthase." *Proceedings of the National Academy of Sciences* 96, no. 12 (1999): 7098–7103.

Liu, Houjun, Junling Zhang, Peter Christie, and Fusuo Zhang. "Influence of iron plaque on uptake and accumulation of Cd by rice (*Oryza sativa* L.) seedlings grown in soil." *Science of the Total Environment* 394, nos. 2–3 (2008): 361–368.

Lombi, Enzo, Fang-Jie Zhao, S. J. Dunham, and S. P. McGrath. "Phytoremediation of heavy metal–contaminated soils: natural hyperaccumulation versus chemically enhanced phytoextraction." *Journal of Environmental Quality* 30, no. 6 (2001): 1919–1926.

Lytle, C. Mel, Farrel W. Lytle, Nancy Yang, Jin-Hong Qian, Drew Hansen, Adel Zayed, and Norman Terry. "Reduction of Cr (VI) to Cr (III) by wetland plants: potential for in situ heavy metal detoxification." *Environmental Science & Technology* 32, no. 20 (1998): 3087–3093.

Macovei, Anca, Lorenzo Ventura, Mattia Donà, Matteo Faè, Alma Balestrazzi, and Daniela Carbonera. "Effects of heavy metal treatments on metallothionein expression profiles in white poplar (*Populus alba* L.) cell suspension cultures." *Analele Universitatii din Oradea, Fascicula Biologie* 1 (2010): 194–198.

Mahajan, Moksh, Faroza Nazir, Badar Jahan, Manzer H. Siddiqui, Noushina Iqbal, and M. Iqbal R. Khan. "Salicylic acid mitigates arsenic stress in rice (*Oryza sativa*) via modulation of nitrogen–sulfur assimilation, ethylene biosynthesis, and defense systems." *Agriculture* 13, no. 7 (2023): 1293.

Margoshes, Marvin, and Bert L. Vallee. "A cadmium protein from equine kidney cortex." *Journal of the American Chemical Society* 79, no. 17 (1957): 4813–4814.

Masion, A., and P. M. Bertsch. "Aluminium speciation in the presence of wheat root cell walls: a wet chemical study." *Plant, Cell & Environment* 20, no. 4 (1997): 504–512.

Mathys, Werner. "The role of malate, oxalate, and mustard oil glucosides in the evolution of zinc-resistance in herbage plants." *Physiologia Plantarum* 40, no. 2 (1977): 130–136.

Memon, Abdul R., Diğdem Aktoprakligil, Aylin Özdemir, and Anastassiia Vertii. "Heavy metal accumulation and detoxification mechanisms in plants." *Turkish Journal of Botany* 25, no. 3 (2001): 111–121.

Mendoza-Cózatl, David G., Emerald Butko, Franziska Springer, Justin W. Torpey, Elizabeth A. Komives, Julia Kehr, and Julian I. Schroeder. "Identification of high levels of phytochelatins, glutathione and cadmium in the phloem sap of Brassica napus. A role for thiol-peptides in the long-distance transport of cadmium and the effect of cadmium on iron translocation." *The Plant Journal* 54, no. 2 (2008): 249–259.

Metwally, Ashraf, Iris Finkemeier, Manfred Georgi, and Karl-Josef Dietz. "Salicylic acid alleviates the cadmium toxicity in barley seedlings." *Plant Physiology* 132, no. 1 (2003): 272–281.

Mhamdi, Amna, Guillaume Queval, Sejir Chaouch, Sandy Vanderauwera, Frank Van Breusegem, and Graham Noctor. "Catalase function in plants: a focus on Arabidopsis mutants as stress-mimic models." *Journal of Experimental Botany* 61, no. 15 (2010): 4197–4220.

Miyasaka, Susan C., J. George Buta, Robert K. Howell, and Charles D. Foy. "Mechanism of aluminum tolerance in snapbeans: root exudation of citric acid." *Plant Physiology* 96, no. 3 (1991): 737–743.

Mourato, Miguel, Rafaela Reis, and Luisa Louro Martins. "Characterization of plant antioxidative system in response to abiotic stresses: a focus on heavy metal toxicity." *Advances in Selected Plant Physiology Aspects* 12 (2012): 1–17.

Nagajyoti, P. C., K. Dtf Lee, and T. V. M. Sreekanth. "Heavy metals, occurrence and toxicity for plants: a review." *Environmental Chemistry Letters* 8 (2010): 199–216.

Nazir, Faroza, Badar Jahan, Noushina Iqbal, Ashish B. Rajurkar, Manzer H. Siddiqui, and M. Iqbal R. Khan. "Methyl jasmonate influences ethylene formation, defense systems, nutrient homeostasis and carbohydrate metabolism to alleviate arsenic-induced stress in rice (*Oryza sativa*)." *Plant Physiology and Biochemistry* 202 (2023a): 107990.

Nazir, Faroza, Badar Jahan, Sarika Kumari, Noushina Iqbal, Mohammed Albaqami, Adriano Sofo, and M. Iqbal R. Khan. "Brassinosteroid modulates ethylene synthesis and antioxidant metabolism to protect rice (*Oryza sativa*) against heat stress-induced inhibition of source–sink capacity and photosynthetic and growth attributes." *Journal of Plant Physiology* 289 (2023b): 154096.

Neill, Steven, Radhika Desikan, and John Hancock. "Hydrogen peroxide signalling." *Current Opinion in Plant Biology* 5, no. 5 (2002): 388–395.

Neumann, Dieter, Olaf Lichtenberger, Detlef Günther, K. Tschiersch, and Lutz Nover. "Heat-shock proteins induce heavy-metal tolerance in higher plants." *Planta* 194 (1994): 360–367.

Neumann, Dieter, Uta Zur Nieden, Olaf Lichtenberger, and Ines Leopold. "How does *Armeria maritima* tolerate high heavy metal concentrations?" *Journal of Plant Physiology* 146, nos. 5–6 (1995): 704–717.

Noctor, Graham, and Christine H. Foyer. "Ascorbate and glutathione: keeping active oxygen under control." *Annual Review of Plant Biology* 49, no. 1 (1998): 249–279.

Nouairi, Issam, Wided Ben Ammar, Nabil Ben Youssef, Douja Daoud Ben Miled, Mohamed Habib Ghorbal, and Mokhtar Zarrouk. "Antioxidant defense system in leaves of Indian mustard (*Brassica juncea*) and rape (*Brassica napus*) under cadmium stress." *Acta Physiologiae Plantarum* 31 (2009): 237–247.

Ouziad, Fouad, Ulrich Hildebrandt, Elmon Schmelzer, and Hermann Bothe. "Differential gene expressions in arbuscular mycorrhizal-colonized tomato grown under heavy metal stress." *Journal of Plant Physiology* 162, no. 6 (2005): 634–649.

Oven, Matjaž, Erwin Grill, Avi Golan-Goldhirsh, Toni M. Kutchan, and Meinhart H. Zenk. "Increase of free cysteine and citric acid in plant cells exposed to cobalt ions." *Phytochemistry* 60, no. 5 (2002): 467–474.

Pagani, María Ayelén, M. Tomas, J. Carrillo, R. Bofill, M. Capdevila, S. Atrian, and Carlos Santiago Andreo. "The response of the different soybean metallothionein isoforms to cadmium intoxication." *Journal of Inorganic Biochemistry* 117 (2012): 306–315.

Pant, Preeti Pandey, A. K. Tripathi, and Vivek Dwivedi. "Effect of heavy metals on some biochemical parameters of sal (*Shorea robusta*) seedling at nursery level, Doon Valley, India." *Journal of Agricultural Sciences* 2, no. 1 (2011): 45–51.

Patra, Manomita, Niladri Bhowmik, Bulbul Bandopadhyay, and Archana Sharma. "Comparison of mercury, lead and arsenic with respect to genotoxic effects on plant systems and the development of genetic tolerance." *Environmental and Experimental Botany* 52, no. 3 (2004): 199–223.

Pence, Nicole S., Paul B. Larsen, Stephen D. Ebbs, Deborah L. D. Letham, Mitch M. Lasat, David F. Garvin, David Eide, and Leon V. Kochian. "The molecular physiology of heavy metal transport in the Zn/Cd hyperaccumulatorThlaspicaerulescens." *Proceedings of the National Academy of Sciences* 97, no. 9 (2000): 4956–4960.

Pich, Axel, Renate Manteuffel, Stefan Hillmer, Günter Scholz, and Wolfgang Schmidt. "Fe homeostasis in plant cells: does nicotianamine play multiple roles in the regulation of cytoplasmic Fe concentration?" *Planta* 213 (2001): 967–976.

Pilon-Smits, Elizabeth, and Marinus Pilon. "Phytoremediation of metals using transgenic plants." *Critical Reviews in Plant Sciences* 21, no. 5 (2002): 439–456.

Pinto, A. P., I. Sim [Otilde] Es, and A. M. Mota. "Cadmium impact on root exudates of sorghum and maize plants: a speciation study." *Journal of Plant Nutrition* 31, no. 10 (2008): 1746–1755.

Pourrut, Bertrand, Geoffrey Perchet, Jérôme Silvestre, Marie Cecchi, Maritxu Guiresse, and Eric Pinelli. "Potential role of NADPH-oxidase in early steps of lead-induced oxidative burst in Vicia faba roots." *Journal of Plant Physiology* 165, no. 6 (2008): 571–579.

Prabhavathi, V., and M. V. Rajam. "Mannitol-accumulating transgenic eggplants exhibit enhanced resistance to fungal wilts." *Plant Science* 173, no. 1 (2007): 50–54.

Qureshi, Javed Anver, David A. Thurman, K. Hardwick, and H. A. Collin. "Uptake and accumulation of zinc, lead and copper in zinc and lead tolerant *Anthoxanthum odoratum* L." *New Phytologist* 100, no. 3 (1985): 429–434.

Rahmaty, Raziyeh, and Jalil Khara. "Effects of vesicular arbuscular mycorrhiza Glomus intraradices on photosynthetic pigments, antioxidant enzymes, lipid peroxidation, and chromium accumulation in maize plants treated with chromium." *Turkish Journal of Biology* 35, no. 1 (2011): 51–58.

Rai, V. K. "Role of amino acids in plant responses to stresses." *Biologiaplantarum* 45, no. 4 (2002): 481–487.

Rascio, Nicoletta, and Flavia Navari-Izzo. "Heavy metal hyperaccumulating plants: how and why do they do it? And what makes them so interesting?" *Plant Science* 180, no. 2 (2011): 169–181.

Rastgoo, Leila, Abbas Alemzadeh, and Alireza Afsharifar. "Isolation of two novel isoforms encoding zinc- and copper-transporting P1B-ATPase from Gouan (*Aeluropus littoralis*)." *Plant Omics Journal* 4, no. 7 (2011): 377–383.

Rauser, Wilfried E. "Structure and function of metal chelators produced by plants: the case for organic acids, amino acids, phytin, and metallothioneins." *Cell Biochemistry and Biophysics* 31 (1999): 19–48.

Remans, Tony, Kelly Opdenakker, Karen Smeets, Dennis Mathijsen, Jaco Vangronsveld, and Ann Cuypers. "Metal-specific and NADPH oxidase dependent changes in lipoxygenase and NADPH oxidase gene expression in *Arabidopsis thaliana* exposed to cadmium or excess copper." *Functional Plant Biology* 37, no. 6 (2010): 532–544.

Riyazuddin, Riyazuddin, Nisha Nisha, Bushra Ejaz, M. Iqbal R. Khan, Manu Kumar, Pramod W. Ramteke, and Ravi Gupta. "A comprehensive review on the heavy metal toxicity and sequestration in plants." *Biomolecules* 12, no. 1 (2021): 43.

Robinson, Nigel J., Catherine M. Procter, Erin L. Connolly, and Mary Lou Guerinot. "A ferric-chelate reductase for iron uptake from soils." *Nature* 397, no. 6721 (1999): 694–697.

Rodríguez-Serrano, María, María C. Romero-Puertas, Imogen Sparkes, Chris Hawes, Luis A. del Río, and Luisa M. Sandalio. "Peroxisome dynamics in Arabidopsis plants under oxidative stress induced by cadmium." *Free Radical Biology and Medicine* 47, no. 11 (2009): 1632–1639.

Romero-Puertas, M. C., M. Rodríguez-Serrano, F. J. Corpas, M. del Gomez, L. A. Del Rio, and L. M. Sandalio. "Cadmium-induced subcellular accumulation of O2·– and H2O2 in pea leaves." *Plant, Cell & Environment* 27, no. 9 (2004): 1122–1134.

Ruscitti, Marcela, María Arango, Marta Ronco, and José Beltrano. "Inoculation with mycorrhizal fungi modifies proline metabolism and increases chromium tolerance in pepper plants (*Capsicum annuum* L.)." *Brazilian Journal of Plant Physiology* 23 (2011): 15–25.

Salazar, F. S., S. Pandey, L. Narro, J. C. Perez, H. Ceballos, S. N. Parentoni, and A. F. C. Bahia Filho. "Diallel analysis of acid-soil tolerant and intolerant tropical maize populations." *Crop Science* 37, no. 5 (1997): 1457–1462.

Salt, David E., Michael Blaylock, Nanda P. B. A. Kumar, Viatcheslav Dushenkov, Burt D. Ensley, Ilan Chet, and Ilya Raskin. "Phytoremediation: a novel strategy for the removal of toxic metals from the environment using plants." *Bio/technology* 13, no. 5 (1995): 468–474.

Salt, David E., Roger C. Prince, Alan J. M. Baker, Ilya Raskin, and Ingrid J. Pickering. "Zinc ligands in the metal hyperaccumulator *Thlaspi caerulescens* as determined using X-ray absorption spectroscopy." *Environmental Science & Technology* 33, no. 5 (1999): 713–717.

Salt, David E., and Wilfried E. Rauser. "MgATP-dependent transport of phytochelatins across the tonoplast of oat roots." *Plant Physiology* 107, no. 4 (1995): 1293–1301.

Sanchez, Diego H., Fernando L. Pieckenstain, Jedrzey Szymanski, Alexander Erban, MariuszBromke, Matthew A. Hannah, Ute Kraemer, Joachim Kopka, and Michael K. Udvardi. "Comparative functional genomics of salt stress in related model and cultivated plants identifies and overcomes limitations to translational genomics." *PloS One* 6, no. 2 (2011): e17094.

Sarry, Jean-Emmanuel, Lauriane Kuhn, Céline Ducruix, Alexandra Lafaye, Christophe Junot, Véronique Hugouvieux, Agnès Jourdain et al. "The early responses of Arabidopsis thaliana cells to cadmium exposure explored by protein and metabolite profiling analyses." *Proteomics* 6, no. 7 (2006): 2180–2198.

Saxena, Mukesh, Rekha Bisht, Suchandra Deb Roy, S. K. Sopory, and Neera Bhalla-Sarin. "Cloning and characterization of a mitochondrial glyoxalase II from *Brassica juncea* that is upregulated by NaCl, Zn, and ABA." *Biochemical and Biophysical Research Communications* 336, no. 3 (2005): 813–819.

Seth, Chandra Shekhar, Tony Remans, Els Keunen, Marijke Jozefczak, Heidi Gielen, Kelly Opdenakker, Nele Weyens, Jaco Vangronsveld, and Ann Cuypers. "Phytoextraction of toxic metals: a central role for glutathione." *Plant, Cell & Environment* 35, no. 2 (2012): 334–346.

Sharma, Pallavi, Ambuj Bhushan Jha, Rama Shanker Dubey, and Mohammad Pessarakli. "Reactive oxygen species, oxidative damage, and antioxidativedefense mechanism in plants under stressful conditions." *Journal of Botany* 2012 (2012): 1–26.

Sharma, Shanti S., and Karl-Josef Dietz. "The relationship between metal toxicity and cellular redox imbalance." *Trends in Plant Science* 14, no. 1 (2009): 43–50.

Shi, Qinghua, Fei Ding, Xiufeng Wang, and Min Wei. "Exogenous nitric oxide protect cucumber roots against oxidative stress induced by salt stress." *Plant Physiology and Biochemistry* 45, no. 8 (2007): 542–550.

Shi, Qinghua, and Zhujun Zhu. "Effects of exogenous salicylic acid on manganese toxicity, element contents and antioxidative system in cucumber." *Environmental and Experimental Botany* 63, nos. 1–3 (2008): 317–326.

Shruti, Mishra, and R. S. Dubey. "Heavy metal uptake and detoxification mechanisms in plants." *International Journal of Agricultural Research* 5, no. 7 (2010): 482–501.

Siddiqui, Manzer H., Mohamed H. Al-Whaibi, and Mohammed O. Basalah. "Role of nitric oxide in tolerance of plants to abiotic stress." *Protoplasma* 248 (2011): 447–455.

Siedlecka, Anna, and Tadeusz BaszyńAski. "Inhibition of electron flow around photosystem I in chloroplasts of Cd-treated maize plants is due to Cd-induced iron deficiency." *Physiologiaplantarum* 87, no. 2 (1993): 199–202.

Singh, Dharmendra, and S. K. Chauhan. "Organic acids of crop plants in aluminium detoxification." *Current Science* (2011): 1509–1515.

Singh, Karam B., Rhonda C. Foley, and Luis Oñate-Sánchez. "Transcription factors in plant defense and stress responses." *Current Opinion in Plant Biology* 5, no. 5 (2002): 430–436.

Singh, Nandita, Lena Q. Ma, Mrittunjai Srivastava, and Bala Rathinasabapathi. "Metabolic adaptations to arsenic-induced oxidative stress in Pteris vittata L and Pteris ensiformis L." *Plant Science* 170, no. 2 (2006): 274–282.

Singla-Pareek, S. L., M. K. Reddy, and S. K. Sopory. "Genetic engineering of the glyoxalase pathway in tobacco leads to enhanced salinity tolerance." *Proceedings of the National Academy of Sciences* 100, no. 25 (2003): 14672–14677.

Solanki, Radha, and Rajesh Dhankhar. "Biochemical changes and adaptive strategies of plants under heavy metal stress." *Biologia* 66, no. 2 (2011): 195–204.

Sytar, Oksana, Abhay Kumar, Dariusz Latowski, Paulina Kuczynska, Kazimierz Strzałka, and Majeti Narasimha Vara Prasad. "Heavy metal-induced oxidative damage, defense reactions, and detoxification mechanisms in plants." *Acta Physiologiae Plantarum* 35 (2013): 985–999.

Szabados, László, and Arnould Savouré. "Proline: a multifunctional amino acid." *Trends in Plant Science* 15, no. 2 (2010): 89–97.

Takács, T., B. Biró, and I. Vörös. "Arbuscular mycorrhizal effect on heavy metal uptake of ryegrass (*Lolium perenne* L.) in pot culture with polluted soils." In W. J. Horst et al. (Eds.). *Plant nutrition. Developments in plant and soil sciences*, pp. 480–481. 2001.

Takahashi, Michiko, Yasuko Terada, Izumi Nakai, Hiromi Nakanishi, Etsuro Yoshimura, Satoshi Mori, and Naoko K. Nishizawa. "Role of nicotianamine in the intracellular delivery of metals and plant reproductive development." *The Plant Cell* 15, no. 6 (2003): 1263–1280.

Takahashi, Shunichi, and Norio Murata. "How do environmental stresses accelerate photoinhibition?" *Trends in Plant Science* 13, no. 4 (2008): 178–182.

Tangahu, B. Voijant, S. Rozaimah Sheikh Abdullah, H. Basri, M. Idris, N. Anuar, and M. Mukhlisin. "Phytotoxicity of wastewater containing lead (Pb) effects *Scirpus grossus*." *International Journal of Phytoremediation* 15, no. 8 (2013): 814–826.

Thakur, Prince, Sanjeev Kumar, Jahid A. Malik, Jens D. Berger, and Harsh Nayyar. "Cold stress effects on reproductive development in grain crops: an overview." *Environmental and Experimental Botany* 67, no. 3 (2010): 429–443.

Thao, Nguyen Phuong, M. Iqbal R. Khan, Nguyen Binh Anh Thu, Xuan Lan Thi Hoang, Mohd Asgher, Nafees A. Khan, and Lam-Son Phan Tran. "Role of ethylene and its cross talk with other signaling molecules in plant responses to heavy metal stress." *Plant Physiology* 169, no. 1 (2015): 73–84.

Theriappan, P., Aditya K. Gupta, and P. Dhasarrathan. "Accumulation of proline under salinity and heavy metal stress in cauliflower seedlings." *Journal of Applied Sciences and Environmental Management* 15, no. 2 (2011).

Thornalley, Paul J. "The glyoxalase system: new developments towards functional characterization of a metabolic pathway fundamental to biological life." *Biochemical Journal* 269, no. 1 (1990): 1.

Thumann, Josef, Erwin Grill, Ernst-L. Winnacker, and Meinhart H. Zenk. "Reactivation of metal-requiring apoenzymes by phytochelatin – metal complexes." *FEBS Letters* 284, no. 1 (1991): 66–69.

Tice, Kathy R., David R. Parker, and Darleen A. DeMason. "Operationally defined apoplastic and symplastic aluminum fractions in root tips of aluminum-intoxicated wheat." *Plant Physiology* 100, no. 1 (1992): 309–318.

Tripathi, B. N., and J. P. Gaur. "Relationship between copper-and zinc-induced oxidative stress and proline accumulation in Scenedesmus sp." *Planta* 219 (2004): 397–404.

Tsuji, Naoki, Nayumi Hirayanagi, Megumi Okada, Hitoshi Miyasaka, Kazumasa Hirata, Meinhart H. Zenk, and Kazuhisa Miyamoto. "Enhancement of tolerance to heavy metals and oxidative stress in *Dunaliella tertiolecta* by Zn-induced phytochelatin synthesis." *Biochemical and Biophysical Research Communications* 293, no. 1 (2002): 653–659.

Turnau, Katarzyna. "Heavy metal content and localization in mycorrhizal Euphorbia cyparissias zinc wastes in southern Poland." *Acta Societatis Botanicorum Poloniae* 67, no. 1 (1998): 105–113.

Turnau, Katarzyna, Przemysław Ryszka, Vivienne Gianinazzi-Pearson, and Diederik Van Tuinen. "Identification of arbuscular mycorrhizal fungi in soils and roots of plants colonizing zinc wastes in southern Poland." *Mycorrhiza* 10 (2001): 169–174.

van de Mortel, Judith E., Henk Schat, Perry D. Moerland, Emiel Ver Loren Van Themaat, S. J. O. E. R. D. Van Der Ent, Hetty Blankestijn, Artak Ghandilyan, Styliani Tsiatsiani, and Mark G. M. Aarts. "Expression differences for genes involved in lignin, glutathione and sulphate metabolism in response to cadmium in *Arabidopsis thaliana* and the related Zn/Cd-hyperaccumulator *Thlaspi caerulescens*." *Plant, Cell & Environment* 31, no. 3 (2008): 301–324.

Van Steveninck, R. F. M., Margaret E. Van Steveninck, Denise R. Fernando, Walter J. Horst, and H. Marschner. "Deposition of zinc phytate in globular bodies in roots of *Deschampsia caespitosa* ecotypes; a detoxification mechanism?" *Journal of Plant Physiology* 131, nos. 3–4 (1987): 247–257.

Vázquez, María Dolores, Charlotte Poschenrieder, Isabel Corrales, and Juan Barceló. "Change in apoplastic aluminum during the initial growth response to aluminum by roots of a tolerant maize variety." *Plant Physiology* 119, no. 2 (1999): 435–444.

Verkleij, J. A. C., F. E. C. Sneller, and H. Schat. "Metallothioneins and phytochelatins: ecophysiological aspects." In *Sulphur in plants*, pp. 163–176. Dordrecht, The Nether lands: Springer Netherlands, 2003.

Vogeli-Lange, Regina, and George J. Wagner. "Subcellular localization of cadmium and cadmium-binding peptides in tobacco leaves: implication of a transport function for cadmium-binding peptides." *Plant Physiology* 92, no. 4 (1990): 1086–1093.

Von Wiren, Nicolaus, Sukhbinder Klair, Suhkibar Bansal, Jean-Francois Briat, Hicham Khodr, Takayuki Shioiri, Roger A. Leigh, and Robert C. Hider. "Nicotianamine chelates both FeIII and FeII. Implications for metal transport in plants." *Plant Physiology* 119, no. 3 (1999): 1107–1114.

Wagner, George J. "Accumulation of cadmium in crop plants and its consequences to human health." *Advances in Agronomy* 51 (1993): 173–212.

Wang, Feijuan, Bin Zeng, Zongxiu Sun, and Cheng Zhu. "Relationship between proline and Hg 2+-induced oxidative stress in a tolerant rice mutant." *Archives of Environmental Contamination and Toxicology* 56 (2009): 723–731.

Wang, Jian, Bill P. Evangelou, Mark T. Nielsen, and George J. Wagner. "Computer-simulated evaluation of possible mechanisms for quenching heavy metal ion activity in plant vacuoles: I. Cadmium." *Plant Physiology* 97, no. 3 (1991): 1154–1160.

Wang, Jian, Bill P. Evangelou, Mark T. Nielsen, and George J. Wagner. "Computer, simulated evaluation of possible mechanisms for sequestering metal ion activity in plant vacuoles: II. Zinc." *Plant Physiology* 99, no. 2 (1992): 621–626.

Weber, Michael, Emiko Harada, Christoph Vess, Edda V. Roepenack-Lahaye, and Stephan Clemens. "Comparative microarray analysis of Arabidopsis thaliana and Arabidopsis halleri roots identifies nicotianamine synthase, a ZIP transporter and other genes as potential metal hyperaccumulation factors." *The Plant Journal* 37, no. 2 (2004): 269–281.

Weyens, Nele, Sascha Truyens, Eline Saenen, Jana Boulet, Joke Dupae, Safiyh Taghavi, Daniel van der Lelie, Robert Carleer, and Jaco Vangronsveld. "Endophytes and their potential to deal with co-contamination of organic contaminants (toluene) and toxic metals (nickel) during phytoremediation." *International Journal of Phytoremediation* 13, no. 3 (2011): 244–255.

Wheeler, D. M., and I. L. Power. "Comparison of plant uptake and plant toxicity of various ions in wheat." *Plant and Soil* 172 (1995): 167–173.

Wong, Hann Ling, Tsuyoshi Sakamoto, Tsutomu Kawasaki, Kenji Umemura, and Ko Shimamoto. "Down-regulation of metallothionein, a reactive oxygen scavenger, by the small GTPase OsRac1 in rice." *Plant Physiology* 135, no. 3 (2004): 1447–1456.

Wuana, Raymond A., and Felix E. Okieimen. "Heavy metals in contaminated soils: a review of sources, chemistry, risks and best available strategies for remediation." *International Scholarly Research Notices* 2011 (2011): 1–20.

Wycisk, Katharina, Eugene J. Kim, Julian I. Schroeder, and Ute Krämer. "Enhancing the first enzymatic step in the histidine biosynthesis pathway increases the free histidine pool and nickel tolerance in *Arabidopsis thaliana*." *FEBS Letters* 578, nos. 1–2 (2004): 128–134.

Xia, Yan, Ying Qi, Yuxiang Yuan, Guiping Wang, Jin Cui, Yahua Chen, Hongsheng Zhang, and Zhenguo Shen. "Overexpression of Elsholtzia haichowensis metallothionein 1 (EhMT1) in tobacco plants enhances copper tolerance and accumulation in root cytoplasm and decreases hydrogen peroxide production." *Journal of Hazardous Materials* 233 (2012): 65–71.

Xu, Jin, Wenying Wang, Jianhang Sun, Yuan Zhang, Qing Ge, Liguo Du, Hengxia Yin, and Xiaojing Liu. "Involvement of auxin and nitric oxide in plant Cd-stress responses." *Plant and Soil* 346 (2011): 107–119.

Yadav, Sudesh Kumar. "Heavy metals toxicity in plants: an overview on the role of glutathione and phytochelatins in heavy metal stress tolerance of plants." *South African Journal of Botany* 76, no. 2 (2010): 167–179.

Yang, Xiao-E., Xiao-Fen Jin, Ying Feng, and Ejazul Islam. "Molecular mechanisms and genetic basis of heavy metal tolerance/hyperaccumulation in plants." *Journal of Integrative Plant Biology* 47, no. 9 (2005): 1025–1035.

Yu, Guo, Jianchu Ma, Pingping Jiang, Jieyue Li, Junyu Gao, Shixuan Qiao, and Zhiyong Zhao. "The mechanism of plant resistance to heavy metal." *IOP Conference Series: Earth and Environmental Science* 310, no. 5 (2019): 052004. IOP Publishing.

Zengin, Fikriye Kirbag, and Sevda Kirbag. "Effects of copper on chlorophyll, proline, protein and abscisic acid level of sunflower (*Helianthus annuus* L.) seedlings." *Journal of Environmental Biology* 28, no. 3 (2007): 561.

Zenk, Meinhart H. "Heavy metal detoxification in higher plants – a review." *Gene* 179, no. 1 (1996): 21–30.

Zhang, Haiyan, Wenzhong Xu, Jiangbo Guo, Zhenyan He, and Mi Ma. "Coordinated responses of phytochelatins and metallothioneins to heavy metals in garlic seedlings." *Plant Science* 169, no. 6 (2005): 1059–1065.

Zhang, Wen-Hao, Peter R. Ryan, and Stephen D. Tyerman. "Malate-permeable channels and cation channels activated by aluminum in the apical cells of wheat roots." *Plant Physiology* 125, no. 3 (2001): 1459–1472.

Zhao, Xiaohong, Liu Han, Jun Xiao, Lingqing Wang, Tao Liang, and Xiaoyong Liao. "A comparative study of the physiological and biochemical properties of tomato (*Lycopersicon esculentum* M.) and maize (*Zea mays* L.) under palladium stress." *Science of the Total Environment* 705 (2020): 135938.

Zhen, Yan, Jin-Liang Qi, Si-Si Wang, Jing Su, Guo-Hua Xu, Ming-Sheng Zhang, L. V. Miao, Xin-Xiang Peng, Dacheng Tian, and Yong-Hua Yang. "Comparative proteome analysis of differentially expressed proteins induced by Al toxicity in soybean." *Physiologia Plantarum* 131, no. 4 (2007): 542–554.

Zheng, Chunfang, Dong Jiang, Fulai Liu, Tingbo Dai, Weicheng Liu, Qi Jing, and Weixing Cao. "Exogenous nitric oxide improves seed germination in wheat against mitochondrial oxidative damage induced by high salinity." *Environmental and Experimental Botany* 67, no. 1 (2009): 222–227.

Zheng, Shao Jian, Jian Feng Ma, and Hideaki Matsumoto. "High aluminum resistance in buckwheat: I. Al-induced specific secretion of oxalic acid from root tips." *Plant Physiology* 117, no. 3 (1998): 745–751.

Zhigang, An, Li Cuijie, Zu Yuangang, Du Yejie, Andreas Wachter, Roland Gromes, and Thomas Rausch. "Expression of BjMT2, a metallothionein 2 from *Brassica juncea*, increases copper and cadmium tolerance in *Escherichia coli* and *Arabidopsis thaliana*, but inhibits root elongation in *Arabidopsis thaliana* seedlings." *Journal of Experimental Botany* 57, no. 14 (2006): 3575–3582.

Zhou, Boru, Wenjing Yao, Shengji Wang, Xinwang Wang, and Tingbo Jiang. "The metallothionein gene, TaMT3, from *Tamarix androssowii* confers Cd2+ tolerance in tobacco." *International Journal of Molecular Sciences* 15, no. 6 (2014): 10398–10409.

Zhou, Zhao Sheng, Kai Guo, Abdelrahman Abdou Elbaz, and Zhi Min Yang. "Salicylic acid alleviates mercury toxicity by preventing oxidative stress in roots of *Medicago sativa*." *Environmental and Experimental Botany* 65, no. 1 (2009): 27–34.

Zhu, Xiao Fang, Zhi Wei Wang, Fang Dong, Gui Jie Lei, Yuan Zhi Shi, Gui Xin Li, and Shao Jian Zheng. "Exogenous auxin alleviates cadmium toxicity in *Arabidopsis thaliana* by stimulating synthesis of hemicellulose 1 and increasing the cadmium fixation capacity of root cell walls." *Journal of Hazardous Materials* 263 (2013): 398–403.

Zhu, Xiao Fang, Cheng Zheng, Yi Ting Hu, T. A. O. Jiang, Y. U. Liu, Ning Yu Dong, Jian Li Yang, and Shao Jian Zheng. "Cadmium-induced oxalate secretion from root apex is associated with cadmium exclusion and resistance in *Lycopersicon esulentum*." *Plant, Cell & Environment* 34, no. 7 (2011): 1055–1064.

2 An Insight into the Role of Plant Growth Regulators in the Mitigation of Abiotic Stresses

Sajad Hussain Shah and Firoz Mohammad

2.1 INTRODUCTION

Abiotic stress refers to adverse environmental conditions that negatively impact the growth, development, and productivity of crop plants. The consequences of abiotic stress on crop plants carry significant implications for global food production and the burgeoning global population. Abiotic stressors, like water scarcity/drought, salinity, heat, temperature, and nutrient imbalances, can pose substantial challenges for agricultural systems, potentially leading to food security concerns and broader socio-economic impacts (Sako et al. 2020; Godoy et al. 2021; Khan et al. 2023a). Reduced crop yields resulting from abiotic stress can lead to food shortages and increased prices. These stressors can manifest various detrimental effects on plants, with the specific impact depending on the type and severity of the stress. They impede growth and physiological processes by hindering cell elongation, disrupting photosynthesis, decreasing water uptake and oxygen availability, inducing ion toxicity, and causing oxidative stress through the accumulation of reactive oxygen species within plant cells. These highly reactive molecules can damage cellular components, including cellular membrane stability, proteins, lipids, and DNA, nutrient imbalances, reduced reproductive success by interfering with flower development, pollen viability, and seed production (Khan et al. 2023a). These implications can impact the plant's long-term survival and dynamics (Petrov et al. 2015; You and Chan 2015; Nadarajah 2020). This often involves up-regulating stress-related genes while down-regulating genes associated with growth and development (Pereira 2016; Saharan et al. 2022). Researchers and farmers are continually exploring and implementing strategies to mitigate the impacts of abiotic stress on plants. Among these approaches, the application of plant growth regulators emerges as an efficient strategy to modulate physico-chemical processes within plants and enhance their endurance under adverse conditions. Plants have evolved a multitude of strategies to cope 1 with environmental challenges, including structural modifications, adjustments in osmotic balance, and alterations in their genetic activity to acclimate to changing conditions. This often involves up-regulating stress-related genes while down-regulating genes associated with growth and development (Pereira 2016; Khan et al. 2021a; Saharan et al. 2022). Researchers and farmers are continually exploring and implementing strategies to mitigate the impacts of abiotic stress on plants (Pereira 2016; Saharan et al. 2022). Among these approaches, the application of plant growth regulators (PGRs) emerges as an efficient strategy that modulates plant's physico-chemical processes and enhances the endurance of plants under adverse conditions. Plant growth regulators such as auxins, brassinosteroids (BR), gibberellins (GA), cytokinins, abscisic acid (ABA), jasmonic acid (JA), ethylene, polyamines, and salicylic acid (SA) are small signaling molecules that control and regulate various aspects of plant growth and development (Shah et al. 2021; Hossain et al. 2022; Khan et al. 2023a). These PGRs frequently interact and respond to environmental cues, influencing the physiological processes of plants. These regulators find wide application in agriculture and horticulture, where they are utilized to manipulate plant growth, enhance crop yields, and regulate plant responses to diverse environmental conditions (Nazir et al. 2023a; Zahid et al. 2023). In this

DOI: 10.1201/9781032630380-2

chapter, an effort has been made to summarize the adverse effects of abiotic stressors on plants, elucidate the mechanism of action of PGRs, and explore their role in mitigating the impacts of abiotic stressors.

2.2 NEGATIVE EFFECTS OF ABIOTIC STRESSORS ON PLANTS

Abiotic stressors have harmful effects on crops, impacting their growth, development, and overall productivity. Some common abiotic stressors and their negative effects on crops have been depicted in Figure 2.1.

2.2.1 Drought Stress

Drought stress poses a significant challenge to plants, profoundly affecting their morphological, physiological, and biochemical processes. As water scarcity becomes more severe, plants initiate a series of adaptive responses to cope with the impending water deficit. One of the initial responses involves the closure of stomata to reduce water loss through transpiration, thereby conserving internal water resources (Farooq et al. 2012; Iqbal et al. 2022; Nazir et al. 2024). However, this protective mechanism comes at a cost, as it hinders carbon dioxide uptake and photosynthesis. With photosynthesis compromised, plants experience a decline in energy production, resulting in reduced growth rates and biomass accumulation. Additionally, water scarcity disrupts nutrient

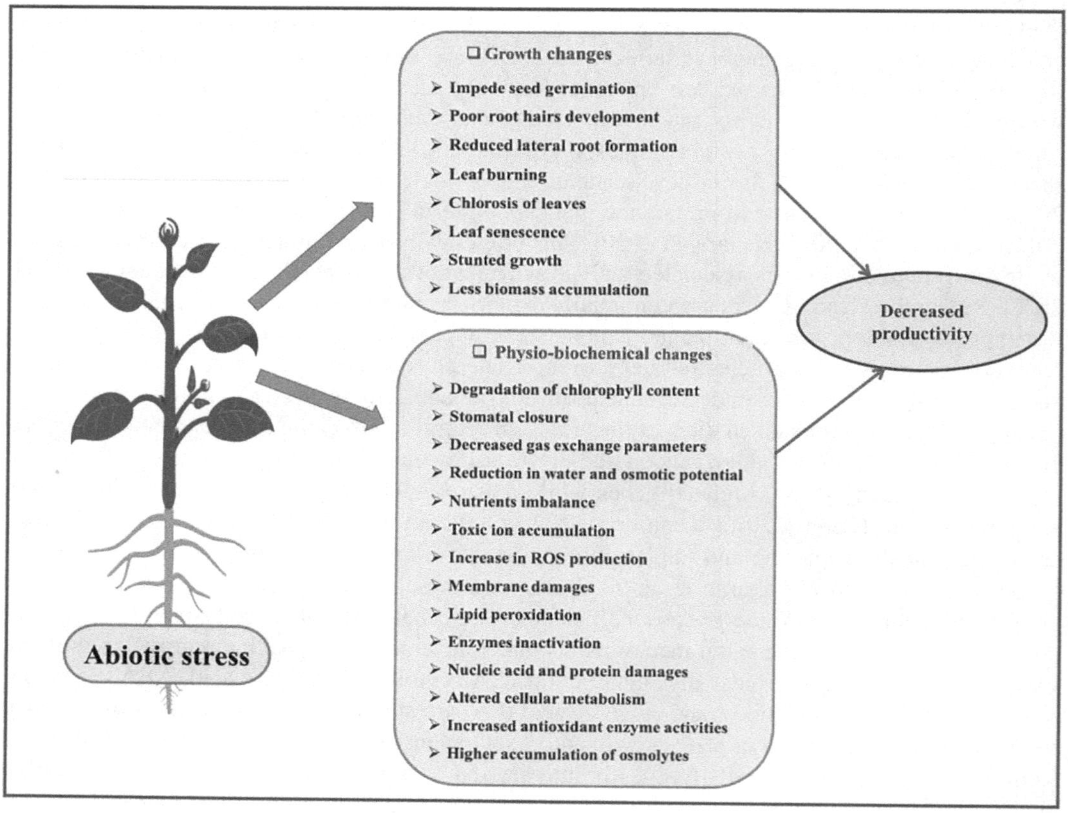

FIGURE 2.1 Common abiotic stressors and their negative effects on plants.

uptake, as many essential minerals depend on water for transport through the plant's vascular system. Consequently, nutrient imbalances further impede metabolic activities and exacerbate overall plant functions (Zargar et al. 2017; Khan et al. 2023b). The prolonged period of drought triggers hormonal changes, with increased levels of ABA, which orchestrates various stress responses. ABA induces the closure of stomata but also activates genes related to stress tolerance (Pirasteh-Anosheh et al. 2016). Drought stress significantly impacts root growth, as the plant allocates resources to deepen and expand its root system in search of water. This adaptive strategy, crucial for survival, comes with trade-offs as it competes with aboveground growth. Drought stress also induces the production of reactive oxygen species (ROS) within plant cells, leading to oxidative stress and damage to cellular structures (Ahmad et al. 2014; Burgess and Huang 2016; Khan et al. 2021b). Additionally, the impact of drought stress extends beyond the physiological realm, influencing plant architecture and development (Khan et al. 2021a). Stunted growth, leaf rolling, and premature senescence are common morphological responses to water scarcity. In reproductive organs, drought stress often results in decreased flower production, impaired pollination, and reduced seed set, ultimately compromising crop yield (Ratnakumar et al. 2016; Per et al. 2017). The severity of these impacts varies among plant species, genotypes, and developmental stages, highlighting the complex and multifaceted nature of drought stress. Crop losses due to drought stress have significant socio-economic implications, affecting food security and livelihoods (Salehi-Lisar and Bakhshayeshan-Agdam 2016).

2.2.2 Salt Stress

Salt stress, a prevalent abiotic stressor, occurs when the concentration of salts, particularly sodium and chloride ions, in the soil surpasses the tolerance levels of plants. This environmental challenge poses significant threats to plant growth and development (Parihar et al. 2015). High salt concentrations in the soil can induce osmotic stress, impeding water uptake by plant roots and leading to dehydration of plant cells. The osmotic imbalance triggers disruptions in cellular processes, compromising turgor pressure and ultimately inhibiting growth (Kumari et al. 2023, 2024). Furthermore, excessive salts can accumulate in plant tissues, causing ion toxicity. Sodium ions, in particular, interfere with vital cellular functions, displacing essential nutrients and leading to nutrient imbalances. This disruption in nutrient homeostasis negatively impacts photosynthesis, respiration, and other metabolic processes critical for plant survival. Additionally, salt stress induces oxidative stress by promoting ROS production, causing damage to cellular structures such as membranes and proteins (Nawaz et al. 2010; Yadav et al. 2019). Plants respond to salt stress through various mechanisms, including the activation of stress-responsive genes, the synthesis of osmolytes, and the enhancement of antioxidant systems. However, prolonged exposure to high salt levels can overwhelm these defense mechanisms, leading to long-term detrimental effects on plant health. Salt stress not only hampers growth and reduces yields but also renders plants more susceptible to other environmental stresses and pathogens. The impact of salt stress is particularly pronounced in arid and semi-arid regions where irrigation practices, compounded by poor drainage, can exacerbate soil salinity. Sustainable agricultural practices, such as precision irrigation, cultivation of salt-tolerant crop varieties through breeding programs, and the use of soil amendments to improve soil structure and reduce salinity, are essential for attenuating the adverse effects of salt stress (Yokoi et al. 2002). Understanding the intricate molecular and physiological responses of plants to salt stress is crucial for developing strategies to ensure food security in regions where salinity is a pervasive challenge (Hernández 2019). As global environmental changes continue, including alterations in precipitation patterns and rising sea levels, the management of salt stress in agriculture becomes increasingly imperative for sustaining crop production and addressing the complex interplay between soil salinity and the well-being of ecosystems.

2.2.3 Heavy Metals

Heavy metal stress is a significant environmental challenge for plants, exerting detrimental effects on their growth, development, and overall physiological well-being. When plants are exposed to elevated concentrations of heavy metals such as lead, cadmium, mercury, and arsenic, these metals can disrupt essential cellular processes. The primary mode of heavy metal toxicity involves interference with enzymatic activities, particularly those associated with essential metabolic pathways (Yadav 2010; Kumar et al. 2022a; Nazir et al. 2023b; Kaur et al. 2023). Heavy metals can bind to sulfhydryl groups in proteins, altering their structure and function. This interference extends to critical processes like photosynthesis, respiration, and nutrient uptake, resulting in a cascade of physiological disruptions. Additionally, heavy metals induce the overproduction of ROS, causing oxidative stress and damaging cellular components such as lipids, proteins, and DNA. The accumulation of heavy metals in plant tissues can also compromise the integrity of cellular membranes and disrupt ion homeostasis. The impact of heavy metal stress is further exacerbated by its ability to disrupt water balance and nutrient absorption in plants, leading to symptoms such as wilting, chlorosis, and stunted growth (Rasheed et al. 2021). Moreover, heavy metal stress extends beyond the individual plant level, as contaminated plants pose risks to ecosystems and human health when consumed. Heavy metals may accumulate in edible plant parts, potentially entering the food chain and posing health hazards (Shahid et al. 2015). Mitigating the impact of heavy metal stress on plants requires a multifaceted approach, including soil remediation techniques, phytoremediation strategies employing metal-accumulating plants, and genetic engineering to develop crop varieties with enhanced tolerance to heavy metals.

2.2.4 Heat and Temperature Stress

Heat and temperature stress exert profound effects on plants, influencing their physiological processes, growth, and overall performance. High temperatures can lead to a cascade of detrimental impacts on plant biology. One of the primary consequences is increased transpiration rates, causing water loss that can result in dehydration and compromised cell turgor pressure (Nazir et al. 2023c). This water stress is often exacerbated by the closure of stomata, responsible for gas exchange, which can limit carbon dioxide uptake for photosynthesis (Bita and Gerats 2013; Ul Hassan et al. 2021). Elevated temperatures can also disrupt photosynthetic machinery, reducing the efficiency of this crucial process and ultimately diminishing crop yields. Furthermore, heat stress can lead to protein denaturation and damage to cellular structures, impacting enzyme activity and disrupting essential metabolic pathways. Plants have evolved various mechanisms to cope with heat stress, including the synthesis of heat shock proteins that aid in protein stabilization, but prolonged exposure to extreme temperatures can overwhelm these protective mechanisms (Wahid et al. 2012; Poór et al. 2022). Additionally, temperature stress can influence the timing of developmental stages, affecting flowering and fruiting patterns, which can have implications for the synchronization of crop growth with seasonal changes. In some cases, high temperatures can lead to the acceleration of flowering, a phenomenon known as heat-induced flowering, potentially affecting the overall quality and quantity of harvested crops (Jha et al. 2014). Moreover, as global temperatures continue to rise due to climate change, the frequency and intensity of heat stress events are expected to increase, posing significant challenges for agriculture.

2.3 MECHANISM OF ACTION OF PLANT GROWTH REGULATORS

Cellular signaling pathways play a pivotal role in mediating the responses of plants to abiotic stress, and various PGRs act as key molecular players in these intricate processes. The mechanism of action of some common PGRs is discussed in Figure 2.2.

Abscisic acid, a crucial phytohormone, stands out as a central regulator in the plant's response to multiple abiotic stresses. ABA signaling involves a complex network of proteins, including receptors

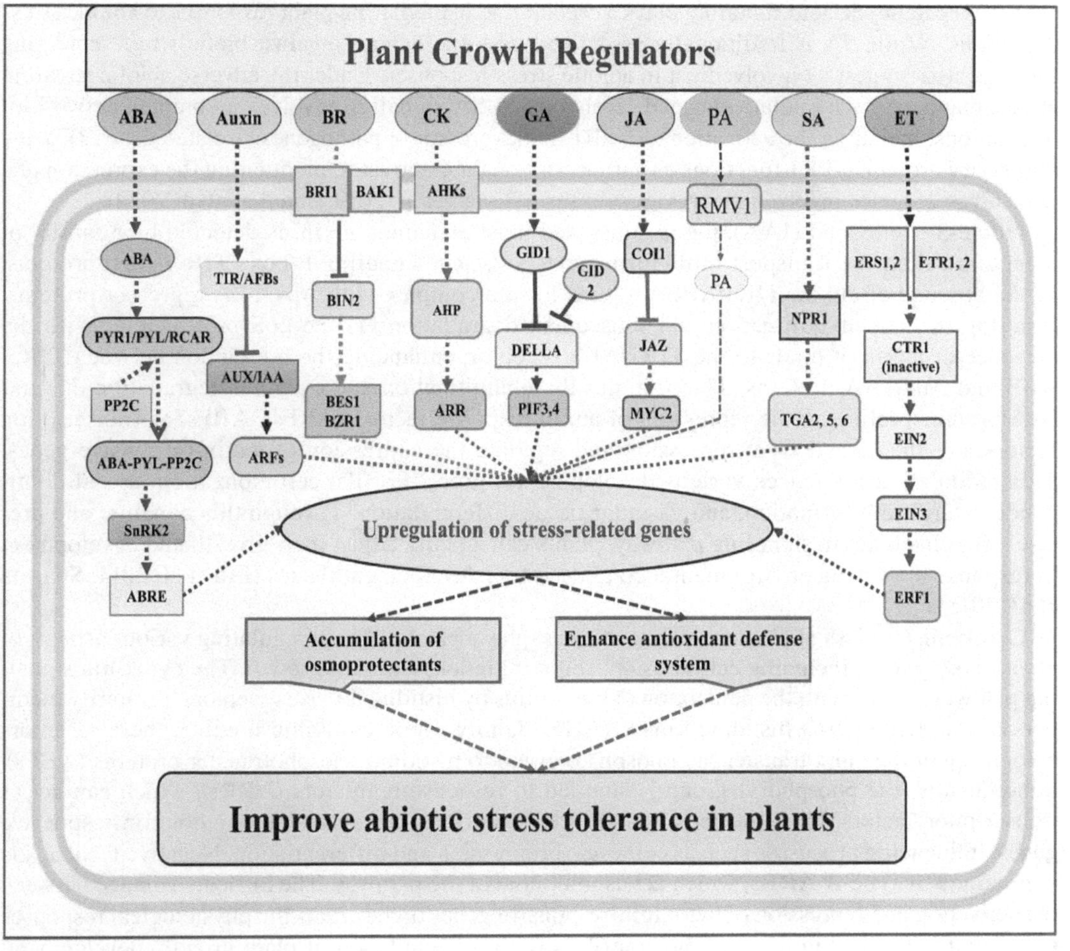

FIGURE 2.2 Mechanism of action of plant growth regulators and their role in abiotic stress mitigation (Shah et al. 2021).

such as PYR/PYL/RCAR, which perceive ABA and initiate a signaling cascade that leads to the activation of downstream effectors, such as SnRK2 protein kinases. These effectors regulate key processes such as stomatal closure, thus reducing water loss, and the activation of stress-responsive genes (Vishwakarma et al. 2017; Kumar et al. 2022b). Ethylene, another important plant hormone, modulates plant responses to various stresses. Ethylene signaling involves receptors such as ETR1 and EIN2, leading to the activation of transcription factors like ERF1. Ethylene regulates diverse physiological processes, such as fruit ripening and senescence, but its role in stress responses involves the induction of adaptive mechanisms to cope with adverse conditions (Sharma et al. 2019; Husain et al. 2020; Khan et al. 2020).

Jasmonic acid signaling in plants is typically associated with responses to biotic stresses, such as herbivory and pathogen attack. However, emerging evidence highlights its involvement in abiotic stress responses as well. Under abiotic stress conditions, plants can induce the synthesis of JA and activate JA signaling pathways. The JA signaling cascade involves the perception of JA by its receptor, COI1, leading to the degradation of JAZ proteins, which normally repress JA-responsive transcription factors. The release of these transcription factors enables the activation of downstream JA-responsive genes involved in various stress-related processes (Ali and Baek 2020; Wang et al. 2021). Salicylic acid signaling plays a crucial role in mediating plant responses to abiotic stress conditions. While SA is traditionally associated with the defense against biotic stress, emerging findings also suggest its involvement in abiotic stress responses. Under the adverse abiotic stressors effect, plants can accumulate elevated levels of SA. SA signaling involves a complex network of interactions, including the activation of NPR1 (nonexpressor of pathogenesis-related genes 1), a key regulatory protein. NPR1 functions as a transcriptional co-activator, modulating the expression of a wide array of genes involved in stress responses (Ahmad et al. 2019; Shaukat et al. 2022).

Indole-3-acetic acid (IAA), the primary and most abundant auxin, is detected by a family of receptors called the transport inhibitor response 1/auxin signaling F-box (TIR1/AFB) proteins. In the absence of auxin, TIR1/AFB proteins form a complex with Aux/IAA repressor proteins, resulting in their ubiquitination and subsequent degradation via the 26S proteasome. With the presence of auxin, it binds to the TIR1/AFB receptor, enhancing the interaction between TIR1/AFB and Aux/IAA proteins. This initiates the ubiquitination and eventual degradation of Aux/IAA proteins, relieving the repression of auxin response factors (ARFs). ARFs, as transcription factors, are then freed from repression and regulate the expression of auxin-responsive genes. This modulation influences various developmental processes like cell elongation, apical dominance, lateral root formation, and vascular tissue differentiation. Through this dynamic and precisely regulated auxin signaling pathway, plants can adeptly adjust their growth and development in response to varying environmental conditions and developmental cues (Liu et al. 2014; Sharma et al. 2015).

Cytokinins, a class of plant growth regulators, play a pivotal role in regulating various aspects of plant development, including cell division, shoot initiation, and root growth. The cytokinin signaling pathway begins with the perception of cytokinins by histidine kinase receptors, primarily members of the *Arabidopsis* histidine kinase (AHK) family. Upon cytokinin binding, these receptors autophosphorylate and transfer the phosphate group to histidine phosphotransfer proteins (AHPs). Subsequently, the phosphoryl group is shuttled to response regulators (ARRs), which can act as transcription factors. In the nucleus, type-B ARRs activate the expression of cytokinin-responsive genes, influencing processes such as cell cycle progression and differentiation. Negative feedback is maintained by type-A ARRs, which act as inhibitors of the pathway. The intricate balance between the activation and repression of cytokinin-responsive genes orchestrates the physiological responses associated with cytokinin signaling, contributing to the regulation of plant growth, development, and stress responses (Pavlů et al. 2018; Li et al. 2021a).

Gibberellins signaling pathway in plants plays a crucial role in modulating growth and development, and its involvement in responses to abiotic stress. The GA signaling pathway involves the binding of GA to its receptor, GID1, forming a complex that targets DELLA proteins for degradation through the 26S proteasome. DELLA proteins act as repressors of GA responses, and their degradation releases the repression, allowing GA-responsive genes to be activated and modulate physiological processes (Banerjee and Roychoudhury 2019; Shah et al. 2023a).

Polyamines, comprising spermidine, spermine, and putrescine, assume a crucial role in plant responses to abiotic stress by functioning as multifaceted signaling molecules. Despite their significance, the signaling pathway linked with these polyamines remains inadequately understood. It is hypothesized that polyamines enter the cell via transporters like resistant to methyl viologen 1 (RMV1) or polyamine uptake transporter 1 (PUT1) and subsequently modulate the expression of relevant genes (Singh et al. 2018; Ali et al. 2020).

Brassinosteroids are steroid hormones that regulate various physiological processes, including growth, development, and stress responses (Nazir et al. 2021). Under abiotic stress, plants activate the BR signaling pathway to enhance stress tolerance. The pathway involves the perception of BR by cell surface receptors, such as brassinosteroid insensitive 1 (BRI1), initiating a signaling cascade that involves several downstream components. This cascade includes the activation of various kinases, such as BR-signaling kinases (BSKs), and the suppression of negative regulators like brassinosteroid insensitive 2 (BIN2). The downstream transcription factors, including brassinazole resistant 1 (BZR1) and BR-suppressor 1 (BES1/BZR2), are subsequently modulated, resulting in the expression of target genes associated with stress adaptation. The BR signaling pathway not only promotes growth and development but also contributes to enhancing stress tolerance by regulating processes such as stomatal closure, antioxidant defense, and the maintenance of cellular homeostasis (Li et al. 2021b; Shah et al. 2023a).

2.4 ROLE OF PLANT GROWTH REGULATORS IN THE MITIGATION OF ABIOTIC STRESS

Plant growth regulators play a crucial role in mitigating abiotic stress in plants, offering a multifaceted approach to enhance stress tolerance and sustaining crop productivity. Abiotic stresses pose significant threats to agricultural systems globally. PGRs function as key molecular players in orchestrating adaptive responses to these stressors by enhancing the antioxidant defense system and accumulation of osmolytes (Khan et al. 2021c; Shah et al. 2021; Altaf et al. 2023). ABA, a central regulator in plants under drought stress, induces stomatal closure, minimizing water loss, and promotes root growth, thereby and improving water uptake efficiency in plants (Trivedi et al. 2016). Cytokinins are known for their contribution to abiotic stress tolerance by enhancing the plant's ability to maintain turgor pressure under water-limiting conditions (Li et al. 2021a). Gibberellins modulate the expression of stress-responsive genes, regulate stomatal conductance, improve water use efficiency, and reduce the impact of abiotic stress. Furthermore, gibberellins have been shown to enhance antioxidant enzyme activities, thereby alleviating oxidative stress, and also modulate root architecture and facilitate resource allocation (Shah et al. 2023a). Ethylene regulates various stress-responsive genes, facilitates antioxidant defense mechanisms, and enhances the plant's ability to withstand unfavorable conditions. It also influences root architecture, water use efficiency, and stomatal closure, contributing to overall stress resilience in plants (Gamalero and Glick 2012). Brassinosteroids contribute to abiotic stress tolerance by promoting antioxidant systems and regulating gene expression (Bhandari and Nailwal 2020).

Moreover, the intricate interplay among these PGRs allows plants to fine-tune their physiological processes, enabling them to direct and adapt to the complex challenges posed by abiotic stressors. For instance, auxins enhance ABA biosynthesis, leading to stomatal closure and reduced water loss. Simultaneously, auxin interacts with cytokinins, influencing shoot growth and promoting stress tolerance (Asghar et al. 2019). ABA and cytokinins exhibit antagonistic effects in stress responses. Cytokinins mitigate stress-induced senescence by inhibiting ABA-regulated processes, promoting cell division, and enhancing antioxidant defense mechanisms (Rowe et al. 2016). GAs antagonize ABA by promoting seed germination and enhancing plant growth, counteracting the inhibitory effects of ABA. Ethylene interacts with GAs to modulate stress responses. Ethylene induces GA biosynthesis, contributing to growth promotion and stress tolerance. Ethylene-ABA crosstalk is critical in balancing growth and stress responses. Ethylene induces ABA biosynthesis, enhancing stress tolerance by regulating stomatal closure, antioxidant defenses, and osmotic adjustment. Ethylene and JA synergistically regulate defense mechanisms against abiotic stress. Ethylene induces JA biosynthesis, promoting the expression of stress-responsive genes and enhancing plant resilience (Munné-Bosch and Müller 2013; Jogawat 2019; Shah et al. 2023b). In summary, the regulation of plant responses to abiotic stress involves intricate crosstalk among PGRs, illustrating the complexity of signaling networks. The synergistic and antagonistic interactions among auxins, cytokinins, GAs, ABA, ethylene, and JA enable plants to adjust growth and defense mechanisms in response to environmental challenges. Understanding these regulatory networks is vital for developing strategies to boost stress tolerance in crops and enhance agricultural sustainability amid changing environmental conditions. Table 2.1 presents significant findings regarding the role of PGRs in alleviating abiotic stresses in crop plants.

TABLE 2.1
Exogenous Application of Plant Growth Regulators Alleviates Effects of Abiotic Stress on Crops

Crop	Type of stress	PGR/s	Dose and mode of application	Plant response	References
Triticum aestivum	Salt	ABA	50 μmol L^{-1} Foliar	Increased plant biomass and proline and sugar content	Li et al. (2020)
Vigna radiata	Cadmium	ABA	15 μM Foliar	Enhanced plant height, fresh and dry weight, antioxidant enzyme activities, and proline content	Leng et al. (2021)
Vicia faba	Salt	IAA	200 ppm Foliar	Improved antioxidant enzyme activities and maintain ion homeostasis	Abdel Latef et al. (2021)
Camellia sinensis	Cadmium	IAA	10 μM Foliar	Influenced plant height, photosynthetic pigment, antioxidant enzyme activities, proline content, and root cell activity	Zhang et al. (2020)
Oryza sativa	Arsenic	IAA, SA, or GA	2 μmol L^{-1} Foliar	Increased shoot length, root elongation, plant biomass, and grain weight	He et al. (2022)
Solanum lycopersicum	Heat	GA_3	75 mg L^{-1} Foliar	Enhanced plant biomass, gas exchange parameters, antioxidant enzyme activities, proline content, and mineral nutrients content	Guo et al. (2022)
Oryza sativa	Antimony	SA	100 μM Foliar	Improved growth traits, photosynthetic rate, and antioxidant defense system	Luo et al. (2021)
Zea mays	Drought	MeJA + SA	20 μM + 2 mM Seed priming and Foliar	Increased growth characters, chlorophyll content, relative water content, proline and sugar content, and antioxidant enzyme activities	Tayyab et al. (2020)
Zea mays	Lead and cadmium	24-EBL	10 μM Seed soaking	Enhanced growth, photosynthetic pigment, proline content, and antioxidant enzyme activities	Shafi et al. (2023)
Brassica juncea	Lead	24-EBL + SA	10^{-7} M + 1 mM Seed priming	Improved dry mass, antioxidant defense system, and osmolyte content	Kohli et al. (2018)
Triticum aestivum	Salt	MeJA + BAP	0.1 μM + 0.04 μM Nutrient solution	Influenced growth rate, antioxidant enzyme activities, and proline content	Avalbaev et al. (2021)
Vigna radiata	Salt	Put, Spd or Spm	0.2 mM Pre-treatment	Increased enzymatic and non-enzymatic antioxidant defense systems, proline content, and ion homeostasis	Nahar et al. (2016)

ABA, abscisic acid; BAP, 6-benzylaminopurine; 24-EBL, 24-Epibrassinolide; GA, gibberellins; IAA, indole-3-acetic acid; MeJA, methyl jasmonate; Put, putrescine; SA, salicylic acid; Spd, spermidine; Spm, spermidine.

2.5 CONCLUSION AND FUTURE PROSPECTS

Plant growth regulators play a pivotal role in mitigating abiotic stress in plants, providing a promising avenue for enhancing crop resilience and productivity. Abiotic stresses, such as drought, salinity, extreme temperatures, and heavy metal toxicity, can significantly impair plant growth and yield. PGRs modulate various physiological and biochemical processes, including the regulation of antioxidant systems and the modulation of stress-responsive gene expression, enabling plants to cope with adverse environmental conditions. Incorporating PGRs into agriculture offers a sustainable approach to alleviate the detrimental impacts of abiotic stress on crops, thereby potentially contributing to global food security. The crosstalk among PGRs represents a sophisticated and adaptive mechanism through which plants integrate signals from multiple hormonal pathways to optimize their responses to adverse environmental conditions. Understanding and harnessing PGRs crosstalk present exciting opportunities to develop crops with enhanced resilience to the increasingly challenging environmental conditions associated with climate change, ultimately contributing to global food security. Looking forward, the future perspective of PGRs in abiotic stress mitigation holds tremendous promise. Advances in biotechnology and molecular biology offer opportunities to engineer crops with enhanced responsiveness to specific PGRs, tailoring them for optimal performance under stress conditions. Additionally, exploring microbial interactions with plants and their influence on PGR synthesis opens avenues for developing biofertilizers that can simultaneously enhance nutrient uptake and stress tolerance. Integrating PGRs with precision agriculture technologies, such as sensors and drones, allows for targeted and efficient application, minimizing input wastage. Continued exploration of natural sources for PGRs and the development of eco-friendly formulations contribute to sustainable agriculture practices. As we navigate a changing climate and increasing demands for food production, the strategic use of plant growth regulators emerges as a key strategy in building resilient and productive agricultural systems.

2.6 AUTHORSHIP CONTRIBUTION

Conceptualization, F.M.; Writing – Original draft preparation, S.H.S. and F.M.; Writing – Review and editing, S.H.S. and F.M. All authors have read and agreed to the published version of the book chapter.

2.7 DECLARATION OF COMPETING INTEREST

The authors declare that they have no known competing financial interests or personal relationships that could have appeared to influence the work reported in this book chapter.

REFERENCES

Abdel Latef, Arafat Abdel Hamed, Md Tahjib-Ul-Arif, and Mohammad Saidur Rhaman. "Exogenous auxin-mediated salt stress alleviation in faba bean (*Vicia faba* L.)." *Agronomy* 11, no. 3 (2021): 547.

Ahmad, Farhan, Ananya Singh, and Aisha Kamal. "Salicylic acid–mediated defense mechanisms to abiotic stress tolerance." In *Plant signaling molecules*, pp. 355–369. Woodhead Publishing, 2019.

Ahmad, Parvaiz, Sumiya Jamsheed, Asiya Hameed, Saima Rasool, Iti Sharma, M. M. Azooz, and Mirza Hasanuzzaman. "Drought stress induced oxidative damage and antioxidants in plants." In *Oxidative damage to plants*, pp. 345–367. Academic Press, 2014.

Ali, Md Sarafat, and Kwang-Hyun Baek. "Jasmonic acid signaling pathway in response to abiotic stresses in plants." *International Journal of Molecular Sciences* 21, no. 2 (2020): 621.

Ali, Rohaya, Sumaya Hassan, Durdana Shah, Nasreena Sajjad, and Eijaz Ahmed Bhat. "Role of polyamines in mitigating abiotic stress." In *Protective chemical agents in the amelioration of plant abiotic stress: biochemical and molecular perspectives*, pp. 291–305. Acdemic Press, 2020.

Altaf, Muhammad Ahsan, Rabia Shahid, Ravinder Kumar, Muhammad Mohsin Altaf, Awadhesh Kumar, Latif Ullah Khan, Muhammad Saqib et al. "Phytohormones mediated modulation of abiotic stress tolerance and potential crosstalk in horticultural crops." *Journal of Plant Growth Regulation* 42, no. 8 (2023): 4724–4750.

Asghar, Muhammad Ahsan, Yan Li, Hengke Jiang, Xin Sun, Bushra Ahmad, Shakeel Imran, Liang Yu, Chunyan Liu, Wenyu Yang, and Junbo Du. "Crosstalk between abscisic acid and auxin under osmotic stress." *Agronomy Journal* 111, no. 5 (2019): 2157–2162.

Avalbaev, Azamat, Chulpan Allagulova, Dilara Maslennikova, Kristina Fedorova, and Farida Shakirova. "Methyl jasmonate and cytokinin mitigate the salinity-induced oxidative injury in wheat seedlings." *Journal of Plant Growth Regulation* 40 (2021): 1741–1752.

Banerjee, Aditya, and Aryadeep Roychoudhury. "The regulatory signaling of gibberellin metabolism and its crosstalk with phytohormones in response to plant abiotic stresses." In *Plant signaling molecules*, pp. 333–339. Woodhead Publishing, 2019.

Bhandari, Sneha, and Tapan K. Nailwal. "Role of brassinosteroids in mitigating abiotic stresses in plants." *Biologia* 75, no. 12 (2020): 2203–2230.

Bita, Craita E., and Tom Gerats. "Plant tolerance to high temperature in a changing environment: scientific fundamentals and production of heat stress-tolerant crops." *Frontiers in Plant Science* 4 (2013): 273.

Burgess, Patrick, and Bingru Huang. "Mechanisms of hormone regulation for drought tolerance in plants." In *Drought stress tolerance in plants, vol 1: physiology and biochemistry*, pp. 45–75. Springer, 2016.

Farooq, Muhammad, M. Hussain, Abdul Wahid, and K. H. M. Siddique. "Drought stress in plants: an overview." In *Plant responses to drought stress: from morphological to molecular features*, pp. 1–33. Springer, 2012.

Gamalero, Elisa, and Bernard R. Glick. "Ethylene and abiotic stress tolerance in plants." In *Environmental adaptations and stress tolerance of plants in the era of climate change*, pp. 395–412. Springer, 2012.

Godoy, Francisca, Karina Olivos-Hernández, Claudia Stange, and Michael Handford. "Abiotic stress in crop species: improving tolerance by applying plant metabolites." *Plants* 10, no. 2 (2021): 186.

Guo, Tianxin, Shaista Gull, Muhammad Moaaz Ali, Ahmed Fathy Yousef, Sezai Ercisli, Hazem M. Kalaji, Arkadiusz Telesiński et al. "Heat stress mitigation in tomato (*Solanum lycopersicum* L.) through foliar application of gibberellic acid." *Scientific Reports* 12, no. 1 (2022): 11324.

He, Yue, Ting Zhang, Yu Sun, Xuejing Wang, Qingqing Cao, Zhigang Fang, Ming Chang, Qingsheng Cai, and Laiqing Lou. "Exogenous IAA alleviates arsenic toxicity to rice and reduces arsenic accumulation in rice grains." *Journal of Plant Growth Regulation* 41, no. 2 (2022): 734–741.

Hernández, Jose Antonio. "Salinity tolerance in plants: trends and perspectives." *International Journal of Molecular Sciences* 20, no. 10 (2019): 2408.

Hossain, Akbar, Biswajit Pamanick, Visha Kumari Venugopalan, Ulkar Ibrahimova, Md Atikur Rahman, Ayaz Latif Siyal, Sagar Maitra, Subhrajyoti Chatterjee, and Tariq Aftab. "Emerging roles of plant growth regulators for plants adaptation to abiotic stress–induced oxidative stress." In *Emerging plant growth regulators in agriculture*, pp. 1–72. Academic Press, 2022.

Husain, Tajammul, Abreeq Fatima, Mohammad Suhel, Samiksha Singh, Anket Sharma, Sheo Mohan Prasad, and Vijay Pratap Singh. "A brief appraisal of ethylene signaling under abiotic stress in plants." *Plant Signaling & Behavior* 15, no. 9 (2020): 1782051.

Iqbal, Noushina, Mehar Fatma, Harsha Gautam, Zebus Sehar, Faisal Rasheed, M. Iqbal R. Khan, Adriano Sofo, and Nafees A. Khan. "Salicylic acid increases photosynthesis of drought grown mustard plants effectively with sufficient-N via regulation of ethylene, abscisic acid, and nitrogen-use efficiency." *Journal of Plant Growth Regulation* 41, no. 5 (2022): 1966–1977.

Jha, Uday Chand, Abhishek Bohra, and Narendra Pratap Singh. "Heat stress in crop plants: its nature, impacts and integrated breeding strategies to improve heat tolerance." *Plant Breeding* 133, no. 6 (2014): 679–701.

Jogawat, Abhimanyu. "Crosstalk among phytohormone signaling pathways during abiotic stress." In *Molecular plant abiotic stress: biology and biotechnology*, pp. 209–220. Academic Press, 2019.

Kaur, Harmanjit, Faroza Nazir, Sofi Javed Hussain, Rupinder Kaur, Ashish B. Rajurkar, Sarika Kumari, Manzer H. Siddiqui, Moksh Mahajan, Sayeda Khatoon, and M. Iqbal R. Khan. "Gibberellic acid alleviates cadmium-induced seed germination inhibition through modulation of carbohydrate metabolism and antioxidant capacity in mung bean seedlings." *Sustainability* 15, no. 4 (2023): 3790.

Khan, M. Iqbal R., Farha Ashfaque, Himanshu Chhillar, Mohammad Irfan, and Nafees A. Khan. "The intricacy of silicon, plant growth regulators and other signaling molecules for abiotic stress tolerance: an entrancing crosstalk between stress alleviators." *Plant Physiology and Biochemistry* 162 (2021b): 36–47.

Khan, M. Iqbal R., Badar Jahan, Mohamed F. AlAjmi, Md Tabish Rehman, Noushina Iqbal, Mohammad Irfan, Zebus Sehar, and Nafees A. Khan. "Crosstalk of plant growth regulators protects photosynthetic performance from arsenic damage by modulating defense systems in rice." *Ecotoxicology and Environmental Safety* 222 (2021a): 112535.

Khan, M. Iqbal R., Sarika Kumari, Faroza Nazir, Risheek Rahul Khanna, Ravi Gupta, and Himanshu Chhillar. "Defensive role of plant hormones in advancing abiotic stress-resistant rice plants." *Rice Science* 30, no. 1 (2023a): 15–35.

Khan, M. Iqbal R., Faroza Nazir, Chirag Maheshwari, Priyanka Chopra, Himanshu Chhillar, and Nese Sreenivasulu. "Mineral nutrients in plants under changing environments: a road to future food and nutrition security." *The Plant Genome* 16, no. 4 (2023b): e20362.

Khan, M. Iqbal R., Sudhakar R. Palakolanu, Priyanka Chopra, Ashish B. Rajurkar, Ravi Gupta, Noushina Iqbal, and Chirag Maheshwari. "Improving drought tolerance in rice: ensuring food security through multi-dimensional approaches." *Physiologia Plantarum* 172, no. 2 (2021c): 645–668.

Khan, M. Iqbal R., Alice Trivellini, Himanshu Chhillar, Priyanka Chopra, Antonio Ferrante, Nafees A. Khan, and Abdelbagi M. Ismail. "The significance and functions of ethylene in flooding stress tolerance in plants." *Environmental and Experimental Botany* 179 (2020): 104188.

Kohli, Sukhmeen Kaur, Neha Handa, Shagun Bali, Saroj Arora, Anket Sharma, Ravdeep Kaur, and Renu Bhardwaj. "Modulation of antioxidative defense expression and osmolyte content by co-application of 24-epibrassinolide and salicylic acid in Pb exposed Indian mustard plants." *Ecotoxicology and Environmental Safety* 147 (2018): 382–393.

Kumar, Sandeep, Mohammad Abass Ahanger, Huda Alshaya, Basit Latief Jan, and Vimala Yerramilli. "Salicylic acid mitigates salt induced toxicity through the modifications of biochemical attributes and some key antioxidants in capsicum annuum." *Saudi Journal of Biological Sciences* 29, no. 3 (2022a): 1337–1347.

Kumar, Sandeep, Sajad Hussain Shah, Yerramilli Vimala, Hanuman Singh Jatav, Parvaiz Ahmad, Yinglong Chen, and Kadambot HM Siddique. "Abscisic acid: metabolism, transport, crosstalk with other plant growth regulators, and its role in heavy metal stress mitigation." *Frontiers in Plant Science* 13 (2022b): 972856.

Kumari, Sarika, Faroza Nazir, Kajal Jain, and M. Iqbal R. Khan. "GABA and potassium modulates defence systems, assimilation of nitrogen and carbon, and yield traits under salt stress in wheat." *Journal of Plant Growth Regulation* 42, no. 10 (2023): 6721–6740.

Kumari, Sarika, Faroza Nazir, Apoorva Singh, Hyda Haroon, Nahida Rashid Khan, Rudra Narayan Sahoo, Mohammed Albaqami, Kadambot H. M. Siddique, and M. Iqbal R. Khan. "γ-aminobutyric acid (GABA) strengthened nutrient accumulation, defense metabolism, growth and yield traits against salt and endoplasmic reticulum stress conditions in wheat plants." *Plant and Soil* (2024): 1–21.

Leng, Yan, Yi Li, Yan-Hua Ma, Li-Fang He, and Shi-Weng Li. "Abscisic acid modulates differential physiological and biochemical responses of roots, stems, and leaves in mung bean seedlings to cadmium stress." *Environmental Science and Pollution Research* 28, no. 5 (2021): 6030–6043.

Li, Si-Min, Hong-Xiang Zheng, Xian-Sheng Zhang, and Na Sui. "Cytokinins as central regulators during plant growth and stress response." *Plant Cell Reports* 40 (2021a): 271–282.

Li, Simin, Hongxiang Zheng, Lin Lin, Fang Wang, and Na Sui. "Roles of brassinosteroids in plant growth and abiotic stress response." *Plant Growth Regulation* 93 (2021b): 29–38.

Li, Xiaoyu, Shuxin Li, Jinghong Wang, and Jixiang Lin. "Exogenous abscisic acid alleviates harmful effect of salt and alkali stresses on wheat seedlings." *International Journal of Environmental Research and Public Health* 17, no. 11 (2020): 3770.

Liu, Liu, Guangyan Guo, Zhijuan Wang, Hongtao Ji, Fupeng Mu, and Xia Li. "Auxin in plant growth and stress responses." In *Phytohormones: a window to metabolism, signaling and biotechnological applications*, pp. 1–35. Springer, 2014.

Luo, Wen-Tao, Li He, Fang Li, and Jiao-Kun Li. "Exogenous salicylic acid alleviates the antimony (Sb) toxicity in rice (*Oryza sativa* L.) seedlings." *Journal of Plant Growth Regulation* 40 (2021): 1327–1340.

Munné-Bosch, Sergi, and Maren Müller. "Hormonal cross-talk in plant development and stress responses." *Frontiers in Plant Science* 4 (2013): 529.

Nadarajah, Kalaivani K. "ROS homeostasis in abiotic stress tolerance in plants." *International Journal of Molecular Sciences* 21, no. 15 (2020): 5208.

Nahar, Kamrun, Mirza Hasanuzzaman, Anisur Rahman, Md Mahabub Alam, Jubayer-Al Mahmud, Toshisada Suzuki, and Masayuki Fujita. "Polyamines confer salt tolerance in mung bean (*Vigna radiata* L.) by reducing sodium uptake, improving nutrient homeostasis, antioxidant defense, and methylglyoxal detoxification systems." *Frontiers in Plant Science* 7 (2016): 1104.

Nawaz, Khalid, Khalid Hussain, Abdul Majeed, Farah Khan, Shahid Afghan, and Kazim Ali. "Fatality of salt stress to plants: morphological, physiological and biochemical aspects." *African Journal of Biotechnology* 9, no. 34 (2010).

Nazir, Faroza, Qazi Fariduddin, Anjuman Hussain, and Tanveer Alam Khan. "Brassinosteroid and hydrogen peroxide improve photosynthetic machinery, stomatal movement, root morphology and cell viability and reduce Cu-triggered oxidative burst in tomato." *Ecotoxicology and Environmental Safety* 207 (2021): 111081.

Nazir, Faroza, Badar Jahan, Noushina Iqbal, Ashish B. Rajurkar, Manzer H. Siddiqui, and M. Iqbal R. Khan. "Methyl jasmonate influences ethylene formation, defense systems, nutrient homeostasis and carbohydrate metabolism to alleviate arsenic-induced stress in rice (*Oryza sativa*)." *Plant Physiology and Biochemistry* 202 (2023a): 107990.

Nazir, Faroza, Badar Jahan, Sarika Kumari, Noushina Iqbal, Mohammed Albaqami, Adriano Sofo, and M. Iqbal R. Khan. "Brassinosteroid modulates ethylene synthesis and antioxidant metabolism to protect rice (*Oryza sativa*) against heat stress-induced inhibition of source–sink capacity and photosynthetic and growth attributes." *Journal of Plant Physiology* 289 (2023b): 154096.

Nazir, Faroza, Peter Poor, Ravi Gupta, Sarika Kumari, Kashif Nawaz, and M. Iqbal R. Khan. "Plant hormone ethylene: a leading edge in conferring drought stress tolerance." *Physiologia Plantarum* 176, no. 1 (2024): e14151.

Nazir, Faroza, Sarika Kumari, Moksh Mahajan, and M. Iqbal R. Khan. "An explicit story of plant abiotic stress resilience: overtone of selenium, plant hormones and other signaling molecules." *Plant and Soil* 486, no. 1 (2023c): 135–163.

Parihar, Parul, Samiksha Singh, Rachana Singh, Vijay Pratap Singh, and Sheo Mohan Prasad. "Effect of salinity stress on plants and its tolerance strategies: a review." *Environmental Science and Pollution Research* 22 (2015): 4056–4075.

Pavlů, Jaroslav, Jan Novák, Vladěna Koukalová, Markéta Luklová, Břetislav Brzobohatý, and Martin Černý. "Cytokinin at the crossroads of abiotic stress signalling pathways." *International Journal of Molecular Sciences* 19, no. 8 (2018): 2450.

Per, Tasir S., Nafees A. Khan, Palakolanu Sudhakar Reddy, Asim Masood, Mirza Hasanuzzaman, M. Iqbal R. Khan, and Naser A. Anjum. "Approaches in modulating proline metabolism in plants for salt and drought stress tolerance: phytohormones, mineral nutrients and transgenics." *Plant Physiology and Biochemistry* 115 (2017): 126–140.

Pereira, Andy. "Plant abiotic stress challenges from the changing environment." *Frontiers in Plant Science* 7 (2016): 1123.

Petrov, Veselin, Jacques Hille, Bernd Mueller-Roeber, and Tsanko S. Gechev. "ROS-mediated abiotic stress-induced programmed cell death in plants." *Frontiers in Plant Science* 6 (2015): 69.

Pirasteh-Anosheh, Hadi, Armin Saed-Moucheshi, Hassan Pakniyat, and Mohammad Pessarakli. "Stomatal responses to drought stress." In *Water stress and crop plants: a sustainable approach*, vol. 1, pp. 24–40. Academic Press, 2016.

Poór, Peter, Kashif Nawaz, Ravi Gupta, Farha Ashfaque, and M. Iqbal R. Khan. "Ethylene involvement in the regulation of heat stress tolerance in plants." *Plant Cell Reports* 41, no. 3 (2022): 675–698.

Rasheed, Adnan, Muhammad Umair Hassan, Shah Fahad, Muhammad Aamer, Maria Batool, Muhammad Ilyas, Fang Shang, Ziming Wu, and Huijie Li. "Heavy metals stress and plants defense responses." In *Sustainable soil and land management and climate change*, pp. 57–82. CRC Press, 2021.

Ratnakumar, Pasala, Md Iqbal Raja Khan, Paramjit Singh Minhas, Mahummad Ansar Farooq, Rafat Sultana, Tasir S. Per, Pallavi P. Deokate, Nafees Ahmad Khan, Yogeshwar Singh, and Jagadish Rane. "Can plant bio-regulators minimize crop productivity losses caused by drought, salinity and heat stress? An integrated review." *Journal of Applied Botony and Food Quality* 89 (2016): 113–125.

Rowe, James H., Jennifer F. Topping, Junli Liu, and Keith Lindsey. "Abscisic acid regulates root growth under osmotic stress conditions via an interacting hormonal network with cytokinin, ethylene and auxin." *New Phytologist* 211, no. 1 (2016): 225–239.

Saharan, Baljeet Singh, Basanti Brar, Joginder Singh Duhan, Ravinder Kumar, Sumnil Marwaha, Vishnu D. Rajput, and Tatiana Minkina. "Molecular and physiological mechanisms to mitigate abiotic stress conditions in plants." *Life* 12, no. 10 (2022): 1634.

Sako, Kaori, Huong Mai Nguyen, and Motoaki Seki. "Advances in chemical priming to enhance abiotic stress tolerance in plants." *Plant and Cell Physiology* 61, no. 12 (2020): 1995–2003.

Salehi-Lisar, Seyed Yahya, and Hamideh Bakhshayeshan-Agdam. "Drought stress in plants: causes, consequences, and tolerance." In *Drought stress tolerance in plants, vol. 1: physiology and biochemistry*, pp. 1–16. Springer, 2016.

Shafi, Zaryab, Mohammad Shahid, Amal Mohamed AlGarawi, Mohammad Tarique Zeyad, Samy A. Marey, Ashraf Atef Hatamleh, Shifa Wang, and Udai B. Singh. "The exogenous application of 24-epibrassinolide

(24-EBL) increases the Cd and Pb resilience in *Zea mays* (L.) by regulating the growth and physiological mechanism." *Applied Biochemistry and Biotechnology* (2023): 1–25.

Shah, Sajad Hussain, Shaistul Islam, Firoz Mohammad, and Manzer H. Siddiqui. "Gibberellic acid: a versatile regulator of plant growth, development and stress responses." *Journal of Plant Growth Regulation* (2023a): 1–22.

Shah, Sajad Hussain, Shaistul Islam, Zubair Ahmad Parrey, and Firoz Mohammad. "Role of exogenously applied plant growth regulators in growth and development of edible oilseed crops under variable environmental conditions: a review." *Journal of Soil Science and Plant Nutrition* 21, no. 4 (2021): 3284–3308.

Shah, Sajad Hussain, Zubair Ahmad Parrey, Shaistul Islam, Firoz Mohammad, Saud Alamri, and Hazem M. Kalaji. "Epibrassinolide application improves *Brassica juncea* L. performance by augmenting gas exchange markers, stomatal behavior, cellular viability, enzymatic activities and yield attributes." *South African Journal of Botany* 162 (2023b): 531–544.

Shahid, Muhammad, Sana Khalid, Ghulam Abbas, Naeem Shahid, Muhammad Nadeem, Muhammad Sabir, Muhammad Aslam, and Camille Dumat. "Heavy metal stress and crop productivity." In *Crop production and global environmental issues*, pp. 1–25. Springer, 2015.

Sharma, Anket, Vinod Kumar, Gagan Preet Singh Sidhu, Rakesh Kumar, Sukhmeen Kaur Kohli, Poonam Yadav, Dhriti Kapoor et al. "Abiotic stress management in plants: role of ethylene." IN *Molecular plant abiotic stress: biology and biotechnology*, pp. 185–208. Academic Press, 2019.

Sharma, Eshan, Raghvendra Sharma, Pratikshya Borah, Mukesh Jain, and Jitendra P. Khurana. "Emerging roles of auxin in abiotic stress responses." In *Elucidation of abiotic stress signaling in plants: functional genomics perspectives*, vol. 1, pp. 299–328. Springer, 2015.

Shaukat, Kanval, Noreen Zahra, Mohammad Bilal Hafeez, Rubina Naseer, Aaliya Batool, Hina Batool, Ali Raza, and Abdul Wahid. "Role of salicylic acid–induced abiotic stress tolerance and underlying mechanisms in plants." In *Emerging plant growth regulators in agriculture*, pp. 73–98. Academic Press, 2022.

Singh, Pratika, Sahana Basu, and Gautam Kumar. "Polyamines metabolism: a way ahead for abiotic stress tolerance in crop plants." In *Biochemical, physiological and molecular avenues for combating abiotic stress tolerance in plants*, pp. 39–55. Academic Press, 2018.

Tayyab, Nimrah, Rabia Naz, Humaira Yasmin, Asia Nosheen, Rumana Keyani, Muhammad Sajjad, Muhammad Nadeem Hassan, and Thomas H. Roberts. "Combined seed and foliar pre-treatments with exogenous methyl jasmonate and salicylic acid mitigate drought-induced stress in maize." *PLoS One* 15, no. 5 (2020): e0232269.

Trivedi, Dipesh Kumar, Sarvajeet Singh Gill, and Narendra Tuteja. "Abscisic acid (ABA): biosynthesis, regulation, and role in abiotic stress tolerance." In *Abiotic stress response in plants*, pp. 315–326. Academic Press, 2016.

Ul Hassan, Mahmood, Tassaduq Rasool, Chandni Iqbal, Adnan Arshad, Muhammad Abrar, Muhammad Mohsin Abrar, Muhammad Habib-ur-Rahman, Mehmood Ali Noor, Alam Sher, and Shah Fahad. "Linking plants functioning to adaptive responses under heat stress conditions: a mechanistic review." *Journal of Plant Growth Regulation* (2021): 1–18.

Vishwakarma, Kanchan, Neha Upadhyay, Nitin Kumar, Gaurav Yadav, Jaspreet Singh, Rohit K. Mishra, Vivek Kumar et al. "Abscisic acid signaling and abiotic stress tolerance in plants: a review on current knowledge and future prospects." *Frontiers in Plant Science* 8 (2017): 161.

Wahid, Abdul, Muhammad Farooq, Iqbal Hussain, Rizwan Rasheed, and Saddia Galani. "Responses and management of heat stress in plants." In *Environmental adaptations and stress tolerance of plants in the era of climate change*, pp. 135–157. Springer, 2012.

Wang, Yun, Salma Mostafa, Wen Zeng, and Biao Jin. "Function and mechanism of jasmonic acid in plant responses to abiotic and biotic stresses." *International Journal of Molecular Sciences* 22, no. 16 (2021): 8568.

Yadav, S. K. "Heavy metals toxicity in plants: an overview on the role of glutathione and phytochelatins in heavy metal stress tolerance of plants." *South African Journal of Botany* 76, no. 2 (2010): 167–179.

Yadav, Shiv Poojan, Reshu Bharadwaj, Harsita Nayak, Rakhi Mahto, Rajesh Kumar Singh, and Saroj Kumar Prasad. "Impact of salt stress on growth, productivity and physicochemical properties of plants: a review." *International Journal of Chemical Studies* 7, no. 2 (2019): 1793–1798.

Yokoi, Shuji, Ray A. Bressan, and Paul Mike Hasegawa. "Salt stress tolerance of plants." *JIRCAS Working Report* 23, no. 1 (2002): 25–33.

You, Jun, and Zhulong Chan. "ROS regulation during abiotic stress responses in crop plants." *Frontiers in Plant Science* 6 (2015): 1092.

Zahid, Ghassan, Sara Iftikhar, Flavien Shimira, Hafiz Muhammad Ahmad, and Yıldız Aka Kaçar. "An overview and recent progress of plant growth regulators (PGRs) in the mitigation of abiotic stresses in fruits: a review." *Scientia Horticulturae* 309 (2023): 111621.

Zargar, Sajad Majeed, Nancy Gupta, Muslima Nazir, Reetika Mahajan, Firdose A. Malik, Nageebul R. Sofi, Asif B. Shikari, and R. K. Salgotra. "Impact of drought on photosynthesis: molecular perspective." *Plant Gene* 11 (2017): 154–159.

Zhang, Chenyu, Qun He, Minghan Wang, Xizhi Gao, Jianjiao Chen, and Chengwen Shen. "Exogenous indole acetic acid alleviates Cd toxicity in tea (*Camellia sinensis*)." *Ecotoxicology and Environmental Safety* 190 (2020): 110090.

3 Interactions between Wheat Root System and Salt Stress Environments

Glimpses of Omics Interventions

Sarika Kumari, Hyda Haroon, Rudra Narayan Sahoo, Kadambot H. M. Siddique, and M. Iqbal R. Khan

3.1 INTRODUCTION

The frequent increase in the severity of environmental stresses, including salt stress, imposes detrimental impacts on crop yield and quality due to climate change (Kumari et al. 2023, 2024). The world's three foremost important staple crops, wheat, rice, and maize, account for a large portion of daily caloric and protein intake (El Sabagh et al. 2021). Among these cereal crops, wheat holds a crucial and significant position due to its ease of domestication and its role as the leading food crop for the global population. Although it currently occupies most of the arable land (38.8%), its productivity is still low despite having comparatively higher grain protein content (12%–15%) than other cereal crops (El Sabagh et al. 2021). Due to the occurrence of salt stress, brought on by climate change, wheat productivity may decline even more. Various climate models have predicted that stressful environments would result in a 6% decrease in wheat production (Farooq et al. 2023). The prevalence of salt stress in these areas is rising, driven by natural and anthropogenic factors. Additionally, in the past two decades, the challenges posed by salt stress have intensified, particularly in arid and semi-arid regions (El Sabagh et al. 2021; Farooq et al. 2023).

Roots are the crucial link between plants and soil, with the root architecture regulating anchoring, uptake of soil nutrients, synthesis and storage of secondary metabolites, and facilitation of other molecular interactions (Lynch 2022). In addition, roots express a wide range of exceedingly variable morphological and physiological traits, including root number, primary root hairs, root tips, total root length, seminal root angle, root thickness, surface area, volume, and fresh and dry weights, reflecting salt tolerance and plant viability (Lynch 2022; Maqbool et al. 2022). Moreover, unique root–intersystemic trait interactions with their genetic and environmental components further provide fascinating prospects for breeding strategies that could improve nutrient uptake and yield in wheat crops (Maqbool et al. 2022).

The root systems are classified based on branch anatomy, root activity, and development in plants. In terms of analyzing root growth, categorization based on development is more common and beneficial, highlighting the differences between monocotyledonous and dicotyledonous root systems. The methodology includes the ontogenetic classification of roots into three distinct categories, namely primary, lateral, and adventitious. Notably, during germination, the primary root is the first root to emerge from the seed in both monocotyledonous and dicotyledonous plants, and it is derived from the embryonic root. For instance, many monocotyledonous plants exhibit a comparable root system development to dicotyledonous plants, with the additional occurrence of nodal roots that create a fibrous adventitious root system. Subsequently, root surface area is formed during post-embryonic root development and is influenced by a plastic genetic program that is adjusted by

DOI: 10.1201/9781032630380-3

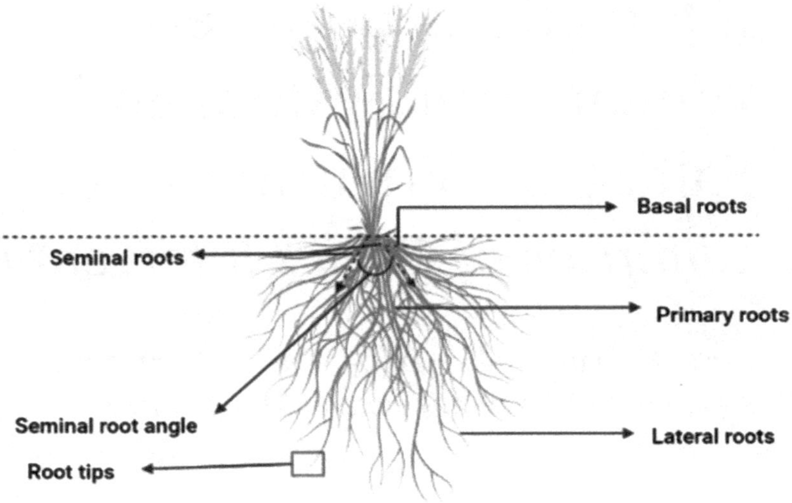

FIGURE 3.1 Representation of a typical wheat root system.

environmental stimuli. Despite the consistent morphology of the root system, the spatial arrangement of the root surface area, including the primary, lateral, and adventitious root number, position, and growth position, is highly variable, even among genetically identical plants (Figure 3.1).

Thus, taking the significance of the root system into consideration, in this chapter we discuss the growth, physiological, and metabolic adaptations in root systems in salt-affected wheat plants. Additionally, this chapter also deals and highlights the important research conducted with the biotechnological interventions to transform root architecture for ameliorating salt stress-induced adversities in wheat plants.

3.2 GROWTH, PHYSIOLOGICAL, AND METABOLIC ADAPTATIONS IN ROOT SYSTEMS OF WHEAT PLANTS UNDER SALT STRESS

Undergoing salt stress has the potential to disrupt the usual dynamic equilibrium within a plant, leading to significant alterations in its root structural configurations. The morphological and physiological functionality of roots could adapt to a variety of environmental conditions, such as fluctuations in water availability, ionic distributions, and nutrient acquisitions (Maqbool et al. 2022). These adaptation processes often involve considerable modifications in the root architecture to withstand salt stress-induced reactions (Figure 3.2; Table 3.1).

In order to counteract these stressors, plants utilize a range of mechanisms to regulate their physiological and metabolic processes, with the ultimate goal of re-establishing cellular homeostasis and promoting optimal growth and development (Khan et al. 2015; Sehar et al. 2021; Wahid et al. 2022). Thus, the investigation into the growth, physiological, and metabolic interactions occurring in wheat roots under salt stress has been the subject of several studies.

It has been reported that physiological adaptations correspond with the root growth management through inducing considerable alterations in root architecture, including epiblema thickness, root hair length, endoderm thickness, and central cylinder diameter, which contribute to the plant's ability to mitigate the impact of salt stress (Terletskaya et al. 2019). Additionally, Dissanayake et al. (2022) found a significant instability in protein translation, source-sink affiliated enzymatic activities, including fructokinase-2 and sucrose synthase in wheat root tips, thus indicating that the compromised protein-building and carbon sink capacity could be exploited to render salt tolerance

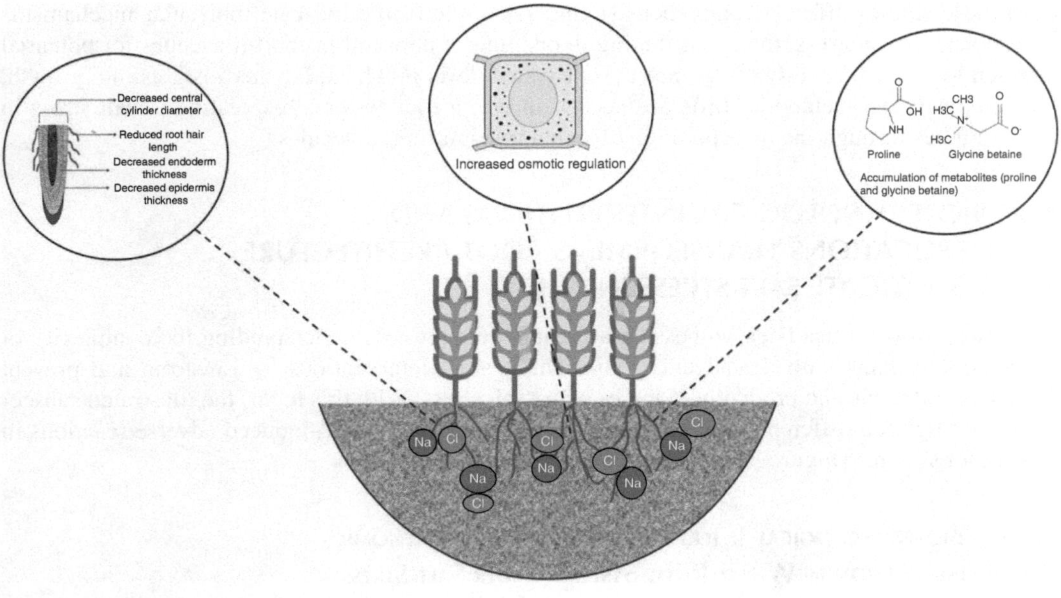

FIGURE 3.2 Growth, physiological, and metabolic adaptations in wheat root systems under salt stress conditions.

TABLE 3.1
Putative growth, physiological, and metabolic adaptations in root systems of wheat plants under salt stress

Wheat genotypes/ varieties/cultivars	Concentrations	Physiological adaptations	References
Triticum dicoccum, *T. polonicum*, *T. compactum*, *T. aestivum*	Sodium chloride (NaCl), 1.68% *w/v*	Induced significant alterations in root architecture, including epiblema thickness, alterations in root hair length, endoderm thickness, and central cylinder diameter.	Terletskaya et al. (2019)
Triticum aestivum	NaCl; 150 mM	Modulated source-sink enzymes (for instance, fructokinase-2, and sucrose synthase) and induced variations in the activation of glycolytic enzymes, including triose-phosphate isomerase and aconitate hydratase.	Dissanayake et al. (2022)
Cang 6001	NaCl; 150 mM	Elevated antioxidant enzymatic activities to scavenge reactive oxygen species (ROS).	Chen et al. (2011)
Triticum boeoticum	NaCl; 250 mM	Increased the accumulation of antioxidants to maintain oxidative balance.	Ahmadi et al. (2018)

in future studies. Additionally, they observed variations in the abundance of glycolytic enzymes, including triose-phosphate isomerase, and aconitate hydratase in wheat root cells, suggesting the impeded energy utilization could serve as a trademark response to increase the salt stress adaptability (Dissanayake et al. 2022). Further, these impaired physiological alterations have led to the anatomical modulations surpassing decreased root growth and respiratory rate in salt-stressed wheat plants. This metabolic response suggested that wheat plants strategically allocated energy to roots while engaging in active synthesis in leaves to foster salt tolerance. The synthesis of information from these studies offers a comprehensive understanding of the intricate molecular mechanisms in response to salt stress, thus contributing in opening of new and insightful avenues for potential interventions to enhance salt tolerance at root system level in wheat. Further insights into growth, physiological, and metabolic shifts are needed in wheat root systems responding to salt stress in future studies through the incorporation of signal transmitting molecules.

3.3 BIOTECHNOLOGICAL INTERVENTIONS AND IMPLICATIONS TRANSFORMING ROOT ARCHITECTURE TO MITIGATE SALT STRESS IN WHEAT

The research world has been witnessing a greater shift towards understanding the complexity of wheat root systems with the advent of biotechnological interventions to transform and prevent growth retardations and productivity losses under salt stress. With this focus, the illustrated subsections shed light on different omics strategies to counteract salt stress-induced adverse reactions in wheat root systems (Figure 3.3).

3.3.1 Biotechnological Interventions Related to Phenomics Implication in Wheat Root Systems under Salt Stress

The recent upsurge in assessing the discipline of phenomics has raised budding opportunities with advanced and novel sensing to monitor the interactions between the root systems and salt environments. These studies have explicitly assessed and provided insights into root-environmental heterogeneity to dissect salt tolerance and capture plant growth (Hu and Schmidhalter 2023). However, research oriented towards wheat root systems under salt stress conditions is limited in the literature, concerning phenomics-mediated studies. Therefore, future investigations are required to understand phenomics studies to unknot wheat root tolerance in response to salt stress conditions.

3.3.2 Biotechnological Interventions Related to Genomics Implication in Wheat Root Systems under Salt Stress

Salt stress environments have been persistently inducing certain inherent transformation in plant genomes, pertaining to the genetic mechanisms of how plants become accustomed to underlying circumstances. For instance, in salt-stressed wheat roots (250 mM sodium chloride; NaCl), a *glutathione sulfur transferase* (*GST*) gene (*TraesCS1D01G207400.1*) was induced, favoring the accumulation of numerous salt-responsive proteins implicated in osmotic and ionic homeostasis, antioxidant metabolism, and heat shock responses under salt stress (Khan et al. 2023). During salt stress, *TaNHX* (sodium-proton antiporter) expressions were differentially expressed in wheat root of the KRL213 variety, as elucidated by Sharma et al. (2023). Among the 10 *TaNHX* genes, five genes, including *TaNHX2* (~14 folds), *TaNHX12* (~9 folds), *TaNHX16* (~7 folds), *TaNHX20* (~10 folds), and *TaNHX23* (~3 folds), showed a considerable induction in the wheat root after 48 hours of salt exposure (150 mM NaCl). This indicates the potential of root systems to deliberately control and mirror the salt resistance, aiding in the regulation of cellular expansion, trafficking of vesicles, and pH regulations in wheat plants (Sharma et al. 2023). In the BARI Gom-25 wheat variety, the introduction of salt stress (100 mM NaCl) up-regulated transcription factors (TFs), including *MYB3*, *MYB4*, and *MYB59* (Sukumaran et al. 2023). Among these, *MYB3* encoded the zinc finger protein (*ZAT11*) and *HHO5* TF, wherein *ZAT11* was up-regulated in root tissues reflecting a salt stress avoidance strategy adapted by wheat plants through reinforcing defense against Na^+ transport through the apoplastic pathway (Sukumaran et al. 2023). These outcomes suggest that uncovering the functional attributes of *MYB* TFs could potentially support wheat root development in response to salt stress.

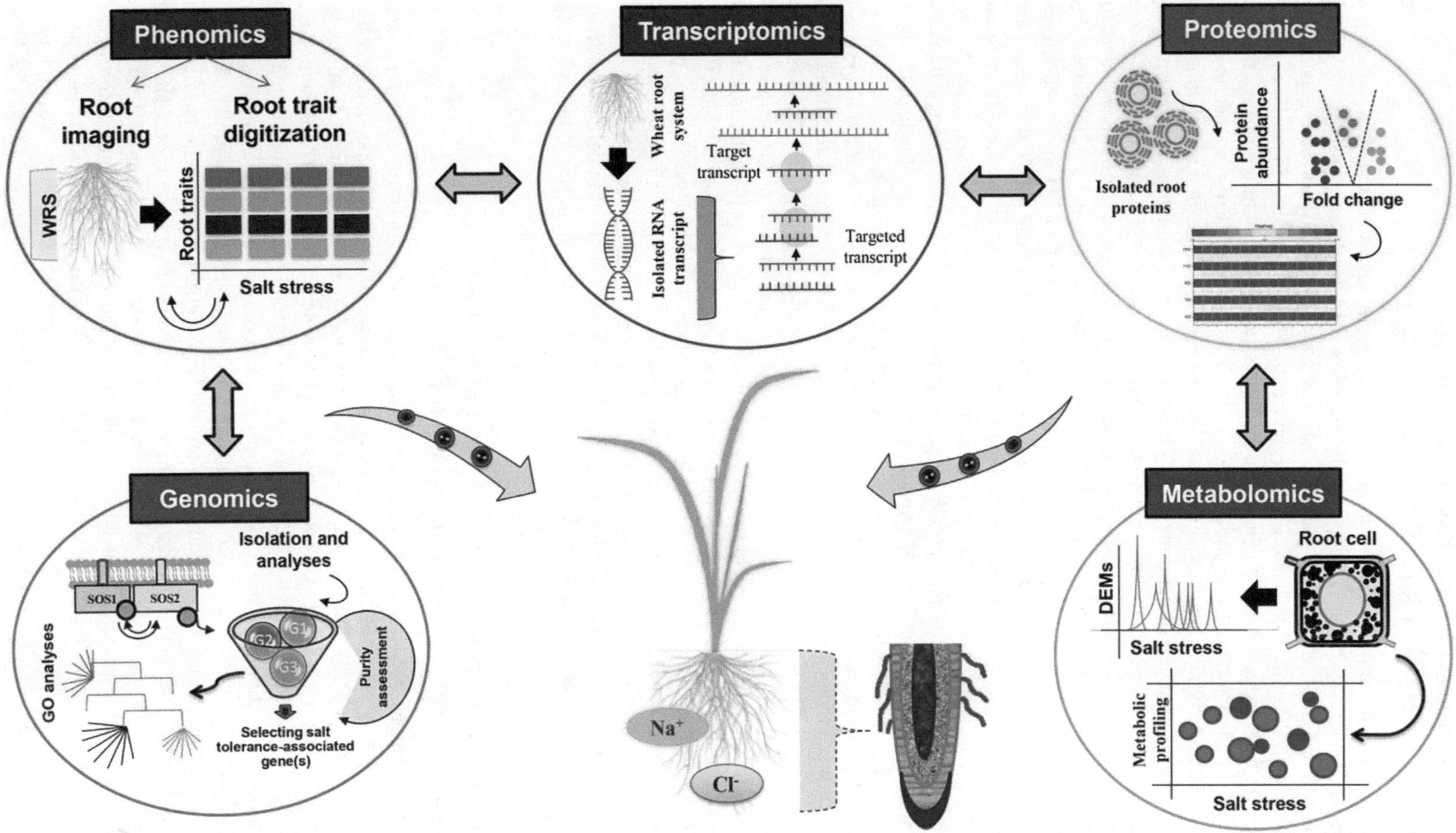

FIGURE 3.3 Wheat root system modulations induced through the involvement of different biotechnological interventions conveyed with omics approaches to ameliorate salt stress-induced adversities. Cl^-, chloride ion; DEFs, differentially expressed metabolites; GO, gene ontology; Na^+, sodium ion; RNA, ribonucleic acid; SOS, salt overly sensitive; WRA, wheat root system.

3.3.3 Biotechnological Interventions Related to Transcriptomics Implication in Wheat Root Systems under Salt Stress

Salt stress considerably imposes a cascade impact on transcript levels of root cells, which lead to their down- and/or up-regulation. This eventually imposes considerable alterations in plant physiological growth and productivity. For instance, when Najran wheat cultivars were exposed to salt stress applied with 200 mM NaCl, the RNA-sequence analysis revealed ~135 million paired-end reads from root cells (Alyahya and Taybi 2023). Additionally, transcriptomic profiling showed the occurrence of 5,829 differentially expressed genes (DEGs) in root cells in response to salt stress conditions in the underlying wheat cultivar. These findings have led to the identification of key physiological pathways, including glutathione (GSH) biosynthesis and secondary metabolism (phenylpropanoids and galactose) of the wheat root system, which might play a decisive role in providing relief from salt-induced adversities. Further, to comprehend the transcript regulations related to salt tolerance in common and synthetic hexaploid (SH) wheat, Nakayama et al. (2022) conducted RNA sequencing in the roots of selected wheat lines, including salt-tolerant SH and common wheat, along with salt-sensitive common wheat, which were exposed to 50 mM NaCl at 0, 6, 24, and 72 hours. This study showed that SH wheat prominently sustained salt tolerance through an unidentified signaling network probably regulating ROS accumulation, *respiratory burst oxidase homologs* (*RBOH*) and *ethylene responsive factors* (*ERFs*) genes. In the case of common wheat, salt tolerance-mediating transporters were involved in enhancing resistibility in response to the underlying stress (Nakayama et al. 2022). Similar observations were reported by Li et al. (2023) in salt-tolerant and sensitive wheat strains during 0, 6, and 24 hours of salt stress treatment (2% NaCl solution), respectively. Further, weighted gene co-expression network analysis (WGCNA) has determined that acute salt responses last more than 6 hours; however, it ends before the completion of 12 hours (Li et al. 2023). These findings were also supported through Kyoto Encyclopedia of Genes and Genomes (KEGG) enrichment analyses aiding in the revelation of DEGs, including phenylpropanoid biosynthesis (ko00940), along with the metabolisms of cysteine (Cys) and methionine (Met) (ko00270), and GSH (ko00480) of wheat root tissues under salt stress. In three-leaf stage wheat plants, the exposure of salt treatment (250 mM NaCl) resulted in DEGs, including *Walls Are Thin 1-related gene* (*TaWAT*), an *aquaporin gene* (*TaAQP*), a *TaGST*, and a *zinc finger gene* (*TaZFP*), which were correlated with root salt tolerance (Chen et al. 2024). These assessments have delineated the peculiar salt endurance mechanisms in wheat lines promising to be effective in future biotechnological programs favoring salt tolerance.

3.3.4 Biotechnological Interventions Related to Proteomics Implication in Wheat Root Systems under Salt Stress

The distinctive responsive behavior of proteome regulations in wheat root tip and mature root in response to salt stress (150 mM NaCl) has been comparatively illustrated in the study conducted by Dissanayake et al. (2022). They showed decreased translation of the enzymes belonging to the glycolytic and tricarboxylic acid (TCA) pathways and adenosine triphosphate (ATP) synthase subunits in the root tips of salt-stressed wheat plants. However, proteins controlling carbon assimilation and energy production were decreased in both wheat root tips and mature root tissues under salt stress (Dissanayake et al. 2022). These results plausibly indicate that translation processes and proteins coordinating with energy metabolism are the limiting responses in the case of the root tips, while the underlying mechanisms are relatively less adverse in mature root tissues of wheat plants exposed to salt stress conditions. Another exploration conducted by Khan et al. (2023) led to the identification of 2436 proteins induced upon salt stress (250 mM NaCl), including 194 differentially abundant proteins (DAPs), which comprises 170 and 24 up- and down-regulated proteins in wheat root. These DAPs, including three peroxidase, six heat shock proteins (HSPs), four GST, four dehydrins, one β-subunit of potassium (K^+) channel, four vacuolar-type ATPase (VATPases), four superoxide dismutases (Cu-Zn SOD), one 14-3-3 protein, three thioredoxins, and two malate dehydrogenases (MDH) proteins were found to be responsible for addressing salt stress through their active induction in wheat root cells (Khan et al. 2023). Among these, 14-3-3 protein abundance in wheat root indicated Na^+ compartmentalization, thus synergistically assisting in coping with salt stress in wheat. Additionally, protein profusion in the KEGG pathway was differentially modulated, reflected through citrate cycle, ribosome formation, oxidative phosphorylation, ribonucleic acid (RNA) transport, amino acid metabolism, photosynthesis, and pentose phosphate, glycolysis, and ubiquitin-proteasome pathways in wheat root (Khan et al. 2023). These studies have highlighted the significance of understanding the proteomic responses of wheat root to improve salt tolerance due to its salt sensitivity and perceptibility leading to significant metabolic and physiologic protection.

3.3.5 Biotechnological Interventions Related to Metabolomics Implication in Wheat Root Systems under Salt Stress

Few studies have essentially targeted wheat root systems to draw a parallel association with salt resistance capability. For instance, salt-tolerant Jimai 22 wheat variety showed 297 differentially expressed metabolites (DEMs) induced in root tissues by salt stress treatment (150 mM NaCl) compared to control (Yue et al. 2021). These DEMs were primarily enriched in pathways of phenylpropanoid biosynthesis and the metabolism of phenylalanine, glycine, serine, threonine, Cys, Met metabolism, and histidine, and thus crucially indicated wheat root-mediated salt adaptability. Likewise, in the Cangmai 6005 (CM6005; wheat variety), 0.5% salt exposure noticeably affected crucial cellular metabolisms, including amino acid and lipid biosynthesis pathways (Wang et al. 2023). Correspondingly, in a study performed by Jiang et al. (2023), wheat root analyses through ultra performance liquid chromatography-tandem mass spectrometer (UPLC-MS/MS) has identified significant modulations in metabolic occurrence, including the derivatives of amino acids, secondary metabolites, organic acids, nucleotides, and lipids. In summary, these reports have shed light on the signature metabolic profiling of the wheat root system under salt stress conditions, which will further aid in optimizing cultivation strategies yielding salt tolerance. However, further explorations are needed to decipher these findings with wheat root system stimulated salt defensive mechanisms to maintain the metabolites conveying carbon and energy expenditure, ionic transportation, and antioxidant biosynthesis.

3.4 CONCLUSION AND FUTURE PERSPECTIVES

The impact of salt stress on wheat root systems is multifaceted, affecting their growth, physiology, and metabolic processes. Since roots serve as the primary organs for nutrient uptake and water absorption, and encounters foremost contact with the stress, they play a critical role in plant responsiveness under salt stress. In response to the underlying stress exposure, wheat plants exhibit myriad modulations in root behavior, thus highlighting their significance in mediating salt tolerance. However, severe salt-induced adverse reactions imperatively affect root sensitivity, leading to wheat crop failures. Several reports have highlighted the potential of wheat root systems, with distinctive modulations in signature molecules and/or responses operating primary and secondary metabolisms. These growth, physiological, and metabolic interactions in wheat roots under salt stress involve a complex network conducting metabolic shifts to retain plant sustenance. Further, the comprehensive understanding of intricate mechanisms through integrating advanced omics technologies, including phenomics, genomics, transcriptomics, proteomics, and metabolomics, provides in-depth examinations of the subsequent events occurring in wheat roots during salt stress. These approaches could lay the foundation for future research and potential interventions that enhance coping mechanisms to impart salt tolerance in wheat crops through the manipulation of root systems. Future studies focusing on the specific gene(s)/transcript(s) and protein(s) involved in root metabolic pathways could aid in conferring salt-acclimatizing properties in wheat plants. This will certainly facilitate the development of biotechnological interventions to selectively breed wheat genotypes/varieties/cultivars/strains with improved root health and salt tolerance, thus creating the operation of sustainable agricultural practices in the future. In addition to this, developing agronomic practices and precision farming techniques through wheat root reprogramming could address the specific needs of mitigating crop yield losses and protect food security for the ever-growing world population. This calls for collaboration between researchers, agronomists, and policymakers to translate scientific findings into practical implications for farmers facing the threats of impeded wheat productivity. Therefore, a concurrent approach facilitating salt resistance is commercially important to secure wheat health and grain production in response to elevated salt-contaminated areas.

3.5 AUTHORSHIP CONTRIBUTION

Conceptualization, M.I.R.K.; Writing – Original draft preparation, S.K., H.H., R.N.S., K.H.M.S., and M.I.R.K.; Writing – Review and editing, M.I.R.K., K.H.M.S., and S.K. All authors have read and agreed to the published version of the book chapter.

3.6 DECLARATION OF COMPETING INTEREST

The authors declare that they have no known competing financial interests or personal relationships that could have appeared to influence the work reported in this book chapter.

ACKNOWLEDGEMENT

MIRK is gratefully acknowledges the SERB-DST grant (SRG/2020/001004).

REFERENCES

Ahmadi, Jafar, Alireza Pour-Aboughadareh, Sedigheh Fabriki-Ourang, Ali-Ashraf Mehrabi, and Kadambot H. M. Siddique. "Screening wild progenitors of wheat for salinity stress at early stages of plant growth: insight into potential sources of variability for salinity adaptation in wheat." *Crop and Pasture Science* 69, no. 7 (2018): 649–658.

Alyahya, Norah, and Tahar Taybi. "Comparative transcriptomic profiling reveals differentially expressed genes and important related metabolic pathways in shoots and roots of a Saudi wheat cultivar (Najran) under salinity stress." *Frontiers in Plant Science* 14 (2023): 1225541.

Chen, Jiating, Lei Zhang, Yingxi Liu, Xinyao Shen, Yujing Guo, Xiaofei Ma, Xiaojun Zhang, et al. "RNA-seq-based WGCNA and association analysis reveal the key regulatory module and genes responding to salt stress in wheat roots." *Plants* 13, no. 2 (2024): 274.

Chen, Lina, Hengxia Yin, Jin Xu, and Xiaojing Liu. "Enhanced antioxidative responses of a salt-resistant wheat cultivar facilitate its adaptation to salt stress." *African Journal of Biotechnology* 10, no. 74 (2011): 16884–16886.

Dissanayake, Bhagya M., Christiana Staudinger, Rana Munns, Nicolas L. Taylor, and A. Harvey Millar. "Distinct salinity-induced changes in wheat metabolic machinery in different root tissue types." *Journal of Proteomics* 256 (2022): 104502.

El Sabagh, Ayman, Mohammad Sohidul Islam, Milan Skalicky, Muhammad Ali Raza, Kulvir Singh, Mohammad Anwar Hossain, Akbar Hossain, W. Mahboob, M. A. Iqbal, D. Ratnasekera, and R. K. Singhal. "Salinity stress in wheat (*Triticum aestivum* L.) in the changing climate: adaptation and management strategies." *Frontiers in Agronomy* 3 (2021): 661932.

Farooq, Ahsan, Nageen Farooq, Haseeb Akbar, Zia Ul Hassan, and Shabbir H. Gheewala. "A critical review of climate change impact at a global scale on cereal crop production." *Agronomy* 13, no. 1 (2023): 162.

Hu, Yuncai, and Urs Schmidhalter. "Opportunity and challenges of phenotyping plant salt tolerance." *Trends in Plant Science* 28, no. 5 (2023): 552–566.

Jiang, Xue-min, Xiang-qian Chen, Hong-yan Li, and Qi-yan Jiang. "Metabolomic analysis of wheat response to salt stress." *Journal of Agricultural Science and Technology* (2023): 43–56.

Khan, M. Iqbal R., Faroza Nazir, Mohd Asgher, Tasir S. Per, and Nafees A. Khan. "Selenium and sulfur influence ethylene formation and alleviate cadmium-induced oxidative stress by improving proline and glutathione production in wheat." *Journal of Plant Physiology* 173 (2015): 9–18.

Khan, Waqif, Adnan Khan, Abd Ullah, Syed Inzimam Ul Haq, Nazim Hassan, Babar Iqbal, Naveed Ahmad, Eman A. Mahmoud, and Hosam O. Elansary. "Insights concerning advancing the agroecological sustainability of salinity tolerance through proteomics profiling of hexaploid wheat (*Triticum aestivum* L.)." *South African Journal of Botany* 158 (2023): 142–148.

Kumari, Sarika, Faroza Nazir, Kajal Jain, and M. Iqbal R. Khan. "GABA and potassium modulates defence systems, assimilation of nitrogen and carbon, and yield traits under salt stress in wheat." *Journal of Plant Growth Regulation* 42, no. 10 (2023): 6721–6740.

Kumari, Sarika, Faroza Nazir, Apoorva Singh, Hyda Haroon, Nahida Rashid Khan, Rudra Narayan Sahoo, Mohammed Albaqami, Kadambot H. M. Siddique, and M. Iqbal R. Khan. "γ-aminobutyric acid (GABA) strengthened nutrient accumulation, defense metabolism, growth and yield traits against salt and endoplasmic reticulum stress conditions in wheat plants." *Plant and Soil* 498, no. 1 (2024): 409–429.

Li, Jianfeng, Xin Gao, Xunji Chen, Zheru Fan, Yueqiang Zhang, Zhong Wang, Jia Shi, C. Wang, H. Zhang, L. Wang, and Q. Zhao. "Comparative transcriptome responses of leaf and root tissues to salt stress in wheat strains with different salinity tolerances." *Frontiers in Genetics* 14 (2023): 1015599.

Lynch, Jonathan P. "Harnessing root architecture to address global challenges." *The Plant Journal* 109, no. 2 (2022): 415–431.

Maqbool, Saman, Muhammad Adeel Hassan, Xianchun Xia, Larry M. York, Awais Rasheed, and Zhonghu He. "Root system architecture in cereals: progress, challenges and perspective." *The Plant Journal* 110, no. 1 (2022): 23–42.

Nakayama, Rio, Mohammad Taheb Safi, Waisuddin Ahmadzai, Kazuhiro Sato, and Kanako Kawaura. "Comparative transcriptome analysis of synthetic and common wheat in response to salt stress." *Scientific Reports* 12, no. 1 (2022): 11534.

Sehar, Zebus, Noushina Iqbal, M. Iqbal R. Khan, Asim Masood, Md Tabish Rehman, Afzal Hussain, Mohamed F. AlAjmi, Altaf Ahmad, and Nafees A. Khan. "Ethylene reduces glucose sensitivity and reverses photosynthetic repression through optimization of glutathione production in salt-stressed wheat (*Triticum aestivum* L.)." *Scientific Reports* 11, no. 1 (2021): 12650.

Sharma, P., S. Mishra, B. Pandey, and G. Singh. Genome-wide identification and expression analysis of the NHX gene family under salt stress in wheat (*Triticum aestivum* L). *Frontiers in Plant Science* 14 (2023): 1266699.

Sukumaran, Selvakumar, Johanna Lethin, Xin Liu, Justyna Pelc, Peng Zeng, Sameer Hassan, and Henrik Aronsson. "Genome-wide analysis of MYB transcription factors in the wheat genome and their roles in salt stress response." *Cells* 12, no. 10 (2023): 1431.

Terletskaya, Nina, Ulshan Duisenbayeva, Aiman Rysbekova, Meruert Kurmanbayeva, and Irina Blavachinskaya. "Architectural traits in response to salinity of wheat primary roots." *Acta Physiologiae Plantarum* 41, no. 9 (2019): 157.

Wahid, Iram, Pratibha Rani, Sarika Kumari, Rafiq Ahmad, Sofi J. Hussain, Saud Alamri, Nirmalya Tripathy, and M. Iqbal R. Khan. "Biosynthesized gold nanoparticles maintained nitrogen metabolism, nitric oxide synthesis, ions balance, and stabilizes the defense systems to improve salt stress tolerance in wheat." *Chemosphere* 287 (2022): 132142.

Wang, Wei, Jinfeng Cao, Sufang Huang, Zhi Wang, Weiwei Wang, Jingwei Zou, Fengzhi Wang, Meng Luo, and Jingjian Zhang. "Integrated transcriptomics and metabolomics analyses provide insights into salt-stress response in germination and seedling stage of wheat (*Triticum aestivum* L.)." *Current Plant Biology* 33 (2023): 100274.

Yue, Jieyu, Yingjie Wang, Jinlan Jiao, and Huazhong Wang. "Comparative transcriptomic and metabolic profiling provides insight into the mechanism by which the autophagy inhibitor 3-MA enhances salt stress sensitivity in wheat seedlings." *BMC Plant Biology* 21 (2021): 1–36.

4 Signalling Molecules
New Players in Optimizing Plant Growth under Stress Conditions

Nitika Kapoor, Roohi Sharma, Deepak Kumar, Shalini Dhiman, Tamanna Bhardwaj, Palak Bakshi, Indu Sharma, Pooja Sharma, Puja Ohri, and Renu Bhardwaj

4.1 INTRODUCTION

Stress in plants refers to the physiological and biochemical responses of plants to adverse environmental conditions or external factors that disrupt their normal growth and development (Zhang et al. 2020). Plants can experience stress from a wide range of sources, including abiotic stressors like extreme temperatures, drought, flooding, salinity, pollutants, and physical damage, as well as biotic stressors like pathogens, herbivores, and competition with other plants (Vandana et al. 2020). Plants have evolved various strategies to cope with stress, including physiological, biochemical, and morphological adaptations (Wahab et al. 2022; Khan et al. 2023a, 2023b). These adaptations can include the production of stress-related proteins, changes in root architecture, altered metabolic pathways such as signalling molecules, and the accumulation of protective compounds like antioxidants (Mittler et al. 2022). Understanding and managing stress in plants is crucial for agriculture and ecosystem health, as stress can reduce crop yields and affect the overall functioning of ecosystems. Researchers and farmers work to develop stress-tolerant plant varieties and implement strategies to mitigate the impact of stress on crops and natural environments. Signalling molecules have been shown to enhance plant growth and development, playing crucial roles in processes, such as seed germination, root growth, flowering, and stress response and senescence, making them important regulators of plant physiology and adaptation (Thakur et al. 2021; Ghosh et al. 2022; Kaya et al. 2023). These signalling molecules are integral to the plant's ability to sense and respond to changes in their environment, helping them adapt to different stresses and conditions. They often function as secondary messengers in signal transduction pathways, regulating gene expression and other cellular processes to maintain plant health and vitality (Di et al. 2015).

Signalling molecules, especially nitric oxide (NO), gamma-aminobutyric acid (GABA), hydrogen sulfide (H_2S), and hydrogen peroxide (H_2O_2) have gained significant attention in recent years for their roles in enhancing plant growth and stress tolerance. These signalling molecules interact with one another and are integrated into complex signalling networks within plants (Fatima et al. 2021; Khan et al. 2021a). The balance and coordination of these molecules are crucial for optimal growth and stress tolerance. They play vital roles in regulating various physiological processes in plants and also in managing abiotic and biotic stresses. SMs are responsible for the integration of different environmental signals and the activation of stress-response networks, thus contributing to the establishment of defence mechanisms and plant resilience (Kohli et al. 2019). Out of these SMs, GABA is a non-proteinogenic amino acid which acts particularly in response to abiotic stresses such as drought, salinity, and oxidative stress in plants (Sita and Kumar 2020). GABA plays diverse roles, including regulating redox status, cytosolic pH, ion transport, osmotic potential, nutrient uptake, metabolic flux, pollen tube growth, cell wall modification, root development, modulating reactive oxygen species (ROS), and plastid functions (Jalil et al. 2019; Ansari et al. 2021; Khan et al. 2021b;

DOI: 10.1201/9781032630380-4

Khanna et al. 2021; Kumari et al. 2023b, 2024). GABA enhances plant traits under stress, improving photosynthesis, osmoprotectant accumulation, and polyamine metabolism (Podlešáková et al. 2019; Kumari et al. 2023a, 2024). It mitigates ROS buildup by activating antioxidant defences, while also modulating gene expression related to signal transduction and stress responses (Kinnersley et al. 2000). Further, H_2O_2 is an ROS that serves as a signalling molecule involved in various cellular processes in plants. H_2O_2 (along with GABA) acts as a secondary messenger in signal transduction pathways (Saxena et al. 2016; Rarata et al. 2019). It possesses several characteristics that make it a plausible candidate for phytohormone status. Oxidative stress represents only one aspect of its actions, as the respiratory chain in mitochondria may be regulated by the ratio of superoxide radical to H_2O_2 and the triggering of apoptosis programmes is believed to be determined by the combination of singlet oxygen plus superoxide radical to hydrogen peroxide (Yan et al. 2020).

H_2S, a colourless and toxic gas with a foul odour, protects plants from oxidative stress, regulates stomatal closure, and influences various physiological processes, including seed germination, premature leaf senescence, and degradation of fruits after harvest in a variety of fruits and vegetables (Ali et al. 2019; Kumari et al. 2023b). In minimal concentrations (30 μmol L^{-1}), it hinders mitochondrial respiration by 50% through cytochrome c oxidase inhibition. Paradoxically, H_2S also acts as a gasotransmitter in plant growth, interacting with plant hormones and other gasotransmitters and ionic signals, while modifying proteins and influencing secondary metabolism (Li et al. 2016). Additionally, NO is also a highly reactive gas, and even as late as 1998, groundbreaking research demonstrating NO as a signalling molecule in plants was published. NO is a biologically active signalling molecule that was initially identified in mammals (Lamattina et al. 2003) where it was found to play an important role in regulating various physiological processes, including immunological modulation, smooth muscle relaxation, brain transmission, apoptosis, and others (Wang et al. 2020). In plants, NO also plays a crucial role in regulating photosynthetic machinery, seed germination, root development, and response to numerous abiotic and biotic stresses by amending gene expression and signalling pathways related to plant defence mechanisms (Asgher et al. 2017; Minguillón et al. 2022).

In this chapter, we aim to encapsulate various dimensions of NO-, GABA-, H_2S-, and H_2O_2-mediated responses in plants. We commence by exploring the sources, properties, biosynthesis and breakdown, and mechanisms of these signalling molecules. We delve into detail on how these signalling molecules affect various processes and their roles in plant growth and development in both favourable and unfavourable environmental settings. Subsequently, we delve into the mechanisms through which plants perceive these molecules, discussing their role within intricate plant signalling networks and their substantial influence on plant growth and development.

4.2 PHYSICO-CHEMICAL PROPERTIES OF SIGNALLING MOLECULES

4.2.1 Nitric Oxide

In plants, nitric oxide is recognized as an incredibly mobile molecule capable of diffusing across membranes and serving as a vital secondary gaseous messenger that plays a pivotal role in plant growth and overall development. Due to these characteristics, NO holds significance as a crucial signalling molecule (Khan et al. 2023b). NO, a small, uncharged molecule with an unpaired electron, exhibits paramagnetic properties, determining its reactive nature with various chemicals. These unpaired electrons, whether present in σ or π orbital, are highly unstable (Fukuto 2022). Not all molecules react with NO, but with radicals it reacts readily because of pairing of two very favourable electrons which are unpaired in nature. O_2•– contains an even number of total electrons, and two of them are unpaired in its most frequent state, namely the ground state. So, in a reaction involving the •NO and O_2, nitrogen dioxide is thought to arise as the first product, which is likewise thought to be a radical molecule (Lehnert 2021; Lee 2022). Certainly, NO_2 acts a nucleophile that reacts quickly with a variety of other compounds that are electron rich and creating other radicals like nitrite anion (NO_2^-). NO radicals also react with $O_2^{\bullet -}$, that is, superoxide anion, which is exceedingly quick, and first-order kinetics form peroxynitrite. Peroxynitrite (ONOO) is not considered as radical but it acts as a moderately effective electron oxidant; nitrosonium ions (NO^+), NO radicals (NO), higher nitrogen oxides (NOx), iron dinitrosyl complexes, and S-nitrosothiols (SNOs), among others, are formed when the free radical •NO combines with different extracellular/intracellular macromolecules. Under biological conditions, the interaction of •NO radical with nonmetals has two main effects: oxidation of cellular nucleophiles and production of the slightly steady anions, namely nitrite and nitrate (Lancaster 2015).

4.2.2 Hydrogen Sulfide

The second gaseous signalling molecule, often known as a gasotransmitter, after carbon monoxide (CO) and •NO, is hydrogen sulphide. It is an inorganic gas which is combustible and water soluble and has a pungent odour that is similar to that of rotten eggs (Hilal 2023). It is a small molecule with a reducing ability. Because of its small size and neutral charge, H_2S can easily transfer across the cell membranes and other organelles without the need of a transporter (Yang et al. 2022). Three forms of H_2S exist in nature: disprotonated, monoanion, and dianion (Figure 4.1). All of them are collectively called as hydrogen sulfide.

4.2.3 Gamma-Aminobutyric Acid

Gamma-aminobutyric acid ($C_4H_9NO_2$) is a non-protein amino acid which is also known as 4-aminobutanoic acid or 4-aminobutyric acid. Numerous microorganisms, plants, and vertebrate animals exhibit it (Lee et al. 2023). Animals' bodies depend on GABA for a variety of physiological processes, including management of blood pressure, neurotransmission, cell signalling, hormone regulation, and brain protection against psychiatric disorders (Ueno et al. 2022). Modifying the architecture of the cell, affecting H^+ and Ca^{2+} concentrations and enzyme configuration to activate glutamate decarboxylase, and regulating the gene and protein expression of enzymes involved in GABA biosynthesis and metabolism are the key components of GABA enrichment (Sun et al. 2022; Ullah et al. 2023).

4.2.4 Hydrogen Peroxide

Hydrogen peroxide, commonly referred to as dioxygenane or dioxygen, is a highly oxidizing substance with an especially foul stench that breaks down into oxygen and water while producing a lot of heat (Andrés et al. 2022). Despite not being flammable, it is a powerful oxidant that can ignite when it comes into touch with organic material or metals like copper, silver, or bronze (Asaeda et al. 2022). Hydrogen peroxide is employed in a variety of human endeavours due to its chemical qualities, including healthcare as an antiseptic, antimicrobial, and antibacterial agent. H_2O_2 is reactive and may adversely impact energy-transforming cell systems at high concentration, for example, by inactivating glyceraldehyde-3-phosphate dehydrogenase which is a glycolytic enzyme (Qureshi et al. 2022). In the Fenton reaction, soluble Fe(II) gives an electron to an H_2O_2 molecule, leading it to break down into hydroxyl radicals, or •OH + OH, which can randomly oxidize almost any organic molecule (Mahaseth and Andrei 2017).

$$\underset{\text{Disprotonated}}{H_2S} \longleftrightarrow \underset{\text{Monoanion}}{HS^- + H^+} \longleftrightarrow \underset{\text{Dianion}}{S^{2-} + 2H^+}$$

FIGURE 4.1 Different isoforms of H_2S.

4.3 BIOSYNTHESIS OF SIGNALLING MOLECULES

4.3.1 Biosynthesis of NO

Animals have well-known systems that control NO production and signalling. Despite evidence suggesting NO's functional presence in plants, its metabolic origin and mechanism of engagement in signalling cascades are still unknown. NO plays a significant role in plant metabolism and can be synthesized through enzymatic (reductive and oxidative) and non-enzymatic pathways (Figure 4.2; Astier et al. 2018; Kumar et al. 2023). The oxidative route entails the oxidation of hydroxylamine, L-arginine, and other PAs to produce NO. The nitrate reductase (NR) and other reductive enzymes found in plasma membrane (PM) and mitochondria, such as nitrite reductase and cytochrome c reductase, are essential for the reductive pathway. Non-enzymatic NO generation results from non-enzymatic reduction of NO_2^-/NO in plant compartments having acidic pH (Singh et al. 2020b).

4.3.1.1 Reductive Pathway of NO Biosynthesis

Nitrite serves as the principal substrate for reductive processes, which are catalyzed by NR, mitochondrial nitrite reduction, and NO-forming nitrite reductase (*NOFNiR*). The major and oldest known enzymatic NO mechanism in plants among these is assumed to be NO generation from nitrite produced by NR (Astier et al. 2021). Due to nitrate being the primary substrate for NR and nitrite reduction comprising only 1% of NR activity under normal circumstances, this pathway generates a small amount of NO. The main source of NR activity is the connection between the cytoplasm and chloroplasts. Additionally, NO_2^-/NO^- reductase (NI-NOR) has the capability to convert NO_2^- to NO by utilizing reduced cytochrome c as an electron acceptor (Mandal et al. 2022; Khan et al. 2023b). When NO_2^- is utilized as an alternative electron acceptor during the synthesis of ATP in mitochondrial complexes III and IV, enzymes also reduce NO_2^- to NO (Graska et al. 2023). This induces hypoxia in plant cells and triggers mitochondrial NO production, which inhibits the activity of cytochrome c oxidase and enhances the energy status of hypoxic cells. In hypoxic cells, cytosolic hemoglobin oxidizes NO produced from mitochondria, ensuring a continual supply of NO_2 for ATP synthesis (Gupta and Igamberdiev 2011; Timilsina et al. 2022). In plant peroxisomes, the reduction

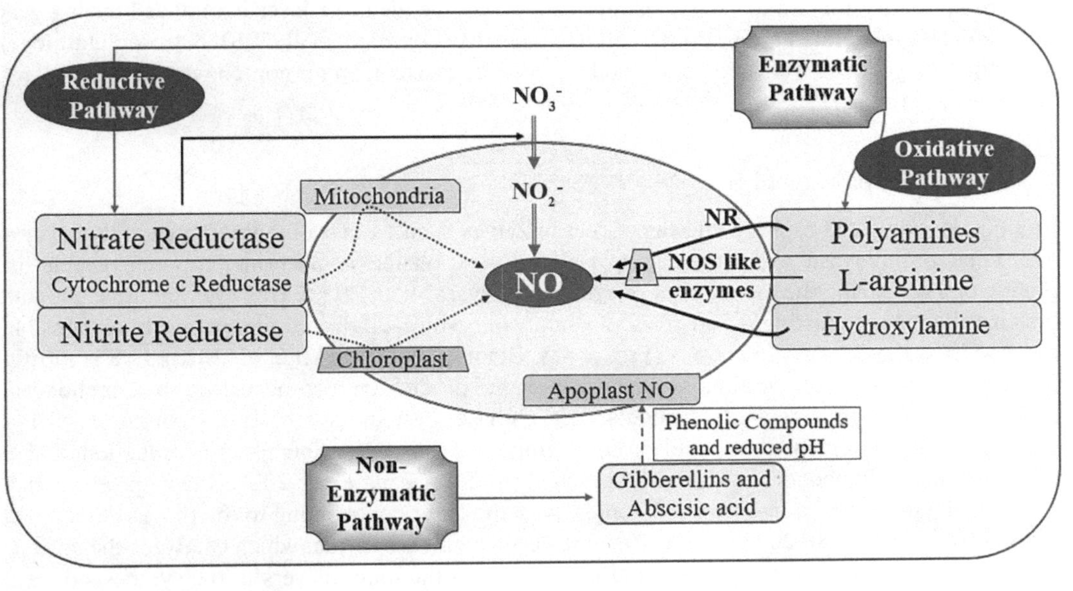

FIGURE 4.2 Biosynthesis of nitric oxide. NO_3^-, Nitrate; NO_2^-, Nitrite; NR, Nitrate reductase; NOS, NO synthase-like enzymes; P, Peroxisomes.

of NO_2^- under hypoxic or anoxic circumstances is another avenue for the formation of reductive NO. NO_2^-/NO decreases the capability of deoxygenated heme-containing proteins in the peroxisomal matrix. Similar reductional NO production in plant PM, cytosol, and endoplasmic reticulum has also been described (Igamberdiev et al. 2010).

4.3.1.2 Oxidative Pathway of NO Biosynthesis

Synthesis similar to oxidative processes in mammals, plants can produce NO by oxidizing N-containing compounds. NO synthase (NOS) proteins and NOS-producing genes have been found in prokaryotes, single-celled eukaryotes, invertebrates, and vertebrates, including mammals, however NOS enzymes have not yet been discovered in higher plants (Shepherd et al. 2022; Kumar and Ohri 2023).

Green algae and vascular plants' leaf peroxisomes and chloroplasts are major sites where L-arginine is converted by enzyme to citrulline and NO (Kolbert et al. 2019). NADPH and the lack of Ca^{2+} are necessary for the oxidation of L-arginine in chloroplasts, whereas NADPH, FAD, FMN, Ca^{2+} and calmodulin are necessary for peroxisomal oxidation (Mir et al. 2022). With the aid of enzymes found in the inter-membrane gap and mitochondrial matrix, L-arginine may also be converted to NO in plant mitochondria (Allagulova et al. 2022). In plant cells, oxidation of polyamines and hydroxylamines also leads to the generation of NO, however the precise mechanism underlying this is yet unknown (Ma et al. 2020). A potential mechanism involves indirect effects of polyamine production on L-arginine metabolism as well as interactions between polyamines and NR catalyzed NO. This mechanism could be used as a substitute for L-arginine-based oxidative NO production (Lopes-Oliveira et al. 2021; Kumar and Ohri 2023).

4.3.1.3 Non-Enzymatic Biosynthesis of NO

Non-enzymatic NO synthesis results from the breakdown of NO from nitrous acid. This necessitates apoplast acidification brought on by low pH brought on by gibberellins and abscisic acid (Santolini et al. 2017; Ciacka et al. 2022). This kind of NO release is encouraged by the acidic apoplast conditions of germination-stage seeds, which also shield them from soil bacteria. In the barley aleurone layer's apoplast, non-enzymatic conversion of NO_2^- to NO has been documented (Fatima et al. 2021). This process was aided by the phenolic chemicals found in the aleurone apoplast. Furthermore, NO breaks up seed dormancy, indicating that both NO_2^- and enzymatic NO production are necessary for healthy seed germination. When the aleurone layer is removed during germination, NO-moderated programmed cell death results (Lombardi et al. 2010) Nitrosoglutathione, which has not yet been well studied, is another potential mechanism for non-enzymatic NO generation (Corpas et al. 2022).

4.3.2 Biosynthesis of H_2S

In animals, pyridoxal-5-phosphate-dependent enzymes such as cystathionine beta-synthase, cystathionine gammalyase, and 3-mercaptopyruvate sulfurtransferase are principally responsible for producing H_2S during the metabolism of cysteine (Shivaraj et al. 2020). However, multiple enzyme systems that can be distinguished from the sulphate assimilation route are primarily responsible for the biosynthesis of H_2S in plants (Figure 4.3; Gotor et al. 2019; Ali et al. 2023). H_2S is mostly produced in chloroplasts in plant systems, with some production also occurring in mitochondria and cytosol. The enzymes beta-cyanoalanine synthase, cysteine synthase, L-cysteine desulfhydrase, D-cysteine desulfhydrase, and carbonic anhydrase are responsible for the production of H_2S in cytosol and mitochondria of plants (Ahmed et al. 2021; Fang et al. 2021). Plants produce H_2S by both enzymatic and non-enzymatic routes, with the latter contributing to the process in a much smaller way (Raza et al. 2022). The L-cysteine desulfhydrase enzyme, which catalyzes the conversion of L-cysteine to pyruvate, NH_4^+, and H_2S, is one of the main players in the synthesis of H_2S in plants. This process makes use of the cofactor pyridoxal phosphate (Santisree et al. 2019). In guard cells of *Arabidopsis*, ABA treatment was discovered to stimulate the expression of *LCD*,

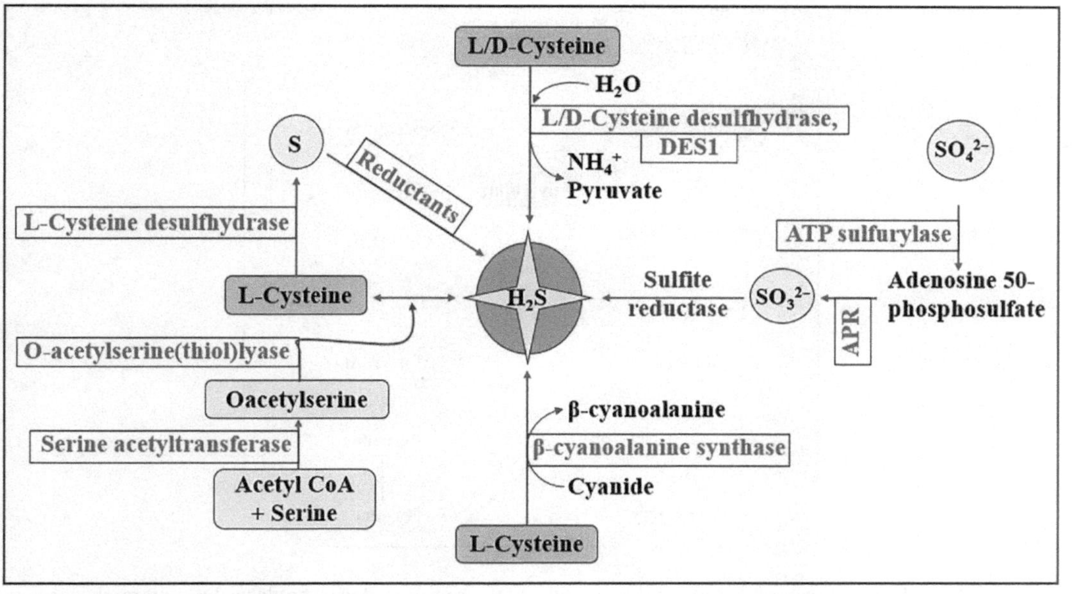

FIGURE 4.3 Biosynthesis of hydrogen sulfide. SO_3^{2-}, sulfite ion; SO_4^{2-}, sulfate ion.

which encodes L-cysteine desulfhydrase. Additionally, L-cysteine desulfhydrase is thought to control L-cysteine homeostasis. Additionally, D-cysteine desulfhydrase also promotes the synthesis of H_2S in plants by converting the substrate D-cysteine into pyruvate, NH_3 and H_2S in the cytoplasm.

Cyanide is hazardous to cells because it binds to cytochrome oxidase or other metalloenzymes, which prevents electron transport in mitochondria and chloroplasts. Higher plants use cyanoalanine synthase to detoxify by converting L-cysteine to cyanide, which then triggers the generation of H_2S. Plant cells, as opposed to animal cells, are less vulnerable to the cytotoxic effects of cyanide (Zhang et al. 2021). This is partly because plant mitochondria have an alternate oxidase that acts as the terminal oxidase in the respiratory chain and has a cyanide-insensitive respiratory route (Vojtovič et al. 2021). Another cytosolic enzyme called cysteine synthase, which is found in chloroplasts and mitochondria, catalyzes the reversible conversion of L-cysteine and acetate to O-acetyl-l-serine and H_2S. Along with these other enzymes, sulfite reductase reduces sulfite (SO_3^{2-}) to H_2S in chloroplast when ferredoxin is present (Hilal et al. 2023). The SO_2 (from the atmosphere) or sulphate (SO_4^{2-}; generated via sulphur nutrition) are the two sources of the SO_3^{2-} that is used as the substrate for this reaction. Adenosine 5′-phosphosulfate is formed when ATP sulfurylase activates SO_4^{2-} in its presence. The adenosine 5′-phosphosulfate resulting from this reaction is metabolized through the adenosine 5′-phosphosulfate reductase pathway, yielding SO_3^{2-}, which subsequently contributes to the production of H_2S in plants (Zhao et al. 2022).

4.3.3 Biosynthesis of GABA

In plants three routes may be used for GABA production (Figure 4.4). One method of GABA production is through the breaking down of polyamines like putrescine and spermidine. Putrescine can be produced either directly from ornithine or indirectly from arginine by the action of the enzyme ornithine decarboxylase and arginine decarboxylase, respectively (Suhel et al. 2023). A second polyamine, spermidine, is created when putrescine reacts with an aminopropyl group generated from decarboxylated S-adenosyl-methionine in the presence of the enzyme spermidine synthase (Shelp et al. 2021). Using diamine oxidase (DO) and polyamine oxidase (PO), respectively, putrescine and

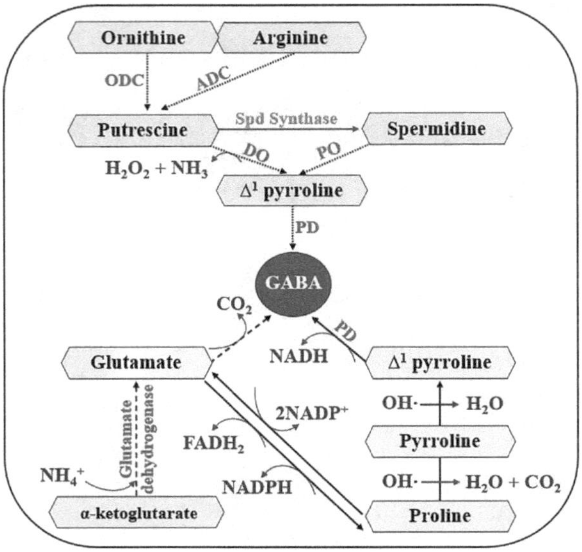

FIGURE 4.4 Biosynthesis of γ-aminobutyric acid. ADC, arginine decarboxylase; DO, diamine oxidase; ODC, ornithine decarboxylase; PO, polyamine oxidase; PD, Δ^1 pyrroline dehydrogenase.

spermidine are further catabolized to Δ^1 pyrroline. Δ^1 pyrroline is then converted into GABA by Δ^1 pyrroline dehydrogenase (PD) (Khan et al. 2021b; Dabravolski et al. 2023). GABA is also produced from glutamate through the GABA shunt pathway in the plant cell's cytoplasm during an irreversible decarboxylation event. GABA produced from glutamate is subsequently transported from the cytosol to the mitochondria where it is transformed into succinyl semi aldehyde (Hasan et al. 2021; Li et al. 2021). Another alternative pathway for GABA biosynthesis arises from non-enzymatic reactions of proline, which become more prevalent in the presence of an excessive buildup of reactive oxygen species. In these reactions, H abstraction from pyrroline leads to the formation of Δ1 pyrroline, which subsequently contributes to GABA biosynthesis (Ramos-Ruiz et al. 2019).

4.3.4 Reactive Oxygen Species

The buildup of reactive oxygen species is a frequent result of abiotic stress, including dryness. Superoxide anion, singlet oxygen, and hydroxyl anion are examples of radical ROS that have been identified. Hydrogen peroxide is an example of a non-radical ROS (Zhanassova et al. 2021). These ROS species have been discovered to induce death by inducing the programmed cell death pathway, oxidizing proteins, damaging nucleic acids, peroxiding lipids, and inhibiting enzymes (Kapoor et al. 2019). The formation of the extremely ROS results from the stepwise reduction of molecular oxygen by high-energy excitation or electron-transfer processes. In plants, these radicals are always produced as a byproduct of multiple metabolic pathways located in various cellular compartments or as a result of the unavoidable leaking of electrons onto molecular oxygen from the electron transport activities of chloroplasts, mitochondria, and PMs (Mansoor et al. 2022). Despite their damaging nature, ROS have been identified as second messengers in numerous cellular activities, including stress tolerance. The precise balance between reactive oxygen species generation and scavenging determines whether ROS will function as harmful or signalling molecules (Martin et al. 2022). Given the multifaceted roles of reactive oxygen species, precise regulation of their levels is essential for cells to prevent oxidative damage rather than completely eradicating them. An effective antioxidant system can scavenge excess ROS or detoxify them (Dumanović et al. 2021; Tyagi et al. 2022).

4.4 TRANSPORTATION OF SIGNALLING MOLECULES

Signalling molecules are essential for the regulation of a wide variety of physiological processes in all species, including plants. This is especially true in the case of plants when they are subjected to external pressures. NO, GABA, H_2S, and H_2O_2 are examples of some of the most important signalling molecules that are found in plant cells. Their transportation and cellular locations along with possible functions inside plant cells have been tabulated under Table 4.1.

NO serves as an essential messenger in plant systems. It is primarily generated through the conversion of nitrite, with processes such as nitrate reductases (NRs; Astier et al. 2018). While some evidence suggests plants might have an NO production method similar to that in animals, the exact enzymes, akin to animal NOS, have not been pinpointed in plants. NO is crucial for protein modifications (Liu and Xue 2021) and impacts various plant hormone signals (Bakshi et al. 2022.). Its role is particularly pronounced in plant parts with low oxygen levels, like roots. The chloroplast, peroxisomes, mitochondria, cytoplasm, PM, and apoplast of plant cells are all possible locations for the formation of nitrogen oxide (Kumar and Ohri 2022). The precise pathways and roles of NO in plants are still under investigation.

Another signalling molecule, GABA, is an amino acid that is not found in proteins but plays an extremely important role in plant stress physiology (Khan et al. 2021b). The enzyme known as glutamate decarboxylase (GAD) is responsible for synthesizing GABA in the cytoplasm from glutamate (Akama et al. 2007). Following its functional role, GABA is broken down by GABA transaminase (GABA-T) into succinic semialdehyde, which ultimately participates in the tricarboxylic acid (TCA) cycle (Michaeli et al. 2011). Certain transporters facilitate the movement of GABA between different parts of the cell, such as from the cytosol to the mitochondria or vacuoles (Khan et al. 2021b). These transporters are responsible for transporting GABA to the location where it undergoes catabolism or is stored. GABA permease (GABA-P) is an enzyme located in mitochondria

TABLE 4.1
Cellular Transportation, Locations and Function Aspects of Signalling Molecules

Most common signalling molecules	Cellular locations	Cellular transportation of signalling molecules	Functional roles inside plant cell	References
H_2S (Hydrogen Sulfide)	Mitochondria, cytoplasm, and maybe other parts of the cell	Diffusion, and perhaps even with the use of transporters	Act as a signalling molecule in stress reactions and root organogenesis	Huang and Yanjie (2023); Yang et al. (2022); Li et al. (2022)
H_2O_2 (Hydrogen Peroxide)	Cytoplasm, peroxisomes, mitochondria, and other locations inside cell	Diffusion, and perhaps also through aquaporins, peroxiporin channels, or some other specialized transporters	Perform functions in the signalling pathways that are associated with the responses to stress, growth, and development	Jahan et al. (2023); Israel et al. (2022)
GABA (Gamma-Aminobutyric Acid)	Cytoplasm, vacuole, and maybe even some more cell locations	Through GABA transporters, aluminum-activated malate transporter (ALMT), with some possibility of diffusion also	Take part in the signalling pathways that are associated with the body's responses to stress, growth, and development	Guo et al. (2023); Heli et al. (2022); Singh et al. (2020a); Seifikalhor et al. (2019)
NO	Cytoplasm, mitochondria, and possibly others	Diffusion, and perhaps even through the use of specialized transporters	Participate in a wide variety of physiological processes, such as growth, development, and the body's response to stress	Allagulova et al. (2023); Gupta et al. (2022); Singhal et al. (2021); Kolbert et al. (2019)

that facilitates the transport of GABA across the plasma membrane in plants. This enzyme belongs to the family of APC transporters (Michaeli et al. 2011). Once GABA is transported into the mitochondrial matrix, it undergoes catabolism. Additionally, in plants, GABA signalling is regulated by the aluminum-activated malate transporter (ALMT), an anion channel exclusive to plants (Ramesh et al. 2015). Its transport and subsequent fluctuations in level can induce various physiological responses in the organism, such as stomatal closure or modulation of ion channel function. GABA also performs a role in cell pH regulation. Both its production and its catabolism can function as a buffer system, assisting plants in overcoming the effects of acidic or alkaline stress. In response to stressors (like temperature, salt, drought, or pathogenic), GABA production and transport are frequently increased (Priya et al. 2019; Khanna et al. 2021). This up-regulation can result in a buildup of GABA, which assists the plant in surviving adverse conditions. Additionally, the metabolism of GABA is connected to the metabolism of polyamines, proline, and many other metabolites (Yong et al. 2017). GABA concentrations in the plant can affect and be affected by how these metabolites are transported and broken down (Khan et al. 2021b). Hence, the transportation of GABA entails a complex process involving its synthesis, movement across cellular compartments, and eventual breakdown. This transportation is indispensable for the myriad roles that GABA plays in plant growth, development, and responses to various stresses. Hydrogen sulfide, as a signalling molecule, has been shown to regulate intricate stomatal movements (Hou et al. 2013; Jin et al. 2017). It is suggested that H_2S may restrict NO production to facilitate stomatal opening (Lisjak et al. 2010). Moreover, H_2S interacts with reactive oxygen species and NO during stomatal closure, thereby constituting an essential component of the abscisic acid (ABA) signalling cascade in guard cells (Scuffi et al. 2014, 2018).

Furthermore, inside healthy plant cells, normal metabolism produces H_2O_2 through chloroplast Mehler reactions, mitochondrial electron transport, and peroxisomal photorespiration. Peroxisomes might additionally contain supplementary H_2O_2-producing systems (Corpas et al. 2001). Abiotic and biological stressors both increase the production of H_2O_2 through the aforementioned pathways, as well as through enzymatic sources such as PM-localized NADPH oxidases or respiratory burst oxidase homologue (RBOH) or cell wall peroxidases (Sagi and Robert 2001; Bolwell et al. 2002; Torres et al. 2002). H_2O_2 has a high capacity for diffusion, which may be aided by its ability to flow through peroxiporin channels within the membrane (Henzler and Ernst 2000). Depending on the overall content of H_2O_2 and the antioxidant state, H_2O_2 might diffuse only across subcellular microdomains when it enters cells from the apoplast or is generated internally (Neill et al. 2002). Signalling proteins, such as protein phosphatases (PP), protein kinases (PK), including a PM histidine kinase and MAPK cascades, transcription factors (TFs), and calcium channels, can be oxidized by H_2O_2 or otherwise modulated by it in one or other ways. These proteins can be found at the PM or elsewhere in the cell. If cytosolic calcium levels are high, calcium-binding proteins like CaM, PKs, and PPs will trigger downstream responses (Neill et al. 2002). TFs and gene expression are affected by reversible protein phosphorylation, leading to downstream signalling in the cytoplasm and nucleus. Oxidation of TFs has the potential to activate them and/or cause nuclear localization. Rop signalling has the ability to regulate H_2O_2 production (Baxter-Burrell et al. 2002).

4.5 ROLE OF SIGNALLING MOLECULES IN PLANTS GROWTH AND DEVELOPMENT

The growth, development, and adaptation of plants, as well as the activation of their antioxidant responses to a variety of environmental stress factors such as extreme temperatures, light, drought, salinity, heavy metals, herbicides, pathogens, and others, depend on signalling molecules (hormones, growth regulators), proteins, amino acids, nucleotides, etc. whose various roles have been tabulated in Table 4.2. One of the most researched plant SMs is GABA. GABA regulates redox status (Jin et al. 2019) and cytosolic pH improves nitrate uptake and its utilization (Zhen et al. 2018), adventitious root development (Xie et al. 2020), and plastid-associated functions (Xiang et al. 2016;

Bashir et al. 2021). The concentration of GABA present in the flower is responsible for pollen tube growth and orientation towards the ovule. Physiological effect of GABA is also reported in seed germination and adventitious root growth. Increase in GABA levels has been reported in soyabean, oats, wheat, rice, and barley (Li et al. 2021).

H_2S confers resistance to a variety of stress conditions, including metal strains, oxidative stress, oxidizing agents, temperature changes, and drought (Zhang et al. 2021). H_2S not only functions as a potent gasotransmitter, but in studies with H_2S donors applied exogenously, it has been found to control plant growth, senescence, and stress responses (Deng et al. 2016; Zhang et al. 2021). For instance, *S. oleracea* seedlings were examined, and interesting observations were reported in which senescent leaves had higher H_2S than the new leaves (Chen et al. 2011). H_2S can post-translationally

TABLE 4.2
Various Roles of Signalling Molecules in Plant Growth and Development

Signalling molecules	Plant species	Role	Reference
NO	*Prunus persica*	Sustains firmness and lowers ethylene production.	Han et al. (2018)
	Litchi chinensis	Promotes reproductive growth, abscission of rudimentary leaves, and expression of the flowering-related genes.	Liu et al. (2015)
	Coriandrum sativum	Simulates germination and seedling growth.	Panngom et al. (2018)
	Brassica juncea	Advances seed germination under copper stress.	Rather et al. (2020)
	Actinidia chinensis	Decreases ethylene production and extends postharvest life.	Zhu et al. (2010)
GABA	*Malus hupehensis* (Chinese crab apple)	Increases biomass and enhanced root growth.	Li et al. (2020)
	Lactuca sativa	Enhances salt tolerance and photosynthetic performance.	Kalhor et al. (2018)
	Trifolium repens	Increases endogenous GABA concentration that improves drought tolerance and homeostasis.	Yong et al. (2017)
	Pak-choi seeds	Significantly enhances the growth and quality of the crop.	Li et al. (2016)
	Tea	Increases contents of free amino acid components.	Zhu et al. (2019)
H_2S	*Phaseolus vulgaris, Pisum sativum, Zea mays*	Improves seed germination, seedling growth, leaf growth, biomass, and photosynthetic traits.	Dooley et al. (2013)
	Jatropha curcas	Increases germination percentage.	Li et al. (2012b)
	Festuca arundinacea	Betters plant growth and developmental metabolism; enhances plant height, biomass, leaf width, number of tillers, turf quality, and photosynthetic activity.	Liu et al. (2019)
	Zea mays	Plays a significant role in stomatal opening, apparently by reducing NO and ABA levels.	Li et al. (2013)
	Glycine max	Increases biomass, nodule number, leaf nitrogen (N) absorption, and N content.	Ali et al. (2014)
H_2O_2	*Oryza sativa*	Enhances cell expansion and root diameter.	Xiong et al. (2015)
	Solanum lycopersicum	Improves chlorophyll content and net photosynthesis rate and its related attributes	Nazir et al. (2019)
	Wheat	Regulates stomatal size and its physiology.	Kumari and Verma (2019)
	Paeonia suffruticosa	Regulates senescence.	Shi et al. (2015a)
	Solanum lycopersicum	Improves the rate of photosynthesis, chlorophyll content, and stomatal conductance.	Khan et al. (2019)

modify proteins and affect secondary metabolism (Li et al. 2016). Several studies have stated that H_2S promotes seed germination. For instance, seeds (bean, pea, wheat, and corn) on exposing to $10-100$ mM H_2S solutions showed an incremental increase in germination rates and seedling size (Dooley et al. 2013). Also, H_2S was reported to promote number and length of adventitious roots in *Ipomoea batata* seedlings while validating its role in promoting number and length of adventitious roots (Zhang et al. 2009). H_2S controls stomatal movement and thus regulates water loss and uptake of carbon dioxide (CO_2) through stomata (Liu and Xue 2021).

NO as a regulator of plant development has been expansively studied. NO is known to modulate cell processes in various ways. It has been reported that it plays a crucial role during seed germination (Pandey et al. 2019). NO inhibits cytochrome oxidase (COX) that limits ROS levels and prevents lipid degradation and DNA damage. Exogenous NO treatment delays ripening, avoids chilling damage, fosters disease resistance, and improves nutritional value. Furthermore, NO content gradually declines during ripening with concurrent protein nitration and nitrosation (Palma et al. 2019). Also, NO treatment fosters maturation of immature sweet pepper fruits resulting in low lipid peroxidation and an increase in antioxidant capacity, which makes the ripening process less severe (González-Gordo et al. 2019).

H_2O_2 at elevated levels results in oxidative stress while at nanomolar levels functions as an SM that controls plant growth and development (Htet et al. 2019). H_2O_2 supplementation in rice plants ($0-100$ mM) enhanced root diameter (Xiong et al. 2015), and H_2O_2 treatment (1–10 mM) in pistachio seedlings had bettered growth attributes and dry weights of shoot and roots (Bagheri et al. 2019). Also, H_2O_2 application promoted chlorophyll content and stomatal conductance (Ashfaque et al. 2014).

4.6 SIGNALLING MOLECULES MEDIATING BIOTIC AND ABIOTIC STRESS MANAGEMENT IN PLANTS

Plants are exposed to various abiotic and biotic stressors during their lifespan. Abiotic stresses mainly include heavy metals, salinity, drought, flooding, temperature, UV-B irradiation, and light, whereas plant pathogens, pests, and herbivores contribute to biotic stress in plants. Plants being sedentary cannot escape these environmental stresses and are adversely affected (Dahuja et al. 2021). Plants have also developed various strategies as defence responses that include complex networks coordinated by a myriad of SMs (Mostofa et al. 2018). For instance, H_2S application improves membrane integrity by reducing oxidative stress in pea plant under nickel toxicity (Shaiek et al. 2023). Similarly, in biotic stress like aphids, foliar application of NO donors, namely *S*-nitrosoglutathione (GSNO) and sodium nitroprusside (SNP), induced the defence response of *Pisum sativum* (Woźniak et al. 2017). Numerous reports elucidating the role of various signalling molecules in various plant species under abiotic and biotic stress have been tabulated in Table 4.3.

SMs have been identified to play an important role in plant defence mechanisms and protection from biotic stressors. For instance, during bacterial and fungal attacks, H_2S regulates the metabolism of glutathione (GSH), induces pathogenesis-related (PR) gene expression, induces post-translational modifications leading to modulation of enzymatic activities, and cross-talks with other signalling molecules and phytohormones (Corpas and Palma 2020; Choudhary et al. 2022). Moreover, H_2S has been demonstrated to effectively mitigate the toxic effects of *Fusarium graminearum* in wheat (Yao et al. 2022). NO has also been shown to regulate plant growth, development, and stress responses. Numerous studies have indicated that the exogenous application of signalling molecules such as NO, H_2S, abscisic acid, glycine-betaine, calcium ions, along with plant growth regulators like brassinosteroids, auxin, ethylene, salicylic acid (SA), jasmonates (JA), and polyamines, among others, has been effective in alleviating both biotic and abiotic stresses in plants (Bhardwaj et al. 2014a, 2014b; Liu and Xue 2021; Sharma et al. 2018, 2021a; Singh et al. 2023; Wang et al. 2021). Under stressful conditions, exogenously applied H_2S can ameliorate abiotic stresses by maintaining redox balance, preventing apoptosis, and mediating the expression of genes and proteins involved in signalling and

TABLE 4.3
Role of Signalling Molecules in Abiotic and Biotic Stress Management

Signalling molecules	Stress	Plant	Genes, enzymes or transcription factor involved	Effects	References
GABA	Drought	*Triticum aestivum*	Activities of phenylalanine ammonia lyase (*PAL*), cinnamic acid 4-hydroxylase (C4H), and 4-coumarate coenzyme A ligase (4CL) were activated	Enhanced activities of antioxidant enzymes, increased level of phenolic compounds	Zhao et al. (2023)
GABA	Drought	*Malus pumila*	Identified 1271 differentially expressed genes (*DEGs*) related to biosynthesis of secondary metabolites, MAPK signalling pathway, glutathione metabolism, and carbon fixation in photosynthetic organisms	Improved plant growth and fruit quality	Cheng et al. (2023)
GABA	Salt	*Vigna radiata*	*SOD*, *APX*, *POD*, *GTR*, *PPO*, (*NR*) nitrate reductase and (*GOGAT*) glutamate synthase activity	Improved growth and biomass, reduced Na^+, H_2O_2, and MDA content, increased photosynthetic pigments, enzymatic antioxidants activities, K^+ and K^+/Na^+ ratio, improved N metabolism and osmolytes content	Ullah et al. (2023)
GABA	Chlorothalonil	*Solanum lycopersicum*	Knockout of *pyruvate-dependent GABA transaminase*, involvement of GABA in CHT metabolism, including 4 *peroxidases*, 5 *glycosyltransferases*, 4 *glutathione S-transferases*, and 4 *ABC transporters*	Improved photosynthetic efficiency, promotes stomatal closure, reduced stomatal conductance	Shan et al. (2022)
GABA and l-Pro chemoperception	*Pseudomonas syringae* pv. tomato	*Solanum lycopersicum*	Genes involved in GABA catabolism	Three genes encoding GABA transaminases *viz.*, *gabT1*, *gabT2*, and *gabT3* and the succinic semialdehyde dehydrogenase gene *gabD* were analyzed for their gene expression. Gene expression of *gabT1*, *gabT2*, and *gabD* was up-regulated by GABA treatment.	Santamaría-Hernando et al. (2022)
H_2S	Salt	*Oryza sativa*	Indole acetic acid (IAA) and 1-aminocyclopropane-1-carboxylate (ACC) deaminase	Improved growth and tolerance against salt stress	Wang et al. (2022)
H_2S	Cd	*Trigonella foenum-graecum*	*APX, CAT*, and *SOD*	Improved morphological attributes, reduced H_2O_2 content and electrolyte leakage. Enhanced activities of antioxidant enzymes.	Javad et al. (2022)

(*Continued*)

TABLE 4.3 (*Continued*)
Role of Signalling Molecule in Abiotic and Biotic Stress Management

Signalling molecules	Stress	Plant	Genes, enzymes or transcription factor involved	Effects	References
H_2S and NO	High temperature	*Oryza sativa*	Photosystem I and II (*psbA* and *psbB*) genes	Increased growth, biomass, leaf water status, photosynthesis, antioxidants, and osmolytes	Gautam et al. (2022)
H_2S	Cd	*Brassica nigra*	Ascorbate and glutathione	Enhanced antioxidants, improved biomass and photosynthesis	Kaur et al. (2022)
NO	Salt	*Triticum aestivum* L.	Up-regulated salt overly sensitive, sodium hydrogen antiporter, aquaporin and osmotin (SOS1, NHX1, AQP, and OSM-34) and down-regulated photosystem II (D2-protein)	Increased photosynthetic efficiency, enhanced accumulation of proline and soluble sugars, up-regulation of antioxidant system	Alnusairi et al. (2021)
$H2O_2$	*Colletotrichum* species	*Vitis vinifera*	Heat shock transcriptional factors (HSFs), *CAT*, and *SOD*	Heat shock transcription factors (HSFs) namely HSF11, HSF12, HSF14, HSF15, HSF16, and HSF17, and gene expression of *CAT* and *SOD* were expressed in grapevines in response to H_2O_2 produced by *Colletotrichum* species	You et al. (2022)
H_2S	Drought and heat	*Hordeum vulgare*	*PPO*, *POD*, *SOD*, *CAT*	Increased biomass, reduced ROS production, MDA, and electrolyte leakage, enhanced enzymatic activities	Naz et al. (2022)
H_2S	Cu	*Artemisia annua*	Carbonic anhydrase (*CA*) and nitrate reductase (*NR*), *CAT*, *POX*, and *SOD*	Reduced lipid peroxidation, enhanced antioxidant activities	Nomani et al. (2022)
H_2S and NO	Waterlogging	*Triticum aestivum*	*CAT* and *POD*	Improved nutrient status, enhanced oxidative defence	Mfarrej et al. (2022)
NO	Cd	*Satureja hortensis*	*CAT* and *POD*	Increased growth, plant pigments, carbohydrates, and peroxidase activity	Azizi et al. (2021)
NO and H_2O_2	As	*Glycine max*	*GSH* and *AsA*	Improved biomass, up-regulation in ascorbate glutathione cycle	Singh et al. (2021)
H_2S	Drought	*Medicago sativa*	*GST17*, *Cu/ZnSOD*, *FeSOD*, *cAPX*, *PIP*, *NR*	Gene expression of antioxidants namely *GST17*, *Cu/ZnSOD*, *FeSOD*, *cAPX*, and NO biosynthesis (*NR*) and aquaporin (*PIP*) were up-regulated during drought stress amelioration	Antoniou et al. (2020)

H_2S and NO	Co	*Triticum aestivum*	*SOD*, *POX*, *APX*, *GR*, *DHAR*, *MDHAR*	Expression of *SOD*, *POX*, *APX*, *glutathione reductase* (*GR*), *monodehydroascorbate reductase* (*MDHAR*), and *dehydroascorbate reductase* (*DHAR*) was upregulated to induce co-stress amelioration. Also, content of ascorbate and glutathione and Rubisco activity was enhanced. Both signalling molecules induced stress alleviation through up-regulation of ascorbate-glutathione cycle (AsA-GSH) in chloroplast.	Ozfidan-Konakci et al. (2020)
H_2O_2	Drought	*Oryza sativa*	*CAT*, *APX*, and *GPOX*	Elevated proline level, improved water status, reduced oxidative damage, and improved photosynthesis	Sohag et al. (2020)
H_2S	High temperature	*Arabidopsis*	ABA-INSENSITIVE 5 (*ASI5*), E3 ligase Constitutive photomorphogenesis 1 (*COP1*), elonghypcotyl 5 (*HY5*), and Aba-insensitive 5 (*ABI5*)	Efflux of nuclear COP1 increased the accumulation of ABI5 and HY5. H_2S enhanced seed germination and thermotolerance *via* the regulation of COP1, HY5, and ABI5 pathways. It retained nuclear COP1 for the degradation of HY5 to induce thermotolerance.	Chen et al. (2019)
H_2S and NO	Al	*Glycine max*	Citrate transporter genes viz., *GmMATE13* and *GmMATE47*	In soybean, H_2S induced Al stress tolerance through interaction with NO. Both the signalling molecules up-regulated the expression of two citrate transporter genes, namely *GmMATE13* and *GmMATE47*, that enhanced citrate secretion, which is a common strategy for the management of Al-induced phytotoxicity. Functional analysis and transgenic *Arabidopsis* revealed that *GmMATE13* and *GmMATE47* induced citrate exudation resulting in amelioration of Al stress.	Wang et al. (2019)
NO and H_2O_2	*Xanthomonas axonopodis*pv. Citri	*Poncirus trifoliata*	*Enhance pseudomonas susceptibility 1* (*PtEPS1*) and *pathogenesis-related gene 4* (*PtPR4*)	*X. axonopodis* infection dramatically decreased root H_2O_2, NO, and SA levels, but increased root CaM and JA concentrations in non-arbuscular mycorrhizal fungi seedlings. Under mycorrhization, there were higher H_2O_2 and CaM levels and lower JA levels in *X. axonopodis*-infected seedlings. Also, gene expression of root *PtEPS1* and *PtPR4* were up-regulated, thus indicated *X. axonopodis* mediated stimulation of mycorrhizal roles in enhancing resistance of citrus canker.	Miao-Miao et al. (2019)

(*Continued*)

TABLE 4.3 (*Continued*)
Role of Signalling Molecule in Abiotic and Biotic Stress Management

Signalling molecules	Stress	Plant	Genes, enzymes or transcription factor involved	Effects	References
H_2S	Alkalinity	*Brassica oleracea*	Glyoxalase (*Gly* I and *Gly* II), *SOD*, and *GSH*	Enhanced glutathione content and enzymes of AsA-GSH cycle	Montesinos-Pereira et al. (2018)
NO	Drought	*Crambe abyssinica*	*APX*, *CAT*, *GR*, and *SOD*	Reduced MDA and reactive oxygen species. Increased proline and chloroplast pigment content.	Batista et al. (2018)
NO	NaCl	*Solanum lycopersicum*	*APX*, *CAT*, *GR*, and *SOD*	Enhanced accumulation of glycine betaine flavonoid, proline. Increased activities of SOD, CAT, Gr, and APX.	Ahmad et al. (2018)
GABA	Cr	*Brassica juncea*	*APOX*, *MDHAR*, *DHAR*, *GPX*, *GS*, *CAT* (*Gly* I and *Gly* II) glyoxalase	Improved growth, increased chlorophyll content, decreased proline and phytochelatins content, up-regulated enzymatic and non-enzymatic antioxidants	Mahmud et al. (2017)
GABA	*Pseudomonas syringae* pv. tomato DC3000	*Solanum lycopersicum*	*T3SS*, *GABApermease* (GabP), and ΔgabP	GABA inhibits *T3SS* in *P. syringae*, reducing HR elicitation in tobacco. *GabP* is the primary GABA uptake mechanism, and *ΔgabP* is insensitive to GABA-mediated T3SS repression. *GabP* plays a vital role in *P. syringae* for nutrition or gene expression regulation.	McCraw et al. (2016)
GABA	Harpin (Bacterial elicitor protein)	*Nicotiana tabaccum*	*Glutamate dehydrogenase* (*GDH*) and *glutamate decarboxylase* (*GAD*) encoding genes	Enhanced transcripts for genes encoding *GDH* and *GAD* were regulated differently in leaves infiltrated with harpin, thereby establishing a connection between harpin-induced responses and GABA in tobacco. Harpin infiltration significantly induced the first components of the GABA signalling pathway, such as GDH, GAD, glutamate, and GABA in tobacco.	Dimlioğlu et al. (2015)
H_2S	*Pseudomonas*	*Arabidopsis thaliana*	*MIR393a/b* and their target genes (*TIR1*, *AFB1*, *AFB2*, and *AFB3*)	Enhanced plant growth by up-regulating auxin signalling	Shi et al. (2015b)
H_2S	*Sclerotinia sclerotiorum*	*Brassica napus*	*GSH*	Modulation in the status of gases was observed that enhanced the growth conditions	Bloem et al. (2012)
H_2S and NO	Cd	*Medicago sativa*	*SOD*, *POX*, and *APX*	Regulation of differential gene expression (transcripts) of antioxidant enzymes such as *SOD*, *peroxidase* (*POX*), and *APX* resulted in the alleviation of Cd-induced stress in alfalfa	Li et al. (2012a)

H_2S and NO	Salt	*Medicago sativa*	*Ascorbate peroxidase* (*APX-1* and *APX-2*), and *Cu/Zn-SOD* (*superoxide dismutase*)	Transcripts of *APX-1* and *APX-2*, and *Cu/Zn-SOD* (*superoxide dismutase*) were activated differentially, and resulted in amelioration of salinity induced oxidative damage. Also, NaHS application was found to enhance the production of NO in alfalfa.	Wang et al. (2012)
H_2S	Drought	*Arabidopsis thaliana*	*Cysteine desulfhydrases* (*CDes*), *l-cysteine desulfhydrase*, and *d-cysteine desulfhydrase*	H_2S application triggered the induction of drought-associated genes, namely *Cysteine desulfhydrases* (*CDes*), *l-cysteine desulfhydrase*, and *d-cysteine desulfhydrase*, degraded cysteine, and enhanced endogenous levels of H_2S to induce resistance against drought	Jin et al. (2011)
GABA	Salt	*Caragana intermedia*	*Expressed sequence tags* (*ESTs*), *1-aminocyclopropane-1-carboxylic acid* (*ACC*) *oxidase* (*ACO*) genes (*CaGR30* and *CaGR31*)	GABA-induced regulation of H_2O_2-generated genes, NADPH oxidase (*CaGR60*), peroxidase (*CaGR61*) and amine oxidase (*CaGR62*) genes at mRNA level that resulted in the inhibition of salt-induced accumulation of H_2O_2. Also, gene expression of *ACO* genes (*Ca*GR30 and *Ca*GR31) was up-regulated.	Shi et al. (2009)

metabolic pathways induced by phytohormones (Kour et al. 2021). Both abiotic and biotic stresses lead to the production of reactive oxygen species through plant Noxes (NADPH oxidase), resulting in oxidative stress in plants (Kaur et al. 2014).

ROS-generating and ROS-scavenging systems regulate the levels of oxidative stress (Sharma et al. 2021b). These systems include enzymes such as ascorbate peroxidases (APX), catalase (CAT), and superoxide dismutase (SOD), and their gene expressions are altered in response to the ROS regulatory network during plant stress management.

It has been observed that exogenous H_2S treatment increases the expression and activities of certain enzymes that alleviate oxidative stress (Liu et al. 2021). It also facilitates the production of another signalling molecule such as H_2O_2 *via* NADPH oxidase but controls ROS content by regulating antioxidant enzymes. Moreover, H_2S increases GSH and GSH-associated genes under chilling stress and inhibits oxidative damage indicators such as malondialdehyde (MDA), proline, and electrolyte leakage. Exogenous application of sodium hydrosulfide (NaHS), an H_2S donor, increases the activities of enzymes and endogenous cysteine and H_2S content (Min et al. 2016). H_2S regulates dynamic sulfur (S) metabolism to promote the production of S-derivatives and S-containing proteins (Liu et al. 2021). It is also reported that H_2S reduces heat stress damage by increasing the synthesis of total sulfhydryl compounds, proteins, and cysteine in tobacco (Chen et al. 2019; Li et al. 2013, 2021). Further, it is found that the exogenous application of H_2S donors helps plants adapt to drought-induced stress by enhancing antioxidant enzymes, reducing MDA content, and promoting stomatal closure to prevent water loss under drought conditions (Thakur and Anand 2021). It also accelerates the photosystem II repair cycle and interacts with secondary messengers and phytohormones.

Various studies on the alleviation of heavy metal toxicities through the application of SMs emphasized the possibility of activation of H_2S by NO in response to cadmium (Cd) stress, and that H_2S is downstream of the NO signal (Li et al. 2012a). This was further demonstrated in a study of Al stress in soybean roots, where NO regulated the expression of two citrate transporter genes, namely *GmMATE13* and *GmMATE47*, to enhance citrate secretion and also controlled the activity of PM H^+-ATPase by regulating H_2S biosynthesis and degradation (Wang et al. 2019). Both H_2S and NO interacted to enhance Al tolerance in soybean roots. Pretreatment with an H_2S donor has been observed to enhance citrate exudation under Al stress, reduce the accumulation of aluminum in the root tips of soybean, and exhibit alleviating effects on root elongation. Conversely, when an H_2S scavenger was applied, it resulted in the opposite effect compared to the donor, thereby underscoring the metal-ameliorating potential of H_2S.

The plant SM GABA plays a vital role in connecting C and N metabolic fluxes and has proven to be an effective tool in plant defence against biotic stressors (Tarkowski et al. 2020). Its versatility and ability to modulate ROS balance in invaded plant tissues make it an essential component of plant immune response against pathogens. GABA has proved to effectively combat various biotic stressors and their connections with other stress-related amino acids like β-aminobutyric acid (BABA), glutamate, and proline. The role of a bacterial elicitor protein called harpin was analyzed through the studies conducted on harpin-infiltrated tobacco and found that it affected growth, photosynthesis, and H_2O_2 levels (Dimlioğlu et al. 2015). GABA accumulation increased and glutamate levels decreased due to the differences in glutamate dehydrogenase (GDH) and GAD enzyme activities. As a result, harpin infiltration induced the first components of the GABA shunt in tobacco and emphasized the potential role of GABA in harpin-elicited biotic stress responses.

In tomato leaf apoplast, GABA is an abundant metabolite, and it is also reported to be produced by *Arabidopsis thaliana* in response to *Pseudomonas syringae* pv. *tomato* DC3000 (McCraw et al. 2016). Exogenous GABA represses the *type III* secretion system (T3SS) in *P. syringae*, resulting in reduced elicitation of the hypersensitive response (HR) in tobacco. GabP provides the primary mechanism for GABA uptake by *P. syringae* and *ΔgabP* is insensitive to GABA-mediated repression of *T3SS* expression. The observation suggests that *GabP* is important for *P. syringae* in a natural setting, either for nutrition or as a mechanism for regulating gene expression. Santamaría-Hernando et al. (2022) emphasized that *P. syringae* pv. infection of tomato plants was alleviated through regulation of genes involved in GABA catabolism through treatment of GABA and l Pro chemoperception (l-Pro). It was found that infected plants had increased accumulation of both GABA and l-Pro levels upon pathogen infection to regulate their defence response, enabling efficient infection by the bacteria.

Furthermore, H_2O_2 can function either as a toxin or as a signal transduction molecule, depending on its concentration (Dikilitas et al. 2020). Priming with H_2O_2 can enhance stress tolerance in plants by regulating reactive oxygen species and gene expression pathways. In a recent study by You et al. (2022), the expression of heat shock factor (HSF) genes in grapevines inoculated with pathogens was analyzed. The results revealed up-regulation of *HSF11*, *HSF12*, *HSF14*, and *HSF15* in response to both high temperatures and pathogen infections. Specifically, *HSF16* and *HSF17* were up-regulated in response to pathogenic infections only. The differential up-regulation of selected *HSF* genes in response to external stress and pathogen inoculation suggests that some *HSFs* are regulated through H_2O_2-mediated pathways in response to pathogens. The impact of various signaling substances on gene expression in the roots of arbuscular mycorrhizal fungi (AMF) *Paraglomus occultum* and non-AMF trifoliate orange (*Poncirus trifoliata*) seedlings following infection by the citrus canker pathogen (*Xanthomonas axonopodis* pv. *Citri*). The results showed that AMF inoculation significantly improved plant height, stem diameter, and leaf number. *X. axonopodis* infection dramatically decreased root H_2O_2, NO, and SA levels, but increased root CaM and JA concentrations in non-AMF seedlings. Under mycorrhization, there were higher H_2O_2 and CaM levels and lower JA levels in *X. axonopodis*-infected seedlings than in non-*X. axonopodis*-infected seedlings. Under non-*X. axonopodis* infection, mycorrhizal treatment reduced root H_2O_2, NO, and SA but increased CaM and JA levels (Miao-Miao et al. (2019). However, under *X. axonopodis* infection, mycorrhizal inoculation distinctly accelerated root H_2O_2, NO, CaM, and SA accumulation, accompanied with up-regulated expression levels of root *enhance pseudomonas susceptibility 1* (*PtEPS1*) and *pathogenesis-related gene 4* (*PtPR4*), indicating that *X. axonopodis* stimulated mycorrhizal roles in enhancing resistance of citrus canker. These findings suggest that citrus plants pre-inoculated with AMF have stronger resistance to *X. axonopodis* infection by increasing signal substrate accumulation and pathogen defence gene expressions.

So, plant diseases caused by diverse pathogens create biotic stress. Plants benefit from stress-alleviating signals and growth regulators specific to the pathogen. For precision agriculture and enhanced crop productivity, optical sensing techniques, including computer vision approaches, can be used for real-time monitoring and identification of diseases that may help in deciding the appropriate signalling molecules for plant stress protection (Sharma et al. 2021c). Moreover, different imaging techniques, such as RGB sensors, IR/thermal sensors, hyperspectral sensors, and multispectral sensors, can be utilized effectively for the quick identification and classification of plant diseases where the quick, reliable, and timely detection of pathogens and diseases might be beneficial for devising the potential strategies to prevent plant disease outbreaks and managing plant biotic stress that can endanger crop production. Furthermore, understanding the mechanism of action of SMs and their interplay can lead to new strategies for crop production and the development of more resistant crop plants.

4.7 CONCLUSION AND FUTURE PROSPECTS

Signalling molecules such as nitric oxide, gamma-aminobutyric acid, hydrogen sulfide, and hydrogen peroxide have emerged as pivotal regulators of both plant growth and stress tolerance. They intricately participate in signalling networks, exerting influence over critical processes such as seed germination, root development, and responses to various environmental stressors. Acting as versatile messengers, these molecules orchestrate biochemical and genetic changes within plants. Ongoing research into their multifaceted roles promises to unveil exciting avenues for enhancing crop productivity and resilience against both abiotic and biotic stressors. Harnessing the potential of these novel players is imperative for fostering sustainable agriculture and ensuring global food security.

4.8 AUTHORSHIP CONTRIBUTION

Conceptualization, R.B.; Writing – Original draft preparation, N.K, R.S, D.K., S.D., T.B., P.B., I.S, P.S., P.O., R.B.; Writing – Review and editing, R.B. All authors have read and agreed to the published version of the book chapter.

4.9 DECLARATION OF COMPETING INTEREST

The authors declare that they have no known competing financial interests or personal relationships that could have appeared to influence the work reported in this book chapter.

REFERENCES

Ahmad, Parvaiz, Mohammad Abass Ahanger, Mohammed Nasser Alyemeni, Leonard Wijaya, Pravej Alam, and Mohammad Ashraf. "Mitigation of sodium chloride toxicity in *Solanum lycopersicum* L. by supplementation of jasmonic acid and nitric oxide." *Journal of Plant Interactions* 13, no. 1 (2018): 64–72.

Ahmed, Mukhtar, Shah Fahad, Muhammad Arif Ali, Sajjad Hussain, Muhammad Tariq, Fariha Ilyas, Shakeel Ahmad et al. "Hydrogen sulfide: a novel gaseous molecule for plant adaptation to stress." *Journal of Plant Growth Regulation* 40 (2021): 2485–2501.

Akama, Kazuhito, and Fumio Takaiwa. "C-terminal extension of rice glutamate decarboxylase (OsGAD2) functions as an autoinhibitory domain and overexpression of a truncated mutant results in the accumulation of extremely high levels of GABA in plant cells." *Journal of Experimental Botany* 58, no. 10 (2007): 2699–2707.

Ali, B., W. J. Song, W. Z. Hu, X. N. Luo, R. A. Gill, J. Wang, and W. J. Zhou. "Hydrogen sulfide alleviates lead-induced photosynthetic and ultrastructural changes in oilseed rape." *Ecotoxicology and Environmental Safety* 102 (2014): 25–33.

Ali, Mohd, Deepak Kumar, Raman Tikoria, Roohi Sharma, Parkirti Parkirti, Vikram Vikram, Kritika Kaushal, and Puja Ohri. "Exploring the potential role of hydrogen sulfide and jasmonic acid in plants during heavy metal stress." *Nitric Oxide* 140–141 (2023): 16–29.

Ali, Sajid, Aamir Nawaz, Shaghef Ejaz, Sakeena Tul-Ain Haider, Muhammad Waqar Alam, and Hafiz Umer Javed. "Effects of hydrogen sulfide on postharvest physiology of fruits and vegetables: an overview." *Scientia Horticulturae* 243 (2019): 290–299.

Allagulova, C. R., A. M. Avalbaev, A. R. Lubyanova, O. V. Lastochkina, and F. M. Shakirova. "Current concepts of the mechanisms of nitric oxide formation in plants." *Russian Journal of Plant Physiology* 69, no. 4 (2022): 61.

Allagulova, Chulpan R., Alsu R. Lubyanova, and Azamat M. Avalbaev. "Multiple ways of nitric oxide production in plants and its functional activity under abiotic stress conditions." *International Journal of Molecular Sciences* 24, no. 14 (2023): 11637.

Alnusairi, Ghalia S. H., Yasser S. A. Mazrou, Sameer H. Qari, Amr A. Elkelish, Mona H. Soliman, Mohamed Eweis, Khaled Abdelaal, Gomaa Abd El-Samad, Mohamed F. M. Ibrahim, and Nihal ElNahhas. "Exogenous nitric oxide reinforces photosynthetic efficiency, osmolyte, mineral uptake, antioxidant, expression of stress-responsive genes and ameliorates the effects of salinity stress in wheat." *Plants* 10, no. 8 (2021): 1693.

Andrés, Celia María Curieses, José Manuel Pérez de la Lastra, Celia Andrés Juan, Francisco J. Plou, and Eduardo Pérez-Lebeña. "Chemistry of hydrogen peroxide formation and elimination in mammalian cells, and its role in various pathologies." *Stresses* 2, no. 3 (2022): 256–274.

Ansari, Mohammad Israil, Syed Uzma Jalil, Shamim Akhtar Ansari, and Mirza Hasanuzzaman. "GABA shunt: a key-player in mitigation of ROS during stress." *Plant Growth Regulation* 94 (2021): 131–149.

Antoniou, Chrystalla, Rafaella Xenofontos, Giannis Chatzimichail, Anastasis Christou, Khosrow Kashfi, and Vasileios Fotopoulos. "Exploring the potential of nitric oxide and hydrogen sulfide (NOSH)-releasing synthetic compounds as novel priming agents against drought stress in *Medicago sativa* plants." *Biomolecules* 10, no. 1 (2020): 120.

Asaeda, Takashi, Mizanur Rahman, Xia Liping, and Jonas Schoelynck. "Hydrogen peroxide variation patterns as abiotic stress responses of *Egeria densa*." *Frontiers in Plant Science* 13 (2022): 855477.

Asgher, Mohd, Tasir S. Per, Asim Masood, Mehar Fatma, Luciano Freschi, Francisco J. Corpas, and Nafees A. Khan. "Nitric oxide signaling and its crosstalk with other plant growth regulators in plant responses to abiotic stress." *Environmental Science and Pollution Research* 24 (2017): 2273–2285.

Ashfaque, Farha, M. Iqbal R. Khan, and Nafees A. Khan. "Exogenously applied H2O2 promotes proline accumulation, water relations, photosynthetic efficiency and growth of wheat (*Triticum aestivum* L.) under salt stress." *Annual Research & Review in Biology* (2014): 105–120.

Astier, Jeremy, Inonge Gross, and Jörg Durner. "Nitric oxide production in plants: an update." *Journal of Experimental Botany* 69, no. 14 (2018): 3401–3411.

Astier, Jeremy, Jordan Rossi, Pauline Chatelain, Agnès Klinguer, Angélique Besson-Bard, Claire Rosnoblet, Sylvain Jeandroz, Valérie Nicolas-Francès, and David Wendehenne. "Nitric oxide production and signalling in algae." *Journal of Experimental Botany* 72, no. 3 (2021): 781–792.

Azizi, Iraj, Behrooz Esmaielpour, and Hamideh Fatemi. "Exogenous nitric oxide on morphological, biochemical and antioxidant enzyme activity on savory (*Satureja hortensis* L.) plants under cadmium stress." *Journal of the Saudi Society of Agricultural Sciences* 20, no. 6 (2021): 417–423.

Bagheri, Mahsa, Mahdiyeh Gholami, and Bahram Baninasab. "Hydrogen peroxide-induced salt tolerance in relation to antioxidant systems in pistachio seedlings." *Scientia Horticulturae* 243 (2019): 207–213.

Bakshi, Palak, Sukhmeen Kaur Kohli, Shagun Bali, Parminder Kaur, Vinod Kumar, Pooja Sharma, Mohammad Ibrahim, Bilal Ahmad Mir, and Renu Bhardwaj. "NO and phytohormones cross-talk in plant defense against abiotic stress." In: Vijay Pratap Singh, Samiksha Singh, Durgesh Kumar Tripathi, Maria C. Romero Puertas, and Luisa Maria Sandalio (Eds.). *Nitric Oxide in Plant Biology* (pp. 573–596). Academic Press, 2022.

Bashir, Rohina, Hafiza Naila Riaz, Sumera Anwar, Nusrat Parveen, Razieh Khalilzadeh, Iqbal Hussain, and Saqib Mahmood. "Morpho-physiological changes in carrots by foliar γ-aminobutyric acid under drought stress." *Brazilian Journal of Botany* 44 (2021): 57–68.

Batista, Priscila Ferreira, Alan Carlos Costa, Caroline Müller, Robson de Oliveira Silva-Filho, Fábia Barbosa da Silva, Andrew Merchant, Giselle Camargo Mendes, and Kelly Juliane Telles Nascimento. "Nitric oxide mitigates the effect of water deficit in *Crambeabyssinica*." *Plant Physiology and Biochemistry* 129 (2018): 310–322.

Baxter-Burrell, Airica, Zhenbiao Yang, Patricia S. Springer, and Julia Bailey-Serres. "RopGAP4-dependent Rop GTPase rheostat control of Arabidopsis oxygen deprivation tolerance." *Science* 296, no. 5575 (2002): 2026–2028.

Bhardwaj, Renu, Indu Sharma, Neha Handa, Dhriti Kapoor, Harpreet Kaur, Vandana Gautam, and Sukhmeen Kohli. "Role of polyamines in stress management." In: N. A. Anjum, S. S. Gill, and R. Gill (Eds.). *Plant Adaptation to Environmental Change: Significance of Amino Acids and Their Derivatives* (pp. 245–265). Wallingford, UK: CABI, 2014a.

Bhardwaj, Renu, Indu Sharma, and Sharma Resham. "Betaines and related osmoprotectants-significance in metabolic engineering of plant stress resistance." In: *Plant Adaptation to Environmental Change: Significance of Amino Acids and Their Derivatives* (pp. 281–299). CABI, 2014b.

Bloem, Elke, Silvia Haneklaus, Jürgen Kesselmeier, and Ewald Schnug. "Sulfur fertilization and fungal infections affect the exchange of H2S and COS from agricultural crops." *Journal of Agricultural and Food Chemistry* 60, no. 31 (2012): 7588–7596.

Bolwell, G. Paul, Laurence V. Bindschedler, Kristopher A. Blee, Vernon S. Butt, Dewi R. Davies, Sarah L. Gardner, Chris Gerrish, and Farida Minibayeva. "The apoplastic oxidative burst in response to biotic stress in plants: a three-component system." *Journal of Experimental Botany* 53, no. 372 (2002): 1367–1376.

Chen, Juan, Fei-Hua Wu, Wen-Hua Wang, Chen-Juan Zheng, Guang-Hui Lin, Xue-Jun Dong, Jun-Xian He, Zhen-Ming Pei, and Hai-Lei Zheng. "Hydrogen sulphide enhances photosynthesis through promoting chloroplast biogenesis, photosynthetic enzyme expression, and thiol redox modification in *Spinacia oleracea* seedlings." *Journal of Experimental Botany* 62, no. 13 (2011): 4481–4493.

Chen, Zhen, Yawen Huang, Wenjuan Yang, Guanxiao Chang, Ping Li, Jiali Wei, Xiaojun Yuan, Jingling Huang, and Xiangyang Hu. "The hydrogen sulfide signal enhances seed germination tolerance to high temperatures by retaining nuclear COP1 for HY5 degradation." *Plant Science* 285 (2019): 34–43.

Cheng, Pengda, Qianyu Yue, Yutian Zhang, Shuang Zhao, Abid Khan, Xinyue Yang, Jieqiang He et al. "Application of γ-aminobutyric acid (GABA) improves fruit quality and rootstock drought tolerance in apple." *Journal of Plant Physiology* 280 (2023): 153890.

Choudhary, A. K., S. Singh, N. Khatri, and R. Gupta. "Hydrogen sulphide: an emerging regulator of plant defencesignalling." *Plant Biology* 24, no. 4 (2022): 532–539.

Ciacka, Katarzyna, Pawel Staszek, Katarzyna Sobczynska, Urszula Krasuska, and Agnieszka Gniazdowska. "Nitric oxide in seed biology." *International Journal of Molecular Sciences* 23, no. 23 (2022): 14951.

Corpas, Francisco J., Juan B. Barroso, and Luis A. del Río. "Peroxisomes as a source of reactive oxygen species and nitric oxide signal molecules in plant cells." *Trends in Plant Science* 6, no. 4 (2001): 145–150.

Corpas, Francisco J., Salvador González-Gordo, and José M. Palma. "NO source in higher plants: present and future of an unresolved question." *Trends in Plant Science* 27, no. 2 (2022): 116–119.

Corpas, Francisco J., and José M. Palma. "H2S signaling in plants and applications in agriculture." *Journal of Advanced Research* 24 (2020): 131–137.

Dabravolski, Siarhei A., and Stanislav V. Isayenkov. "The role of the γ-aminobutyric acid (GABA) in plant salt stress tolerance." *Horticulturae* 9, no. 2 (2023): 230.

Dahuja, Anil, Ranjeet R. Kumar, Akshay Sakhare, Archana Watts, Bhupinder Singh, Suneha Goswami, Archana Sachdev, and Shelly Praveen. "Role of ATP-binding cassette transporters in maintaining plant homeostasis under abiotic and biotic stresses." *Physiologia Plantarum* 171, no. 4 (2021): 785–801.

Deng, Yun-Quan, Jing Bao, Fang Yuan, Xue Liang, Zhong-Tao Feng, and Bao-Shan Wang. "Exogenous hydrogen sulfide alleviates salt stress in wheat seedlings by decreasing Na+ content." *Plant Growth Regulation* 79 (2016): 391–399.

Di, Dong-Wei, Caiguo Zhang, and Guang-Qin Guo. "Involvement of secondary messengers and small organic molecules in auxin perception and signaling." *Plant Cell Reports* 34 (2015): 895–904.

Dikilitas, Murat, Eray Simsek, and Aryadeep Roychoudhury. "Modulation of abiotic stress tolerance through hydrogen peroxide." In: *Protective Chemical Agents in the Amelioration of Plant Abiotic Stress: Biochemical and Molecular Perspectives* (pp. 147–173). John Wiley & Sons, 2020.

Dimlioğlu, Gizem, Zeycan Akcan Daş, Melike Bor, Filiz Özdemir, and İsmail Türkan. "The impact of GABA in harpin-elicited biotic stress responses in *Nicotiana tabaccum*." *Journal of Plant Physiology* 188 (2015): 51–57.

Dooley, Frederick D., Suven P. Nair, and Peter D. Ward. "Increased growth and germination success in plants following hydrogen sulfide administration." *Plos One* 8, no. 4 (2013): e62048.

Dumanović, Jelena, Eugenie Nepovimova, Maja Natić, Kamil Kuča, and Vesna Jaćević. "The significance of reactive oxygen species and antioxidant defense system in plants: a concise overview." *Frontiers in Plant Science* 11 (2021): 552969.

Fang, Pingping, Ting Sun, Yonggang Wang, Yanfei Ding, Arun Kumar Pandey, Cheng Zhu, and Pei Xu. "Plant gasotransmitters: light molecules interplaying with heavy metals." *Reviews in Environmental Science and Bio/Technology* 20 (2021): 31–53.

Fatima, Abreeq, Tajammul Husain, Mohammad Suhel, Sheo Mohan Prasad, and Vijay Pratap Singh. "Implication of nitric oxide under salinity stress: the possible interaction with other signaling molecules." *Journal of Plant Growth Regulation* 41, no. 1 (2021): 163–177.

Fukuto, Jon M., Cristina Perez-Ternero, Jessica Zarenkiewicz, Joseph Lin, Adrian J. Hobbs, and John P. Toscano. "Hydropersulfides (RSSH) and nitric oxide (NO) signaling: possible effects on S-nitrosothiols (RS-NO)." *Antioxidants* 11, no. 1 (2022): 169.

Gautam, Harsha, Mehar Fatma, Zebus Sehar, Iqbal R. Mir, and Nafees A. Khan. "Hydrogen sulfide, ethylene, and nitric oxide regulate redox homeostasis and protect photosynthetic metabolism under high temperature stress in rice plants." *Antioxidants* 11, no. 8 (2022): 1478.

Ghosh, U. K., M. N. Islam, M. N. Siddiqui, X. Cao, and M. A. R. Khan. "Proline, a multifaceted signalling molecule in plant responses to abiotic stress: understanding the physiological mechanisms." *Plant Biology* 24, no. 2 (2022): 227–239.

González-Gordo, Salvador, Rocío Bautista, M. Gonzalo Claros, Amanda Cañas, José M. Palma, and Francisco J. Corpas. "Nitric oxide-dependent regulation of sweet pepper fruit ripening." *Journal of Experimental Botany* 70, no. 17 (2019): 4557–4570.

Gotor, Cecilia, Irene García, Ángeles Aroca, Ana M. Laureano-Marín, Lucía Arenas-Alfonseca, Ana Jurado-Flores, Inmaculada Moreno, and Luis C. Romero. "Signaling by hydrogen sulfide and cyanide through post-translational modification." *Journal of Experimental Botany* 70, no. 16 (2019): 4251–4265.

Graska, Jakub, Justyna Fidler, Marta Gietler, Beata Prabucka, Małgorzata Nykiel, and Mateusz Labudda. "Nitric oxide in plant functioning: metabolism, signaling, and responses to infestation with ecdysozoa parasites." *Biology* 12, no. 7 (2023): 927.

Guo, Zhujuan, Junqing Gong, Shuitian Luo, Yixin Zuo, and Yingbai Shen. "Role of gamma-aminobutyric acid in plant defense response." *Metabolites* 13, no. 6 (2023): 741.

Gupta, Kapuganti J., and Abir U. Igamberdiev. "The anoxic plant mitochondrion as a nitrite: NO reductase." *Mitochondrion* 11, no. 4 (2011): 537–543.

Gupta, Kapuganti Jagadis, Vemula Chandra Kaladhar, Teresa B. Fitzpatrick, Alisdair R. Fernie, Ian Max Møller, and Gary J. Loake. "Nitric oxide regulation of plant metabolism." *Molecular Plant* 15, no. 2 (2022): 228–242.

Han, Shuai, Hongfang Cai, Xiujuan An, Chen Huan, Xiaoqin Wu, Li Jiang, Mingliang Yu, Ruijuan Ma, and Zhifang Yu. "Effect of nitric oxide on sugar metabolism in peach fruit (cv. Xiahui 6) during cold storage." *Postharvest Biology and Technology* 142 (2018): 72–80.

Hasan, Md Mahadi, Nadiyah M. Alabdallah, Basmah M. Alharbi, Muhammad Waseem, Guangqian Yao, Xu-Dong Liu, Hany G. Abd El-Gawad et al. "GABA: a key player in drought stress resistance in plants." *International Journal of Molecular Sciences* 22, no. 18 (2021): 10136.

Heli, Z., C. Hongyu, B. Dapeng, T. Yee Shin Z. Yejun, Z. Xi, and W. Yingying (2022). "Recent advances of γ-aminobutyric acid: physiological and immunity function, enrichment, and metabolic pathway." *Frontiers in Nutrition* 9: 1076223.

Henzler, Tobias, and Ernst Steudle. "Transport and metabolic degradation of hydrogen peroxide in *Chara corallina*: model calculations and measurements with the pressure probe suggest transport of H2O2 across water channels." *Journal of Experimental Botany* 51, no. 353 (2000): 2053–2066.

Hilal, Bisma, Tanveer Ahmad Khan, and Qazi Fariduddin. "Recent advances and mechanistic interactions of hydrogen sulfide with plant growth regulators in relation to abiotic stress tolerance in plants." *Plant Physiology and Biochemistry* 196 (2023): 1065–1083.

Hou, Zhihui, Lanxiang Wang, Jing Liu, Lixia Hou, and Xin Liu. "Hydrogen sulfide regulates ethylene-induced stomatal closure in Arabidopsis thaliana." *Journal of Integrative Plant Biology* 55, no. 3 (2013): 277–289.

Htet, Yamin, Zhuomin Lu, Sunia A. Trauger, and Andrew G. Tennyson. "Hydrogen peroxide as a hydride donor and reductant under biologically relevant conditions." *Chemical Science* 10, no. 7 (2019): 2025–2033.

Huang, Jingjing, and Yanjie Xie. "Hydrogen sulfide signaling in plants." *Antioxidants & Redox Signaling* 39, nos. 1–3 (2023): 40–58.

Igamberdiev, Abir U., Natalia V. Bykova, Jay K. Shah, and Robert D. Hill. "Anoxic nitric oxide cycling in plants: participating reactions and possible mechanisms." *Physiologia Plantarum* 138, no. 4 (2010): 393–404.

Israel, David, Seong Hee Lee, Thomas Matthew Robson, and Janusz Jerzy Zwiazek. "Plasma membrane aquaporins of the PIP1 and PIP2 subfamilies facilitate hydrogen peroxide diffusion into plant roots." *BMC Plant Biology* 22, no. 1 (2022): 1–15.

Jahan, Ajmat, M. Masroor A. Khan, Bilal Ahmad, Khan Bilal Mukhtar Ahmed, Ram Prakash Pandey, and Mohd Gulfishan. "Hydrogen peroxide: regulator of plant development and abiotic stress response." In: Mohammad Faizan, Shamsul Hayat and S. Maqbool Ahmed. (Eds.). *Reactive Oxygen Species: Prospects in Plant Metabolism* (pp. 213–228). Singapore: Springer Nature Singapore, 2023.

Jalil, Syed Uzma, M. Iqbal R. Khan, and Mohammad Israil Ansari. "Role of GABA transaminase in the regulation of development and senescence in *Arabidopsis thaliana*." *Current Plant Biology* 19 (2019): 100119.

Javad, S., A. A. Shah, M. Ramzan, R. Sardar, T. Javed, A. A. Al-Huqail, H. M. Ali et al. "Hydrogen sulphide alleviates cadmium stress in *Trigonella foenum-graecum* by modulating antioxidant enzymes and polyamine content." *Plant Biology* 24, no. 4 (2022): 618–626.

Jin, Xiaoqing, Tao Liu, Jiaojiao Xu, Zixing Gao, and Xiaohui Hu. "Exogenous GABA enhances muskmelon tolerance to salinity-alkalinity stress by regulating redox balance and chlorophyll biosynthesis." *BMC Plant Biology* 19, no. 1 (2019): 1–15.

Jin, Zhuping, Jiejie Shen, Zengjie Qiao, Guangdong Yang, Rui Wang, and Yanxi Pei. "Hydrogen sulfide improves drought resistance in *Arabidopsis thaliana*." *Biochemical and Biophysical Research Communications* 414, no. 3 (2011): 481–486.

Jin, Zhuping, Zhiqing Wang, Qingxia Ma, Limin Sun, Liping Zhang, Zhiqiang Liu, Danmei Liu, Xuefeng Hao, and Yanxi Pei. "Hydrogen sulfide mediates ion fluxes inducing stomatal closure in response to drought stress in *Arabidopsis thaliana*." *Plant and Soil* 419 (2017): 141–152.

Kalhor, Maryam Seifi, Sasan Aliniaeifard, Mehdi Seif, Elahe Javadi Asayesh, Françoise Bernard, Batool Hassani, and Tao Li. "Enhanced salt tolerance and photosynthetic performance: implication of ɤ-amino butyric acid application in salt-exposed lettuce (*Lactuca sativa* L.) plants." *Plant Physiology and Biochemistry* 130 (2018): 157–172.

Kapoor, Dhriti, Simranjeet Singh, Vijay Kumar, Romina Romero, Ram Prasad, and Joginder Singh. "Antioxidant enzymes regulation in plants in reference to reactive oxygen species (ROS) and reactive nitrogen species (RNS)." *Plant Gene* 19 (2019): 100182.

Kaur, Gurpreet, Ashutosh Sharma, Kunchur Guruprasad, and Pratap Kumar Pati. "Versatile roles of plant NADPH oxidases and emerging concepts." *Biotechnology Advances* 32, no. 3 (2014): 551–563.

Kaur, H., S. J. Hussain, A. A. Al-Huqail, M. H. Siddiqui, A. A. Al-Huqail, and M. I. R. Khan. "Hydrogen sulphide and salicylic acid regulate antioxidant pathway and nutrient balance in mustard plants under cadmium stress." *Plant Biology* 24, no. 4 (2022): 660–669.

Kaya, Cengiz, Ferhat Ugurlar, Muhammed Ashraf, and Parvaiz Ahmad. "Salicylic acid interacts with other plant growth regulators and signal molecules in response to stressful environments in plants." *Plant Physiology and Biochemistry* 196 (2023): 431–443.

Khan, Ekhlaque A., Sabistan Aftab, and Mirza Hasanuzzaman. "Unraveling the importance of Nitric Oxide in plant-microbe interaction." *Plant Stress* (2023a): 100258.

Khan, M. Iqbal R., Farha Ashfaque, Himanshu Chhillar, Mohammad Irfan, and Nafees A. Khan. "The intricacy of silicon, plant growth regulators and other signaling molecules for abiotic stress tolerance: an entrancing crosstalk between stress alleviators." *Plant Physiology and Biochemistry* 162 (2021a): 36–47.

Khan, M. Iqbal R., Syed Uzma Jalil, Priyanka Chopra, Himanshu Chhillar, Antonio Ferrante, Nafees A. Khan, and Mohammad Israil Ansari. "Role of GABA in plant growth, development and senescence." *Plant Gene* 26 (2021b): 100283.

Khan, Murtaza, Tiba Nazar Ibrahim Al Azzawi, Sajid Ali, Byung-Wook Yun, and Bong-Gyu Mun. "Nitric oxide, a key modulator in the alleviation of environmental stress-mediated damage in crop plants: a meta-analysis." *Plants* 12, no. 11 (2023b): 2121.

Khan, Tanveer Alam, Mohammad Yusuf, Aqeel Ahmad, Zoobia Bashir, Taiba Saeed, Qazi Fariduddin, Shamsul Hayat, Hans-Peter Mock, and Tingquan Wu. "Proteomic and physiological assessment of stress sensitive and tolerant variety of tomato treated with brassinosteroids and hydrogen peroxide under low-temperature stress." *Food Chemistry* 289 (2019): 500–511.

Khanna, Risheek Rahul, Badar Jahan, Noushina Iqbal, Nafees A. Khan, Mohamed F. AlAjmi, Md Tabish Rehman, and M. Iqbal R. Khan. "GABA reverses salt-inhibited photosynthetic and growth responses through its influence on NO-mediated nitrogen-sulfur assimilation and antioxidant system in wheat." *Journal of Biotechnology* 325 (2021): 73–82.

Kinnersley, Alan M., and Frank J. Turano. "Gamma aminobutyric acid (GABA) and plant responses to stress." *Critical Reviews in Plant Sciences* 19, no. 6 (2000): 479–509.

Kohli, Sukhmeen Kaur, Kanika Khanna, Renu Bhardwaj, Elsayed Fathi Abd_Allah, Parvaiz Ahmad, and Francisco J. Corpas. "Assessment of subcellular ROS and NO metabolism in higher plants: multifunctional signaling molecules." *Antioxidants* 8, no. 12 (2019): 641.

Kolbert, Z. S., Juan Bautista Barroso, Renaud Brouquisse, Francisco J. Corpas, KapugantiJagadis Gupta, Christian Lindermayr, Gary J. Loake et al. "A forty year journey: the generation and roles of NO in plants." *Nitric Oxide* 93 (2019): 53–70.

Kour, Jaspreet, Kanika Khanna, Pooja Sharma, Arun Dev Singh, Indu Sharma, Priya Arora, Pardeep Kumar et al. "Hydrogen sulfide and phytohormones crosstalk in plant defense against abiotic stress." In: Samiksha Singh, Vijay Pratap Singh, Durgesh Kumar Tripathi, Sheo Kumar Tripathi, Devendra Kumar Chauhan and Nawal Kishore Dubey (Eds.). *Hydrogen Sulfide in Plant Biology* (pp. 267–302). Academic Press, 2021.

Kumar, Deepak, Rajesh Kumari Manhas, and Puja Ohri. "Deciphering the growth promoting and stress curtailing role of nitric oxide in Meloidogyne incognita infested *Solanum lycopersicum* seedlings." *Scientia Horticulturae* 319 (2023): 112147.

Kumar, Deepak, and Puja Ohri. "Say 'NO' to plant stresses: Unravelling the role of nitric oxide under abiotic and biotic stress." *Nitric Oxide* 130 (2022): 36–57.

Kumari, Santosh, and Vipin Kumar Verma. "Hydrogen peroxide mediates suberization, root thickness and stomatal movement in wheats." *International Journal of Current Microbiology and Applied Sciences* 8 (2019): 2448–2454.

Kumari, Sarika, Faroza Nazir, Kajal Jain, and M. Iqbal R. Khan. "GABA and potassium modulates defence systems, assimilation of nitrogen and carbon, and yield traits under salt stress in wheat." *Journal of Plant Growth Regulation* 42, no. 10 (2023a): 6721–6740.

Kumari, Sarika, Harmanjit Kaur, Ayushi Jain, Sofi J. Hussain, Manzer H. Siddiqui, and M. Iqbal R. Khan. "Hydrogen sulfide modulates ascorbate-glutathione system, osmolytes production, nutrient content and yield responses under salt stress in wheat." *South African Journal of Botany* 160 (2023b): 295–308.

Kumari, Sarika, Faroza Nazir, Apoorva Singh, Hyda Haroon, Nahida Rashid Khan, Rudra Narayan Sahoo, Mohammed Albaqami, Kadambot H. M. Siddique, and M. Iqbal R. Khan. "γ-Aminobutyric acid (GABA) strengthened nutrient accumulation, defense metabolism, growth and yield traits against salt and endoplasmic reticulum stress conditions in wheat plants." *Plant and Soil* 498, no. 1 (2024): 409–429.

Lamattina, Lorenzo, Carlos García-Mata, Magdalena Graziano, and Gabriela Pagnussat. "Nitric oxide: the versatility of an extensive signal molecule." *Annual Review of Plant Biology* 54, no. 1 (2003): 109–136.

Lancaster Jr, Jack R. "Nitric oxide: a brief overview of chemical and physical properties relevant to therapeutic applications." *Future Science OA* 1, no. 1 (2015).

Lee, I. J. "Nitric oxide modulates *Glycine max* L. growth and physio-molecular responses during flooding stress." *Annals of Agricultural & Crop Sciences* 7, no. 3 (2022): 1–18.

Lee, X. Y., J. S. Tan, and L. H. Cheng. "Gamma aminobutyric acid (GABA) enrichment in plant-based food – a mini review." *Food Reviews International* 39, no. 8 (2023): 5864–5885.

Lehnert, Nicolai, Eunsuk Kim, Hai T. Dong, Jill B. Harland, Andrew P. Hunt, Elizabeth C. Manickas, Kady M. Oakley, John Pham, Garrett C. Reed, and Victor Sosa Alfaro. "The biologically relevant coordination chemistry of iron and nitric oxide: electronic structure and reactivity." *Chemical Reviews* 121, no. 24 (2021): 14682–14905.

Li, Hua, Hongyu Chen, Lulu Chen, and Chenyang Wang. "The role of hydrogen sulfide in plant roots during development and in response to abiotic stress." *International Journal of Molecular Sciences* 23, no. 3 (2022): 1024.

Li, Le, Yanqin Wang, and Wenbiao Shen. "Roles of hydrogen sulfide and nitric oxide in the alleviation of cadmium-induced oxidative damage in alfalfa seedling roots." *Biometals* 25, no. 3 (2012a): 617–631.

Li, Li, Na Dou, Hui Zhang, and Chunxia Wu. "The versatile GABA in plants." *Plant Signaling & Behavior* 16, no. 3 (2021): 1862565.

Li, Yuxing, Boyang Liu, Yuxiao Peng, Chenlu Liu, Xiuzhi Zhang, Zhijun Zhang, Wei Liang, Fengwang Ma, and Cuiying Li. "Exogenous GABA alleviates alkaline stress in *Malus hupehensis* by regulating the accumulation of organic acids." *Scientia Horticulturae* 261 (2020): 108982.

Li, Zhong-Guang, Xiao-Jiao Ding, and Pei-Fang Du. "Hydrogen sulfide donor sodium hydrosulfide-improved heat tolerance in maize and involvement of proline." *Journal of Plant Physiology* 170, no. 8 (2013): 741–747.

Li, Zhong-Guang, Ming Gong, and Ping Liu. "Hydrogen sulfide is a mediator in H2O2-induced seed germination in *Jatropha Curcas*." *Acta Physiologiae Plantarum* 34 (2012b): 2207–2213.

Li, Zhong-Guang, Xiong Min, and Zhi-Hao Zhou. "Hydrogen sulfide: a signal molecule in plant cross-adaptation." *Frontiers in Plant Science* 7 (2016): 1621.

Lisjak, M., N. Srivastava, T. Teklic, L. Civale, K. Lewandowski, I. Wilson, M. E. Wood, M. Whiteman, and J. T. Hancock. "A novel hydrogen sulfide donor causes stomatal opening and reduces nitric oxide accumulation." *Plant Physiology and Biochemistry* 48, no. 12 (2010): 931–935.

Liu, Hai, Jicheng Wang, Jianhao Liu, Tong Liu, and Shaowu Xue. "Hydrogen sulfide (H2S) signaling in plant development and stress responses." *Abiotech* 2 (2021): 32–63.

Liu, Hai, and Shaowu Xue. "Interplay between hydrogen sulfide and other signaling molecules in the regulation of guard cell signaling and abiotic/biotic stress response." *Plant Communications* 2, no. 3 (2021).

Liu, Wei, Zhidan Xiao, Xiuli Bao, Xiaoyan Yang, Jing Fang, and Xu Xiang. "Identifying litchi (*Litchi chinensis* Sonn.) cultivars and their genetic relationships using single nucleotide polymorphism (SNP) markers." *PLoS One* 10, no. 8 (2015): e0135390.

Liu, Y. H., X. H. Zhang, B. W. Liu, B. Ao, Q. Liu, S. Y. Wen, and Y. F. Xu. "Hydrogen sulfide regulates photosynthesis of tall fescue under low-light stress." *Photosynthetica* 57, no. 3 (2019).

Lombardi, Lara, Nello Ceccarelli, Piero Picciarelli, Carlo Sorce, and Roberto Lorenzi. "Nitric oxide and hydrogen peroxide involvement during programmed cell death of *Sechium edule* nucellus." *Physiologia Plantarum* 140, no. 1 (2010): 89–102.

Lopes-Oliveira, Patrícia Juliana, Halley Caixeta Oliveira, Zsuzsanna Kolbert, and Luciano Freschi. "The light and dark sides of nitric oxide: multifaceted roles of nitric oxide in plant responses to light." *Journal of Experimental Botany* 72, no. 3 (2021): 885–903.

Ma, Ming, David Wendehenne, Laurent Philippot, Robert Hänsch, Emmanouil Flemetakis, Bin Hu, and Heinz Rennenberg. "Physiological significance of pedospheric nitric oxide for root growth, development and organismic interactions." *Plant, Cell & Environment* 43, no. 10 (2020): 2336–2354.

Mahaseth, Tulip, and Andrei Kuzminov. "Potentiation of hydrogen peroxide toxicity: from catalase inhibition to stable DNA-iron complexes." *Mutation Research/Reviews in Mutation Research* 773 (2017): 274–281.

Mahmud, Jubayer A. L., Mirza Hasanuzzaman, Kamrun Nahar, Anisur Rahman, Md Shahadat Hossain, and Masayuki Fujita. "γ-Aminobutyric acid (GABA) confers chromium stress tolerance in *Brassica juncea* L. by modulating the antioxidant defense and glyoxalase systems." *Ecotoxicology* 26 (2017): 675–690.

Mandal, Mamun, Manisha Sarkar, Azmi Khan, Moumita Biswas, Antonio Masi, Randeep Rakwal, Ganesh Kumar Agrawal, Amrita Srivastava, and Abhijit Sarkar. "Reactive Oxygen Species (ROS) and Reactive Nitrogen Species (RNS) in plants–maintenance of structural individuality and functional blend." *Advances in Redox Research* 5 (2022): 100039.

Mansoor, Sheikh, Owais Ali Wani, Jafar K. Lone, Sweeta Manhas, Navneet Kour, Pravej Alam, Ajaz Ahmad, and Parvaiz Ahmad. "Reactive oxygen species in plants: from source to sink." *Antioxidants* 11, no. 2 (2022): 225.

Martin, R. Emily, Anthony E. Postiglione, and Gloria K. Muday. "Reactive oxygen species function as signaling molecules in controlling plant development and hormonal responses." *Current Opinion in Plant Biology* 69 (2022): 102293.

McCraw, S. L., D. H. Park, R. Jones, M. A. Bentley, A. Rico, R. G. Ratcliffe, N. J. Kruger, A. Collmer, and G. M. Preston. "GABA (γ-aminobutyric acid) uptake via the GABA permease GabP represses virulence gene expression in *Pseudomonas syringae* pv. tomato DC3000." *Molecular Plant-Microbe Interactions* 29, no. 12 (2016): 938–949.

Mfarrej, M. F. B., X. Wang, M. Hamzah Saleem, I. Hussain, R. Rasheed, M. Arslan Ashraf, M. Iqbal, M. Sohaib Chattha, and M. Nasser Alyemeni. "Hydrogen sulphide and nitric oxide mitigate the negative impacts of waterlogging stress on wheat (*Triticum aestivum* L.)." *Plant Biology* 24, no. 4 (2022): 670–683.

Miao-Miao, Xie, Yi-Can Zhang, Li-Ping Liu, Ying-Ning Zou, Qiang-Sheng Wu, and Kamil Kuča. "Mycorrhiza regulates signal substance levels and pathogen defense gene expression to resist citrus canker." *Notulae Botanicae Horti Agrobotanici Cluj-Napoca* 47, no. 4 (2019): 1161–1167.

Michaeli, Simon, Aaron Fait, Kelly Lagor, Adriano Nunes-Nesi, Nicole Grillich, Ayelet Yellin, Dana Bar et al. "A mitochondrial GABA permease connects the GABA shunt and the TCA cycle, and is essential for normal carbon metabolism." *The Plant Journal* 67, no. 3 (2011): 485–498.

Min, Yang, Bao-ping Qin, Ping Wang, Mei-ling Li, Lu-lu Chen, Lei-tai Chen, Ai-qing Sun, Zhen-lin Wang, and Yan-ping Yin. "Foliar application of sodium hydrosulfide (NaHS), a hydrogen sulfide (H2S) donor, can protect seedlings against heat stress in wheat (*Triticum aestivum* L.)." *Journal of Integrative Agriculture* 15, no. 12 (2016): 2745–2758.

Minguillón, Samuel, Manuel A. Matamoros, DeqiangDuanmu, and Manuel Becana. "Signaling by reactive molecules and antioxidants in legume nodules." *New Phytologist* 236, no. 3 (2022): 815–832.

Mir, Iqbal R., Harsha Gautam, Naser A. Anjum, Asim Masood, and Nafees A. Khan. "Calcium and nitric oxide signaling in plant cadmium stress tolerance: a cross talk." *South African Journal of Botany* 150 (2022): 387–403.

Mittler, Ron, Sara I. Zandalinas, Yosef Fichman, and Frank Van Breusegem. "Reactive oxygen species signalling in plant stress responses." *Nature Reviews Molecular Cell Biology* 23, no. 10 (2022): 663–679.

Montesinos-Pereira, D., A. de la Torre-González, B. Blasco, and J. M. Ruiz. "Hydrogen sulphide increase the tolerance to alkalinity stress in cabbage plants (*Brassica oleracea* L. 'Bronco')." *Scientia Horticulturae* 235 (2018): 349–356.

Mostofa, Mohammad Golam, Ajit Ghosh, Zhong-Guang Li, Md Nurealam Siddiqui, Masayuki Fujita, and Lam-Son Phan Tran. "Methylglyoxal – a signaling molecule in plant abiotic stress responses." *Free Radical Biology and Medicine* 122 (2018): 96–109.

Naz, R., F. Gul, S. Zahoor, A. Nosheen, H. Yasmin, R. Keyani, M. Shahid et al. "Interactive effects of hydrogen sulphide and silicon enhance drought and heat tolerance by modulating hormones, antioxidant defence enzymes and redox status in barley (*Hordeum vulgare* L.)." *Plant Biology* 24, no. 4 (2022): 684–696.

Nazir, Faroza, Anjuman Hussain, and Qazi Fariduddin. "Hydrogen peroxide modulate photosynthesis and antioxidant systems in tomato (*Solanum lycopersicum* L.) plants under copper stress." *Chemosphere* 230 (2019): 544–558.

Neill, Steven, Radhika Desikan, and John Hancock. "Hydrogen peroxide signalling." *Current Opinion in Plant Biology* 5, no. 5 (2002): 388–395.

Nomani, L., A. Zehra, S. Choudhary, K. I. Wani, M. Naeem, M. H. Siddiqui, M. M. A. Khan, and T. Aftab. "Exogenous hydrogen sulphide alleviates copper stress impacts in *Artemisia annua* L.: Growth, antioxidant metabolism, glandular trichome development and artemisinin biosynthesis." *Plant Biology* 24, no. 4 (2022): 642–651.

Ozfidan-Konakci, Ceyda, Evren Yildiztugay, Fevzi Elbasan, Mustafa Kucukoduk, and Ismail Turkan. "Hydrogen sulfide (H2S) and nitric oxide (NO) alleviate cobalt toxicity in wheat (*Triticum aestivum* L.) by modulating photosynthesis, chloroplastic redox and antioxidant capacity." *Journal of Hazardous Materials* 388 (2020): 122061.

Palma, José M., Luciano Freschi, Marta Rodríguez-Ruiz, Salvador González-Gordo, and Francisco J. Corpas. "Nitric oxide in the physiology and quality of fleshy fruits." *Journal of Experimental Botany* 70, no. 17 (2019): 4405–4417.

Pandey, Sonika, Aprajita Kumari, Manu Shree, Vinod Kumar, Pooja Singh, Chellapilla Bharadwaj, Gary J. Loake, Swarup K. Parida, Shyam Kumar Masakapalli, and Kapuganti Jagadis Gupta. "Nitric oxide accelerates germination via the regulation of respiration in chickpea." *Journal of Experimental Botany* 70, no. 17 (2019): 4539–4555.

Panngom, Kamonporn, Thanyarat Chuesaard, Natchathorn Tamchan, Thitima Jiwchan, Kantirat Srikongsritong, and Gyungsoon Park. "Comparative assessment for the effects of reactive species on seed germination, growth and metabolisms of vegetables." *Scientia Horticulturae* 227 (2018): 85–91.

Podlešáková, Kateřina, Lydia Ugena, Lukáš Spíchal, Karel Doležal, and Nuria De Diego. "Phytohormones and polyamines regulate plant stress responses by altering GABA pathway." *New Biotechnology* 48 (2019): 53–65.

Priya, Manu, Lomeshwar Sharma, Ramanpreet Kaur, H. Bindumadhava, Ramkrishnan M. Nair, K. H. M. Siddique, and Harsh Nayyar. "GABA (γ-aminobutyric acid), as a thermo-protectant, to improve the reproductive function of heat-stressed mungbean plants." *Scientific Reports* 9, no. 1 (2019): 7788.

Qureshi, Muhammad Kamran, Piotr Gawroński, Sana Munir, Sunita Jindal, and Pavel Kerchev. "Hydrogen peroxide-induced stress acclimation in plants." *Cellular and Molecular Life Sciences* 79, no. 2 (2022): 129.

Ramesh, Sunita A., Stephen D. Tyerman, Bo Xu, Jayakumar Bose, Satwinder Kaur, Vanessa Conn, Patricia Domingos et al. "GABA signalling modulates plant growth by directly regulating the activity of plant-specific anion transporters." *Nature Communications* 6, no. 1 (2015): 7879.

Ramos-Ruiz, Roberto, Felix Martinez, and Gertrude Knauf-Beiter. "The effects of GABA in plants." *Cogent Food & Agriculture* 5, no. 1 (2019): 1670553.

Rarata, Grzegorz, Mitja Vahčič, David Anderson, Michael Berglund, Dimitris Kyprianou, Giovanni Emma, and Gabriela Diaconu. "Hydrogen peroxide (H2O2): a review of its use in homemade explosives." *MateriałyWysokoenergetyczne* 11, no. 2 (2019): 5–13.

Rather, Bilal A., Iqbal R. Mir, Asim Masood, Naser A. Anjum, and Nafees A. Khan. "Nitric oxide pre-treatment advances seed germination and alleviates copper-induced photosynthetic inhibition in Indian mustard." *Plants* 9, no. 6 (2020): 776.

Raza, A., J. Tabassum, M. S. Mubarik, S. Anwar, N. Zahra, Y. Sharif, M. B. Hafeez, C. Zhang, Francisco J. Corpas, and H. Chen. "Hydrogen sulfide: an emerging component against abiotic stress in plants." *Plant Biology* 24, no. 4 (2022): 540–558.

Sagi, Moshe, and Robert Fluhr. "Superoxide production by plant homologues of the gp91phox NADPH oxidase. Modulation of activity by calcium and by tobacco mosaic virus infection." *Plant Physiology* 126, no. 3 (2001): 1281–1290.

Santamaría-Hernando, Saray, Álvaro López-Maroto, Clara Galvez-Roldán, Martí Munar-Palmer, Elizabet Monteagudo-Cascales, José-Juan Rodríguez-Herva, Tino Krell, and Emilia López-Solanilla. "*Pseudomonas syringae* pv. tomato infection of tomato plants is mediated by GABA and l-Pro chemoperception." *Molecular Plant Pathology* 23, no. 10 (2022): 1433–1445.

Santisree, Parankusam, Srivani S. Adimulam, Pradeepreddy Bommineni, Pooja Bhatnagar-Mathur, and Kiran K. Sharma. "Hydrogen sulfide in plant abiotic stress tolerance: Progress and perspectives." In: *Reactive Oxygen, Nitrogen and Sulfur Species in Plants: Production, Metabolism, Signaling and Defense Mechanisms* (pp. 743–775). John Wiley & Sons, 2019.

Santolini, Jérôme, François André, Sylvain Jeandroz, and David Wendehenne. "Nitric oxide synthase in plants: where do we stand?" *Nitric Oxide* 63 (2017): 30–38.

Saxena, Ina, Sandhya Srikanth, and Zhong Chen. "Cross talk between H2O2 and interacting signal molecules under plant stress response." *Frontiers in Plant Science* 7 (2016): 570.

Scuffi, Denise, Consolación Álvarez, Natalia Laspina, Cecilia Gotor, Lorenzo Lamattina, and Carlos García-Mata. "Hydrogen sulfide generated by L-cysteine desulfhydrase acts upstream of nitric oxide to modulate abscisic acid-dependent stomatal closure." *Plant Physiology* 166, no. 4 (2014): 2065–2076.

Scuffi, Denise, Thomas Nietzel, Luciano M. Di Fino, Andreas J. Meyer, Lorenzo Lamattina, Markus Schwarzländer, Ana M. Laxalt, and Carlos García-Mata. "Hydrogen sulfide increases production of NADPH oxidase-dependent hydrogen peroxide and phospholipase D-derived phosphatidic acid in guard cell signaling." *Plant Physiology* 176, no. 3 (2018): 2532–2542.

Seifikalhor, Maryam, Sasan Aliniaeifard, Batool Hassani, Vahid Niknam, and Oksana Lastochkina. "Diverse role of γ-aminobutyric acid in dynamic plant cell responses." *Plant Cell Reports* 38 (2019): 847–867.

Shaiek, Oumayma, Yathreb Mahjoubi, Oussama Kharbech, Ahmed Debez, Abdelilah Chaoui, and Wahbi Djebali. "Seed pretreatment by CaCl2 and NaHS alleviates oxidative stress, preserves membrane integrity, and stimulates hydrogen sulfide biosynthesis in pea (*Pisum sativum* L.) under nickel stress." *Journal of Soil Science and Plant Nutrition* (2023): 1–14.

Shan, Qing, Minghui Liu, Rui Li, Qinghua Shi, Yan Li, and Biao Gong. "γ-Aminobutyric acid (GABA) improves pesticide detoxification in plants." *Science of the Total Environment* 835 (2022): 155404.

Sharma, Ashutosh, Pooja Sharma, Rahul Kumar, Vikas Sharma, Renu Bhardwaj, and Indu Sharma. "Role of reactive oxygen species in the regulation of abiotic stress tolerance in legumes." In: Vijay Pratap Singh, Samiksha Singh, Durgesh Kumar Tripathi, Shoe Mohan Prasad, Renu Bhardwaj, and Devendra Kumar Chauhan (Eds.). *Abiotic Stress and Legumes* (pp. 217–243). Academic Press, 2021b.

Sharma, Deepka, Karuna Dhiman, Sadaf Jan, Indu Sharma, Vandana Gautam, and Reena Kumari. "Mechanism of salicylic acid synthesis and homeostasis in plants." In: Dhriti Kapoor, Vandana Gautam, and Renu Bhardwaj (Eds.). *Salicylic Acid Contribution in Plant Biology Against a Changing Environment* (pp. 1–37). Nova Science Publishers, 2021a.

Sharma, Indu, Renu Bhardwaj, Vandana Gautam, Ravdeep Kaur, and Ashutosh Sharma. "Brassinosteroids: occurrence, structure and stress protective activities." *Frontiers in Natural Product Chemistry* 4 (2018): 204–239.

Sharma, Indu, Aditi Sharma, Inderjit Singh, Rahul Kumar, Yogesh Kumar, and Ashutosh Sharma. "Plant disease detection using image sensors: a step towards precision agriculture." In: Jyoti Moy Chatterjee, Abhishek Kumar, Pramod Singh Rathore and Vishal Jain (Eds.). *Internet of Things and Machine Learning in Agriculture* (pp. 89–130). Nova Science Publishers, 2021c.

Shelp, Barry J., Morteza Soleimani Aghdam, and Edward J. Flaherty. "γ-Aminobutyrate (GABA) regulated plant defense: mechanisms and opportunities." *Plants* 10, no. 9 (2021): 1939.

Shepherd, Mark, Daniela Giordano, Cinzia Verde, and Robert K. Poole. "The evolution of nitric oxide function: from reactivity in the prebiotic earth to examples of biological roles and therapeutic applications." *Antioxidants* 11, no. 7 (2022): 1222.

Shi, Haitao, Tiantian Ye, Ning Han, Hongwu Bian, Xiaodong Liu, and Zhulong Chan. "Hydrogen sulfide regulates abiotic stress tolerance and biotic stress resistance in Arabidopsis." *Journal of Integrative Plant Biology* 57, no. 7 (2015a): 628–640.

Shi, Jiang, Guoan Shi, and Zhen Tian. "Effect of exogenous hydrogen peroxide or ascorbic acid on senescence in cut flowers of tree peony (*Paeonia suffruticosa* Andr.)." *The Journal of Horticultural Science and Biotechnology* 90, no. 6 (2015b): 689–694.

Shi, S. Q., Z. Shi, Z. P. Jiang, L. W. Qi, X. M. Sun, C. X. Li, J. F. Liu, W. F. Xiao, and S. G. Zhang. "Effects of exogenous GABA on gene expression of *Caragana intermedia* roots under NaCl stress: regulatory roles for H2O2 and ethylene production." *Plant, Cell & Environment* 33, no. 2 (2009): 149–162.

Shivaraj, Sheelavanta M., Sanskriti Vats, Javaid A. Bhat, Priyanka Dhakte, Vinod Goyal, Praveen Khatri, Surbhi Kumawat et al. "Nitric oxide and hydrogen sulfide crosstalk during heavy metal stress in plants." *Physiologia Plantarum* 168, no. 2 (2020): 437–455.

Singh, Ankur, and Aryadeep Roychoudhury. "Role of γ-aminobutyric acid in the mitigation of abiotic stress in plants." In *Protective Chemical Agents in the Amelioration of Plant Abiotic Stress: Biochemical and Molecular Perspectives* (pp. 413–423). John Wiley & Sons, 2020a.

Singh, Arun Dev, Jaspreet Kour, Pardeep Kumar, Neerja Sharma, Priyanka Sharma, Isha Madaan, Indu Sharma, Nitika Kapoor, Amrit Pal Singh, and Renu Bhardwaj. "Role of jasmonates in regulating physiological and molecular aspects of plants under abiotic stress." In: Sharma Anket, Sangeeta Pandey, Renu Bharrdwaj, Bingsong Zheng, and Durgesh Kumar Tripathi (Eds.). *The Role of Growth Regulators and Phytohormones in Overcoming Environmental Stress* (pp. 137–173). Academic Press, 2023.

Singh, Samiksha, Tajammul Husain, Bishwajit Kumar Kushwaha, Mohd Suhel, Abreeq Fatima, Vipul Mishra, Sani Kumar Singh et al. "Regulation of ascorbate-glutathione cycle by exogenous nitric oxide and hydrogen peroxide in soybean roots under arsenate stress." *Journal of Hazardous Materials* 409 (2021): 123686.

Singh, Simranjeet, Vijay Kumar, Dhriti Kapoor, Sanjay Kumar, Satyender Singh, Daljeet Singh Dhanjal, Shivika Datta et al. "Revealing on hydrogen sulfide and nitric oxide signals co-ordination for plant growth under stress conditions." *Physiologia Plantarum* 168, no. 2 (2020b): 301–317.

Singhal, Rajesh Kumar, Hanuman Singh Jatav, Tariq Aftab, Saurabh Pandey, Udit Nandan Mishra, Jyoti Chauhan, Subhash Chand et al. "Roles of nitric oxide in conferring multiple abiotic stress tolerance in plants and crosstalk with other plant growth regulators." *Journal of Plant Growth Regulation* 40 (2021): 2303–2328.

Sita, Kumari, and Vaneet Kumar. "Role of Gamma Amino Butyric Acid (GABA) against abiotic stress tolerance in legumes: a review." *Plant Physiology Reports* 25 (2020): 654–663.

Sohag, Abdullah Al Mamun, Md Tahjib-Ul-Arif, Marian Brestic, Sonya Afrin, Md Arif Sakil, Md Tahmeed Hossain, Mohammad Anowar Hossain, and Md Afzal Hossain. "Exogenous salicylic acid and hydrogen peroxide attenuate drought stress in rice." *Plant, Soil and Environment* 66, no. 1 (2020): 7–13.

Suhel, Mohammad, Tajammul Husain, Aparna Pandey, Samiksha Singh, Nawal Kishore Dubey, Sheo Mohan Prasad, and Vijay Pratap Singh. "An appraisal of ancient molecule GABA in abiotic stress tolerance in plants, and its crosstalk with other signaling molecules." *Journal of Plant Growth Regulation* 42, no. 2 (2023): 614–629.

Sun, Yu, Arshad Mehmood, Maurizio Battino, Jianbo Xiao, and Xiumin Chen. "Enrichment of gamma-aminobutyric acid in foods: from conventional methods to innovative technologies." *Food Research International* (2022): 111801.

Tarkowski, Łukasz P., Santiago Signorelli, and Monica Höfte. "γ-Aminobutyric acid and related amino acids in plant immune responses: emerging mechanisms of action." *Plant, Cell & Environment* 43, no. 5 (2020): 1103–1116.

Thakur, Meenakshi, and Anjali Anand. "Hydrogen sulfide: an emerging signaling molecule regulating drought stress response in plants." *Physiologia Plantarum* 172, no. 2 (2021): 1227–1243.

Timilsina, Arbindra, Wenxu Dong, Mirza Hasanuzzaman, Binbin Liu, and Chunsheng Hu. "Nitrate–nitrite–nitric oxide pathway: a mechanism of hypoxia and anoxia tolerance in plants." *International Journal of Molecular Sciences* 23, no. 19 (2022): 11522.

Torres, Miguel Angel, Jeffery L. Dangl, and Jonathan DG Jones. "Arabidopsis gp91phox homologues AtrbohD and AtrbohF are required for accumulation of reactive oxygen intermediates in the plant defense response." *Proceedings of the National Academy of Sciences* 99, no. 1 (2002): 517–522.

Tyagi, Shaily, Anshuman Shah, Kesiraju Karthik, Maniraj Rathinam, Vandna Rai, Nidhee Chaudhary, and Rohini Sreevathsa. "Reactive oxygen species in plants: an invincible fulcrum for biotic stress mitigation." *Applied Microbiology and Biotechnology* 106, no. 18 (2022): 5945–5955.

Ueno, Fumihiko, Shinichiro Nakajima, Yusuke Iwata, Shiori Honda, Edgardo Torres-Carmona, Wanna Mar, Sakiko Tsugawa et al. "Gamma-aminobutyric acid (GABA) levels in the midcingulate cortex and clozapine response in patients with treatment-resistant schizophrenia: a proton magnetic resonance spectroscopy (1H-MRS) study." *Psychiatry and Clinical Neurosciences* 76, no. 11 (2022): 587–594.

Ullah, Abd, Iftikhar Ali, Javaria Noor, Fanjiang Zeng, Sami Bawazeer, Sayed M. Eldin, Muhammad Ahsan Asghar et al. "Exogenous γ-aminobutyric acid (GABA) mitigated salinity-induced impairments in mungbean plants by regulating their nitrogen metabolism and antioxidant potential." *Frontiers in Plant Science* 13 (2023): 1081188.

Vandana, Priyam, Dinesh Singh, Guru Dayal Ram Guru, and Sangeeta Srivastava. "Plant response to the abiotic and biotic stresses: a review." *Journal of Pharmacognosy and Phytochemistry* 9, no. 6S (2020): 457–461.

Vojtovič, Daniel, Lenka Luhová, and Marek Petřivalský. "Something smells bad to plant pathogens: Production of hydrogen sulfide in plants and its role in plant defence responses." *Journal of Advanced Research* 27 (2021): 199–209.

Wahab, Abdul, Gholamreza Abdi, Muhammad Hamzah Saleem, Baber Ali, Saqib Ullah, Wadood Shah, Sahar Mumtaz, Ghulam Yasin, Crina Carmen Muresan, and Romina Alina Marc. "Plants' physio-biochemical and phyto-hormonal responses to alleviate the adverse effects of drought stress: a comprehensive review." *Plants* 11, no. 13 (2022): 1620.

Wang, Chunlei, Yuzheng Deng, Zesheng Liu, and Weibiao Liao. "Hydrogen sulfide in plants: Crosstalk with other signal molecules in response to abiotic stresses." *International Journal of Molecular Sciences* 22, no. 21 (2021): 12068.

Wang, Gang, Bowen Li, Danliu Peng, Hongzhi Zhao, Mingyang Lu, Lishuang Zhang, Jiali Li, Songhao Zhang, Chunfeng Guan, and Jing Ji. "Combined application of H2S and a plant growth promoting strain JIL321 regulates photosynthetic efficacy, soil enzyme activity and growth-promotion in rice under salt stress." *Microbiological Research* 256 (2022): 126943.

Wang, Huahua, Fang Ji, Yangyang Zhang, Junjie Hou, Wenwen Liu, Junjun Huang, and Weihong Liang. "Interactions between hydrogen sulphide and nitric oxide regulate two soybean citrate transporters during the alleviation of aluminium toxicity." *Plant, Cell & Environment* 42, no. 8 (2019): 2340–2356.

Wang, X.-X., B. Zhang, R. Xia, and Q.-Y. Jia. "Inflammation, apoptosis and autophagy as critical players in vascular dementia." *European Review for Medical & Pharmacological Sciences* 24, no. 18 (2020).

Wang, Yanqin, Le Li, Weiti Cui, Sheng Xu, Wenbiao Shen, and Ren Wang. "Hydrogen sulfide enhances alfalfa (*Medicago sativa*) tolerance against salinity during seed germination by nitric oxide pathway." *Plant and Soil* 351 (2012): 107–119.

Woźniak, Agnieszka, Magda Formela, Piotr Bilman, Katarzyna Grześkiewicz, Waldemar Bednarski, Łukasz Marczak, Dorota Narożna et al. "The dynamics of the defense strategy of pea induced by exogenous nitric oxide in response to aphid infestation." *International Journal of Molecular Sciences* 18, no. 2 (2017): 329.

Xiang, Lixia, Lipan Hu, Weinan Xu, Ai Zhen, Liang Zhang, and Xiaohui Hu. "Exogenous γ-aminobutyric acid improves the structure and function of photosystem II in muskmelon seedlings exposed to salinity-alkalinity stress." *PLoS One* 11, no. 10 (2016): e0164847.

Xie, Tiantian, Jing Ji, Wei Chen, Jianyun Yue, Changjian Du, Jiacheng Sun, Lanzhen Chen, Zeping Jiang, and Shengqing Shi. "GABA negatively regulates adventitious root development in poplar." *Journal of Experimental Botany* 71, no. 4 (2020): 1459–1474.

Xiong, Jie, Yongjie Yang, Guanfu Fu, and Longxing Tao. "Novel roles of hydrogen peroxide (H2O2) in regulating pectin synthesis and demethylesterification in the cell wall of rice (*Oryza sativa*) root tips." *New Phytologist* 206, no. 1 (2015): 118–126.

Yan, Jialong, Jinyong Jiang, Lu He, and Linxi Chen. "Mitochondrial superoxide/hydrogen peroxide: an emerging therapeutic target for metabolic diseases." *Free Radical Biology and Medicine* 152 (2020): 33–42.

Yang, Zhifeng, Xiaoyu Wang, Jianrong Feng, and Shuhua Zhu. "Biological functions of hydrogen sulfide in plants." *International Journal of Molecular Sciences* 23, no. 23 (2022): 15107.

Yao, Yuanyuan, Wenjie Kan, Pengfei Su, Yan Zhu, Wenling Zhong, Jinfeng Xi, Dacheng Wang, Caiguo Tang, and Lifang Wu. "Hydrogen sulphide alleviates Fusarium Head Blight in wheat seedlings." *PeerJ* 10 (2022): e13078.

Yong, Bin, Huan Xie, Zhou Li, Ya-Ping Li, Yan Zhang, Gang Nie, Xin-Quan Zhang et al. "Exogenous application of GABA improves PEG-induced drought tolerance positively associated with GABA-shunt, polyamines, and proline metabolism in white clover." *Frontiers in Physiology* 8 (2017): 1107.

You, Young Jun, Soon Young Ahn, and Hae Keun Yun. "Heat shock transcriptional factors (HSFs) are expressed in response to hydrogen peroxide production in grapevines inoculated with *Colletotrichum* Species." *Horticulture, Environment, and Biotechnology* 63, no. 5 (2022): 735–745.

Zhanassova, Kuralay, Assylay Kurmanbayeva, Bakhytgul Gadilgereyeva, Roza Yermukhambetova, Nurgul Iksat, Ulbike Amanbayeva, Assemgul Bekturova, Zhanerke Tleukulova, Rustem Omarov, and Zhaksylyk Masalimov. "ROS status and antioxidant enzyme activities in response to combined temperature and drought stresses in barley." *Acta Physiologiae Plantarum* 43, no. 114 (2021).

Zhang, Heng, Yang Zhao, and Jian-Kang Zhu. "Thriving under stress: how plants balance growth and the stress response." *Developmental Cell* 55, no. 5 (2020): 529–543.

Zhang, Hua, Jun Tang, Xiao-Ping Liu, Yun Wang, Wei Yu, Wei-Yan Peng, Fang Fang, Dai-Fu Ma, Zhao-Jun Wei, and Lan-Ying Hu. "Hydrogen sulfide promotes root organogenesis in *Ipomoea batatas*, *Salix matsudana* and *Glycine max*." *Journal of Integrative Plant Biology* 51, no. 12 (2009): 1086–1094.

Zhang, Jing, Mingjian Zhou, Heng Zhou, Didi Zhao, Cecilia Gotor, Luis C. Romero, Jie Shen. "Hydrogen sulfide, a signaling molecule in plant stress responses." *Journal of Integrative Plant Biology* 63, no. 1 (2021): 146–160.

Zhao, Qiuyan, Yan Ma, Xianqing Huang, Lianjun Song, Ning Li, Mingwu Qiao, Tiange Li, Dan Hai, and Yongxia Cheng. "GABA application enhances drought stress tolerance in wheat seedlings (*Triticum aestivum* L.)." *Plants* 12, no. 13 (2023): 2495.

Zhao, Ya-Qin, Liang-Yi Zhao, Shun-Qing Hu, Yuan-Yuan Hou, Yong-Hua Zheng, and Peng Jin. "Hydrogen sulfide: a luminous future in the postharvest preservation of fruits and vegetables." *Food Materials Research* 2, no. 1 (2022).

Zhen, Ai, Zhi Zhang, Xiaoqing Jin, Tao Liu, Wenqi Ren, and Xiaohui Hu. "Exogenous GABA application improves the NO3−-N absorption and assimilation in Ca (NO3) 2-treated muskmelon seedlings." *Scientia Horticulturae* 227 (2018): 117–123.

Zhu, Shuhua, Lina Sun, and Jie Zhou. "Effects of different nitric oxide application on quality of kiwifruit during 20 C storage." *International Journal of Food Science & Technology* 45, no. 2 (2010): 245–251.

Zhu, Xujun, Jieren Liao, Xingli Xia, Fei Xiong, Yue Li, Jiazhi Shen, Bo Wen, Yuanchun Ma, Yuhua Wang, and Wanping Fang. "Physiological and iTRAQ-based proteomic analyses reveal the function of exogenous γ-aminobutyric acid (GABA) in improving tea plant (*Camelliasinensis* L.) tolerance at cold temperature." *BMC Plant Biology* 19, no. 1 (2019): 1–20.

5 Insight into the Perception and Signaling of Brassinosteroids in Plants under Drought and Heavy Metal Stresses

Bisma Hilal and Qazi Fariduddin

5.1 INTRODUCTION

As sessile organisms, plants are inevitably exposed to diverse environmental stressors, whether they stem from abiotic or biotic factors present in their surroundings. Among abiotic stresses, factors such as temperature extremes (heat, chilling, and freezing), water (drought, waterlogging), salt, heavy metals, light (intense and weak), radiation (UV-A/B), and so on pose significant threats to global agriculture (Vardhini and Anjum, 2015; Per et al., 2017; Hasanuzzaman et al., 2018; Khan et al., 2022; Nazir et al., 2023a). These stresses disrupt fundamental life processes, affecting physiological and molecular functions. Due to the immobile nature of the plants, their survival predominantly relies on promptly perceiving stress stimuli and swiftly executing responses to counter the adverse effects of stress (Planas-Riverola et al., 2019; Kumari et al., 2023). Recently, brassinosteroids (BRs) have gained significant attention within the phytohormone family due to their multifaceted roles in plant growth and development. They interact with various factors, including reactive oxygen species (ROS) and metabolites, to accelerate the plant's response to stressful stimuli (Xia et al., 2009; Yusuf et al., 2011; Nazir et al., 2021). BRs are plant steroid hormones, discovered in 1979, include more than 70 compounds found in plants, from lower to higher plants, and are now being designated as a new class of plant hormones regulating a wide range of plant processes including cell division and elongation, xylem differentiation, photomorphogenesis, and reproduction along with complex plant stress responses (Grove et al., 1979; Bishop and Yokota, 2001; Bajguz and Tretyn, 2003; Han et al., 2023). 24-epibrassinolide (BR27), castasterone, and brassinolide are a few examples of BRs (Mandava, 1988; Zullo and Kohout, 2004).

However, brassinolide is identified as the most potent BR, which was isolated from over 200 kilograms of rapeseed (*Brassica napus*) pollen, and its structure was elucidated through X-ray analysis (Grove et al., 1979). The growth-enhancing impact of the unrefined lipid extract derived from rapeseed pollen had been previously observed nine years before in a well-known bean second-internode bioassay (Mitchell et al., 1970). Furthermore, it has been shown that *in vitro* cultures of *Arabidopsis thaliana* L., BRs promote the proliferation of callus tissue (Hu et al., 2000). Moreover, it has been widely accepted that exogenous supplementation of BRs along with overexpression of the BR biosynthetic enzyme DWF4 both promoted drought stress tolerance (Krishna, 2003; Kagale et al., 2007; Sahni et al., 2016). Also, the inhibition of the BRI1 receptor leads to drought-resistant characteristics (Feng et al., 2015; Ye et al., 2017). Multiple research findings have also indicated that the application of brassinosteroids enhances the ability of plants to withstand water stress (Krishna, 2003).

It has also been widely studied that BRs play a crucial role in mitigating heavy metal stress in plants. They exert their protective effects through various mechanisms like (a) triggering the production of H_2O_2, (b) enhancing the defensive antioxidant system to counteract ROS, (c) increasing

DOI: 10.1201/9781032630380-5

the expression of MAPK, and (d) mitigating metal toxicity by elevating the levels of potassium and sodium ions, proline, antioxidants, and osmolytes (Rajewska et al., 2016). Brassinazole, a specific inhibitor of BRs, and the bioactive BR, 24-brassinolide (EBL), were employed to investigate the potential of EBL in alleviating metal toxicity, and it has been found that EBL restores the photosynthetic apparatus and defense mechanisms of *Arabidopsis thaliana* in response to heavy metal stress (Wu et al., 2019; Liaqat et al., 2020). BRs are also involved in maintaining cellular homeostasis in plants by reinstating CO_2 uptake and strengthening antioxidant capacity, thereby counteracting the detrimental impacts of heavy metals (Yusuf et al., 2011; Ahammed et al., 2020). Therefore, in this chapter we discuss in detail the recent advances in molecular mechanism of BR sensing, biosynthesis, and how BR alleviates the drought and heavy metal stress in plants as well as the crosstalk of BR with other plant growth regulators to prevent heavy metal–induced damage. Moreover, special emphasis is laid on signaling of BR in response to drought and heavy metal stress.

5.2 BIOSYNTHESIS OF BRASSINOSTEROIDS

Understanding the biosynthesis pathways of brassinosteroids and their metabolism is crucial for comprehending how plants regulate their internal levels of active brassinosteroids to support their optimal growth and development. In this regard, most of the enzymes associated with BR biosynthesis have been identified through various genetic and biochemical studies. However, further research is needed to fully elucidate the molecular mechanism of BR biosynthesis and release, providing a deeper understanding of the regulation of its synthesis, catalysis, and conjugation. These activities were explained in a coherent manner by Japanese scientists who systematically deciphered the biosynthetic pathways pertaining to BRs within plant cells (Fujioka and Sakurai, 1997). They elucidated that the synthesis of brassinolide (28-epibrassinolide) was linked to two main pathways, the early and late C6 oxidation pathways (Fujioka et al., 1998). Just as steroidal hormones in animals and insects originate from cholesterol, brassinosteroids are synthesized from specific plant sterol compounds. Within these plant sterols, campesterol or related compounds serve as the biosynthetic precursors for producing brassinolide, the most biologically active among over 50 natural brassinosteroids. During the biosynthetic pathway, this BR-specific precursor, campesterol, undergoes an initial conversion into campestanol (Figure 5.1). Subsequently, campestanol is then converted into castasterone (CS) through either early C-6 oxidation or late C-6 oxidation pathways, also known as campestanol-dependent pathways. Enzymes involved in this biosynthetic process include DET2, DWF4, cytochrome monooxygenases (Cyp P450), ROT3, BAS1, BR oxidase 2 (BRox2), constitutive photomorphogenesis and dwarfism (CPD), and BR oxidase 6 (BR6ox1) (Kour et al., 2021). The final step involves the conversion of CS to brassinolide (BL) and requires the enzyme Cyp450 (CYP85A2), which also plays a crucial role in the rate-limiting step for converting 6-deoxycatasterone into castasterone and eventually into brassinolide (Kour et al., 2021). Moreover, comprehensive understanding of both early and late C-6 oxidation pathways has been verified across various plant species. For instance, the conversion of 6-oxocampestanol to cathasterone was affirmed in cultured *C. roseus* cells, providing strong support for the pathway from campestanol to BL through early C-6 oxidation. Additionally, the sequence of campestanol → 6-deoxocathasterone → 6-deoxoteasterone in the late C-6 oxidation pathway has been established in *Arabidopsis* (Fujioka and Yokota, 2003). Evidence suggests that both early and late C-6 oxidation pathways are common across plant species like rice, pea, and zinnia. However, in tomato and tobacco, the late C-6 oxidation pathway appears to be predominant, as their endogenous BRs consist solely of members from this pathway (Fujioka and Yokota, 2003).

Subsequent phases of C_{28}-BR biosynthesis may take place through two parallel pathways known as the late or early C22 oxidation pathways (Bajguz et al., 2020). The late C-22 oxidation pathway starts with the reduction of 24-methylenecholesterol to campesterol. Campesterol then undergoes a 5,4-isomerization reaction to form (24R)-ergostan-4-en-3β-one, which is further converted to (24R)-5α-ergostan-3-one through DET2-mediated 5a-reduction, eventually leading to campestanol

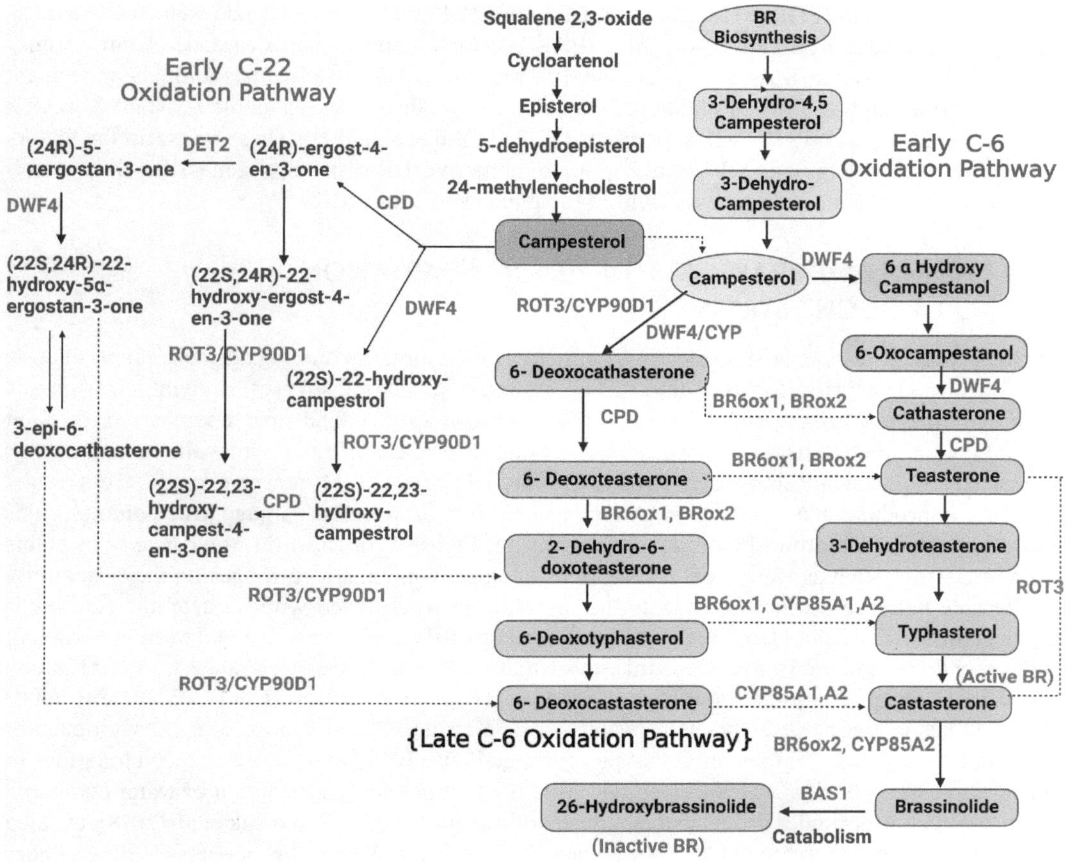

FIGURE 5.1 A schematic depiction of the brassinosteroid biosynthetic pathways, outlining early C-6 oxidation pathway, late C-6 oxidation pathway, and early C-22 oxidation pathway, modified from Kour et al. (2021).

(CN). In the early C-22 oxidation pathway, C-22a hydroxylation of 24-methylene CR results in the synthesis of 22-hydroxy-24-methylene CR. This C22α hydroxylation reaction is catalyzed by C-22α hydroxylase, encoded by the DWF4 gene which belongs to the P450 cytochrome family (Bajguz et al., 2020). Subsequent reactions in this pathway are similar to those in the late C-22 oxidation pathway, leading to the synthesis of 22-hydroxy forms of the corresponding compounds. Therefore, DWF4 can initiate a C-22 hydroxylation pathway by acting on various intermediates upstream, branching the pathway at campesterol (Kour et al., 2021).

5.3 SENSING OF BR

BRs are sensed by a leucine rich-repeat (LRR)–receptor-like-kinase (RLK) named brassinosteroid insensitive 1 (BRI1), localized on the plasma membrane. The isolation and cloning of BRI1 followed the discovery of numerous recessive mutant alleles at a single locus, identified through rigorous screenings aimed at finding mutants that displayed insensitivity to BRs either in root inhibition (Clouse et al., 1996) or hypocotyl elongation (Li and Chory, 1997). Typically, the BRI1 protein comprises three primary domains, each serving a distinct role in perceiving and activating the BR receptor: a substantial extracellular domain, a minor transmembrane domain, and an intracellular kinase domain. The hormone BR binds directly to a specific 93-amino acid segment situated in the extracellular portion of the membrane-bound BRI1 protein (Sun et al., 2013). This direct binding prompts the formation of a heterodimer consisting of BRI1 and brassinosteroid insensitive 1-associated receptor kinase 1 (BAK1, also called SERK3), which then initiates a cascade of intracellular phosphorylation events (Russinova et al., 2004). This cascade ultimately leads to the enhancement of both activity and stability of plant-specific transcription factors brassinazole resistant 1 (BZR1; Wang et al., 2002) and BRI1-EMS suppressor 1 (BES1; Yin et al., 2002). These transcription factors play a direct role in regulating the transcription of numerous BR-responsive genes, thereby controlling various developmental processes within the plant (Sun et al., 2010).

5.4 THE ROLE OF BRASSINOSTEROIDS IN ADAPTATION TO DROUGHT STRESS

Drought stress poses the most serious environmental constraint, threatening agricultural production and world food security (Khan et al., 2023a). Among other abiotic stresses, drought stress causes the maximum yield losses (Rukundo et al., 2014). Its broad-spectrum adverse repercussions on plant physiology and metabolism include stomatal closure, disruption of the photosynthetic machinery, decreased efficiency of Rubisco activity, and Calvin cycle enzymes. Moreover, drought stress results in excessive production of reactive oxygen species, leading to oxidative damage to cell biomolecules and ultimately cell death (Khan et al., 2023a, 2023b; Figure 5.3). However, the capacity of plants to endure stress, such as water deficit conditions and heavy metal stress, hinges on their capability to alternate between promoting growth and inhibiting it when faced with unfavorable conditions (Feng et al., 2016; Bechtold and Field, 2018). Exogenous BR supplementation and overexpression of BR biosynthetic enzymes have been linked to enhanced drought stress tolerance, while inhibition of the BRI1 receptor confers drought-resistant traits (Feng et al., 2015; Ye et al., 2017). BR application enhances tolerance of plants to water stress (Krishna, 2003). An experiment with cucumber (*Cucumis sativus*) demonstrated that plants treated with BR27 exhibited enhanced tolerance to dehydration. Specifically, leaves treated with this BR retained a higher amount of water compared to untreated controls under drought stress (Pustovoitova et al., 2001). Khamsuk et al. (2018) revealed that exogenous foliar treatment of BRs enhanced light use efficiency in pepper seedlings under drought stress, thereby increasing endurance to dehydration. This intervention also shows potential in mitigating the long-term consequences of drought (Nawaz et al., 2017). Recent study by Naveen et al. (2021) evaluated that a brassinolide strengthens the physio-biochemical parameters by improving the antioxidant system and photosynthetic efficiency in *Brassica juncea* plants. Moreover, it has also been revealed that exogenous application of BR27 increased wheat yield by about 20% under drought stress in field conditions (Janeczko et al., 2010). The positive influence of brassinosteroid treatment could be due to its ability to enhance the assimilation of CO_2 and leaf water economy, proline content, anthocyanins, and soluble phenolics and enables plants to resist drought stress (Farooq et al., 2009). Furthermore, Ali et al. (2008) found the similar result that this hormone increased the total grain yield and 100 grain weight under salt stress. Barros Junior et al. (2021) concluded that BR27 attenuated the negative effects of drought stress by enhancing the antioxidant enzyme activities, PSII efficiency, chloroplastic pigments, and electron flux, as well as net photosynthetic rate in *Eucalyptus urophylla* plants under water deficit. In a recent study, it has been demonstrated that pretreatment with BR27 had a protective impact on wheat plants, as indicated by the sustained growth rate and the decrease in lipid peroxidation and electrolyte leakage in wheat plant tissues under drought stress. This protective effect is attributed to the capability of BR27 to decrease the stress-induced buildup of ROS and regulate the activity of antioxidant enzymes. Additionally, BR pretreatment promoted proline accumulation and enriched the barrier properties of cell walls in seedlings by accelerating lignin deposition (Avalbaev et al., 2024). Findings of Gholami et al. (2023) stated that application of EBL improved total chlorophyll content, seed yield, pod weight, and number of pods per plant, as well as decreased malondialdehyde content in cowpea under water deficit stress. Similarly, Raza et al. (2023) also reveals that application of BR significantly increase the growth, physiological, and yield attributes of wheat under drought stress, which could be related to enhanced leaf area index, relative water contents, chlorophyll contents, and stomatal conductance in wheat plants. In line with this, findings of Lalarukh et al. (2022) further corroborated that BR application substantially increased vegetative growth parameters and photosynthetic pigments and upraised the levels of enzymatic and non-enzymatic antioxidants while it decreased stress-related attributes in wheat under drought stress. Systematic analysis at the molecular level mediated by BRs have uncovered that overexpression of BRL3 receptor confers drought tolerance without penalizing overall growth by inducing the accumulation of osmoprotectant metabolites including proline and sugars and differential expression of genes in the vascular tissues (Fàbregas et al., 2018). In another study, treating pepper plants with BR improves their ability to withstand drought by enhancing their photosynthetic system, maintaining leaf hydration, and boosting antioxidant defenses under challenging conditions (Kaya et al., 2019). Sorghum plants treated with BR exhibited better root system development and taller stature, indicating the potential of BR in mitigating water deficit at the field level (Dias De Souza et al., 2023). The primary mechanism underlying sorghum's drought tolerance is its higher stomatal sensitivity, which swiftly reduces transpiration and mitigates plant dehydration during water deficit conditions. Therefore, these studies highlight the potential of BR application as a valuable approach for improving crop resilience in water-limited environments.

5.5 BR SIGNALING UNDER DROUGHT STRESS

The potential of a plant to adapt to stress, like deficit water conditions, heavy metal stress, and salt stress, depends on its capability between activation and repression of growth under adverse conditions (Feng et al., 2016; Bechtold and Field, 2018). Under drought stress conditions, the signaling of BRs is known to be affected, leading to alterations in plant growth and development (Hayes, 2019). One of the key mechanisms by which BRs alleviate the impacts of drought stress is through their interaction with other hormones, such as abscisic acid. The pathway of ABA, as emphasized in studies by Yoshida et al. (2014) and Zhu et al. (2017), plays a key role in controlling how plants respond to environmental stresses. However, recent evidence indicates that BRs also play a vital role in balancing normal growth and adaptability to environmental stresses either by interacting with the ABA pathway or functioning independently. Various mechanisms have been suggested to describe the involvement of BR signaling in mediating stress tolerance which includes regulation of stress-related genes, upregulation of the antioxidant system, and elevated levels of osmoprotectants (Xia et al., 2009; Yuan et al., 2010; Todorova et al., 2016; Ye et al., 2017; Lima and Lobato, 2017; Fàbregas et al., 2018; Cui et al., 2019; Hayes, 2019; Ahmad Lone et al., 2022; Sukpitak et al., 2024). Advances in comprehending the relationship between drought and BR have unveiled many mechanisms that inhibit BR activity during drought stress.

Recent advances in understanding the relationship between drought and BR has revealed numerous mechanisms that inhibit BR activity during drought stress. It has been observed that BR and ABA perform antagonistic interplay across various stages, ranging from receptor complexes to subsequent transcription factors, and the main BR signaling components involved in this crosstalk are BIN2 and BZR1 (Cai et al., 2014; Nolan et al., 2017). BIN2 inhibits BR signaling and boosts ABA-induced stress responses by phosphorylating SnRK2, leading to the activation of ABA-responsive genes. Furthermore, exogenous BR application suppresses the ABA-triggered increase of *RD26*, a gene responsible for activating stress-responsive gene expression (Cai et al., 2014). This reciprocal inhibition between BR signaling and ABA-responsive transcription factors plays a crucial role in balancing plant growth and drought resilience (Figure 5.2). *RD26*, additionally, is directly influenced by BES1 and is suppressed by BR during drought conditions, and in turn, *RD26* regulates the transcription of genes controlled by BES1 to check the function of BR (Ye et al., 2017). Transcription factors WRKY46, 54, and 70 directly interact with BES1 to stimulate BR-mediated plant growth while inhibiting the expression of drought-responsive genes to decline drought tolerance (Chen and Yin, 2017). It has been also suggested by Chen et al. (2017) that BIN2 phosphorylates and undermines the WRKY54 to diminish its impact on BES1-regulated BR response. Recent studies have also unveiled that BR

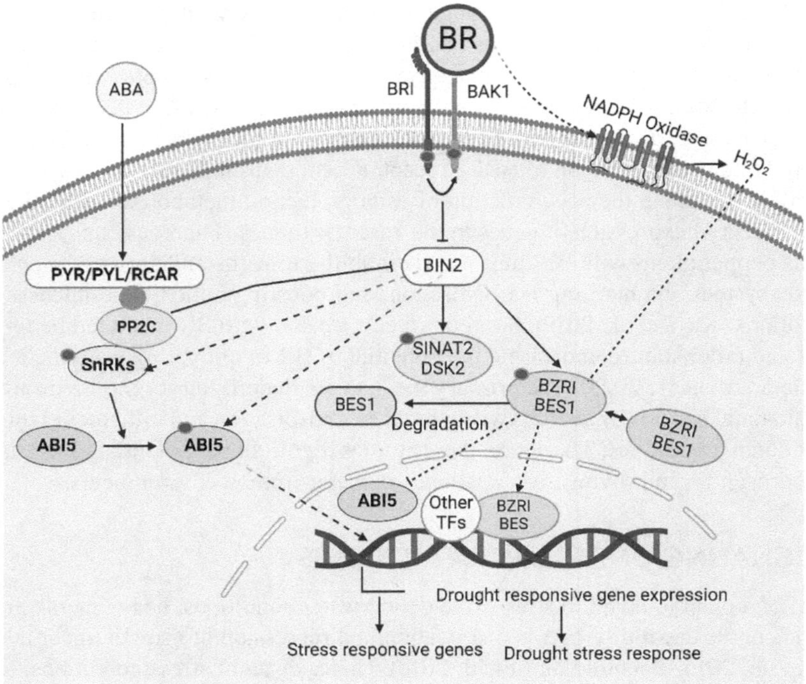

FIGURE 5.2 Schematic illustration of crosstalk of BR-ABA under drought stress.

signaling, via BIN2, participates with autophagy pathways to regulate plant growth and survival in response to adverse conditions such as drought stress (Trevor Nolan et al., 2017). In this process, BIN2 phosphorylates and activates the ubiquitin receptor protein DSK2, which then interacts with BES1, leading to its degradation through autophagy. Different concentrations of BR regulate the opening and closing of stomata through distinct mechanisms: at low concentrations, BR activates respiratory burst oxidase homolog (RBOH), increasing H_2O_2 levels and promoting stomatal opening, whereas at high concentrations, it induces H_2O_2 accumulation and reduces K^+ uptake in guard cells, triggering abscisic acid (ABA) activation which thereby causes stomatal closure (Daszkowska-Golec and Szarejko, 2013). BR binds to the specific receptor brassinosteroid insensitivel (BRI1) on the plasma membrane, leading to increased Ca^{2+} influx that activates NADPH oxidase, enhancing ROS production via binding with the EF-motif of RBOH (Xia et al., 2015). Alternatively, BR also upregulates mitogen-activated protein kinase (MAPK) to scavenge ROS, and stimulates nitric oxide (NO) production, enhancing the antioxidative defense system (Xia et al., 2015). Taken together these studies underscore the intricate nature of BR-mediated responses to drought stress.

ABA is detected by receptors known as PYR/PYL/RCAR, leading to the activation of SnRKs through phosphorylation, thus releasing them from inhibition by PP2C. These activated SnRKs subsequently phosphorylate downstream transcription factors like ABI5, which govern the expression of various genes involved in responding to stress. BIN2 suppresses BR signaling and can phosphorylate and activate both SnRKs and ABI5 directly, whereas PP2C can deactivate BIN2. Furthermore, BZR1 directly targets ABI5, inhibiting its transcription and consequently suppressing the expression of stress-responsive genes. Additionally, BIN2 acts as a negative regulator of BR signaling, suppressing BES1-regulated BR response. BIN2 phosphorylates and activates the ubiquitin receptor protein DSK2, initiating its interaction with BES1, thereby promoting its degradation. Conversely, BR binds to the specific receptor BRI1 on the plasma membrane, triggering the production of H_2O_2 as a signaling molecule via NADPH oxidase. This, in turn, activates BZRI/BES1-regulated BR response.

5.6 THE ROLE OF BRASSINOSTEROIDS IN ADAPTATION TO HEAVY METAL STRESS

Soil contamination by metals poses a significant challenge in numerous industrialized and developing nations, leading to various health issues in humans (Ishihara et al., 2001). This contamination, often originating from tainted sewage sludge, permeates cultivated soils and is absorbed by plants, subsequently spreading throughout the food chain. Consequently, comprehending how plants safeguard themselves and sequester toxins emerges as a crucial issue for the forthcoming decades to mitigate the levels and impact of metal pollutants. Heavy metals are inadvertently absorbed by low-specificity transporters in plants. Once within the cell, they disrupt numerous biological processes, notably impeding growth and causing leaf discoloration.

This disturbance extends to nutrient uptake, triggering the deterioration of photosynthetic apparatus and disrupting water balance by inducing stomatal closure (Fagioni et al., 2009). Additionally, heavy metals disturb the activities of various enzymes, particularly those reliant on metallic ions as cofactors, such as zinc-binding proteins. Heavy metal stress strongly induces the ROS production due to the dysregulation of redox control mechanisms and thereby leads the oxidative damage to the plant cells. The impact of adverse conditions such as drought and heavy metal stress is illustrated in Figure 5.3. However, extensive research has been conducted on the influence of heavy metals on plant physiology, resulting in the identification of detoxification pathways, and BRs are one of the important signaling molecules involved in the detoxification of the toxicities generated by heavy metals in plants (Nazir et al., 2023b). Plants have evolved intricate signaling systems to react and adapt to environmental cues such as drought, temperature, salinity stress, and heavy metal stress. They synthesize a diverse range of hormones which not only govern growth and development but also play important roles in responding to environmental stresses (Bari and Jones, 2009). BRs have been studied to regulate stress responses as depicted in Figure 5.3 under heavy metal exposure in plants. They mitigate the heavy metal stress by stimulating the synthesis of phytochelatins, aiding metal ion complexation, and boosting the activity of antioxidant enzymes to reduce the accumulation of ROS (Sharma et al., 2011).

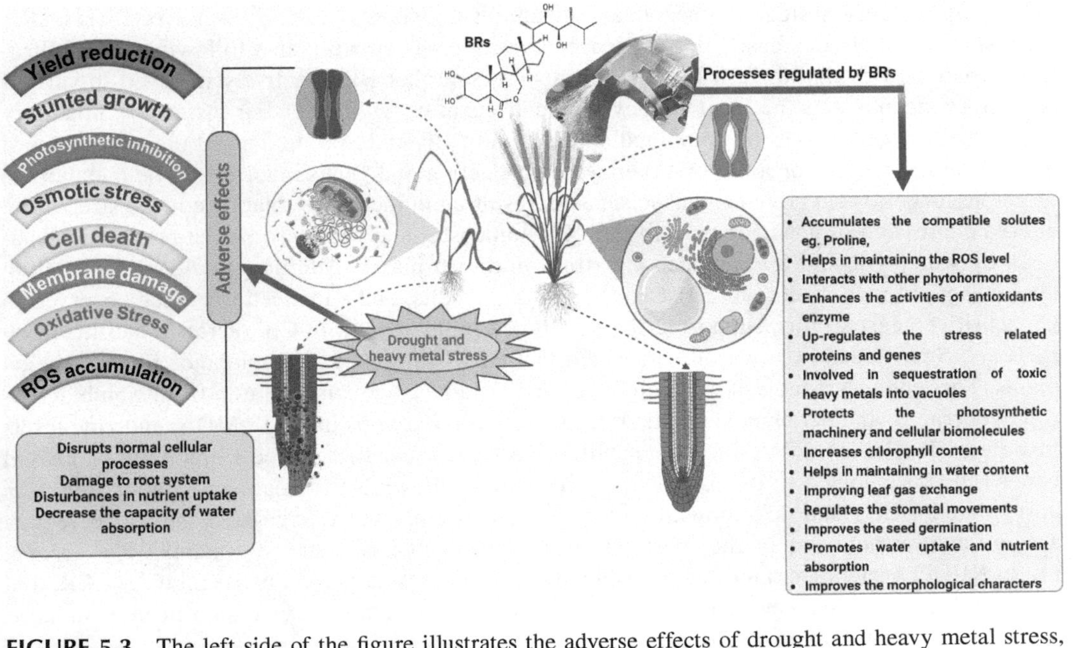

FIGURE 5.3 The left side of the figure illustrates the adverse effects of drought and heavy metal stress, while the right side depicts the processes regulated by brassinosteroids to these environmental challenges.

Comparable advantages are found in *Brassica juncea* plants facing cobalt salt exposure, as BR treatment enhances antioxidant defenses and improves growth (Arora et al., 2012). Alam et al. (2020) reported that the exogenous application of 24-homobrassinolide (HBL) or BR27 to mustard plants exposed to Cd stress neutralized the adverse effects generated by Cd stress that reflect in better photosynthetic performance and increase in proline content. BR promotes growth, ascorbate-glutathione cycle, and glyoxalase system in brown mustard under Cd stress. HBL, a class of BR effectively reduced Pb toxicity and improved resistance in turnip seedlings to Pb contaminated soil by improving photosynthesis and nutritional content (Khan et al., 2023c). It has also been studied that BR in combination with jasmonate improves growth and tolerance by reducing oxidative stress, ROS, lipid peroxidation, and electrolyte leakage while also increasing antioxidant enzymes such as SOD, POD, CAT, APX, and GR, along with non-antioxidants (flavonols, tocopherols), proline content, and plant metal chelating capacity in bamboo plants (Emamverdian et al., 2024). Similarly, individually and in combination, the exogenous supplementation of BR27 and Si mitigated the negative impacts of Cd by enhancing growth, biochemical indicators, nutrient absorption, accumulation of osmolytes, and the antioxidative defense and glyoxalase systems in *Pisum sativum* seedlings (Jan et al., 2018). The application of BRs also significantly reduced chromium (Cr) toxicity by decreasing the uptake and accumulation as well as translocation of Cr within soybean plant tissues. Additionally, it enhanced seed germination rates, photosynthetic functions, plant growth, and biomass, while also preventing the inhibition of nutrient uptake under Cr stress. Moreover, BRs stimulated both enzymatic and non-enzymatic antioxidative defense mechanisms, facilitated the sequestration of ions, reduced the excessive production of ROS, and decreased electrolyte leakage in response to Cr-induced toxicity. Furthermore, BR enhanced the tolerance of soybean seedlings to Cr stress by regulating the expression of stress-related genes involved in Cr accumulation and translocation (Basit et al., 2022). BR either individually or in combination with H_2O_2 enhanced the efficiency of photosynthesis and altered various features such as chloroplast structure, stomatal activity, root morphology, cell viability, and the production of antioxidants and proline (Jan et al., 2018). These modifications increased the ability of tomato plants to withstand copper (Cu) stress. The application of BR to plants under phenanthrene (PHE) and Cd stress led to the mitigation of photosynthetic suppression and oxidative stress. This was achieved by boosting the activities of antioxidant enzymes and associated transcript levels of the antioxidant defense system and secondary metabolism. Moreover, the presence of PHE and/or Cd residues notably decreased in both leaves and roots of tomato plants following BR application (Ahammed et al., 2013). It has also been suggested that BR is actively involved in various regulatory mechanisms that deal with cytotoxic levels of heavy metals. For instance, Choudhary et al. (2012) revealed that the combined application of BR and Spd influenced the expression of genes responsible for polyamine (PA) enzyme production and genes involved in the metabolism of indole-3-acetic acid (IAA) and abscisic acid, resulting in increased tolerance to Cu stress. The altered expression of genes associated with Cu homeostasis appeared to be the primary mechanism by which BR and Spd reduced Cu stress in radish plants. Additional evidence, including ion leakage, in vivo detection of H_2O_2, comet assay results, and enhanced tolerance observed in Cu-sensitive yeast strains, supported the significant improvement in Cu tolerance facilitated by EBR and Spd. Recently, it was also revealed that BR27 improved zinc tolerance in watermelon plants. This enhancement was associated with BR27-induced enhancement in fresh weight, chlorophyll content, and net photosynthetic rate, as well as reductions in H_2O_2, MDA, and zinc levels in watermelon leaves. RNA deep sequencing (RNA-seq) revealed 350 differentially expressed genes (DEGs) involved in the response to zinc stress following BR treatment, comprising 175 upregulated DEGs and 175 downregulated DEGs. Among the upregulated DEGs, there was a significant enrichment in the 'phenylpropanoid biosynthesis' pathway (map00940) according to KEGG enrichment analysis. Notably, the gene expression levels of PAL, 4CL, CCR, and CCoAOMT, which are key genes in the phenylpropanoid pathway, were significantly induced after BR treatment (Liu et al., 2023).

5.7 BR SIGNALING UNDER HEAVY METAL STRESS

The signaling of BRs under heavy metal stress in plants is a complex process that involves intricate molecular mechanisms, which has not been fully elucidated yet. Heavy metal stress disrupts various physiological processes in plants, including growth, development, and metabolism (Hilal et al., 2023). BRs, a class of plant hormones, play a crucial role in mediating plant responses to environmental stresses, including heavy metal toxicity (Kour et al., 2021). Under heavy metal stress, BR signaling pathways are activated, triggering a cascade of molecular events aimed at mitigating the adverse effects of heavy metals on plant growth and survival. BRs alter cell membrane permeability, thereby reducing the uptake of heavy metals, and also trigger the activation of a set of defensive enzymes. They promote the synthesis of stress proteins by influencing anti-stress genes through increased ATPase expression (Madhan et al., 2014). Exogenous treatment of BRs stimulates a transient increase in H_2O_2 levels, activating MAPK and subsequently leading the production of NADPH oxidase (Jiang et al., 2012). Xia et al. (2014) reported that a lower dose (0.1 μM) of BR induced stomatal opening, whereas a higher dose (1.0 μM) of BR promoted stomatal closure. These processes lead to the upregulation of stress-responsive genes, stress proteins, and defensive enzymes, as well as the regulation of plant responses which ultimately mitigate the effects of metal stress (Yin et al., 2016). Moreover, bioactive BRs such as HBL and EBL are also involved in phytoremediation, assisting plants in removing toxic metals (Barbafieri and Tassi, 2011). Therefore, BRs aid plants in conferring metal tolerance, consequently enhancing crop yield and quality (Vriet et al., 2013).

5.8 CROSSTALK OF BRS WITH OTHER PLANT GROWTH REGULATORS TO REGULATE STRESS RESPONSES

BRs have been implicated in various physiological processes in plants, including sugar metabolism and transport. Studies have shown that BRs can enhance the retention and uptake of sucrose in plants. This effect may be attributed to alterations in the activity of proton pump enzymes like H^+-ATPase, as shown by higher *Vmax*-values for sucrose uptake (Petzold et al., 1992). Additionally, studied have found that BRs, similar to other plant hormones such as gibberellic acid (GA_3) and indole-3-acetic acid (IAA), play a role in the unloading of 14C-labeled compounds in plants. Specifically, experiments conducted on *Vicia faba* have demonstrated that treatment with BRs, in combination with GA_3 or IAA, promotes the transport of 14C-labeled compounds to the apical sink region of the plant (Petzold et al., 1992). The regulation of transpiration rate and water loss in plants is governed by the opening and closing of stomata. Consequently, stomata are pivotal in safeguarding plants against water stress and pathogens. The process of stomatal opening and closing depends on reversible changes in turgor pressure and osmotically driven water movement within the guard cells. This essential mechanism is initiated by a variety of internal and external stimuli. However, the involvement of BR in the regulation of stomatal movement has been documented in various plant species (Acharya and Assmann, 2009). For instance, treatment of BR to the epidermis of *Vicia faba* induces stomatal closure while hindering stomatal opening (Haubrick and Assmann, 2006). However, when *Pinus banksiana* seedlings were pretreated with HBL, it was observed that stomatal closure was inhibited under water stress conditions (Rajasekaran and Blake, 1999). The impact of exogenous BR on stomatal response in tomato was revealed to be contingent upon the concentration of BR. Lower concentrations of BR prompted stomatal opening, whereas higher concentrations prompted stomatal closure (Xia et al., 2014). It has been found that cytokinin and auxin augment ethylene (ET) production, thereby facilitating stomatal closure. However, while cytokinin and auxin play a role, it's important to note that ET, rather than ABA, is crucial for BR induced stomatal closure (Suhita et al., 2004; Shi et al., 2015). BR triggers the upregulation of *ACS5* and *ACS9* genes, consequently promoting ethylene synthesis. This activation subsequently stimulates the Gα protein, which in turn facilitates AtrbohF-dependent H_2O_2 production. The resultant increase in H_2O_2 levels leads to the accumulation of NO catalyzed by Nia1. Collectively, the activation of these signaling components is essential for inducing stomatal closure (Shi et al., 2015).

5.9 CONCLUSION AND FUTURE PROSPECTS

BRs play a crucial role in regulating plant growth under normal conditions and under challenging environmental cues such as drought and heavy metal stress. BRs alleviate drought and heavy metal stress in plants by acting as a potent signaling molecule and enhancing antioxidant defense systems. Additionally, BRs regulate ion homeostasis and induce osmotic adjustment, contributing to plant resilience against drought and heavy metal–induced osmotic stress and oxidative damage. Therefore, it may be concluded that modifying the BR signaling pathway could be a valuable strategy for developing crops that are better suited to drought and heavy metal stress conditions. However, our understanding of the precise functions of BR signaling during stress remains general, and research into the specific regulatory mechanisms in different contexts and at different times is still in its infancy. Further investigation is needed to gain a more detailed understanding of how the BR pathway operates. With this understanding, it could be possible to enhance plant growth rates and improve their ability to adapt to environmental challenges by selectively altering BR signaling in specific tissues. For instance, it might be possible to create plants that are more tolerant to drought and heavy metal stress without affecting their overall growth. Such an approach could be crucial for meeting the food needs of a rapidly growing global population, particularly as the demand for increased plant yield in challenging environments continues to rise.

5.10 AUTHORSHIP CONTRIBUTION

Conceptualization, Q.F.; Writing – Original draft preparation, B.H. and Q.F.; Writing – Review and editing, B.H. and Q.F. All authors have read and agreed to the published version of the book chapter.

5.11 DECLARATION OF COMPETING INTEREST

The authors declare that they have no known competing financial interests or personal relationships that could have appeared to influence the work reported in this book chapter.

ACKNOWLEDGEMENT

Bisma Hilal gratefully acknowledges the financial support in the form of the Senior Research Fellowship rendered by Maulana Azad National Fellowship for Minority students under Ref. No. (191620103675) at Aligarh Muslim University, Aligarh, India.

REFERENCES

Acharya, B.R., Assmann, S.M., 2009. Hormone interactions in stomatal function. *Plant Mol. Biol.* 69, 451–462. https://doi.org/10.1007/S11103-008-9427-0

Ahammed, G.J., Choudhary, S.P., Chen, S., Xia, X., Shi, K., Zhou, Y., Yu, J., 2013. Role of brassinosteroids in alleviation of phenanthrene–cadmium co-contamination-induced photosynthetic inhibition and oxidative stress in tomato. *J. Exp. Bot.* 64, 199–213. https://doi.org/10.1093/JXB/ERS323

Ahammed, G.J., Li, X., Liu, A., Chen, S., 2020. Brassinosteroids in plant tolerance to abiotic stress. *J. Plant Growth Regul.* 39, 1451–1464. https://doi.org/10.1007/S00344-020-10098-0

Ahmad Lone, W., Majeed, N., Yaqoob, U., John, R., 2022. Exogenous brassinosteroid and jasmonic acid improve drought tolerance in *Brassica rapa* L. genotypes by modulating osmolytes, antioxidants and photosynthetic system. *Plant Cell Rep.* 41, 603–617. https://doi.org/10.1007/S00299-021-02763-9

Alam, P., Kohli, S.K., Al Balawi, T., Altalayan, F.H., Alam, P., Ashraf, M., Bhardwaj, R., Ahmad, P., 2020. Foliar application of 24-epibrassinolide improves growth, ascorbate-glutathione cycle, and glyoxalase system in brown mustard (*Brassica juncea* (L.) Czern.) under cadmium toxicity. *Plants* 9, 1487. https://doi.org/10.3390/PLANTS9111487

Ali, Q., Athar, H.U.R., Ashraf, M., 2008. Modulation of growth, photosynthetic capacity and water relations in salt stressed wheat plants by exogenously applied 24-epibrassinolide. *Plant Growth Regul.* 56, 107–116. https://doi.org/10.1007/S10725-008-9290-7/METRICS

Arora, P., Bhardwaj, R., Kanwar, M.K., 2012. Effect of 24-epibrassinolide on growth, protein content and antioxidative defense system of *Brassica juncea* L. subjected to cobalt ion toxicity. *Acta Physiol. Plant.* 34, 2007–2017. https://doi.org/10.1007/S11738-012-1002-2/METRICS

Avalbaev, A., Fedyaev, V., Lubyanova, A., Yuldashev, R., Allagulova, C., 2024. 24-Epibrassinolide reduces drought-induced oxidative stress by modulating the antioxidant system and respiration in wheat seedlings. *Plants (Basel, Switzerland)* 13. https://doi.org/10.3390/PLANTS13020148

Bajguz, A., Chmur, M., Gruszka, D., 2020. Comprehensive overview of the brassinosteroid biosynthesis pathways: substrates, products, inhibitors, and connections. *Front. Plant Sci.* 11, 535077. https://doi.org/10.3389/FPLS.2020.01034/BIBTEX

Bajguz, A., Tretyn, A., 2003. The chemical characteristic and distribution of brassinosteroids in plants. *Phytochemistry* 62, 1027–1046. https://doi.org/10.1016/S0031-9422(02)00656-8

Barbafieri, M., Tassi, E., 2011. Brassinosteroids for phytoremediation application. *Brassinosteroids A Cl. Plant Horm.* 403–437. https://doi.org/10.1007/978-94-007-0189-2_16

Bari, R., Jones, J.D.G., 2009. Role of plant hormones in plant defence responses. *Plant Mol. Biol.* 69, 473–488. https://doi.org/10.1007/S11103-008-9435-0

Barros Junior, U.O., Lima, M.D.R., Alsahli, A.A., Lobato, A.K.S., 2021. Unraveling the roles of brassinosteroids in alleviating drought stress in young *Eucalyptus urophylla* plants: implications on redox homeostasis and photosynthetic apparatus. *Physiol. Plant.* 172, 748–761. https://doi.org/10.1111/PPL.13291

Basit, F., Bhat, J.A., Hu, J., Kaushik, P., Ahmad, A., Guan, Y., Ahmad, P., 2022. Brassinosteroid supplementation alleviates chromium toxicity in soybean (*Glycine max* L.) via reducing its translocation. *Plants* 11, 2292. https://doi.org/10.3390/PLANTS11172292/S1

Bechtold, U., Field, B., 2018. Molecular mechanisms controlling plant growth during abiotic stress. *J. Exp. Bot.* 69, 2753–2758. https://doi.org/10.1093/JXB/ERY157

Bishop, G.J., Yokota, T., 2001. Plants steroid hormones, brassinosteroids: current highlights of molecular aspects on their synthesis/metabolism, transport, perception and response. *Plant Cell Physiol.* 42, 114–120. https://doi.org/10.1093/PCP/PCE018

Cai, Z., Liu, J., Wang, H., Yang, C., Chen, Y., Li, Y., Pan, S., Dong, R., Tang, G., De Dios Barajas-Lopez, J., Fujii, H., Wang, X., 2014. GSK3-like kinases positively modulate abscisic acid signaling through phosphorylating subgroup III SnRK2s in Arabidopsis. *Proc. Natl. Acad. Sci. U. S. A.* 111, 9651–9656. https://doi.org/10.1073/PNAS.1316717111/-/DCSUPPLEMENTAL/PNAS.201316717SI.PDF

Chen, J., Nolan, T.M., Ye, H., Zhang, M., Tong, H., Xin, P., Chu, J., Chu, C., Li, Z., Yina, Y., 2017. Arabidopsis WRKY46, WRKY54, and WRKY70 transcription factors are Involved in brassinosteroid-regulated plant growth and drought responses. *Plant Cell* 29, 1425–1439. https://doi.org/10.1105/TPC.17.00364

Chen, J., Yin, Y., 2017. WRKY transcription factors are involved in brassinosteroid signaling and mediate the crosstalk between plant growth and drought tolerance. *Plant Signal. Behav.* 12. https://doi.org/10.1080/15592324.2017.1365212

Choudhary, S.P., Oral, H.V., Bhardwaj, R., Yu, J.Q., Tran, L.S.P., 2012. Interaction of brassinosteroids and polyamines enhances copper stress tolerance in *Raphanus sativus*. *J. Exp. Bot.* 63, 5659–5675. https://doi.org/10.1093/JXB/ERS219

Clouse, S.D., Langford, M., McMorris, T.C., 1996. A brassinosteroid-insensitive mutant in *Arabidopsis thaliana* exhibits multiple defects in growth and development. *Plant Physiol.* 111, 671–678. https://doi.org/10.1104/PP.111.3.671

Cui, X.Y., Gao, Y., Guo, J., Yu, T.F., Zheng, W.J., Liu, Y.W., Chen, J., Xu, Z.S., Ma, Y.Z., 2019. BES/BZR transcription factor TaBZR2 positively regulates drought responses by activation of *TaGST1*. *Plant Physiol.* 180, 605–620. https://doi.org/10.1104/PP.19.00100

Daszkowska-Golec, A., Szarejko, I., 2013. Open or close the gate – stomata action under the control of phytohormones in drought stress conditions. *Front. Plant Sci.* 4, 44747. https://doi.org/10.3389/FPLS.2013.00138/BIBTEX

Dias De Souza, A.C., Stefhane, B., Reis, S., Fernandes Bacin, D., Robson De Oliveira, L., Silva, W., Santos, D., Borges, L.P., Cardoso, N.C., Matos, S., 2023. Importance of brassinosteroids for mitigating water stress in sorghum. *AJCS* 17, 1835–2707. https://doi.org/10.21475/ajcs.23.17.09.p3950

Emamverdian, A., Khalofah, A., Pehlivan, N., Zia-ur-Rehman, M., Li, Y., Zargar, M., 2024. Exogenous application of jasmonates and brassinosteroids alleviates lead toxicity in bamboo by altering biochemical and physiological attributes. *Environ. Sci. Pollut. Res. Int.* 31, 7008–7026. https://doi.org/10.1007/S11356-023-31549-7

Fàbregas, N., Lozano-Elena, F., Blasco-Escámez, D., Tohge, T., Martínez-Andújar, C., Albacete, A., Osorio, S., Bustamante, M., Riechmann, J.L., Nomura, T., Yokota, T., Conesa, A., Alfocea, F.P., Fernie, A.R., Caño-Delgado, A.I., 2018. Overexpression of the vascular brassinosteroid receptor BRL3 confers drought resistance without penalizing plant growth. *Nat. Commun.* 91(9), 1–13. https://doi.org/10.1038/s41467-018-06861-3

Fagioni, M., D'Amici, G.M., Timperio, A.M., Zolla, L., 2009. Proteomic analysis of multiprotein complexes in the thylakoid membrane upon cadmium treatment. *J. Proteome Res.* 8, 310–326. https://doi.org/10.1021/PR800507X

Farooq, M., Wahid, A., Basra, S.M.A., Islam-Ud-Din, 2009. Improving water relations and gas exchange with brassinosteroids in rice under drought stress. *J. Agron. Crop Sci.* 195, 262–269. https://doi.org/10.1111/J.1439–037X.2009.00368.X

Feng, W., Lindner, H., Robbins, N.E., Dinneny, J.R., 2016. Growing out of stress: the role of cell- and organ-scale growth control in plant water-stress responses. *Plant Cell* 28, 1769–1782. https://doi.org/10.1105/TPC.16.00182

Feng, Y., Yin, Y., Fei, S., 2015. Down-regulation of BdBRI1, a putative brassinosteroid receptor gene produces a dwarf phenotype with enhanced drought tolerance in *Brachypodium distachyon*. *Plant Sci.* 234, 163–173. https://doi.org/10.1016/J.PLANTSCI.2015.02.015

Fujioka, S., Noguchi, T., Yokota, T., Takatsuto, S., Yoshida, S., 1998. Brassinosteroids in *Arabidopsis thaliana*. *Phytochemistry* 48, 595–599. https://doi.org/10.1016/S0031-9422(98)00065-X

Fujioka, S., Sakurai, A., 1997. Brassinosteroids. *Nat. Prod. Rep.* 14, 1–10. https://doi.org/10.1039/NP9971400001

Fujioka, S., Yokota, T., 2003. Biosynthesis and metabolism of brassinosteroids. *Annu. Rev. Plant Biol.* 54, 137–64. https://doi.org/10.1146/annurev.arplant.54.031902.134921

Gholami, F., Amerian, M.R., Asghari, H.R., Ebrahimi, A., 2023. Assessing the effects of 24-epibrassinolide and yeast extract at various levels on cowpea's morphophysiological and biochemical responses under water deficit stress. *BMC Plant Biol.* 23, 1–20. https://doi.org/10.1186/S12870-023-04548-6/FIGURES/2

Grove, M.D., Spencer, G.F., Rohwedder, W.K., Mandava, N., Worley, J.F., Warthen, J.D., Steffens, G.L., Flippen-Anderson, J.L., Cook, J.C., 1979. Brassinolide, a plant growth-promoting steroid isolated from Brassica napus pollen. *Nature* 281, 216–217. https://doi.org/10.1038/281216A0

Han, C., Wang, L., Lyu, J., Shi, W., Yao, L., Fan, M., Bai, M.Y. 2023. Brassinosteroid signaling and molecular crosstalk with nutrients in plants. *J. Genetics Genom.* 50(8), 541–553.

Hasanuzzaman, M., Nahar, K., Anee, T.I., Khan, M.I.R., Fujita, M. 2018. Silicon-mediated regulation of antioxidant defense and glyoxalase systems confers drought stress tolerance in *Brassica napus* L. *S. Afr. J. Botany* 115, 50–57.

Haubrick, L.L., Assmann, S.M., 2006. Brassinosteroids and plant function: some clues, more puzzles. *Plant. Cell Environ.* 29, 446–457. https://doi.org/10.1111/J.1365-3040.2005.01481.X

Hayes, S., 2019. BRacing for water stress: brassinosteroid signaling promotes drought survival in wheat. *Plant Physiol.* 180, 18–19. https://doi.org/10.1104/PP.19.00314

Hilal, B., Khan, T.A., Fariduddin, Q., 2023. Recent advances and mechanistic interactions of hydrogen sulfide with plant growth regulators in relation to abiotic stress tolerance in plants. *Plant Physiol. Biochem.* 196, 1065–1083. https://doi.org/10.1016/J.PLAPHY.2023.03.006

Hu, Y., Bao, F., Li, J., 2000. Promotive effect of brassinosteroids on cell division involves a distinct CycD3-induction pathway in Arabidopsis. *Plant J.* 24, 693–701. https://doi.org/10.1046/J.1365-313X.2000.00915.X

Ishihara, T., Kobayashi, E., Okubo, Y., Suwazono, Y., Kido, T., Nishijyo, M., Nakagawa, H., Nogawa, K., 2001. Association between cadmium concentration in rice and mortality in the Jinzu River basin, Japan. *Toxicology* 163, 23–28. https://doi.org/10.1016/S0300-483X(01)00367-5

Jan, S., Alyemeni, M.N., Wijaya, L., Alam, P., Siddique, K.H., Ahmad, P., 2018. Interactive effect of 24-epibrassinolide and silicon alleviates cadmium stress via the modulation of antioxidant defense and glyoxalase systems and macronutrient content in *Pisum sativum* L. seedlings. *BMC Plant Biol.* 18. https://doi.org/10.1186/S12870-018-1359-5

Janeczko, A., Biesaga-Kościelniak, J., Oklešt'ková, J., Filek, M., Dziurka, M., Szarek-Łukaszewska, G., Kościelniak, J., 2010. Role of 24-epibrassinolide in wheat production: physiological effects and uptake. *J. Agron. Crop Sci.* 196, 311–321. https://doi.org/10.1111/J.1439-037X.2009.00413.X

Jiang, Y.P., Cheng, F., Zhou, Y.H., Xia, X.J., Mao, W.H., Shi, K., Chen, Z., Yu, J.Q., 2012. Cellular glutathione redox homeostasis plays an important role in the brassinosteroid-induced increase in CO_2 assimilation in *Cucumis sativus*. *New Phytol.* 194, 932–943. https://doi.org/10.1111/J.1469-8137.2012.04111.X

Kagale, S., Divi, U.K., Krochko, J.E., Keller, W.A., Krishna, P., 2007. Brassinosteroid confers tolerance in *Arabidopsis thaliana* and *Brassica napus* to a range of abiotic stresses. *Planta* 225, 353–364. https://doi.org/10.1007/S00425-006-0361-6

Kaya, C., Ashraf, M., Wijaya, L., Ahmad, P., 2019. The putative role of endogenous nitric oxide in brassinosteroid-induced antioxidant defence system in pepper (*Capsicum annuum* L.) plants under water stress. *Plant Physiol. Biochem.* 143, 119–128. https://doi.org/10.1016/J.PLAPHY.2019.08.024

Khamsuk, O., Sonjaroon, W., Suwanwong, S., Jutamanee, K., Suksamrarn, A., 2018. Effects of 24-epibrassinolide and the synthetic brassinosteroid mimic on chili pepper under drought. *Acta Physiol. Plant.* 40. https://doi.org/10.1007/S11738-018-2682-Z

Khan, M., Ahmed, S., Yasin, N.A., Sardar, R., Hussaan, M., Gaafar, A.R.Z., Haider, F.U., 2023c. 28-Homobrassinolide primed seed improved lead stress tolerance in *Brassica rapa* L. through modulation of physio-biochemical attributes and nutrient uptake. *Plants* 12, 3528. https://doi.org/10.3390/PLANTS12203528

Khan, T.A., Saleem, M., Fariduddin, Q., 2023b. Melatonin influences stomatal behavior, root morphology, cell viability, photosynthetic responses, fruit yield, and fruit quality of tomato plants exposed to salt stress. *J. Plant Growth Regul.* 42, 2408–2432. https://doi.org/10.1007/s00344-022-10713-2

Khan, T.A., Saleem, M., Fariduddin, Q., 2022. Recent advances and mechanistic insights on melatonin-mediated salt stress signaling in plants. *Plant Physiol. Biochem.* 188, 97–107. https://doi.org/10.1016/j.plaphy.2022.08.007

Khan, T.A., Saleem, M., Fariduddin, Q., 2023a. Differential drought stress tolerance in five varieties of tomato (*Solanum lycopersicum* L.): an evaluation of photosynthesis and antioxidant system. *J. Soil Sci. Plant Nutr.* 23, 2810–2831. https://doi.org/10.1007/S42729-023-01236-0/METRICS

Kour, J., Kohli, S.K., Khanna, K., Bakshi, P., Sharma, P., Singh, A.D., Ibrahim, M., Devi, K., Sharma, N., Ohri, P., Skalicky, M., Brestic, M., Bhardwaj, R., Landi, M., Sharma, A., 2021. Brassinosteroid signaling, crosstalk and, physiological functions in plants under heavy metal stress. *Front. Plant Sci.* 12, 608061. https://doi.org/10.3389/FPLS.2021.608061/BIBTEX

Krishna, P., 2003. Brassinosteroid-mediated stress responses. *J. Plant Growth Regul.* 22, 289–297. https://doi.org/10.1007/S00344-003-0058-Z

Kumari, S., Nazir, F., Maheshwari, C., Kaur, H., Gupta, R., Siddique, K.H.M., Khan, M.I.R., 2023. Plant hormones and secondary metabolites under environmental stresses: Shedding light on defense molecules. *Plant Physiol. Biochem.* 108238.

Lalarukh, I., Amjad, S.F., Mansoora, N., Al-Dhumri, S.A., Alshahri, A.H., Almutari, M.M., Alhusayni, Fatimah S., Al-Shammari, W.B., Poczai, P., Abbas, M.H.H., Elghareeb, D., Kubra, K. tul, Abdelhafez, A.A., 2022. Integral effects of brassinosteroids and timber waste biochar enhances the drought tolerance capacity of wheat plant. *Sci. Reports* 121(12), 1–10. https://doi.org/10.1038/s41598-022-16866-0

Li, J., Chory, J., 1997. A putative leucine-rich repeat receptor kinase involved in brassinosteroid signal transduction. *Cell* 90, 929–938. https://doi.org/10.1016/S0092-8674(00)80357-8

Liaqat, S., Umar, S., Saffeullah, P., Iqbal, N., Siddiqi, T.O., Khan, M.I.R., 2020. Protective effect of 24-epibrassinolide on barley plants growing under combined stress of salinity and potassium deficiency. *J. Plant Growth Regul.* 39, 1543–1558.

Lima, J.V., Lobato, A.K.S., 2017. Brassinosteroids improve photosystem II efficiency, gas exchange, antioxidant enzymes and growth of cowpea plants exposed to water deficit. *Physiol. Mol. Biol. Plants* 23, 59–72. https://doi.org/10.1007/S12298-016-0410-Y/TABLES/4

Liu, X., Zhu, Q., Liu, W., Zhang, J., 2023. Exogenous brassinosteroid enhances zinc tolerance by activating the phenylpropanoid biosynthesis pathway in *Citrullus lanatus* L. *Plant Signal. Behav.* 18. https://doi.org/10.1080/15592324.2023.2186640

Madhan, M., Mahesh, K., Seeta, S., Rao, R., 2014. Effect of 24-epibrassinolide on aluminium stress induced inhibition of seed germination and seedling growth of *Cajanus cajan* (L.) Millsp. *Int. J. Multidiscip. Curr. Res.* 2.

Mandava, N.B., 1988. Plant growth-promoting brassinosteroids. *Annu. Rev. Plant Biol.* 39, 23–52. https://doi.org/10.1146/ANNUREV.PP.39.060188.000323

Mitchell, J.W., Mandava, N., Worley, J.F., Plimmer, J.R., Smith, M.V., 1970. Brassins – a new family of plant hormones from rape pollen. *Nature* 2255237(225), 1065–1066. https://doi.org/10.1038/2251065a0

Naveen, N., Kumari, N., Avtar, R., Jattan, M., Ahlawat, S., Rani, B., Malik, K., Sharma, A., Singh, M., 2021. Evaluation of effect of brassinolide in *Brassica juncea* leaves under drought stress in field conditions. *Horticulturae* 7. https://doi.org/10.3390/HORTICULTURAE7110514

Nawaz, F., Naeem, M., Zulfiqar, B., Akram, A., Ashraf, M.Y., Raheel, M., Shabbir, R.N., Hussain, R.A., Anwar, I., Aurangzaib, M., 2017. Understanding brassinosteroid-regulated mechanisms to improve stress tolerance in plants: a critical review. *Environ. Sci. Pollut. Res.* 2419(24), 15959–15975. https://doi.org/10.1007/S11356-017-9163-6

Nazir, F., Fariduddin, Q., Hussain, A., Khan, T.A., 2021. Brassinosteroid and hydrogen peroxide improve photosynthetic machinery, stomatal movement, root morphology and cell viability and reduce Cu-triggered oxidative burst in tomato. *Ecotoxicol. Environ. Saf.* 207, 111081.

Nazir, F., Jahan, B., Iqbal, N., Rajurkar, A.B., Siddiqui, M.H., Khan, M.I.R., 2023a. Methyl jasmonate influences ethylene formation, defense systems, nutrient homeostasis and carbohydrate metabolism to alleviate arsenic-induced stress in rice (*Oryza sativa*). *Plant Physiol. Biochem.* 202, 107990.

Nazir, F., Jahan, B., Kumari, S., Iqbal, N., Albaqami, M., Sofo, A., Khan, M.I.R., 2023b. Brassinosteroid modulates ethylene synthesis and antioxidant metabolism to protect rice (*Oryza sativa*) against heat stress-induced inhibition of source–sink capacity and photosynthetic and growth attributes. *J. Plant Physiol.* 289, 154096.

Nolan, T., Chen, J., Yin, Y., 2017. Cross-talk of brassinosteroid signaling in controlling growth and stress responses. *Biochem. J.* 474, 2641–2661. https://doi.org/10.1042/BCJ20160633

Per, T.S., Khan, N.A., Reddy, P.S., Masood, A., Hasanuzzaman, M., Khan, M.I.R., Anjum, N.A., 2017. Approaches in modulating proline metabolism in plants for salt and drought stress tolerance: phytohormones, mineral nutrients and transgenics. *Plant Physiol. Biochem.* 115, 126–140.

Petzold, U., Peschel, S., Dahse, I., Adam, G., 1992. Stimulation of source-applied 14C-sucrose export in *Vicia faba* plants by brassinosteroids, GA3 and IAA. *Acta Bot. Neerl.* 41, 469–479. https://doi.org/10.1111/J.1438-8677.1992.TB00517.X

Planas-Riverola, A., Gupta, A., Betegoń-Putze, I., Bosch, N., Ibanẹs, M., Cano-Delgado, A.I., 2019. Brassinosteroid signaling in plant development and adaptation to stress. *Development* 146. https://doi.org/10.1242/DEV.151894/48995

Pustovoitova, T.N., Zhdanova, N.E., Zholkevich, V.N., 2001. Epibrassinolide increases plant drought resistance. *Dokl. Biochem. Biophys.* 376, 36–38. https://doi.org/10.1023/A:1018852110393

Rajasekaran, L.R., Blake, T.J., 1999. New plant growth regulators protect photosynthesis and enhance growth under drought of jack pine seedlings. *J. Plant Growth Regul.* 18, 175–181. https://doi.org/10.1007/PL00007067

Rajewska, I., Talarek, M., Bajguz, A., 2016. Brassinosteroids and response of plants to heavy metals action. *Front. Plant Sci.* 7, 197845. https://doi.org/10.3389/FPLS.2016.00629/BIBTEX

Raza, M.A.S., Ibrahim, M.A., Ditta, A., Iqbal, R., Aslam, M.U., Muhammad, F., Ali, S., Çiğ, F., Ali, B., Muhammad Ikram, R., Muzamil, M.N., Rahman, M.H. ur, Alwahibi, M.S., Elshikh, M.S., 2023. Exploring the recuperative potential of brassinosteroids and nano-biochar on growth, physiology, and yield of wheat under drought stress. *Sci. Reports* 131(13), 1–13. https://doi.org/10.1038/s41598-023-42007-2

Rukundo, P., Betaw, H.G., Ngailo, S., Balcha, F., 2014. Assessment of drought stress tolerance in root and tuber crops. *African J. Plant Sci.* 8, 214–224. https://doi.org/10.5897/AJPS2014.1169

Russinova, E., Borst, J.W., Kwaaitaal, M., Caño-Delgado, A., Yin, Y., Chory, J., De Vries, S.C., 2004. Heterodimerization and endocytosis of *Arabidopsis brassinosteroid* receptors BRI1 and AtSERK3 (BAK1). *Plant Cell* 16, 3216–3229. https://doi.org/10.1105/TPC.104.025387

Sahni, S., Prasad, B.D., Liu, Q., Grbic, V., Sharpe, A., Singh, S.P., Krishna, P., 2016. Overexpression of the brassinosteroid biosynthetic gene DWF4 in *Brassica napus* simultaneously increases seed yield and stress tolerance. *Sci. Rep.* 6. https://doi.org/10.1038/SREP28298

Sharma, I., Pati, P.K., Bhardwaj, R., 2011. Effect of 28-homobrassinolide on antioxidant defence system in *Raphanus sativus* L. under chromium toxicity. *Ecotoxicology* 20, 862–874. https://doi.org/10.1007/S10646-011-0650-0

Shi, C., Qi, C., Ren, H., Huang, A., Hei, S., She, X., 2015. Ethylene mediates brassinosteroid-induced stomatal closure via Gα protein-activated hydrogen peroxide and nitric oxide production in Arabidopsis. *Plant J.* 82, 280–301. https://doi.org/10.1111/TPJ.12815

Suhita, D., Raghavendra, A.S., Kwak, J.M., Vavasseur, A., 2004. Cytoplasmic alkalization precedes reactive oxygen species production during methyl jasmonate- and abscisic acid-induced stomatal closure. *Plant Physiol.* 134, 1536–1545. https://doi.org/10.1104/PP.103.032250

Sukpitak, C., Munné-Bosch, S., Seraypheap, K., 2024. Brassinosteroids alleviate postharvest water deficit stress in cut Dendrobium "Khao Sanan" orchid through modulation of protective mechanisms, osmotic regulation, and endogenous hormonal levels. *Postharvest Biol. Technol.* 211, 112832. https://doi.org/10.1016/J.POSTHARVBIO.2024.112832

Sun, Y., Fan, X.Y., Cao, D.M., Tang, W., He, K., Zhu, J.Y., He, J.X., Bai, M.Y., Zhu, S., Oh, E., Patil, S., Kim, T.W., Ji, H., Wong, W.H., Rhee, S.Y., Wang, Z.Y., 2010. Integration of brassinosteroid signal transduction with the transcription network for plant growth regulation in arabidopsis. *Dev. Cell* 19, 765–777. https://doi.org/10.1016/J.DEVCEL.2010.10.010

Sun, Y., Han, Z., Tang, J., Hu, Z., Chai, C., Zhou, B., Chai, J., 2013. Structure reveals that BAK1 as a co-receptor recognizes the BRI1-bound brassinolide. *Cell Res.* 2311(23), 1326–1329. https://doi.org/10.1038/cr.2013.131

Todorova, D., Talaat, N.B., Katerova, Z., Alexieva, V., Shawky, B.T., 2016. Polyamines and brassinosteroids in drought stress responses and tolerance in plants. In *Water Stress Crop Plants A Sustain. Approach*, Vol. 2, pp. 608–627. https://doi.org/10.1002/9781119054450.CH35

Trevor Nolan, A.M., Brennan, B., Yang, M., Bassham, D.C., Walley, J., Yin Correspondence, Y., 2017. Selective autophagy of BES1 mediated by DSK2 balances plant growth and survival. *Dev. Cell* 41, 33–46.e7. https://doi.org/10.1016/j.devcel.2017.03.013

Vardhini, B.V., Anjum, N.A., 2015. Brassinosteroids make plant life easier under abiotic stresses mainly by modulating major components of antioxidant defense system. *Front. Environ. Sci.* 2, 128493. https://doi.org/10.3389/FENVS.2014.00067/BIBTEX

Vriet, C., Eugenia, R., Reuzeau, C., 2013. From squalene to brassinolide: the steroid metabolic and signaling pathways across the plant kingdom. *Mol. Plant* 6, 1738–1757. https://doi.org/10.1093/MP/SST096

Wang, Z.Y., Nakano, T., Gendron, J., He, J., Chen, M., Vafeados, D., Yang, Y., Fujioka, S., Yoshida, S., Asami, T., Chory, J., 2002. Nuclear-localized BZR1 mediates brassinosteroid-induced growth and feedback suppression of brassinosteroid biosynthesis. *Dev. Cell* 2, 505–513. https://doi.org/10.1016/S1534-5807(02)00153-3

Wu, C., Li, F., Xu, H., Zeng, W., Yu, R., Wu, X., Shen, L., Liu, Y., Li, J., 2019. The potential role of brassinosteroids (BRs) in alleviating antimony (Sb) stress in *Arabidopsis thaliana*. *Plant Physiol. Biochem.* 141, 51–59. https://doi.org/10.1016/J.PLAPHY.2019.05.011

Xia, X.J., Gao, C.J., Song, L.X., Zhou, Y.H., Shi, K., Yu, J.Q., 2014. Role of H_2O_2 dynamics in brassinosteroid-induced stomatal closure and opening in *Solanum lycopersicum*. *Plant. Cell Environ.* 37, 2036–2050. https://doi.org/10.1111/PCE.12275

Xia, X.J., Huang, L.F., Zhou, Y.H., Mao, W.H., Shi, K., Wu, J.X., Asami, T., Chen, Z., Yu, J.Q., 2009. Brassinosteroids promote photosynthesis and growth by enhancing activation of Rubisco and expression of photosynthetic genes in *Cucumis sativus*. *Planta* 230, 1185–1196. https://doi.org/10.1007/S00425-009-1016-1/METRICS

Xia, X.J., Zhou, Y.H., Shi, K., Zhou, J., Foyer, C.H., Yu, J.Q., 2015. Interplay between reactive oxygen species and hormones in the control of plant development and stress tolerance. *J. Exp. Bot.* 66, 2839–2856. https://doi.org/10.1093/JXB/ERV089

Ye, H., Liu, S., Tang, B., Chen, J., Xie, Z., Nolan, T.M., Jiang, H., Guo, H., Lin, H.Y., Li, L., Wang, Y., Tong, H., Zhang, M., Chu, C., Li, Z., Aluru, M., Aluru, S., Schnable, P.S., Yin, Y., 2017. RD26 mediates crosstalk between drought and brassinosteroid signalling pathways. *Nat. Commun.* 8. https://doi.org/10.1038/NCOMMS14573

Yin, Y., Wang, Z.Y., Mora-Garcia, S., Li, J., Yoshida, S., Asami, T., Chory, J., 2002. BES1 accumulates in the nucleus in response to brassinosteroids to regulate gene expression and promote stem elongation. *Cell* 109, 181–191. https://doi.org/10.1016/S0092-8674(02)00721-3

Yin, Y.L., Zhou, Y., Zhou, Y.H., Shi, K., Zhou, J., Yu, Y., Yu, J.Q., Xia, X.J., 2016. Interplay between mitogen-activated protein kinase and nitric oxide in brassinosteroid-induced pesticide metabolism in *Solanum lycopersicum*. *J. Hazard. Mater.* 316, 221–231. https://doi.org/10.1016/J.JHAZMAT.2016.04.070

Yoshida, T., Mogami, J., Yamaguchi-Shinozaki, K., 2014. ABA-dependent and ABA-independent signaling in response to osmotic stress in plants. *Curr. Opin. Plant Biol.* 21, 133–139. https://doi.org/10.1016/J.PBI.2014.07.009

Yuan, G.F., Jia, C.G., Li, Z., Sun, B., Zhang, L.P., Liu, N., Wang, Q.M., 2010. Effect of brassinosteroids on drought resistance and abscisic acid concentration in tomato under water stress. *Sci. Hortic. (Amsterdam).* 126, 103–108. https://doi.org/10.1016/J.SCIENTA.2010.06.014

Yusuf, M., Fariduddin, Q., Ahmad, A., 2011. 28-Homobrassinolide mitigates boron induced toxicity through enhanced antioxidant system in Vigna radiata plants. *Chemosphere* 85, 1574–1584. https://doi.org/10.1016/J.CHEMOSPHERE.2011.08.004

Zhu, Y., Wang, B., Tang, K., Hsu, C.C., Xie, S., Du, H., Yang, Y., Tao, W.A., Zhu, J.K., 2017. An *Arabidopsis nucleoporin* NUP85 modulates plant responses to ABA and salt stress. *PLOS Genet.* 13, e1007124. https://doi.org/10.1371/JOURNAL.PGEN.1007124

Zullo, M.A.T., Kohout, L., 2004. Semisystematic nomenclature of brassinosteroids. *Plant Growth Regul.* 42, 15–28. https://doi.org/10.1023/B:GROW.0000014898.30414.33/METRICS

6 Current Prospects in Enhancing Productivity and Nutritional Security of Crop Plants Exposed to Salt Stress

Rohit Chhabra and Aditi Bisht

6.1 INTRODUCTION

Salinity, an abiotic stressor, imposes constraints on crop productivity through the escalation of salinization in cultivable land, leading to an anticipated reduction of 30% incrop yield by the year 2050 (Ebrahimi and Bhatla 2011; Kumari et al. 2023, 2024). According to Tuteja (2007), high salinity stress is a significant environmental stressor that negatively affects crop productivity on a minimum of 20% of irrigated land globally. It is estimated that almost 30% of the global food supply is cultivated on land that is irrigated (Schultz et al. 2005). Approximately 10% of the earth's land surface, equivalent to 20.3×108 hectares, is influenced by salinity. Soil salinization not only results in elevated salt levels within the soil, but also renders the land unsuitable for agricultural purposes. The problem of salt stress is prevalent throughout various continents and diverse climatic conditions. In contrast, the problem of salt stress is more commonly observed in arid and semi-arid regions than in humid climates, when the annual precipitation is insufficient to offset evapotranspiration rates (Corwin 2021). Salinity stressors impact approximately 12,781 million hectares (Mha) of land on a global scale (Schultz et al. 2005). It is noteworthy to observe that a significant proportion of the whole world land area, approximately 85%, experiences only mild to moderate levels of salinity, but the remaining 15% faces severe to extreme constraints on agricultural productivity (Shelden et al. 2016). The nations in which notable salt-affected soils are present encompass, among others, India (27%), Pakistan (28%), Israel (13%), Australia (20%), China (15%), Iraq (50%), and Egypt (30%). According to Corwin (2021), a significant number of countries, specifically 75, have been identified as experiencing issues related to salt (Vij and Tyagi 2007). Salt-affected soils play a crucial role in the ecological composition of arid or semi-arid regions (Vij and Tyagi 2007). In the Indian context, it is observed that the extent of salt-affected soils covers around 6.73 million hectares (Mha). Notably, the states of Gujarat, Uttar Pradesh, Maharashtra, West Bengal, and Rajasthan collectively contribute to over 75% of the overall saline and sodic soil area in the country. Specifically, Gujarat accounts for 2.23 Mha, Uttar Pradesh for 1.37 Mha, Maharashtra for 0.61 Mha, West Bengal for 0.44 Mha, and Rajasthan for 0.38 Mha. The majority of salt-affected habitats have a high occurrence of water with low quality, characterized by salinity and sodicity (Hao et al. 2021). The north-western dry regions of Rajasthan, Haryana, and Punjab in India are significantly affected by the issue of poor-quality water. Based on estimations, it is projected that the current extent of salt-affected soils in the country, which is around 6.73 million hectares, will nearly triple to reach 20 million hectares by the year 2050 (Gu et al. 2004).

The issue of water of substandard quality is expected to escalate in the next few years as a result of projected expansion in irrigated regions and the extensive utilization of natural resources to meet the food and other essential needs of a burgeoning population (Vij and Tyagi 2007). Salinity

DOI: 10.1201/9781032630380-6

stress can induce alterations in plant morphology, physiology, and metabolism, exhibiting potential variability across distinct cultivars of the identical crop species (Wang et al. 2009; Kumari et al. 2023). Plants exhibit various phenological, physiological, and biochemical adaptations in response to salinity, including modifications in osmotic effects, specific ion toxicity, and nutritional problems (De Leon et al. 2017). Salinity, as a prevailing factor, induces water and ion imbalances, disrupts the integrity of the cell membrane, impairs the capacity to detoxify reactive oxygen species (ROS), diminishes the activity of antioxidant enzymes, and suppresses photosynthetic activity (Wang et al. 2009; Kumari et al. 2024). The creation of ROS as a result of salinity exposure gives rise to oxidative stress, which adversely affects macromolecules including DNA, proteins, and lipids, hence disrupting crucial cellular processes in plants. The exposure to high levels of salt induces both oxidative and osmotic stress, thus impacting various fundamental physiological processes including germination, growth, photosynthetic pigment production, photosynthesis, protein synthesis, nucleic acid synthesis, lipid metabolism, and secondary metabolite synthesis (Allakhverdiev et al. 2002). Enhancing the salt tolerance of crops is of utmost significance in a world affected by salinization, characterized by escalating populations, diminishing crop productivity, and a reduction in arable lands (Yadav et al. 2011).

A range of techniques are employed to mitigate the adverse impacts of salt stress on plant physiology (De Leon et al. 2017). These strategies encompass traditional breeding techniques that have facilitated the advancement of crops with enhanced tolerance to salinity (Ondrasek et al. 2011). The application of several chemical compounds and microbial inoculants, such as plant growth-promoting rhizobacteria and arbuscular mycorrhizal fungi, has been found to have the potential to alleviate the negative impacts of salt stress. The reduction of salinity stress can be facilitated by the exogenous application of certain plant growth substances like phytohormones, signalling molecules, amino acids, and sugars (Hasanuzzaman et al. 2020; Nazir et al. 2024). Due to the inherent constraints associated with existing methodologies, there is a clear need for an alternate approach that offers economic viability and cost-effectiveness, hence rendering it suitable for adoption by farmers. Thus, this review may provide valuable insights for interdisciplinary investigations aimed at evaluating the ecological implications of salt stress (Parida and Das 2005). The review also sheds light on the various aspects of stress at physiological and biochemical processes of the host plant and the perplexing nature of plant stress responses. Additionally, the study highlighted the key aspects of salinity stress and its consequences on plant physiology and compiled extensive research findings on cellular, metabolic, and genetic mechanisms involved in the response to salt stress.

6.2 EFFECT OF SALINITY ON MORPHO-PHYSIOLOGICAL TRAITS

6.2.1 Seed Germination and Vigour

Seed germination is one of the critical vital phases in the plant's life cycle, ultimately determining yield. However, this phase is highly vulnerable to salt stress. Salinity is a major stress responsible for the delay in germination and the reduction in seed germination percentage (Ondrasek et al. 2011). High salt levels in the soil cause a reduction in the osmotic potential of soil water, which ultimately reduces water uptake by dry seeds. Also, the uptake of excessive Na^+ and Cl^- ions causes ionic stress and ion toxicity, disrupting critical biochemical processes like nucleic acid and protein metabolism, energy production, and respiration. Salinity decreases water potential, protein synthesis, and reserved food in the germinating seeds of many plants, such as broccoli and cauliflower (Park et al. 2016). Salinity causes hormonal imbalance, especially in abscisic acid and gibberellin, during the seed germination phase, resulting in delayed germination and sometimes even inhibition of seed germination. Salinity enhances the production of reactive oxygen species (Hao et al. 2021). At the same time, reduced ROS scavenging damages essential macromolecules such as proteins, nucleic acids, lipids, and carbohydrates in the cells, resulting in the inhibition of the seed germination process (Farooq et al. 2015). Seed vigor is a physiological property that governs the ability of a seed to produce a seedling rapidly in soil and the extent to which that seed tolerates a range of environmental factors (Chhabra et al. 2019). It is assumed that seeds with high vigor are more tolerant to environmental stress. Salinity not only causes ionic and osmotic stress but also oxidative stress. The higher the germination velocity of a seed in less time, the better the quality and the higher the seed vigor (Wang et al. 2010). The relationship between various seedling growth parameters and seed yield and yield components under saline conditions has been reported in rice by Wang et al. (2010).

6.2.2 Seedling Length and Biomass

Decreased seedling length may be attributed to a decrease in water absorption in the presence of sodium chloride, followed by hydrolysis of the limited food reserves of the storage tissues, as well as the deficient translocation of these storage reserve tissues for the development of the embryonic axis (Khan and Panda 2008). The length of the root and shoot decreases with the increase in salinity level, which may be due to the plants slowing down their water absorption (Kaya et al. 2003). During salinity stress, less affected roots provide sufficient water and nutrients. Better root growth is an essential adaptive response in crops (Shelden et al. 2016). Radicle length was more suppressed than plumule by salinity at all salt concentration levels in different rice cultivars (Anbumalarmathi and Mehta 2013). A decrease in root length and shoot length in *Brassica napus* may be due to salt toxicity and a disproportion in nutrient absorption by the seedling, as suggested by Bybordi and Tabatabaei (2009). The increased NaCl concentration markedly reduced the leaf thickness and diameter of vessels, thus reducing the lamina length as seen in sugar beet genotypes (Skorupa et al. 2019). The reduction in shoot and root lengths is due to decreased physiological activities resulting from water and nutrient stress occurring under high salinity stress. These processes may cause the shoot to remain small. Salinity leads to disturbances in plant metabolism, which consequently leads to a reduction in plant growth and productivity (Shafi et al. 2009).

An increase in salinity creates a stressful situation and evokes the antioxidant defence system of plants. High salt concentrations in nutrient mediums cause stunted growth in plants, which might be due to a reduction in the rate of leaf surface expansion and a decrease in the fresh weight of the stem in a well-grown plant (Chartzoulakis and Klapaki 2000; Takemura et al. 2002). Salt stress has an inhibitory effect on fresh and dry weight production. Reduction in fresh weights in response to salt stress has been reported for crops such as pepper (Chartzoulakis and Klapaki 2000), melon (Sivritepe et al. 2005) and tomato (Yurtseven et al. 2005). NaCl-imposed stress in wheat significantly reduced dry weight by reducing shoot and root dry weights in wheat genotypes (Li et al. 2011). Mahmood and Athar (2003) suggested that as salt concentration increases in the medium, *Panicum turgidum* absorbs less water, causing physiological desiccation of plants. This type of restriction on water absorption might be responsible for the decrease in fresh weight. When these conditions were continuous, weight loss occurred in bean plants (Shabala et al. 2000), as noticed in all genotypes and at all concentration levels. A reduction in fresh weight may be attributed to a decrease in the growth and development of plants caused by reduced uptake of mineral nutrients dissolved in water and thus altered metabolism. A decline in seedling fresh weight with increasing salinity has been reported in maize (Hoque et al. 2015).

6.2.3 Photosynthetic Attributes

Plants produce their food through one of the important metabolic processes, photosynthesis, which helps accumulate dry matter and productivity (Maggio et al. 2007). Salt stress disrupts and limits photosynthesis either by stomatal closure, leading to reduced intercellular carbon dioxide concentration, or by non-stomatal limitations (Ran et al. 2021). Short-term salt effects on photosynthesis are visible within a few hours to one or two days, stopping plant carbon assimilation. The long-term effect of salt on photosynthesis occurs after a few days (Parida and Das 2005). Salt stress affects the ultrastructure of a chloroplast. Disruption of thylakoid structure, increment in size and number of pellets of the plastid, decreased amount of starch, aggregation of chloroplast without stroma, and grana were evident under salt stress (El-Banna et al. 2022). Many factors reduce photosynthesis under salinity: dehydration of membranes causing a reduction in CO_2 permeability; osmotic stress leading to a decrease in the osmotic potential of water in the soil that limits the availability of water to plants; disruption of the photosynthetic electron transport system; and changes in enzyme activities (Nawaz et al. 2020). Also, the ion toxicity of salts reduces the uptake of essential nutrients, limiting photosynthesis and generating reactive oxygen species. Salinity affects receptors of photosystem II (PSII) of light-dependent reactions, which reduces the density of active reaction centres (Nakhaie et al. 2022). Plant pigments harvest sunlight and offer photosynthetic protection in the light reaction of photosynthesis (Friedland et al. 2019). Leaf photosynthetic pigments are vital for photosynthesis and gross primary production. The chlorophyll content represents the plant's photosynthetic capacity (Shah et al. 2017). Carotenoids composed of carotenes and xanthophylls are other major photosynthetic pigments. Carotenoids are an essential structural component of a photosynthetic antenna that harvests light energy (Sudhir and Murthy 2004). Besides contributing to the photosynthesis process, carotenoids act as antioxidants and are demonstrated to be a scavenger of ROS (Allakhverdiev et al. 2002). Carotenoids also protect reaction centres by dissipating excess light energy absorbed during photosynthesis (Kao et al. 2003). The chlorophyll a, chlorophyll b, total chlorophyll, and carotenoids were reduced under salinity in salt-sensitive lines of *Medicago truncatula* (Najar et al. 2019). Over-accumulation of Na^+ ions in leaf tissue causes changes in chlorophyll pigments and limits their synthesis, leading to leaf chlorosis in soybean plants (Sadak et al. 2020).

6.2.4 Effect of Salinity on Oxidative Stress and Membrane Damage

Plants constantly produce reactive oxygen species during photosynthesis and respiration, which can cause oxidative damage to cells (Zhao et al. 2021). Thus, plants must scavenge these ROS to protect plant cells from oxidative damage. Plants have two ROS-scavenging mechanisms involving enzymatic and non-enzymatic antioxidants (Bohnert et al. 2001). Under natural conditions, there is a balance between ROS production and its scavenging in plant cells. However, this balance is disturbed under different environmental stresses, resulting in oxidative stress (Hussain et al. 2013). ROS are oxygen radicals or their derivatives such as hydroxyl radical, superoxide, singlet oxygen, hydrogen peroxide, ozone, hypoiodous acid, hypochlorous acid, hypobromous acid, perhydroxy radical, peroxyl, peroxy radical, semiquinone, and carbonate (Hasanuzzaman et al. 2020). Under extreme environmental conditions such as salinity, ROS are produced in different cell organelles such as the chloroplast, mitochondria, endoplasmic reticulum, peroxisomes, plasma membrane, and cell wall, damaging the cell membrane and cell organelles. Salt stress induces osmotic and ionic stress and an imbalance of nutrients, and the associated secondary effects generate excess ROS. Salinity-induced ROS are toxic and can cause detrimental effects on lipids, proteins, and nucleic acids. ROS induce changes in these biomolecules' functions, resulting in physiological and biochemical alterations in cells, ultimately leading to oxidative stress (Hasanuzzaman et al. 2020).

Plant membranes are an important biological barrier for plant cells, protecting the internal contents of cells and organelles from abiotic and biotic stresses. These membranes are made of lipids and proteins and play an essential role in transporting substances, transmitting energy, and transducing signals (Momeni et al. 2021). Additionally, selective permeability of cell membranes for ions helps regulate ion homeostasis inside a cell. Salinity-induced changes in the structure and functions of membrane lipids are observed in both halophytes and glycophytes. Salinity-induced ROS overproduction causes oxidative stress by increasing relative permeability and decreasing the fluidity of membranes (Acosta-Motos et al. 2017). It also affects selectivity, flow rate, and transportation of ions and alters the properties of membrane proteins, signalling molecules, and signal transduction. Moreover, it also causes osmotic stress due to the exosmosis of many electrolytes. Malondialdehyde (MDA), one of the leading lipid peroxidation products, can alter and inactivate the proteins and enzymes present on the plasma membrane, thereby altering membrane structure and function. MDA content is also the primary indicator of plasma membrane damage under salinity (Hussain et al. 2013). Momeni et al. (2023) have reported significantly increased MDA content in salt-sensitive durum wheat genotypes compared to salt-tolerant types. ROS accumulation, membrane damage, and imbalanced ion homeostasis hamper the rate of protein synthesis, resulting in the accumulation of amino acids and toxic substances inside the cell. Amino acids such as isoleucine, ornithine, and arginine are converted to toxic polyamines such as butane diamine and glutamine. The accumulation of these toxic substances is very harmful to the growth and development of plants (Maggio et al. 2007).

6.2.5 Effect of Salinity on Crop Yield and Productivity

About 30 crop species, such as rice, sorghum, barley, maize, wheat, sugarcane, potato, sugar beet, soybean, sweet potato, cotton, and mungbean, provide 90% of plant-based human food. At moderate salinity conditions (EC 4–8 dS/m), 50%–80% yield loss was observed in these crops (Shavrukov 2013). If salinity increases in the initial phase of the plant‘s life cycle, plant growth, development, yield, and production are significantly reduced, compromising the quality and quantity of the harvest. Loss of crop yield due to salinity has an economic impact on agriculture and bio-based industries (Allbed and Kumar 2013). It has been reported that soil salinity in irrigated agricultural lands causes a loss of over 27 billion US dollars per year. Various studies have reported differences in yield loss at different salinity levels. At 8 to 10 dS/m salinity, 55%, 28%, and 15% yield loss were observed in corn, wheat, and cotton. The 18 ds/m salinity causes a 55% yield loss in cotton. In mungbean, pod parameters such as the number of pods per plant, average pod length, fresh and dry weight of pods per plant, and seed parameters such as the number of seeds per plant, the weight of 100 seeds, and seed yield per plant decreased with increasing salinity (Hussain et al. 2013). Less-green leaves, poor leaf expansion, a smaller number of leaves, and leaf senescence result in less photosynthetic activity, which may be the reason behind the reduction in yield under salinity. Attributes like spike length, the number of spikelets, spike/1000 grain weight, and yield in wheat were reduced with increasing salt concentration, and such an immediate reduction in yield under salinity is due to osmotic and ionic stress (Acosta-Motos et al. 2017).

6.3 PLANT INHERENT TOLERANCE MECHANISMS

The term "salinity tolerance" can be defined as "an inherent capability of host plants to withstand the exposure of high concentrations of salts in the rhizosphere or in the leaves." On the basis of their inhabitants, plants can be categorized into two groups: halophytes (which grow in saline conditions) and glycophytes (which grow in non-saline habitats). In order to overcome the negative impacts of salinity, plants develop various physiological and biochemical strategies, which are discussed as follows.

6.3.1 Ion Homeostasis and Compartmentalization

The maintenance of ion homeostasis by the uptake and compartmentalization of ions is essential and crucial for the normal growth of plants in salinity-stressed environments. Both halophytes and glycophytes cannot tolerate a higher salt level in their cytoplasm; hence, the surplus amount of salt is either compartmentalized into the vacuole or transported to the older plant tissues, which ultimately are sacrificed (Bressan et al. 1998). Some ions, like nitrogen, potassium, and calcium, are essential for the growth and development of the plant. Although these ions are present in the soil, they cannot enter plant cells due to competition with other ions with high concentrations. In plants, ionic homeostasis is crucially essential for growth and development associated responses (Jouyban 2012). Cell membranes play a vital role in maintaining a low concentration and the transportation of ions. Ions are transported inside plants through membrane proteins such as ion channel proteins, symporters, and antiporters on the cell membrane. NaCl is the main salt present in saline soil; hence, the study of the exclusion of Na^+ from roots, the long-distance Na^+ transport mechanism, and the compartmentalization of Na^+ ions is a crucial area of research to understand ion homeostasis at the tissue and cell level. Na^+ toxicity in a cell is controlled through Na^+ and H^+ antiporters on the vacuolar membrane. These antiporters help sequester excess Na^+ in the vacuole under salinity. Vacuolar Na^+ sequestration prevents the increase of Na^+ and maintains the Na^+/K^+ ratio in the cytosol. Two main antiporters are present on vacuolar membranes: vacuolar-type H^+-ATPase (also called V-ATPase) and vacuolar pyrophosphate (V-PPase) (Almeida et al. 2017). NHX1 Na^+, K^+/H^+ exchanger, Na^+, and K^+/H^+ exchanger are a few of the best-known transporters for Na^+ sequestration in vacuoles. The salt overly sensitive (SOS) signalling pathway maintains ion homeostasis under salinity (Park et al. 2016). It comprises plasma membrane SOS3, SOS2, and SOS1-like Na^+ and H^+ exchangers. These SOS exchangers extrude Na^+ from the root cells to maintain ion homeostasis and increase salt tolerance. Under normal conditions, K^+ concentrations are maintained up to 100 mM in the cell cytoplasm. Maintaining this cytosolic K^+ concentration for plant growth and development is crucial (Zhang et al. 2017). However, this K^+ homeostasis is disturbed under salinity, altering critical physiological processes dependent on K^+ concentration. Under salinity, Na^+ concentration is increased in soil; hence, because of similar charges, Na^+ competes with K^+ for the same transporter, which ultimately reduces K^+ uptake (Hasanuzzaman et al. 2020). It is well known that Ca^{2+} also helps during salinity stress, which reduces the toxic effect of NaCl stress by facilitating higher K^+ and Na^+ selectivity. Cytosolic Ca^{2+} increases under salt stress and is transported from the apoplast and intracellular compartments. The resultant increase in Ca^{2+} initiates signal transduction, which reduces salt stress on plant growth and development (Zhang et al. 2017).

6.3.2 Organic Solutes

Plants accumulate many cellular osmoprotectants under stressful environmental conditions, which play an essential role in plant defence mechanisms. These osmolytes are small, organic, low-molecular-weight, electrically neutral, and highly water-soluble compounds. Moreover, osmoprotectants are nontoxic to plants at high concentrations under unfavourable environmental conditions due to their involvement in intracellular metabolisms operating under stress conditions (Hasanuzzaman et al. 2020). Three main types of osmolytes are recognized: (a) amino acids (proline, gamma-aminobutyric acid (GABA), hydroxyproline, pipecolic acid, and polyamines); (b) betaines and associated molecules (glycine betaine, proline betaine, and alanine betaine); and (c) sugars (sucrose, trehalose, sorbitol, fructan, and raffinose) and polyols (mannitol and inositol) (Hasanuzzaman et al. 2020). These osmolytes accumulate during stress conditions for the cell's survival and maintain the osmotic balance between the cell's cytoplasm and surroundings. Besides osmotic osmoregulation, osmoprotectants also contribute to scavenging reactive oxygen species, stabilizing enzymes or proteins, protecting membrane integrity, maintaining cellular redox balance and cell turgor pressure, replacing inorganic ions, mitigating ion toxicity, and protecting cellular components. Moreover, osmoprotectants regulate protein folding, photosynthetic upregulation, and activation of defence-related genes under various stress conditions (Butt et al. 2016).

6.3.2.1 Proline

Proline is one of the significant osmolytes accumulated in the cytoplasm under salinity. It has a strong ability to hydrate. Its hydrophilic end can bind to the water molecule, while its hydrophobic ends binds to proteins. Proteins attached to proline can bind more water molecules, thus protecting proteins from dehydration and degradation under stress (El Moukhtari et al. 2020; Khan et al. 2015, 2020). Proline synthesis increases and its degradation decreases under a stressful environment. It acts as a signalling molecule that contributes to stabilizing membranes, proteins, and enzymes. Proline upregulates membrane proteins to maintain membrane integrity, scavenges ROS, and maintains the balance of ion homeostasis under salinity. Many studies have shown that proline increases water uptake, enhances the activity of antioxidant machinery, and reduces the accumulation of toxic ions under salinity. Proline also acts as a molecular chaperone that can maintain protein folding and confirmation. It can also buffer cytosolic pH and balance the redox condition (Khedr et al. 2003). Proline can mitigate the effect of stress in two ways: (a) by accumulating an excess of proline under stress by upregulating its biosynthesis, which can serve as an osmolyte, chaperone, and direct scavenger of ROS, and (b) by activating the metabolic flux of proline, which is linked to other mechanisms in plants. This proline metabolic flux helps maintain cellular energy and the NADP+/NADPH balance, and it activates signal transduction that promotes cell survival (Sui et al. 2015).

6.3.2.2 Glycine Betaine

Glycine betaine, a quaternary ammonium compound, mitigates the negative effects of different abiotic stresses in animals, fungi, algae, cyanobacteria, bacteria, and members of the Poaceae and Chenopodiaceae families. Most importantly, as an osmolyte under salinity, it protects higher plants from stress damage (Ashraf and Foolad 2007; Khan et al. 2012). Glycine betaine has a unique structure that can bind with the hydrophilic and hydrophobic ends of enzymes and proteins in plants. Thus, it prevents enzymes and proteins from denaturing, retains membrane integrity and the osmotic balance, and scavenges ROS under stress conditions (Ali et al. 2020). In flowering plants, glycine betaine is synthesized in the chloroplast, which helps protect thylakoid membranes, enzymes, and proteins in photosynthetic machinery under stress conditions (Ali et al. 2020). Glycine betaine protects against membrane damage and increases the antioxidant defence system, enhancing photosynthetic activity and plant growth. Glycine betaine regulates the osmotic balance in stressed plants and helps ion transporters maintain normal functioning. Hence, glycine betaine is considered a protective compound that discriminates Na^+ against K^+ under salinity. Glycine betaine is thought to increase the efficiency of vacuoles found in root cells to accumulate and store more Na^+ ions under salinity (Hu and Schmidhalter 2004).

6.3.2.3 Amino Acids

It is important for plants to be able to handle high salt levels, and free amino acids play a big part in this. Protein amino acids like glycine, alanine, proline, isoleucine, leucine, valine, arginine, glutamine, and asparagine, and non-protein amino acids like pipecolic acid, citrulline, ornithine, and gamma-aminobutyric acid, build up in cells when they are stressed and dry out (Khedr et al. 2003). Proline is one of the major amino acids that increase under salinity (Watanabe et al. 2000).

6.3.2.4 Sugars and Sugar Alcohols

Plants undergo dehydration or desiccation under salinity due to osmotic stress. Sugar and sugar alcohols maintain the osmotic equilibrium of cells under dehydration (Lv et al. 2019). Plants synthesize and accumulate various monosaccharides, disaccharides, and other sugars such as glucose, fructose, sucrose, fructans, and starch to minimize salinity-induced osmotic stress. Sugar molecules replace their hydroxyl group with water molecules under dehydration and maintain a hydrophilic structure in their hydrated orientation (Liu and Zhu 1997). Under severe desiccation, these sugars may substitute for water-bound macromolecules, thereby maintaining hydrogen bands. Thus, they prevent protein folding and membrane disturbance. Plants also synthesize sugar alcohols such as mannitol, sorbitol, and inositol under salinity stress to improve salt tolerance through osmotic adjustment. They reduce stress by enhancing growth, scavenging ROS, maintaining cell turgor, and sequestering sodium inside vacuoles (Dkhil and Denden 2010).

6.3.3 Modulation of the Antioxidant Defence System

Plants produce an excess of ROS under salinity, which can damage the cell. Hence, to scavenge ROS, plants have adapted an antioxidant defence mechanism (Ahmad et al. 2021a, 2021b; Kaur et al. 2022). This mechanism mainly consists of two types of antioxidants: (1) enzymatic antioxidants such as catalase, ascorbate peroxidase, glutathione peroxidase, superoxide dismutase, monodehydroascorbate reductase, dehydroascorbate reductase, glutathione S-transferase, and glutathione reductase and (2) non-enzymatic antioxidants such as carotenoids, alkaloids, flavonoids, glutathione, ascorbic acid, and α-tocopherol (Serrano et al. 1999).

6.3.3.1 Enzymatic Antioxidants

Superoxide dismutase (SOD) is plants' most critical and effective omnipresent metal-containing enzymatic antioxidant. It forms the first line of defence against oxidative stress and thus increases plant stress tolerance (Temizgul et al. 2016). SOD dismutates the superoxide radical into hydrogen peroxide and oxygen. SOD has three types of isozymes based on SOD-binding metal ions: (a) Cu/Zn-SOD localized in chloroplast, cytoplasm, and peroxisomes; (b) Fe-SOD localized in chloroplast; and (c) Mn-SOD localized in mitochondria. Butt et al. (2021) observed increased SOD activity in chilli plants under salinity. Catalase (CAT) was the first enzyme discovered to play a role in the antioxidant defence system. In angiosperms, three CAT genes are studied that express isozymes: CAT1, CAT2, and CAT3. These isozymes are mainly found in peroxisomes and glyoxysomes and were elevated under salinity (Li et al. 2019). Peroxidase (POD) is a family of isozymes, heme-containing monomeric glycoproteins, that oxidize various molecules by utilizing hydrogen peroxide. It acts as an electron acceptor and is converted into $2H_2O$. POD is localized in the cell wall, cytoplasm, vacuoles, and organelles of plant cells.

Ascorbate peroxidase (APX) has a higher affinity for hydrogen peroxide than SOD and CAT. It is an integral part of the ascorbic acid (AsA) and glutathione (GSH) cycles. In an APX-catalyzed reaction, ascorbic acid acts as a reducing agent, which reduces H_2O_2 to $2H_2O$. There are four isozymes of APX found in different organelles: the cytosol, mitochondria, peroxisome thylakoid, and the stroma of plant cells (Mushtaq et al. 2023). Al Kharusi et al. (2019) observed an increase in APX activity in salt-tolerant compared to salt-susceptible date palm varieties under salinity.

Other antioxidant enzymes, such as monodehydroascorbate reductase (MDHAR), dehydroascorbate reductase (DHAR), and glutathione reductase (GR), are not directly involved in ROS scavenging. However, they are involved in AsA and GSH regeneration in the AsA and GSH cycles. This AsA-GSH cycle ultimately detoxifies hydrogen peroxide and reduces oxidative damage (Hasanuzzaman et al. 2019). Glutathione peroxidase (GPX) protects the cell from oxidative stress by using glutathione thioperoxide with the help of glutathione S-transferase (GST) (Agarwal et al. 2010).

6.3.3.2 Non-Enzymatic Antioxidants

Ascorbic acid, or vitamin C, is a low-molecular-weight, water-soluble antioxidant molecule and a major substrate for detoxifying ROS under stress and non-stress conditions. It is one of the most potent antioxidants due to its regenerative nature (Akram et al. 2017). AsA acts like a co-enzyme that donates electrons to scavenge ROS. It also regenerates vitamin E, which acts as an antioxidant (Hasanuzzaman et al. 2019). The AsA-GSH pair regulates many developmental processes in plants by manipulating oxidative metabolism (Akram et al. 2017). Noreen et al. (2021) reported improvement in barley's antioxidant defence system after foliar fertigation of ascorbic acid under salinity. Glutathione (γ-glutamyl-cysteinyl-glycine) is a low-molecular-weight, water-soluble thiol compound. It plays a vital role in regulating intracellular defence through direct or indirect detoxification of ROS (Hasanuzzaman et al. 2019). It also plays a significant role in the AsA-GSH cycle, detoxifying ROS (Hasanuzzaman et al. 2019). Al Kharusi et al. (2019) reported enhanced glutathione accumulation under salinity in a salt-tolerant date palm cultivar compared to a salt-sensitive one.

Carotene is a lipid-soluble antioxidant pigment present in chloroplasts. It primarily acts as a light-harvesting accessory pigment and is also known to detoxify ROS in plants. Under stressed conditions, carotenoids react with triplet chlorophyll (3Chl*) or activated chlorophyll molecules, thus preventing the formation of singlet oxygen and other ROS and thus protecting the photosynthetic apparatus. Ben-Abdallah et al. (2019) reported increased carotenoid content in *Solanum villosum* at 100 mM salt stress. Moreover, they found enhanced expressions of genes related to carotenoid production, such as phytoene synthase 1, phytoene synthase 2, and beta-lycopene cyclase. Alpha-tocopherol (vitamin E) is also known to scavenge the ROS derived from photosynthesis, mainly singlet oxygen and hydroxyl radicle. It scavenges lipid peroxidation radicals and prevents lipid peroxidation in the thylakoid membrane, thus protecting plants from oxidative damage under stress. Its level is regulated by plant sensitivity and the severity of stress (Munné-Bosch 2005). Other non-enzymatic antioxidants, such as flavonoid, mannitol, and proline, also help scavenge ROS in plants (Gill and Tuteja 2010).

6.4 STRATEGIES TO OVERCOME SALINITY STRESS

Various methods are used to overcome the negative effects of salt stress in plants. These include conventional breeding methods that led to the development of salt-tolerant crops. The use of various chemicals, microbial inoculants such as plant growth-promoting rhizobacteria (PGPR), arbuscular mycorrhizal fungi (AMF), and the like can also mitigate the effect of salt stress (Bhat et al. 2020; Bisht and Garg 2022a, 2022b). Plant growth-promoting rhizobacteria like *Rhizobium*, *Azospirillum*, *Pseudomonas*, and *Flavobacterium* are also used to reduce the effect of stress-induced changes. Exogenous application of certain osmoprotectants and growth regulators may also help reduce saline stress. Biotechnological and functional genomics methods include molecular marker-based crop improvement, gene silencing, gene mutation, gene knockout, and a transgenic approach. Because of the limitations of these methods, an alternative method that would be economical and cost-effective to be adapted by the farmers would be beneficial. Seed priming is the best method to overcome salinity stress and improve plant yield.

6.4.1 Effect of Exogenously Applied Plant Growth Substances

Seed priming is an important technology to improve the stress resilience in plants exposed to various environmental stresses, including salinity (Hu et al. 2022). The term seed priming can be defined as "pre-treatment of seeds before sowing, in order to improve the rate and uniformity of germination." Seed germination is a complex process that involves different metabolic events that mobilize reserve food in the seed to an activation phase during which plumule and radicle emerge. It occurs in three phases when dry seeds are kept in water: imbibition, lag phase, and radicle emergence through testa. Since imbibition and lag phase are already completed in primed seeds, seeds can complete the germination process once water is provided. Priming decreases the time required for cellular activities such as enzyme, DNA, RNA, and protein synthesis (Rhaman et al. 2020). It also reduces cell membrane damage during storage or drying. Seed priming is a low-cost and effective physiological and biochemical process that stimulates seed germination under abiotic stress by using different priming agents. It includes controlled hydration of seeds, which improves and triggers pre-germination metabolic events, but radicle emergence is not allowed (Momeni et al. 2023). However, the efficacy of priming technique depends on the method of treatment as well its duration, dose of priming agent, and nature of plant species. Not only the seed priming, but also the concept of priming can be extended to seedlings and the whole plants. Therefore, plant priming is a technique which leads to a physiological state that enables the plants to respond more quickly to unfavourable environmental circumstances. Different priming techniques are developed based on the nature of priming agents to provide better seed germination (Hu et al. 2021).

Salt stress has also been reported to be alleviated by the application of different chemicals. Some of these are natural metabolites which are synthesized by plants in small quantity, but if supplied exogenously, they can modulate the growth and metabolism in host cells and thereby impart stress tolerance. These substances include plant growth regulators (salicylic acids, polyamines, jasmonic acid, melatonin), polyamines, nitric oxide, amino acids, ascorbic acid glutathione, and so on (Naeem et al. 2020). Pre-treatment of crop plants before a stress encounter imparts tolerance towards adverse conditions, since it contributes to plants responding more quickly than the untreated ones do. Most of the studies have established the positive role of chemical priming in plant tolerance (Zheng et al. 2018). The type, concentration, and mechanisms of various chemical compounds in alleviating the effects of salt stress are discussed, in detail, in Table 6.1.

TABLE 6.1
Detailed List Depicting the Mechanism of Action of Different Exogenously Applied Plant Growth Substances on Amelioration of Salt-Induced Stress Responses

Chemical strategies	Concentration	Plant species	Mechanism	References
Salicylic acid	1.0 mM	*Oryza sativa*	Enhanced growth, yield, nutrient status, and antioxidant potential	Jini and Joseph (2017)
	0.5, 1.0 mM	*Lycopersicon esculentum*	Improved fruit length, fruit diameter, number of fruits per plant, and level of osmolytes	Naeem et al. (2020)
	1.0 mM	*Zea mays*	Increased photosynthesis-related parameter, reduced membrane permeability and ROS generation, and enhanced the activity of catalase, ascorbate peroxidise	Tahjib-Ul-Arif et al. (2018)
	0.5 mM	*Dianthus superbus*	Higher leaf biomass, soluble sugars and proteins, upregulated the expression of *P5CS* and *MYB* genes	Zheng et al. (2018)
	0.5 mM	*Dianthus superbus*	Increased growth, antioxidant enzymes activity, chloroplast development	Ma et al. (2017)
	0.25, 0.5, 1.0 mM	*Vicia faba*	Improved seed germination and salinity tolerance	Anaya et al. (2018)
	0.5 mM	*Brassica juncea*	Higher photosynthetic activity and activities in enzymes of ascorbate-glutathione pathway	Nazar et al. (2015)
	0.5 mM	*Cicer arietinum*	Higher photosynthetic efficacy, arrested nodular senescence, and reduced ROS generation	Kaur et al. (2022)
	200 ppm	*Triticum aestivum*	Lowered ROS and osmolyte accumulation and enhanced photosynthetic pigments, antioxidant content	El-Hawary et al. (2023)
Melatonin	50, 100, 150 μM	*Citrus trifoliata*	Improved plant biomass, stomatal conductance, photosynthetic rate and efficacy	Hu et al. (2022)
	100, 200 μM	*Cucumis sativus*	Increased protein content, nutrient acquisition, and antioxidative activities	Brengi et al. (2022)
	200, 300, 400 μM	*Oryza sativa*	Promoted plant growth, reduced lipid peroxidation of membrane, and improved the activities of antioxidant enzymes	Wei et al. (2022)
	50, 100 μM	*Solanum lycopersicum*	Lowered membrane damage and oxidative burden and accelerated the activities of SOD, CAT, POX	Khan et al. (2023)

(Continued)

TABLE 6.1 (*Continued*)
Detailed List Depicting the Mechanism of Action of Different Exogenously Applied Plant Growth Substances on Amelioration of Salt-Induced Stress Responses

Chemical strategies	Concentration	Plant species	Mechanism	References
	30, 60, 90 μM	*Zea mays*	Increased plant growth, photosynthetic efficacy, and SOD, POD, CAT activities and reduced ROS generation	Ahmad et al. (2021b)
	1, 10, 50, 100, and 500 μM	*Medicago sativa*	Reduced ROS generation and increased antioxidative enzymes activity	Cen et al. (2020)
	10, 50 μM	*Cucumis melo*	Improved growth and chlorophyll content as well as bettered biochemical parameters	Castañares and Bouzo (2019)
	50, 100, and 500 μM	*Citrullus lanatus*	Protected plants from oxidative injury and improved light absorption as well as electron transport in PSII	Li et al. (2017)
	100 μM	*Beta vulgaris*	Increased accumulation of osmolytes and strengthened the activities of enzymes of ascorbate-glutathione system	Zhang et al. (2021)
	0.2 mM	*Lactuca sativa*	Increased ascorbic acid, antioxidant enzymes, and total phenol content and decreased H_2O_2 content	Sardar et al. (2023)
SNP (NO donor)	0.1 mM	*Triticum aestivum*	Enhanced plant growth and development, decreased Na level and oxidative stress	Kaya et al. (2023)
	10 μM	*Crocus sativus*	Improved accumulation of osmolytes and induced antioxidants activity and synthesis of secondary metabolites	Babaei et al. (2021)
	70 μM	*Lactuca sativa*	Enhanced nutrient uptake, maintained photosynthetic rate, and established phytohormone balance	Compos et al. (2019)
	0.1 mM	*Pisum sativum*	Improved radical length and biomass accumulation and reduced ROS accumulation by increasing the activity of antioxidant enzymes	Yadu et al. (2017); Dadasoglu et al. (2021)
	3, 6 mg L^{-1}	*Zea mays*	Reduced ROS and Na^+ but improved N, P, K, and Ca uptake	Kaya et al. (2015)
	100 μM	*Piper nigrum*	Increased all the growth traits and decreased dry weight of leaves, stomatal conductivity, and proline level	Badem and Söylemez (2022)
	50 μM, 100 μM	*Triticum aestivum*	Increased nutrient uptake, osmolytes accumulation, and antioxidants activity and reduced Na^+ uptake	Alnusairi et al. (2021)
	100 μM	*Glycine max*	Activated antioxidant enzymes activity and reduced H_2O_2 level	Simaei et al. (2011)
Brassinosteroids	10^{-11}, 10^{-9}, and 10^{-7} M	*Oryza sativa*	Reduced MDA content and modulated the expression of oxidative stress marker genes (*OsBRI1*, *OsDWF4*) and salt-responsive gene (*SalT*)	Sharma et al. (2013)

TABLE 6.1 (*Continued*)
Detailed List Depicting the Mechanism of Action of Different Exogenously Applied Plant Growth Substances on Amelioration of Salt-Induced Stress Responses

Chemical strategies	Concentration	Plant species	Mechanism	References
24-epibrassinolide	1 μM	*Solanum lycopersicum*	Increased chlorophyll and carotenoid synthesis and improved stomatal conductance, net photosynthetic rate, and overall growth	Ahanger et al. (2020)
Epi-brassinolide	0.1–100 μg/L	*Solanum tuberosum*	Enhanced proline content and antioxidant enzymes activity and reduced ROS accretion	Hu et al. (2016)
Brassinolide	0.0001, 0.01, and 1 μM	*Zea mays*	Balanced the ionic status, reduced oxidative damage, and enhanced the activities of antioxidative enzymes	Rattan et al. (2020)
24-epibrassinolide, 28-homobrassionolide	0.05, 0.2, and 1.0 mg/L	*Malus hupehensis*	Strengthened the antioxidative enzymes activity and promoted the accumulation of osmolytes	Su et al. (2020)
Brassinolide	0 and 50 nM	*Eucalyptus urophylla*	Improved nutrient uptake, photosynthetic efficacy, and antioxidative potential	de Oliveira et al. (2019)
24-epibrassinolide	100 mM	*Glycine max*	Triggered the accumulation of antioxidants and osmolytes, and improved growth	Soliman et al. (2020)
24-epibrassinolide	0, 0.5, 1.5, and 2.5 mg l^{-1}	*Mentha piperita*	Increased essential oil content and plant growth and improved antioxidant enzymes	Çoban and Baydar (2016)
Brassinolide	1, 2 mg l^{-1}	*Vitis vinifera*	Modified leaf structure and various organelles ultrastructure, protected them from salinity stress by increasing the accumulation of osmolytes	El-Banna et al. (2022)
Polyamines				
Spermidine	0.1 and 1 mM	*Pistacia vera*	Imparted salt tolerance by increasing the antioxidant enzymes activity and reducing Na/K ratio	Kamiab et al. (2014)
Putrescine, spermidine, and spermine	0.15, 0.30, 0.45, 0.60 mM	*Zoysia japonica*	Decreased MDA and H_2O_2 levels and improved the activity of polyamine biosynthetic enzymes and antioxidant enzymes	Li et al. (2016)
Spermidine	1 mM	*Poa pratensis*	Improved chlorophyll content and proline content and reduced membrane leakage and Na content	Puyang et al. (2016)
Spermidine	0.25 mM	*Brassica napus*	Elevated CO_2 assimilation and Rubisco activity, chlorophyll, and proline content and modulated the transcript encoding GR, APX, CAT, SOD, and GR	ElSayed et al. (2022)
Spermidine and spermine	0.5, 1.0, 1.5, and 2.0 mM	*Setaria italica*	Efficient antioxidative system, reduced ROS accumulation	Rathinapriya et al. (2020)
Putrescine, spermidine	10 μM	*Oryza sativa*	Improved growth, yield, floral morphogenesis, number of fertile tillers per plant, and CO_2 assimilation	Ndayiragije and Lutts (2007)

(Continued)

TABLE 6.1 (*Continued*)
Detailed List Depicting the Mechanism of Action of Different Exogenously Applied Plant Growth Substances on Amelioration of Salt-Induced Stress Responses

Chemical strategies	Concentration	Plant species	Mechanism	References
Putrescine	1 mM	*Cajanus cajan*	Improved nutrient status and nitrogen fixation efficacy and modulated the activity of putrescine catabolic enzymes	Garg and Sharma (2019)
Compatible solutes				
Proline	50, 100 mM	*Zea mays*	Restricted Na and Cl uptake, improved plant growth, nutrient uptake, and light absorption capacity	Mosaad et al. (2020); Naz et al. (2023)
	25, 50 mM	*Oryza sativa*	Increased plant growth, yield, chlorophyll content, and ascorbate content	Bhusan et al. (2016)
	50, 100 mM	*Triticum aestivum*	Higher root and shoot growth, photosynthetic pigment, and osmolytes accumulation	Mahboob et al. (2016)
	5 mM	*Phaseolus vulgaris*	Reduced oxidative burden by improving the activities of antioxidant enzymes	Abdelhamid et al. (2013)
	50 mM	*Capsicum annuum*	Significant improvement in growth, yield, and photosynthetic pigments	Jamil et al. (2018)
	50, 75, 100 mg l^{-1}	*Calendula officinalis*	Higher growth and physiological and biochemical parameters	El-Shawa et al. (2020)
	10 and 30 mM	*Aloe vera*	Reduced Na accumulation, increased water status, improved PSII functioning and oxygen evolving complex efficacy	Nakhaie et al. (2022)
	10, 20 mM	*Pisum sativum*	Improved growth, yield, nutrient uptake, soluble proteins, and catalase activity	Shahid et al. (2022b)
	10, 20, 30, 40 mM	*Medicago sativa*	Greater K^+/Na^+ ratio, upregulated the expression of transcripts encoding antioxidants (Cu/Zn-SOD, APX) and ion homeostasis (HKT1, SOS1, NHX1), etc.	Guo et al. (2022)
	25, 50 mM	*Allium cepa*	Increased growth indices, water use efficiency, chlorophyll content, stomatal conductance, and membrane stability index	Rady et al. (2018)
Glycine betaine	2.5, 5.0, 7.5 mM	*Gossypium hirsutum*	Increased photosynthetic and transpiration rate, intracellular CO_2 level, and stomatal conductance	Hamani et al. (2021)
	25, 50 mM	*Phaseolus vulgaris*	Increased enzymatic and non-enzymatic activities, water status, and photosynthetic pigment	Sofy et al. (2020)
	50 mM	*Oryza sativa*	Enhanced photosynthetic pigments, stomatal conductance, water content, and yield	Hafez et al. (2021)
	5, 25, 50 mM	*Glycine max*	Reduced MDA and Na^+ level and increased SOD and CAT activity	Malekzadeh (2015)

TABLE 6.1 (*Continued*)
Detailed List Depicting the Mechanism of Action of Different Exogenously Applied Plant Growth Substances on Amelioration of Salt-Induced Stress Responses

Chemical strategies	Concentration	Plant species	Mechanism	References
	5, 10, 25 mM	*Lactuca sativa*	Increased dry matter production and increased total antioxidants and total phenolics	Yildirim et al. (2015); Shams et al. (2016)
	0.05 mM; 0.2 mM	*Triticum aestivum*	Improved salt tolerance by upregulating the ascorbate-glutathione cycle	Kaya et al. (2023)
Sodium hydrosulfide (H_2S)	25, 50, 75, 100 µM	*Cucumis sativus*	Lowered membrane permeability, increased relative water content, improved photosynthetic properties	Turan et al. (2022)
	25, 50, 75, 100 µM	*Solanum lycopersicum*	Modulated mineral uptake and hormone content, increased photosynthetic activity	Yildirim et al. (2023)
	1 mM	*Reaumuria soongorica*	Reduced ROS level, increased activity of antioxidant enzymes and osmolytes	Liu et al. (2023)
	20, 50 µM	*Triticum aestivum*	Improved activity of antioxidant enzymes and increased concentrations of endogenous NO, hydrogen sulphide, and osmolytes	Alamer (2023)
	50, 100 µM	*Phaseolus vulgaris*	Increased growth and lowered oxidative species production due to higher activity of antioxidative enzymes	Dawood et al. (2022)
	50, 100 µM	*Oryza sativa*	Achieved homeostasis Na^+/K^+ ratio, imparted higher redox balance, and enhanced the activities of ROS as well as methylglyoxal detoxifying enzymes	Mostofa et al. (2015)
	25, 50, 100 µM	*Solanum melongena*	Improved photosynthetic attributes and achieved higher osmolyte accumulation antioxidant enzymes activity	Ekinci et al. (2021)

6.4.2 Microbial Symbionts

Recently, microbial inoculants have been widely used to alleviate stress-induced responses in plants due to their eco-friendly and cost-effective nature. Arbuscular mycorrhizal fungi and plant growth-promoting rhizobacteria are vital components of the rhizosphere, which help in promoting growth and productivity of plants growing under abiotic stressed condition, especially under salinity (Sapre et al. 2018). The major strategies employed by these microbes to improve plant growth, including nutrient cycling, mineral solubilization, and improvement of soil enzymatic activity, ultimately make nutrients available to the plants (Kang et al. 2014a). Furthermore, microbial supplements improve the resilience of host plants to unfavourable environmental constraints by modulating various physiological and metabolic processes such as developing a well-proliferated root system, improving photosynthetic and transpiration attributes, altering the ultrastructure of various cell organelles, enhancing nutrient and water acquisition, and increasing the activities of defence-related enzymes, and so on (Khan et al. 2016). However, diverse root exudates of the host, which act as a substrate for the synthesis of diverse active compounds, impact the activity of these soil microbes. Under salinity-stressed conditions, AMF and PGPR can also stimulate the growth of host by improving growth-related attributes and nutrient (especially P) uptake, modulating osmolytes and phytohormone synthesis, inducing transcriptional as well as translational machinery, regulating the antioxidant defence, and so on. The detailed account of microbe-induced stress resistance in host plants are given in Table 6.2.

TABLE 6.2
Detailed List Depicting the Mechanism and Effect of Various Microbial Inoculants on Amelioration of Salt-Induced Stress Responses

Biological strategies	Name of PGPR	Plant species	Mechanism	References
PGPR	*Bacillus pumilus*	*Oryza sativa*	Enhanced plant growth, improved activity of antioxidative enzymes, and reduced Na+ accumulation	Khan et al. (2016)
	Pseudomonas putida	*Glycine max*	Increased shoot length, fresh weight, and chlorophyll content and improved level of phytohormones (abscisic acid, salicylic acid)	Kang et al. (2014a)
	Bacillus aquimaris	*Zea mays*	Promoted plant growth, increased relative water and chlorophyll content, increased accumulation of osmolytes and activities of SOD, CAT, POX, and APOX enzymes	Li and Jiang (2017)
	Burkholdera cepacia, *Promicromonospora* sp., *Acinetobacter calcoaceticus*	*Cucumis sativus*	Improved plant biomass, chlorophyll content, and water potential and reduced electrolytic leakage and upregulation of stress-responsive phytohormones	Kang et al. (2014b)
	Klebsiella sp.	*Avena sativa*	Higher dry weight, relative water content, and proline accumulation, reduced membrane damage, and modulated the expression of rbcL and WRKY1 genes	Sapre et al. (2018)
	Bacillus magaterium	*Lycopersicon esculentum*	Increased plant growth and photosynthetic efficacy, reduced ROS and ethylene production, and altered metabolic profile of the plants	Akram et al. (2019)
	Paenibacillus yonginensis	*Panax ginseng*	Improved nutrient availability and osmolytes accumulation and increased the synthesis of ABA biosynthetic enzymes	Sukweenadhi et al. (2014)

TABLE 6.2 (*Continued*)
Detailed List Depicting the Mechanism and Effect of Various Microbial Inoculants on Amelioration of Salt-Induced Stress Responses

Biological strategies	Name of PGPR	Plant species	Mechanism	References
	Enterobacter sp.	*Arabidopsis thaliana, Lycopersicon esculentum, Abelmoschus esculentus*	Increased fresh and dry weight, plant height, and expression of salinity responsive genes RD29A;B, DREB2b, RAB18 and increased activities of antioxidative enzymes	Kim et al. (2014); Habib et al. (2016)
	Chryseobacterium gleum	*Triticum aestivum*	Improved fresh and dry weights, amino acid, phenolics, and flavonoids content and reduced proline content, Na+ uptake	Bhise et al. (2017)
	Leclercia adecarboxylata	*Solanum lycopersicum*	Improved plant growth and osmolytes accumulation and lowered level of endogenous abscisic acid	Kang et al. (2019)
	Pseudomonas sp.	*Brassica juncea*	Enhanced seed germination, root, and shoot dry weight and achieved high IAA and ALA production	Phour and Sindhu (2020)
	Bacillus subtilis, B. safensis	*Zea mays*	Increased chlorophyll, proline, and soluble sugar content and activities of defence enzymes	Misra and Chauhan (2020)
	Pseudomonas fluorescence, Bacillus pumilus, Exiguobacterium aurantiacum	*Triticum aestivum*	Improved fresh and dry biomass, water and osmotic potential, and nutrient accumulation	Nawaz et al. (2020)
	Priestia aryabhattai	*Triticum aestivum*	Increased germination, vigour indices, and leaf pigments and reduced membrane damage and electrolytic leakage	Shahid et al. (2022a)
	Trichoderma harzianum, Bacillus amyloliquefacien	*Triticum aestivum*	Improved nutrient uptake and increased expression of *TaHKT-1* and *TaNHX1* genes in roots which maintained Na+/K+ balance	Singh et al. (2021)
	Enterobacter cloacae, Bacillus cereus, Bacillus megaterium	*Triticum aestivum*	Improved growth and enzymatic activities and reduced electrolytic leakage, MDA and H_2O_2 content	Shankar and Prasad (2023)
AMF	*Rhizophagus irregularis, Funneliformis mosseae*	*Arachis hypogaea*	Increased photosynthetic rate, plant height, osmolytes synthesis, and relative leaf water content and reduced MDA level due to higher activity of antioxidant enzymes	Qin et al. (2021)

(*Continued*)

TABLE 6.2 (*Continued*)
Detailed List Depicting the Mechanism and Effect of Various Microbial Inoculants on Amelioration of Salt-Induced Stress Responses

Biological strategies	Name of PGPR	Plant species	Mechanism	References
	Rhizoglomus irregular	*Zea mays*	Increased biomass and nutrient content and reduced Na^+ accumulation in plant tissues	Moreira et al. (2020)
	Glomus mosseae	*Zea mays*	Improved growth parameters, nutrient uptake, activities of antioxidant enzymes and contents of oleic acid, palmitoleic acid, linolenic acid	Ndiate et al. (2021); Wang et al. (2021)
	Funneliformis mosseae, *Septoglomus deserticola*, *Acaulospora lacunose*	*Eclipta prostrata*	Improved growth, proline, polyphenol, and phenolic content	Duc et al. (2021)
	Glomus sp., *Sclerocystis* sp., *Acaulospora* sp.	*Phoenix dactylifera*	Increased nutrient uptake, chlorophyll content, water potential, and antioxidant enzymes reduced H_2O_2 level, lipid peroxidation	Ait-El-Mokhtar et al. (2020)
	Funneliformis mosseae, *Claroideoglomus etunicatum*	*Dracocephalum moldavica*	Increased polyphenol, flavonoids, and non-enzymatic antioxidant activity and reduced cell death	Alizadeh et al. (2021)
	Funneliformis mosseae	*Glycyrrhiza glabra*	Improved growth, P and K content, and promoted proline accumulation and higher K^+/Na^+ ratio	Amanifar et al. (2019)
	G. mosseae, *G. intraradices*, *G. hoi*	*Citrus aurantium*	Increased growth, chlorophyll content, water status, and gas exchange capacity and lowered ROS levels	Hadian-Deljou et al. (2020)
	Glomus deserticola	*Ocimum basilicum*	Increased growth and productivity, chlorophyll content, water use efficiency	Elhindi et al. (2017)
	Claroideoglomus claroideum	*Lactuca sativa*	Increased proline synthesis, SOD, CAT, and APOX activities and reduced synthesis of phenolic compounds and oxidative species	Santander et al. (2020)

6.4.3 Genetic Interventions in Improving Salt Resilience

Recently, various conventional techniques such as plant breeding and omics-based approaches have been adopted to improve stress tolerance in various crop plants (Hu et al. 2022). In breeding practices, breeders make genetic alterations in crop plants at intra-specific, inter-generic, and site-specific levels to enhance the tolerance of cultivars. For example, rice varieties such as Pokkali, Nona Bokra, and Bura Rara exhibit better salinity tolerance than other prominent varieties (Das et al. 2015). Among these, Pokkali is well known for contributing salinity tolerance genes in a breeding program (De Leon et al. 2017). Furthermore, genome-wide association studies (GWAS) have been explored to identify specific regions in the genome of plants that are responsible for abiotic stress tolerance, including salinity. Hu et al. (2021) performed GWAS at the adult stage of wheat using a panel of about 191 accessions to establish 11 quantitative trait loci for salinity tolerance related attribute. Besides this, advances in high-throughput approaches for next generation sequencing (NGS) provide new insights into identifying salt-responsive genes in crops through transcriptomics and genomics. Overall, a system biology approach that integrates all omics techniques is essential for unravelling the mechanisms involved in salinity defence across different plant species.

6.5 CONCLUSION AND FUTURE PROSPECTS

Salinity stress exerts a significant negative impact on crop output due to its influence on many physiological processes such as photosynthesis, respiration, nutritional assimilation, and hormone balance. The presence of salinity also indirectly harms plants by increasing the production of reactive oxygen species when the plants are under stress. These reactive oxygen species can then cause damage to important macromolecules such as lipids, proteins, and nucleic acids, further limiting crop productivity. To enhance the salt tolerance of plants, it is crucial to identify the fundamental components of the plant salt tolerance network. In recent years, the fields of genomics, transcriptomics, proteomics, and metabolomics have been effectively utilized to investigate various aspects of salt tolerance in plants, yielding promising results. The components encompass a diverse array of genes, such as those involved in the SOS signalling network, as well as transcription factors like HLH and MYB. Additionally, proteins and metabolites such as osmolytes, phytohormones, and lipids are also included. These components have the potential to be utilized in the genetic engineering of plants with enhanced salt tolerance. The implementation of a multidisciplinary strategy holds significant potential in enhancing plant and crop yield to address the growing food requirements of an expanding global population.

6.6 AUTHORSHIP CONTRIBUTION

Conceptualization, A.B.; Writing – Original draft preparation, R.C. and A.B.; Writing – Review and editing, R.C. and A.B. All authors have read and agreed to the published version of the book chapter.

6.7 DECLARATION OF COMPETING INTEREST

The authors declare that they have no known competing financial interests or personal relationships that could have appeared to influence the work reported in this book chapter.

REFERENCES

Abdelhamid, Magdi T., Mostafa M. Rady, Ashraf Sh Osman, and Mohamed A. Abdalla. "Exogenous application of proline alleviates salt-induced oxidative stress in *Phaseolus vulgaris* L. plants." *The Journal of Horticultural Science and Biotechnology* 88, no. 4 (2013): 439–446.

Acosta-Motos, Jose Ramón, Maria Fernanda Ortuño, et al. "Plant responses to salt stress: adaptive mechanisms." *Agronomy* 7, no. 1 (2017): 18.

Agarwal, Parinita, Pradeep K. Agarwal, Arvind J. Joshi, Sudhir K. Sopory, and Malireddy K. Reddy. "Overexpression of PgDREB2A transcription factor enhances abiotic stress tolerance and activates downstream stress-responsive genes." *Molecular Biology Reports* 37, no. (2010): 1125–1135.

Ahanger, Mohammad Abass, Rayees Ahmad Mir, Mohammed Nasser Alyemeni, and Parvaiz Ahmad. "Combined effects of brassinosteroid and kinetin mitigates salinity stress in tomato through the modulation of antioxidant and osmolyte metabolism." *Plant Physiology and Biochemistry* 147, no. (2020): 31–42.

Ahmad, F., Aisha Kamal, Ananya Singh, F. Ashfaque, S. Alamri, M. H. Siddiqui, and M. I. R. Khan. "Seed priming with gibberellic acid induces high salinity tolerance in *Pisum sativum* through antioxidants, secondary metabolites and up-regulation of antiporter genes." *Plant Biology* 23 (2021a): 113–121.

Ahmad, Shakeel, Wenwen Cui, Muhammad Kamran, et al. "Exogenous application of melatonin induces tolerance to salt stress by improving the photosynthetic efficiency and antioxidant defense system of maize seedling." *Journal of Plant Growth Regulation* 40, no. (2021b): 1270–1283.

Ait-El-Mokhtar, Mohamed, Marouane Basla, et al. "Alleviation of detrimental effects of salt stress on date palm (*Phoenix dactylifera* L.) by the application of arbuscular mycorrhizal fungi and/or compost." *Frontiers in Sustainable Food Systems* 4, no. (2020): 131.

Akram, Nudrat A., Fahad Shafiq, and Muhammad Ashraf. "Ascorbic acid-a potential oxidant scavenger and its role in plant development and abiotic stress tolerance." *Frontiers in Plant Science* (2017): 613.

Akram, Waheed, Hina Aslam, Sajid Rashid Ahmad, et al. "*Bacillus megaterium* strain A12 ameliorates salinity stress in tomato plants through multiple mechanisms." *Journal of Plant Interactions* 14, no. 1 (2019): 506–518.

Al Kharusi, Latifa, Rashid Al Yahyai, and Mahmoud W. Yaish. "Antioxidant response to salinity in salt-tolerant and salt-susceptible cultivars of date palm." *Agriculture* 9, no. 1 (2019): 8.

Alamer, Khalid H. "Exogenous hydrogen sulfide supplementation alleviates the salinity-stress-mediated growth decline in wheat (*Triticum aestivum* L.) by modulating tolerance mechanisms." *Plants* 12, no. 19 (2023): 3464.

Ali, Shafaqat, Zohaib Abbas, Mahmoud F. Seleiman, et al. "Glycine betaine accumulation, significance and interests for heavy metal tolerance in plants." *Plants* 9, no. 7 (2020): 896.

Alizadeh, Sevda, Syamak Fallahi Gharagoz, Latifeh Pourakbar, Sina Siavash Moghaddam, and Masoomeh Jamalomidi. "Arbuscular mycorrhizal fungi alleviate salinity stress and alter phenolic compounds of Moldavian balm." *Rhizosphere* 19 (2021): 100417.

Allakhverdiev, Suleyman I., Yoshitaka Nishiyama, Sachio Miyairi, et al. "Salt stress inhibits the repair of photodamaged photosystem II by suppressing the transcription and translation of psbA genes in Synechocystis." *Plant Physiology* 130, no. 3 (2002): 1443–1453.

Allbed, Amal, and Lalit Kumar. "Soil salinity mapping and monitoring in arid and semi-arid regions using remote sensing technology: a review." *Advances in Remote Sensing* 2013 (2013).

Almeida, Diego M., M. Margarida Oliveira, and Nelson JM Saibo. "Regulation of Na+ and K+ homeostasis in plants: towards improved salt stress tolerance in crop plants." *Genetics and Molecular Biology* 40, no. 1 suppl 1 (2017): 326–345.

Alnusairi, Ghalia S. H., Yasser S. A. Mazrou, Sameer H. Qari, et al. "Exogenous nitric oxide reinforces photosynthetic efficiency, osmolyte, mineral uptake, antioxidant, expression of stress-responsive genes and ameliorates the effects of salinity stress in wheat." *Plants* 10, no. 8 (2021): 1693.

Amanifar, Setareh, Maryam Khodabandeloo, Ehsan Mohseni Fard, Mohammad Sadegh Askari, and Mohsen Ashrafi. "Alleviation of salt stress and changes in glycyrrhizin accumulation by arbuscular mycorrhiza in liquorice (Glycyrrhiza glabra) grown under salinity stress." *Environmental and Experimental Botany* 160 (2019): 25–34.

Anaya, Fatima, Rachid Fghire, Said Wahbi, and Kenza Loutfi. "Influence of salicylic acid on seed germination of *Vicia faba* L. under salt stress." *Journal of the Saudi Society of Agricultural Sciences* 17, no. 1 (2018): 1–8.

Anbumalarmathi, J., and Preeti Mehta. "Effect of salt stress on germination of indica rice varieties." *European Journal of Biological Science* 6, no. 1 (2013): 1–6.

Ashraf, M. F. M. R., and Majid R. Foolad. "Roles of glycine betaine and proline in improving plant abiotic stress resistance." *Environmental and Experimental Botany* 59, no. 2 (2007): 206–216.

Babaei, Samareh, Vahid Niknam, and Mehrdad Behmanesh. "Comparative effects of nitric oxide and salicylic acid on salinity tolerance in saffron (*Crocus sativus*)." *Plant Biosystems-An International Journal Dealing with all Aspects of Plant Biology* 155, no. 1 (2021): 73–82.

Badem, Adalet, and Selçuk Söylemez. "Effects of nitric oxide and silicon application on growth and productivity of pepper under salinity stress." *Journal of King Saud University-Science* 34, no. 6 (2022): 102189.

Ben-Abdallah, Saoussen, Walid Zorrig, Lisa Amyot, et al. "Potential production of polyphenols, carotenoids and glycoalkaloids in *Solanum villosum* Mill. under salt stress." *Biologia* 74 (2019): 309–324.

Bhat, Mujtaba Aamir, Vijay Kumar, Mudasir Ahmad Bhat, et al. "Mechanistic insights of the interaction of plant growth-promoting rhizobacteria (PGPR) with plant roots toward enhancing plant productivity by alleviating salinity stress." *Frontiers in Microbiology* 11 (2020): 1952.

Bhise, Komal K., Prashant K. Bhagwat, and Padma Babulal Dandge. "Synergistic effect of *Chryseobacterium gleum* sp. SUK with ACC deaminase activity in alleviation of salt stress and plant growth promotion in *Triticum aestivum* L." *3 Biotech* 7 (2017): 1–13.

Bhusan, Dhiman, Debasish Kumar Das, Mahmud Hossain, Yoshiyuki Murata, and Md Anamul Hoque. "Improvement of salt tolerance in rice (*Oryza sativa* L.) by increasing antioxidant defense systems using exogenous application of proline." *Australian Journal of Crop Science* 10, no. 1 (2016): 50–56.

Bisht, Aditi, and Neera Garg. "AMF modulated rhizospheric microbial enzyme activities and their impact on sulphur assimilation along with thiol metabolism in pigeonpea under Cd stress." *Rhizosphere* 21 (2022a): 100478.

Bisht, Aditi, and Neera Garg. "AMF species improve yielding potential of Cd stressed pigeonpea plants by modulating sucrose-starch metabolism, nutrients acquisition and soil microbial enzymatic activities." *Plant Growth Regulation* 96, no. 3 (2022b): 409–430.

Bohnert, Hans J., Patricia Ayoubi, Chris Borchert, et al. "A genomics approach towards salt stress tolerance." *Plant Physiology and Biochemistry* 39, nos. 3–4 (2001): 295–311.

Brengi, Sary H., E. M. Abd Allah, and Ibrahim A. Abouelsaad. "Effect of melatonin or cobalt on growth, yield and physiological responses of cucumber (*Cucumis sativus* L.) plants under salt stress." *Journal of the Saudi Society of Agricultural Sciences* 21, no. 1 (2022): 51–60.

Bressan, Ray A., Paul M. Hasegawa, and José M. Pardo. "Plants use calcium to resolve salt stress." *Trends in Plant Science* 3, no. 11 (1998): 411–412.

Butt, Madiha, C. M. Ayyub, M. Amjad, and Rashid Ahmad. "Proline application enhances growth of chilli by improving physiological and biochemical attributes under salt stress." *Pakistan Journal of Agricultural Sciences* 53, no. 1 (2016).

Butt, Madiha, Abdul Sattar, Tahira Abbas, et al. "Morpho-physiological and biochemical attributes of Chili (*Capsicum annum* L.) genotypes grown under varying salinity levels." *Plos One* 16, no. 11 (2021): e0257893.

Bybordi, Ahmad, and Jalal Tabatabaei. "Effect of salinity stress on germination and seedling properties in canola cultivars (*Brassica napus* L.)." *Notulae Botanicae Horti Agrobotanici Cluj-Napoca* 37, no. 2 (2009): 71–76.

Campos, Fernanda V., Juraci A. Oliveira, Mayara G. Pereira, and Fernanda S. Farnese. "Nitric oxide and phytohormone interactions in the response of *Lactuca sativa* to salinity stress." *Planta* 250 (2019): 1475–1489.

Castañares, José Luis, and Carlos Alberto Bouzo. "Effect of exogenous melatonin on seed germination and seedling growth in melon (*Cucumis melo* L.) under salt stress." *Horticultural Plant Journal* 5, no. 2 (2019): 79–87.

Cen, Huifang, Tingting Wang, Huayue Liu, Danyang Tian, and Yunwei Zhang. "Melatonin application improves salt tolerance of alfalfa (*Medicago sativa* L.) by enhancing antioxidant capacity." *Plants* 9, no. 2 (2020): 220.

Chartzoulakis, K., and G. Klapaki. "Response of two greenhouse pepper hybrids to NaCl salinity during different growth stages." *Scientia Horticulturae* 86, no. 3 (2000): 247–260.

Chhabra, Rohit, Shabnam, and Tavisha Singh. "Seed aging, storage and deterioration: an irresistible physiological phenomenon." *Agricultural Reviews* 40, no. 3 (2019): 234–238.

Çoban, Özkan, and Nilgün Göktürk Baydar. "Brassinosteroid effects on some physical and biochemical properties and secondary metabolite accumulation in peppermint (*Mentha piperita* L.) under salt stress." *Industrial Crops and Products* 86 (2016): 251–258.

Corwin, Dennis L. "Climate change impacts on soil salinity in agricultural areas." *European Journal of Soil Science* 72, no. 2 (2021): 842–862.

Dadasoglu, Esin, Melek Ekinci, Raziye Kul, Mostafakamal Shams, Metin Turan, and Ertan Yildirim. "Nitric oxide enhances salt tolerance through regulating antioxidant enzyme activity and nutrient uptake in pea." *Legume Research-An International Journal* 44, no. 1 (2021): 41–45.

Das, Priyanka, Kamlesh K. Nutan, Sneh L. Singla-Pareek, and Ashwani Pareek. "Understanding salinity responses and adopting 'omics-based' approaches to generate salinity tolerant cultivars of rice." *Frontiers in Plant Science* 6 (2015): 712.

Dawood, Mona F. A., Mahmoud R. Sofy, Heba I. Mohamed, Ahmed R. Sofy, and Huwida A. A. Abdel-kader. "Hydrogen sulfide modulates salinity stress in common bean plants by maintaining osmolytes and regulating nitric oxide levels and antioxidant enzyme expression." *Journal of Soil Science and Plant Nutrition* 22, no. 3 (2022): 3708–3726.

De Leon, Teresa B., Steven Linscombe, and Prasanta K. Subudhi. "Identification and validation of QTLs for seedling salinity tolerance in introgression lines of a salt tolerant rice landrace 'Pokkali'." *PloS One* 12, no. 4 (2017): e0175361.

de Oliveira, Victor Pereira, Michael Douglas Roque Lima, Breno Ricardo Serrão da Silva, Bruno Lemos Batista, and Allan Klynger da Silva Lobato. "Brassinosteroids confer tolerance to salt stress in *Eucalyptus urophylla* plants enhancing homeostasis, antioxidant metabolism and leaf anatomy." *Journal of Plant Growth Regulation* 38 (2019): 557–573.

Dkhil, Besma Ben, and Mounir Denden. "Salt stress induced changes in germination, sugars, starch and enzyme of carbohydrate metabolism in *Abelmoschus esculentus* L. (Moench.) seeds." *African Journal of Agricultural Research* 5, no. 12 (2010): 1412–1418.

Duc, Nguyen Hong, Au Trung Vo, Imane Haddidi, Hussein Daood, and Katalin Posta. "Arbuscular mycorrhizal fungi improve tolerance of the medicinal plant *Eclipta prostrata* (L.) and induce major changes in polyphenol profiles under salt stresses." *Frontiers in Plant Science* 11 (2021): 612299.

Ebrahimi, Reza, and Satish C. Bhatla. "Effect of sodium chloride levels on growth, water status, uptake, transport, and accumulation pattern of sodium and chloride ions in young sunflower plants." *Communications in Soil Science and Plant Analysis* 42, no. 7 (2011): 815–831.

Ekinci, Melek, Ertan Yildirim, and Metin Turan. "Ameliorating effects of hydrogen sulfide on growth, physiological and biochemical characteristics of eggplant seedlings under salt stress." *South African Journal of Botany* 143 (2021): 79–89.

El Moukhtari, Ahmed, Cécile Cabassa-Hourton, Mohamed Farissi, and Arnould Savouré. "How does proline treatment promote salt stress tolerance during crop plant development?" *Frontiers in Plant Science* 11 (2020): 1127.

El-Banna, Mostafa F., Arwa Abdulkreem A. L.-Huqail, Saad Farouk, Bassam E. A. Belal, Mosaad A. El-Kenawy, and Ahmed F. Abd El-Khalek. "Morpho-physiological and anatomical alterations of salt-affected thompson seedless grapevine (*Vitis vinifera* L.) to brassinolide spraying." *Horticulturae* 8, no. 7 (2022): 568.

El-Hawary, Mohammed Mohammed, Omnia S. M. Hashem, and Mirza Hasanuzzaman. "Seed priming and foliar application with ascorbic acid and salicylic acid mitigate salt stress in wheat." *Agronomy* 13, no. 2 (2023): 493.

Elhindi, Khalid M., Ahmed Sharaf El-Din, and Abdallah M. Elgorban. "The impact of arbuscular mycorrhizal fungi in mitigating salt-induced adverse effects in sweet basil (*Ocimum basilicum* L.)." *Saudi Journal of Biological Sciences* 24, no. 1 (2017): 170–179.

ElSayed, Abdelaleim I., Azza H. Mohamed, Mohammed Suhail Rafudeen, Ahmad A. Omar, Mohamed F. Awad, and Elsayed Mansour. "Polyamines mitigate the destructive impacts of salinity stress by enhancing photosynthetic capacity, antioxidant defense system and upregulation of calvin cycle-related genes in rapeseed (*Brassica napus* L.)." *Saudi Journal of Biological Sciences* 29, no. 5 (2022): 3675–3686.

El-Shawa, Ghada M. R., Eman M. Rashwan, and Kh A. A. Abdelaal. "Mitigating salt stress effects by exogenous application of proline and yeast extract on morpho-physiological, biochemical and anatomical characters of calendula plants." *Scientific Journal of Flowers and Ornamental Plants* 7, no. 4 (2020): 461–482.

Farooq, Muhammad, Mubshar Hussain, Abdul Wakeel, and Kadambot H. M. Siddique. "Salt stress in maize: effects, resistance mechanisms, and management. A review." *Agronomy for Sustainable Development* 35 (2015): 461–481.

Friedland, N., S. Negi, T. Vinogradova-Shah, et al. "Fine-tuning the photosyntheticlight harvesting apparatus for improved photosynthetic efficiency and biomass yield." *Scientific Reports* 9, no. 1 (2019): 13028.

Garg, Neera, and Amrita Sharma. "Role of putrescine (put) in imparting salt tolerance through modulation of put metabolism, mycorrhizal and rhizobial symbioses in *Cajanus cajan* (L.) Millsp." *Symbiosis* 79, no. 1 (2019): 59–74.

Gill, Sarvajeet Singh, and Narendra Tuteja. "Reactive oxygen species and antioxidant machinery in abiotic stress tolerance in crop plants." *Plant Physiology and Biochemistry* 48, no. 12 (2010): 909–930.

Gu, Ruisheng, Sandra Fonseca, László G. Puskás, et al. "Transcript identification and profiling during salt stress and recovery of *Populus euphratica*." *Tree Physiology* 24, no. 3 (2004): 265–276.

Guo, Shuaiqi, Xuxia Ma, Wenqi Cai, et al. "Exogenous proline improves salt tolerance of alfalfa through modulation of antioxidant capacity, ion homeostasis, and proline metabolism." *Plants* 11, no. 21 (2022): 2994.

Habib, Sheikh Hasna, Hossain Kausar, and Halimi Mohd Saud. "Plant growth-promoting rhizobacteria enhance salinity stress tolerance in okra through ROS-scavenging enzymes." *BioMed Research International* 2016 (2016).

Hadian-Deljou, M., M. Esna-Ashari, and A. Mirzaie-asl. "Alleviation of salt stress and expression of stress-responsive gene through the symbiosis of arbuscular mycorrhizal fungi with sour orange seedlings." *Scientia Horticulturae* 268 (2020): 109373.

Hafez, Emad M., Salah M. Gowayed, Yasser Nehela, Raghda M. Sakran, Asmaa MS Rady, Abdelmoniem Awadalla, Alaa El-Dein Omara, and Bassam F. Alowaiesh. "Incorporated biochar-based soil amendment and exogenous glycine betaine foliar application ameliorate rice (Oryza sativa L.) tolerance and resilience to osmotic stress." *Plants* 10, no. 9 (2021): 1930.

Hamani, Abdoul Kader Mounkaila, Shuang Li, Jinsai Chen, et al. "Linking exogenous foliar application of glycine betaine and stomatal characteristics with salinity stress tolerance in cotton (*Gossypium hirsutum* L.) seedlings." *BMC Plant Biology* 21, no. 1 (2021): 1–12.

Hao, Shanhu, Yiran Wang, Yunxiu Yan, Yuhang Liu, Jingyao Wang, and Su Chen. "A review on plant responses to salt stress and their mechanisms of salt resistance." *Horticulturae* 7, no. 6 (2021): 132.

Hasanuzzaman, Mirza, M. H. M. Borhannuddin Bhuyan, Taufika Islam Anee, et al. "Regulation of ascorbate-glutathione pathway in mitigating oxidative damage in plants under abiotic stress." *Antioxidants* 8, no. 9 (2019): 384.

Hasanuzzaman, Mirza, M. H. M. Borhannuddin Bhuyan, Faisal Zulfiqar, et al. "Reactive oxygen species and antioxidant defense in plants under abiotic stress: revisiting the crucial role of a universal defense regulator." *Antioxidants* 9, no. 8 (2020): 681.

Hoque, Mohammad M. Ibne, Zheng Jun, and Wang Guoying. "Evaluation of salinity tolerance in maize (*Zea mays* L.) genotypes at seedling stage." *Journal of BioScience & Biotechnology* 4, no. 1 (2015).

Hu, Ceng-hong, Ying Zheng, Cui-ling Tong, and De-jian Zhang. "Effects of exogenous melatonin on plant growth, root hormones and photosynthetic characteristics of trifoliate orange subjected to salt stress." *Plant Growth Regulation* 97, no. 3 (2022): 551–558.

Hu, Pan, Qi Zheng, Qiaoling Luo, Wan Teng, Hongwei Li, Bin Li, and Zhensheng Li. "Genome-wide association study of yield and related traits in common wheat under salt-stress conditions." *BMC Plant Biology* 21 (2021): 1–20.

Hu, Yueqing, Shitou Xia, Yi Su, et al. "Brassinolide increases potato root growth in vitro in a dose-dependent way and alleviates salinity stress." *BioMed Research International* 2016 (2016).

Hu, Yuncai, and Urs Schmidhalter. "Limitation of salt stress to plant growth." *HOCK, E. Plant Toxicology* 4 (2004): 191–224.

Hussain, Saddam, Abdul Khaliq, Amar Matloob, Muhammad Ashfaq Wahid, and Irfan Afzal. "Germination and growth response of three wheat cultivars to NaCl salinity." *Soil Environ* 32, no. 1 (2013): 36–43.

Jamil, Moazzam, Muhammad Ali Kharal, Maqshoof Ahmad, Ghulam Hassan Abbasi, Farheen Nazli, Azhar Hussain, and Muhammad Fakhar-u-Zaman Akhtar. "Inducing salinity tolerance in red pepper (*Capsicum annuum* L.) through exogenous application of proline and L-tryptophan." *Soil & Environment* 37, no. 2 (2018): 160–168.

Jini, D., and B. Joseph. "Physiological mechanism of salicylic acid for alleviation of salt stress in rice." *Rice Science* 24, no. 2 (2017): 97–108.

Jouyban, Zeinolabedin. "The effects of salt stress on plant growth." *Technical Journal of Engineering and Applied Sciences* 2, no. 1 (2012): 7–10.

Kamiab, Fereshteh, Alireza Talaie, Masood Khezri, and Amanallah Javanshah. "Exogenous application of free polyamines enhance salt tolerance of pistachio (*Pistacia vera* L.) seedlings." *Plant Growth Regulation* 72 (2014): 257–268.

Kang, Sang-Mo, Abdul Latif Khan, Muhammad Waqas, et al. "Plant growth-promoting rhizobacteria reduce adverse effects of salinity and osmotic stress by regulating phytohormones and antioxidants in *Cucumis sativus*." *Journal of Plant Interactions* 9, no. 1 (2014a): 673–682.

Kang, Sang-Mo, Ramalingam Radhakrishnan, Abdul Latif Khan, et al. "Gibberellin secreting rhizobacterium, *Pseudomonas putida* H-2–3 modulates the hormonal and stress physiology of soybean to improve the plant growth under saline and drought conditions." *Plant Physiology and Biochemistry* 84 (2014b): 115–124.

Kang, Sang-Mo, Raheem Shahzad, Saqib Bilal, et al. "Indole-3-acetic-acid and ACC deaminase producing *Leclercia adecarboxylata* MO1 improves *Solanum lycopersicum* L. growth and salinity stress tolerance by endogenous secondary metabolites regulation." *BMC Microbiology* 19, no. 1 (2019): 1–14.

Kao, W.-Y., T.-T. Tsai, and C.-N. Shih. "Photosynthetic gas exchange and chlorophyll a fluorescence of three wild soybean species in response to NaCl treatments." *Photosynthetica* 41 (2003): 415–419.

Kaur, Harmanjit, Sofi J. Hussain, Gursharan Kaur, et al. "Salicylic acid improves nitrogen fixation, growth, yield and antioxidant defence mechanisms in chickpea genotypes under salt stress." *Journal of Plant Growth Regulation* 41, no. 5 (2022): 2034–2047.

Kaya, Cengiz, Muhammad Ashraf, Osman Sönmez, Atilla Levent Tuna, and Salih Aydemir. "Exogenously applied nitric oxide confers tolerance to salinity-induced oxidativestress in two maize (*Zea mays* L.) cultivars differing in salinity tolerance." *Turkish Journal of Agriculture and Forestry* 39, no. 6 (2015): 909–919.

Kaya, Cengiz, Ferhat Ugurlar, Muhammed Ashraf, Pravej Alam, and Parvaiz Ahmad. "Nitric oxide and hydrogen sulfide work together to improve tolerance to salinity stress in wheat plants by upraising the AsA-GSH cycle." *Plant Physiology and Biochemistry* 194 (2023): 651–663.

Kaya, Mehmet Demir, Arif Ipek, and Ahmet Öztürk. "Effects of different soil salinity levels on germination and seedling growth of safflower (*Carthamus tinctorius* L.)." *Turkish Journal of Agriculture and Forestry* 27, no. 4 (2003): 221–227.

Khan, Alamgir, Xue Qiang Zhao, M. Tariq Javed, et al. "*Bacillus pumilus* enhances tolerance in rice (*Oryza sativa* L.) to combined stresses of NaCl and high boron due to limited uptake of Na+." *Environmental and Experimental Botany* 124 (2016): 120–129.

Khan, M. I. R., N. Iqbal, A. Masood, and N. A. Khan. "Variation in salt tolerance of wheat cultivars: role of glycinebetaine and ethylene." *Pedosphere* 22, no. 6 (2012): 746–754.

Khan, M., R. Iqbal, Badar Jahan, Mohamed F. AlAjmi, Md Tabish Rehman, and Nafees A. Khan. "Ethephon mitigates nickel stress by modulating antioxidant system, glyoxalase system and proline metabolism in Indian mustard." *Physiology and Molecular Biology of Plants* 26 (2020): 1201–1213.

Khan, M., R. Iqbal, Faroza Nazir, Mohd Asgher, Tasir S. Per, and Nafees A. Khan. "Selenium and sulfur influence ethylene formation and alleviate cadmium-induced oxidative stress by improving proline and glutathione production in wheat." *Journal of Plant Physiology* 173 (2015): 9–18.

Khan, M. H., and S. K. Panda. "Alterations in root lipid peroxidation and antioxidative responses in two rice cultivars under NaCl-salinity stress." *Acta Physiologiae Plantarum* 30 (2008): 81–89.

Khan, Tanveer Ahmad, Mohd Saleem, and Qazi Fariduddin. "Melatonin influences stomatal behavior, root morphology, cell viability, photosynthetic responses, fruit yield, and fruit quality of tomato plants exposed to salt stress." *Journal of Plant Growth Regulation* 42, no. 4 (2023): 2408–2432.

Khedr, Abdel Hamid A., Mohammad A. Abbas, Amal A. Abdel Wahid, W. Paul Quick, and Gaber M. Abogadallah. "Proline induces the expression of salt-stress-responsive proteins and may improve the adaptation of *Pancratium maritimum* L. to salt-stress." *Journal of Experimental Botany* 54, no. 392 (2003): 2553–2562.

Kim, Kangmin, Ye-Jin Jang, Sang-Myeong Lee, Byung-Taek Oh, Jong-Chan Chae, and Kui-Jae Lee. "Alleviation of salt stress by Enterobacter sp. EJ01 in tomato and Arabidopsis is accompanied by up-regulation of conserved salinity responsive factors in plants." *Molecules and Cells* 37, no. 2 (2014): 109.

Kumari, Sarika, Faroza Nazir, Kajal Jain, and M. Iqbal R. Khan. "GABA and potassium modulates defence systems, assimilation of nitrogen and carbon, and yield traits under salt stress in wheat." *Journal of Plant Growth Regulation* 42, no. 10 (2023): 6721–6740.

Kumari, Sarika, Faroza Nazir, Apoorva Singh, Hyda Haroon, Nahida Rashid Khan, Rudra Narayan Sahoo, Mohammed Albaqami, Kadambot H. M. Siddique, and M. Iqbal R. Khan. "γ-Aminobutyric acid (GABA) strengthened nutrient accumulation, defense metabolism, growth and yield traits against salt and endoplasmic reticulum stress conditions in wheat plants." *Plant and Soil* (2024): 1–21.

Li, Hao, Jingjing Chang, Hejie Chen, et al. "Exogenous melatonin confers salt stress tolerance to watermelon by improving photosynthesis and redox homeostasis." *Frontiers in Plant Science* 8 (2017): 295.

Li H. Q, and Jiang X. W. Inoculation with plant growth-promoting bacteria (PGPB) improves salt tolerance of maize seedling. *Russian Journal of Plant Physiology* 64 (2017): 235–41.

Li, Jin-Ting, Zong-Bo Qiu, Xiao-Wei Zhang, and Lin-Song Wang. "Exogenous hydrogen peroxide can enhance tolerance of wheat seedlings to salt stress." *Acta Physiologiae Plantarum* 33 (2011): 835–842.

Li, Ping, Hong Yang, Lu Wang, et al. "Physiological and transcriptome analyses reveal short-term responses and formation of memory under drought stress in rice." *Frontiers in Genetics* 10 (2019): 55.

Li, Shucheng, Han Jin, and Qiang Zhang. "The effect of exogenous spermidine concentration on polyamine metabolism and salt tolerance in Zoysiagrass (*Zoysia japonica* Steud) subjected to short-term salinity stress." *Frontiers in Plant Science* 7 (2016): 1221.

Liu, Hanghang, Peifang Chong, Zehua Liu, Xinguang Bao, and Bingbing Tan. "Exogenous hydrogen sulfide improves salt stress tolerance of *Reaumuria soongorica* seedlings by regulating active oxygen metabolism." *PeerJ* 11 (2023): e15881.

Liu, Jiping, and Jian-Kang Zhu. "Proline accumulation and salt-stress-induced gene expression in a salt-hypersensitive mutant of Arabidopsis." *Plant Physiology* 114, no. 2 (1997): 591–596.

Lv, Xiaoyan, Sixue Chen, and Yuguang Wang. "Advances in understanding the physiological and molecular responses of sugar beet to salt stress." *Frontiers in Plant Science* 10 (2019): 1431.

Ma, Xiaohua, Jian Zheng, Xule Zhang, Qingdi Hu, and Renjuan Qian. "Salicylic acid alleviates the adverse effects of salt stress on *Dianthus superbus* (Caryophyllaceae) by activating photosynthesis, protecting morphological structure, and enhancing the antioxidant system." *Frontiers in Plant Science* 8 (2017): 600.

Maggio, Albino, Giampaolo Raimondi, Armando Martino, and Stefania De Pascale. "Salt stress response in tomato beyond the salinity tolerance threshold." *Environmental and Experimental Botany* 59, no. 3 (2007): 276–282.

Mahboob, Wajid, Muhammad Athar Khan, and M. U. Shirazi. "Induction of salt tolerance in wheat (*Triticum aestivum* L.) seedlings through exogenous application of proline." *Pakistan Journal of Botany* 48, no. 3 (2016): 861–867.

Mahmood, S., and H. Athar. "Germination and growth of *Panicum turgidum* provenance under saline conditions." *Pakistan Journal of Biological Sciences (Pakistan)* 6, no. 2 (2003).

Malekzadeh, Parviz. "Influence of exogenous application of glycinebetaine on antioxidative system and growth of salt-stressed soybean seedlings (*Glycine max* L.)." *Physiology and Molecular Biology of Plants* 21 (2015): 225–232.

Misra, Sankalp, and Puneet Singh Chauhan. "ACC deaminase-producing rhizosphere competent Bacillus spp. mitigate salt stress and promote *Zea mays* growth by modulating ethylene metabolism." *3 Biotech* 10, no. 3 (2020): 119.

Momeni, Mohammad Mehdi, Mansoor Kalantar, and Mahdi Dehghani-Zahedani. "Physiological, biochemical and molecular responses of durum wheat under salt stress." *Plant Genetic Resources* 19, no. 2 (2021): 93–103.

Momeni, Mohammad Mehdi, Mansoor Kalantar, and Mehdi Dehghani-Zahedani. "H2O2 seed priming improves tolerance to salinity stress in durum wheat." *Cereal Research Communications* 51, no. 2 (2023): 391–401.

Moreira, Helena, Sofia I. A. Pereira, Alberto Vega, Paula M. L. Castro, and Ana P. G. C. Marques. "Synergistic effects of arbuscular mycorrhizal fungi and plant growth-promoting bacteria benefit maize growth under increasing soil salinity." *Journal of Environmental Management* 257 (2020): 109982.

Mosaad, Ibrahim S. M., Ayman H. I. Serag, Mohamed Moustafa-Farag, and Ali K. Seadh. "Effect of exogenous proline application on maize yield and the optimum rate of mineral nitrogen under salinity stress." *Journal of Plant Nutrition* 43, no. 3 (2020): 354–370.

Mostofa, Mohammad G., Daisuke Saegusa, Masayuki Fujita, and Lam-Son Phan Tran. "Hydrogen sulfide regulates salt tolerance in rice by maintaining Na+/K+ balance, mineral homeostasis and oxidative metabolism under excessive salt stress." *Frontiers in Plant Science* 6 (2015): 1055.

Munné-Bosch, Sergi. "The role of α-tocopherol in plant stress tolerance." *Journal of Plant Physiology* 162, no. 7 (2005): 743–748.

Mushtaq, Naveed Ul, Khalid M. Alghamdi, Seerat Saleem, et al. "Exogenous zinc mitigates salinity stress by stimulating proline metabolism in proso millet (*Panicum miliaceum* L.)." *Frontiers in Plant Science* 14 (2023): 1053869.

Naeem, Mohammad, Abdul Basit, Imran Ahmad, Heba I. Mohamed, and Humaira Wasila. "Effect of salicylic acid and salinity stress on the performance of tomato plants." *Gesunde Pflanzen* 72, no. 4 (2020).

Najar, Rim, Samir Aydi, Sameh Sassi-Aydi, Abdennabi Zarai, and Chedly Abdelly. "Effect of salt stress on photosynthesis and chlorophyll fluorescence in *Medicago truncatula*." *Plant Biosystems-An International Journal Dealing with all Aspects of Plant Biology* 153, no. 1 (2019): 88–97.

Nakhaie, Ali, Ghader Habibi, and Atousa Vaziri. "Exogenous proline enhances salt tolerance in acclimated Aloe vera by modulating photosystem II efficiency and antioxidant defense." *South African Journal of Botany* 147 (2022): 1171–1180.

Nawaz, Aniqa, Muhammad Shahbaz, Asadullah, Asma Imran, Muhammad U. Marghoob, Muhammad Imtiaz, and Fathia Mubeen. "Potential of salt tolerant PGPR in growth and yield augmentation of wheat (*Triticum aestivum* L.) under saline conditions." *Frontiers in Microbiology* 11 (2020): 2019.

Naz, Maliha, Sadam Hussain, Imran Ashraf, and Muhammad Farooq. "Exogenous application of proline and phosphorus help improving maize performance under salt stress." *Journal of Plant Nutrition* 46, no. 10 (2023): 2342–2350.

Nazar, Rahat, Shahid Umar, and Nafees A. Khan. "Exogenous salicylic acid improves photosynthesis and growth through increase in ascorbate-glutathione metabolism and S assimilation in mustard under salt stress." *Plant Signaling & Behavior* 10, no. 3 (2015): e1003751.

Nazir, Faroza, Laiba Khan, Shahid Umar, and M. Iqbal R. Khan. "Omics-based strategies for improving salt tolerance in rice." In *Current Omics Advancement in Plant Abiotic Stress Biology*, pp. 63–70. Academic Press, 2024.

Ndayiragije, Alexis, and Stanley Lutts. "Long term exogenous putrescine application improves grain yield of a salt-sensitive rice cultivar exposed to NaCl." *Plant and Soil* 291 (2007): 225–238.

Ndiate, Ndiaye Ibra, Qudsia Saeed, Fasih Ullah Haider, Cai Liqun, Jackson Nkoh Nkoh, and Adnan Mustafa. "Co-application of biochar and arbuscular mycorrhizal fungi improves salinity tolerance, growth and lipid metabolism of maize (*Zea mays* L.) in an alkaline soil." *Plants* 10, no. 11 (2021): 2490.

Noreen, Sibgha, Maham Sultan, Muhammad Salim Akhter, et al. "Foliar fertigation of ascorbic acid and zinc improves growth, antioxidant enzyme activity and harvest index in barley (*Hordeum vulgare* L.) grown under salt stress." *Plant Physiology and Biochemistry* 158 (2021): 244–254.

Ondrasek, Gabrijel, Zed Rengel, and Szilvia Veres. "Soil salinisation and salt stress in crop production." In *Abiotic Stress in Plants-Mechanisms and Adaptations*, pp. 171–190. InTech, 2011. ISBN: 978-953-307-394-1

Parida, Asish Kumar, and Anath Bandhu Das. "Salt tolerance and salinity effects on plants: a review." *Ecotoxicology and Environmental Safety* 60, no. 3 (2005): 324–349.

Park, Hee Jin, Woe-Yeon Kim, and Dae-Jin Yun. "A new insight of salt stress signaling in plant." *Molecules and Cells* 39, no. 6 (2016): 447.

Phour, Manisha, and Satyavir Singh Sindhu. "Amelioration of salinity stress and growth stimulation of mustard (*Brassica juncea* L.) by salt-tolerant Pseudomonas species." *Applied Soil Ecology* 149 (2020): 103518.

Puyang, Xuehua, Mengying An, Lixin Xu, Liebao Han, and Xunzhong Zhang. "Protective effect of exogenous spermidine on ion and polyamine metabolism in Kentucky bluegrass under salinity stress." *Horticulture, Environment, and Biotechnology* 57 (2016): 11–19.

Qin, Wenjie, Hengyu Yan, Bingyin Zou, et al. "Arbuscular mycorrhizal fungi alleviate salinity stress in peanut: evidence from pot-grown and field experiments." *Food and Energy Security* 10, no. 4 (2021): e314.

Rady, Mohamed O. A., Wael M. Semida, Taia A. Abd El-Mageed, Khaulood A. Hemida, and Mostafa M. Rady. "Up-regulation of antioxidative defense systems by glycine betaine foliar application in onion plants confer tolerance to salinity stress." *Scientia Horticulturae* 240 (2018): 614–622.

Ran, Xin, Xiao Wang, Xiaokuan Gao, Haiyong Liang, Bingxiang Liu, and Xiaoxi Huang. "Effects of salt stress on the photosynthetic physiology and mineral ion absorption and distribution in white willow (*Salix alba* L.)." *PloS One* 16, no. 11 (2021): e0260086.

Rathinapriya, Periyasamy, Subramani Pandian, Kasinathan Rakkammal, et al. "The protective effects of polyamines on salinity stress tolerance in foxtail millet (*Setaria italica* L.), an important C4 model crop." *Physiology and Molecular Biology of Plants* 26 (2020): 1815–1829.

Rattan, Amandeep, Dhriti Kapoor, Nitika Kapoor, Renu Bhardwaj, and Anket Sharma. "Brassinosteroids regulate functional components of antioxidative defense system in salt stressed maize seedlings." *Journal of Plant Growth Regulation* 39 (2020): 1465–1475.

Rhaman, Mohammad Saidur, Shahin Imran, Farjana Rauf, et al. "Seed priming with phytohormones: An effective approach for the mitigation of abiotic stress." *Plants* 10, no. 1 (2020): 37.

Sadak, Mervat Sh, Asmaa R. Abd El-Hameid, Faten S. A. Zaki, Mona G. Dawood, and Mohamed E. El-Awadi. "Physiological and biochemical responses of soybean (*Glycine max* L.) to cysteine application under sea salt stress." *Bulletin of the National Research Centre* 44 (2020): 1–10.

Santander, Christian, Antonieta Ruiz, Susana García, Ricardo Aroca, Jonathan Cumming, and Pablo Cornejo. "Efficiency of two arbuscular mycorrhizal fungal inocula to improve saline stress tolerance in lettuce plants by changes of antioxidant defense mechanisms." *Journal of the Science of Food and Agriculture* 100, no. 4 (2020): 1577–1587.

Sapre, Swapnil, Iti Gontia-Mishra, and Sharad Tiwari. "*Klebsiella sp.* confers enhanced tolerance to salinity and plant growth promotion in oat seedlings (*Avena sativa*)." *Microbiological Research* 206 (2018): 25–32.

Sardar, Hasan, Zubair Khalid, Muhammad Ahsan, et al. "Enhancement of salinity stress tolerance in lettuce (*Lactuca sativa* L.) via foliar application of nitric oxide." *Plants* 12, no. 5 (2023): 1115.

Schultz, Bart, C. D. Thatte, and V. K. Labhsetwar. "Irrigation and drainage. Main contributors to global food production." *Irrigation and Drainage: The Journal of the International Commission on Irrigation and Drainage* 54, no. 3 (2005): 263–278.

Serrano, Ramón, Jose M. Mulet, Gabino Rios, et al. "A glimpse of the mechanisms of ion homeostasis during salt stress." *Journal of Experimental Botany* (1999): 1023–1036.

Shabala, Sergey, Olga Babourina, and Ian Newman. "Ion-specific mechanisms of osmoregulation in bean mesophyll cells." *Journal of Experimental Botany* 51, no. 348 (2000): 1243–1253.

Shafi, Mohammad, Jehan Bakht, Mohammad Jaffar Hassan, Mohammad Raziuddin, and Guoping Zhang. "Effect of cadmium and salinity stresses on growth and antioxidant enzyme activities of wheat (*Triticum aestivum* L.)." *Bulletin of Environmental Contamination and Toxicology* 82 (2009): 772–776.

Shah, Syed Haleem, Rasmus Houborg, and Matthew F. McCabe. "Response of chlorophyll, carotenoid and SPAD-502 measurement to salinity and nutrient stress in wheat (*Triticum aestivum* L.)." *Agronomy* 7, no. 3 (2017): 61.

Shahid, Mohammad, Mohammad Tarique Zeyad, Asad Syed, et al. "Stress-tolerant endophytic isolate *Priestia aryabhattai* BPR-9 modulates physio-biochemical mechanisms in wheat (*Triticum aestivum* L.) for enhanced salt tolerance." *International Journal of Environmental Research and Public Health* 19, no. 17 (2022a): 10883.

Shahid, Sadia, Muhammad Shahbaz, Muhammad Faisal Maqsood, et al. "Proline-induced modifications in morpho-physiological, biochemical and yield attributes of pea (*Pisum sativum* L.) cultivars under salt stress." *Sustainability* 14, no. 20 (2022b): 13579.

Shams, Mostafakamal, Ertan Yildirim, Melek Ekinci, et al. "Exogenously applied glycine betaine regulates some chemical characteristics and antioxidative defence systems in lettuce under salt stress." *Horticulture, Environment, and Biotechnology* 57 (2016): 225–231.

Shankar, Ajay, and Vishal Prasad. "Potential of desiccation-tolerant plant growth-promoting rhizobacteria in growth augmentation of wheat (*Triticum aestivum* L.) under drought stress." *Frontiers in Microbiology* 14 (2023): 1017167.

Sharma, Isha, Erwan Ching, Shivani Saini, Renu Bhardwaj, and Pratap Kumar Pati. "Exogenous application of brassinosteroid offers tolerance to salinity by altering stress responses in rice variety Pusa Basmati–1." *Plant Physiology and Biochemistry* 69 (2013): 17–26.

Shavrukov, Yuri. "Salt stress or salt shock: which genes are we studying?" *Journal of Experimental Botany* 64, no. 1 (2013): 119–127.

Shelden, Megan C., Daniel A. Dias, Nirupama S. Jayasinghe, Antony Bacic, and Ute Roessner. "Root spatial metabolite profiling of two genotypes of barley (*Hordeum vulgare* L.) reveals differences in response to short-term salt stress." *Journal of Experimental Botany* 67, no. 12 (2016): 3731–3745.

Simaei, M., R. A. Khavarinejad, S. Saadatmand, F. Bernard, and H. Fahimi. "Interactive effects of salicylic acid and nitric oxide on soybean plants under NaCl salinity." *Russian Journal of Plant Physiology* 58 (2011): 783–790.

Singh, Udai B., Deepti Malviya, Shailendra Singh, et al. "Salt-tolerant compatible microbial inoculants modulate physio-biochemical responses enhance plant growth, Zn biofortification and yield of wheat grown in saline-sodic soil." *International Journal of Environmental Research and Public Health* 18, no. 18 (2021): 9936.

Sivritepe, H. Özkan, Nuray Sivritepe, Atilla Eriş, and Ece Turhan. "The effects of NaCl pre-treatments on salt tolerance of melons grown under long-term salinity." *Scientia Horticulturae* 106, no. 4 (2005): 568–581.

Skorupa, Monika, Marcin Gołębiewski, Katarzyna Kurnik, et al. "Salt stress vs. salt shock-the case of sugar beet and its halophytic ancestor." *BMC Plant Biology* 19 (2019): 1–18.

Sofy, Mahmoud R., Nevien Elhawat, and Tarek Alshaal. "Glycine betaine counters salinity stress by maintaining high K+/Na+ ratio and antioxidant defense via limiting Na+ uptake in common bean (*Phaseolus vulgaris* L.)." *Ecotoxicology and Environmental Safety* 200 (2020): 110732.

Soliman, Mona, Amr Elkelish, Trabelsi Souad, Haifa Alhaithloul, and Muhammad Farooq. "Brassinosteroid seed priming with nitrogen supplementation improves salt tolerance in soybean." *Physiology and Molecular Biology of Plants* 26 (2020): 501–511.

Su, Qiufang, Xiaodong Zheng, Yike Tian, and Caihong Wang. "Exogenous brassinolide alleviates salt stress in *Malus hupehensis* Rehd. by regulating the transcription of NHX-Type Na+ (K+)/H+ antiporters." *Frontiers in Plant Science* 11 (2020): 38.

Sudhir, P., and S. D. S. Murthy. "Effects of salt stress on basic processes of photosynthesis." *Photosynthetica* 42 (2004): 481–486.

Sui, Na, Zhen Yang, Mingli Liu, and Baoshan Wang. "Identification and transcriptomic profiling of genes involved in increasing sugar content during salt stress in sweet sorghum leaves." *BMC Genomics* 16, no. 1 (2015): 1–18.

Sukweenadhi, Johan, Yeon-Ju Kim, Kwang Je Lee, et al. "*Paenibacillus yonginensis* sp. nov., a potential plant growth promoting bacterium isolated from humus soil of Yongin forest." *Antonie van Leeuwenhoek* 106 (2014): 935–945.

Tahjib-Ul-Arif, Md, Md Nurealam Siddiqui, Abdullah Al Mamun Sohag, et al. "Salicylic acid-mediated enhancement of photosynthesis attributes and antioxidant capacity contributes to yield improvement of maize plants under salt stress." *Journal of Plant Growth Regulation* 37 (2018): 1318–1330.

Takemura, Taro, Nobutaka Hanagata, Zvy Dubinsky, and Isao Karube. "Molecular characterization and response to salt stress of mRNAs encoding cytosolic Cu/Zn superoxide dismutase and catalase from *Bruguiera gymnorrhiza*." *Trees* 16 (2002): 94–99.

Temizgul, Ridvan, Mahmut Kaplan, Rukiye Kara, and Semih Yilmaz. "Effects of salt concentrations on antioxidant enzyme activity of grain sorghum." *Current Trends in Natural Sciences* 5, no. 9 (2016): 171–178.

Turan, Metin, Melek Ekinci, Raziye Kul, Fatma G. Boynueyri, and Ertan Yildirim. "Mitigation of salinity stress in cucumber seedlings by exogenous hydrogen sulfide." *Journal of Plant Research* 135, no. 3 (2022): 517–529.

Tuteja, Narendra. "Mechanisms of high salinity tolerance in plants." *Methods in Enzymology* 428 (2007): 419–438.

Vij, Shubha, and Akhilesh K. Tyagi. "Emerging trends in the functional genomics of the abiotic stress response in crop plants." *Plant Biotechnology Journal* 5, no. 3 (2007): 361–380.

Wang, Hao, Tingting An, Di Huang, Runjin Liu, et al. "Arbuscular mycorrhizal symbioses alleviating salt stress in maize is associated with a decline in root-to-leaf gradient of Na+/K+ ratio." *BMC Plant Biology* 21, no. 1 (2021): 1–15.

Wang, Xuchu, Pengxiang Fan, Hongmiao Song, Xianyang Chen, Xiaofang Li, and Yinxin Li. "Comparative proteomic analysis of differentially expressed proteins in shoots of Salicornia europaea under different salinity." *Journal of Proteome Research* 8, no. 7 (2009): 3331–3345.

Wang, Zhou-fei, Jian-fei Wang, Yong-mei Bao, Fu-hua Wang, and Hong-sheng Zhang. "Quantitative trait loci analysis for rice seed vigor during the germination stage." *Journal of Zhejiang University Science B* 11 (2010): 958–964.

Watanabe, Shin, Katsumi Kojima, Yuji Ide, and Satohiko Sasaki. "Effects of saline and osmotic stress on proline and sugar accumulation in *Populus euphratica* in vitro." *Plant Cell, Tissue and Organ Culture* 63 (2000): 199–206.

Wei, Long, Haiyan Zhao, Baoxiang Wang, et al. "Exogenous melatonin improves the growth of rice seedlings by regulating redox balance and ion homeostasis under salt stress." *Journal of Plant Growth Regulation* 41, no. 6 (2022): 2108–2121.

Yadav, Sangeeta, Mohd Irfan, Aqil Ahmad, and Shamsul Hayat. "Causes of salinity and plant manifestations to salt stress: a review." *Journal of Environmental Biology* 32, no. 5 (2011): 667.

Yadu, Shrishti, Teman Lal Dewangan, Vibhuti Chandrakar, and S. Keshavkant. "Imperative roles of salicylic acid and nitric oxide in improving salinity tolerance in *Pisum sativum* L." *Physiology and Molecular Biology of Plants* 23 (2017): 43–58.

Yildirim, Ertan, Melek Ekinci, Metin Turan, Atilla Dursun, Raziye Kul, and Fazilet Parlakova. "Roles of glycine betaine in mitigating deleterious effect of salt stress on lettuce (*Lactuca sativa* L.)." *Archives of Agronomy and Soil Science* 61, no. 12 (2015): 1673–1689.

Yildirim, Ertan, Melek Ekinci, Metin Turan, Selda Ors, and Atilla Dursun. "Physiological, morphological and biochemical responses of exogenous hydrogen sulfide in salt-stressed tomato seedlings." *Sustainability* 15, no. 2 (2023): 1098.

Yurtseven, E., G. D. Kesmez, and A. Ünlükara. "The effects of water salinity and potassium levels on yield, fruit quality and water consumption of a native central anatolian tomato species (*Lycopersicon esculantum*)." *Agricultural Water Management* 78, nos. 1–2 (2005): 128–135.

Zhang, Pengfei, Lei Liu, Xin Wang, et al. "Beneficial effects of exogenous melatonin on overcoming salt stress in sugar beets (*Beta vulgaris* L.)." *Plants* 10, no. 5 (2021): 886.

Zhang, Pengfei, Masateru Senge, and Yanyan Dai. "Effects of salinity stress at different growth stages on tomato growth, yield, and water-use efficiency." *Communications in Soil Science and Plant Analysis* 48, no. 6 (2017): 624–634.

Zhao, Shuangshuang, Qikun Zhang, Mingyue Liu, Huapeng Zhou, Changle Ma, and Pingping Wang. "Regulation of plant responses to salt stress." *International Journal of Molecular Sciences* 22, no. 9 (2021): 4609.

Zheng, Jian, Xiaohua Ma, Xule Zhang, Qingdi Hu, and Renjuan Qian. "Salicylic acid promotes plant growth and salt-related gene expression in *Dianthus superbus* L. (Caryophyllaceae) grown under different salt stress conditions." *Physiology and Molecular Biology of Plants* 24, no. 2 (2018): 231–238.

7 Plant Stress Hormones
Fostering Drought Stress Resilience

Moksh Mahajan, Sayeda Khatoon, Sneha Prakash, and M. Iqbal R. Khan

7.1 INTRODUCTION

The global population is steadily increasing and is anticipated to grow by more than 5 billion by 2050 (Mottaleb et al. 2023). Meeting the food demands of this expanding population requires a substantial increase in both the quantity and the quality of food production, estimated to be up to 56% by 2050 (Van Dijk et al. 2021). Abiotic stresses such as heat, cold, salinity, and drought have a profound impact on plant growth, causing significant reductions in crop productivity and widespread losses (Chaudhry and Sidhu 2022). In response to these stresses, plants activate various physio-biochemical and molecular mechanisms that affect their morphology and cellular processes. Drought stress, specifically resulting from a lack of water precipitation, creates soil water deficits, leading to a decrease in groundwater levels and potentially impeding the growth and survival of plants (Khan et al. 2021). The primary defence mechanism against water loss is the plant cuticle, an extracellular lipid polymer located in the epidermis, safeguarding all aboveground plant organs (Xue et al. 2017). Plant species lack a well-defined immune system, but they have signalling mechanisms to ensure optimized growth and secure overall development. Plant stress hormones such as ethylene and abscisic acid (ABA) serve as central key hormones to link and reprogram complex developmental mechanisms and adaptive stress responses (Parwez et al. 2022). ABA is an essential key regulator that regulates many aspects of plant growth and development, including embryo maturation, seed dormancy and germination, floral induction, and root growth (Chen et al. 2020b). ABA also helps reduce the adverse effects related to stresses in plants, such as drought, thereby enabling plant species to adapt to a continuously changing environment and maintain physiological processes under drought stress (Ullah et al. 2018). Plant physiological processes can be regulated by modulating the expression of various ABA-responsive genes responsible for the opening and closing of the stomatal aperture (Hsu et al. 2021). Drought stress can have detrimental effects on plants by disrupting various activities, including carbon assimilation, leaf gas exchange, and turgor pressure, and causing increased oxidative damage. Additionally, it can impact on ionic balance, enzymatic activity, stem extension, leaf size, and root proliferation, ultimately leading to a reduction in yield (Manavalan and Nguyen 2017; Mukarram et al. 2021; Wahid et al. 2020). Plants, being sessile, cannot avoid both biotic and abiotic stresses. While lacking a well-defined immune system, they employ signalling mechanisms to optimize growth and ensure overall fitness.

The acquisition of ethylene plays a crucial role in plants, especially in response to varying levels of drought exposure, affecting all stages of plant development (Salvi et al. 2021). Moreover, increased ethylene production under drought stress has detrimental effects on growth rate, photochemical efficiency, and grain yield by hastening leaf and flower abscission and senescence (Habben et al. 2014; Nazir et al. 2024), suggesting a dual function of ethylene in plants. Drought stress influences source-sink dynamics and osmotic adjustments by primarily controlling ethylene production in reproductive tissues such as floral, pedicel, calyx, and fruit (Albacete et al. 2014). The external application of ethylene, including ethephon, has a positive impact on plants, enhancing their drought stress tolerance (Arraes et al. 2015). This positive effect is achieved through the regulation of cuticular wax production, crucial for maintaining water levels and membrane stability essential for plant

DOI: 10.1201/9781032630380-7

growth (Yu et al. 2017). Ethylene signalling is vital for alleviating challenges induced by drought in plants, as it regulates various physiological and metabolic processes. This includes the balance between managing oxidative stress and antioxidant systems (Khan et al. 2016; Sehar et al. 2021; Czékus et al. 2022; Rehman et al. 2022). Furthermore, ethylene contributes to the modulation of carbohydrate metabolism and the supply of photosynthates, proline metabolism, and the production of dry matter (Man et al. 2011; Cui et al. 2015; Khan et al. 2019), ultimately enhancing crop productivity (Ojuederie et al. 2019). In essence, ethylene plays a crucial role in fine-tuning plant growth and facilitating tolerance to drought stress.

The goal of this chapter is to advance our thorough comprehension of the crucial components of ethylene and ABA's involvement in improving plants' resistance to drought stress. It emphasizes the significance of ethylene and ABA signalling components in controlling plant growth and metabolic responses, such as source-sink dynamics, root architecture, and osmotic and oxidative stress metabolism, in order to regulate plant defence and impart drought stress resistance. Moreover, the mechanisms underpinning signal transduction involved in ethylene and ABA signalling have also been detailed to broaden our understanding of drought tolerance.

7.2 PLANT STRESS HORMONES

7.2.1 Ethylene Synthesis under Drought Stress

Ethylene, an endogenous gaseous plant hormone, plays a major role in regulating plant developmental and physiological processes ranging from germination to senescence. It plays major roles in seed germination, fruit ripening, nutrient allocation, and senescence (Iqbal et al. 2017). Ethylene easily diffuses through cells and binds to ethylene receptors to trigger ethylene responses (Gautam et al. 2024). Ethylene's function was first identified in dark-grown pea seedlings which displayed a set of morphological features known as triple response (Chen et al. 2021a).

Biosynthesis of ethylene comprises a cascade of three enzyme-catalyzed reactions. Ethylene is derived from methionine, an amino acid where methionine is first converted into *S*-adenosylmethionine (SAM) by SAM synthetases. SAM is further converted into 1-amino cyclopropane-1-carboxylic acid (ACC) with the help of ACC synthetases (ACS; Rodrigues et al. 2014). Another by-product MTA (5′-methylthioadenosine) is formed in this step. MTA plays a major role in reforming methionine for further formation of ethylene (Pattyn et al. 2021). Finally, ACC is converted to ethylene via ACC oxidases (ACO) (Singh et al. 2022). SAM is also involved in the biosynthesis of polyamines, which also plays a major role in stress responses (Ali et al. 2020).

Ethylene, in addition to influencing major stages of a plant's growth, also plays a critical role in stress defence (Baharudin and Osman 2023). ACS is the major rate-limiting enzyme in ethylene production, making it a prime target for studying stress-mediated regulation (Park et al. 2021). ACC and ACS were up-regulated under drought stress in soybean (*Glycine max*; Arraes et al. 2015). Enhanced expression of ACS enzyme corresponds to increased tolerance in *Thymus kotschyanus* to drought stress (Ashrafi et al. 2022). In *Arabidopsis* (*Arabidopsis thaliana*), a total of 12 ACS genes have been identified. In addition to their role in the formation of ACC, these ACS genes also play a distinctive role in plant growth (Ahmadizadeh et al. 2020). Depending upon the sequence of its C-terminal region ACS protein can be divided into three types. Type I ACS proteins are those which undergo phosphorylation by mitogen-activated protein kinases (*AtMPK3–6*) and calcium dependent kinases (*AtCDPK2*), type II which undergo phosphorylation by calcium dependent protein kinases (CPK), and type III which do not undergo phosphorylation by either kinase. Moreover, secondary messengers like calcium and phytohormones can regulate ACSs. When there is no putative signal present, 26S proteasome degrades type II ACS by abridging it. Ethylene overproducer (ETO) and ETO-like (EOL) proteins help in the degradation process by tagging the C-terminal of ACS for degradation. MPK3–6 kinases protect the type I ACS degradation by phosphorylizing the C-terminal thereby resulting in enhanced ethylene production under stressed conditions (Arraes et al. 2015).

Under drought stress, *phaco1* and *phaco3* mutant petunia were found less tolerant in comparison to wild plants due to less ethylene production (Naing et al. 2022). Similarly in rose (*Rosa siinensis*), *RhACS1* and *RhACS2* expression was enhanced under drought stress in roots and decreased under rehydration of the plants (Liu et al. 2013). Dalal et al. (2018) suggested that in wheat plants, up-regulation of ethylene biosynthesis pathway genes, such as ACO and 2-oxoglutarate/FeII (2OGFeII) oxygenases, enhanced ethylene to encourage root development under drought stress. Furthermore, in rice seedlings up-regulation of acireductone dioxygenase 1 (OsARD1), which enhances ethylene biosynthesis, exhibited an enhanced root growth, thereby aiding in drought tolerance (Liang et al. 2019). Also, in rice (*Oryza sativa*) ETO1 mutants with an insensitive ethylene overproducer 1-like (OsETOL1) protein aids in combating drought stress as plants over-expressing OsETOL1 exhibits lower ethylene production and thus shows susceptibility to drought stress conditions (Du et al. 2014). In pearl millet (*Pennisetum glaucum*) PRLT2/89–33, a drought-tolerant variety, the expression of ACO was found to enhance under drought stress conditions (Shivhare et al. 2020). This highlights the importance of understanding changes in ethylene biosynthesis under drought stress to undergo biochemical alterations to cope with drought stress.

7.2.2 Abscisic Acid Synthesis under Drought Stress

The main mechanism through which water exchanges with the atmosphere is through leaves. The major benefit of ABA production in leaves, especially in dry spells, is that leaves can stop excessive water potential reductions. Research has shown that ABA is rapidly synthesized in leaves as a result of exposure to high vapour pressure deficit (VPD), which allows angiosperm species to efficiently close their stomata in a short amount of time (Liu et al. 2022a). Up until a certain point, there is a strong correlation between increases in foliar ABA and a drop in leaf water potential, which is directly linked to bulk leaf turgor loss. It is generally agreed upon that a substantial increase in foliar ABA biosynthesis occurs when the water potential at turgor loss occurs, but it is still unclear whether turgor fluctuations or changes in cell volume serve as the primary cues for up-regulating the enzymes that produce ABA (Cao et al. 2020). Furthermore, because guard cells and nearby mesophyll cells have lower water potentials during transpiration, it is expected that these cells will produce small amounts of ABA as transpirational demand increases, resulting in stomatal closure before the leaf turgor loss point (Li et al. 2021).

A precise coordination between ABA production, catabolism, conjugation, and transportation is required to regulate ABA levels (Zhang et al. 2021b). The ABA biosynthesis route and the identification of key enzymes have been elucidated by means of initial biochemical investigations, followed by genetic and physiological analysis of ABA-deficient mutants (Ablazov et al. 2020). It has been discovered that environmental stimuli, particularly drought, have an impact on specific genes that encode the ABA metabolic enzymes, suggesting that ABA metabolism is sensitive to feedback regulation.

With two exceptions, plastids are the primary sites of ABA synthesis. Here, xanthoxin is transformed into ABA during the last two stages of the process. Isopentenyl pyrophosphate (IPP), the biological isoprene unit, is the starting point of the "indirect" pathway to ABA production (Jia et al. 2022). In a two-step epoxidation process mediated by ZEP/AtABA1, zeaxanthin is transformed to trans-violaxanthin; over-expressing ZEP in transgenic plants improves drought tolerance. The next stage is the oxidative cleavage of 9-cis-violaxanthin and/or 9-cis-neoxanthin, which is catalyzed by NCED (9-cis epoxycarotenoid dioxygenase). This produces the 15C molecule xanthoxin, which is thought to be the rate-limiting step in plastids. Rapid activation is suggested by the rapid induction of NCED gene expression, particularly NCED3 in response to a water deficit. Xanthoxin is subsequently transferred to the cytosol and transformed to ABA via an intermediate, abscisic aldehyde, via SDR/AtABA2, followed by oxidation to ABA by aldehyde oxidase / ascorbate oxidase (AAO/AO) (Mercado-Reyes et al. 2024).

Enzymes involved in ABA production are more prevalent in parenchyma cells found in vascular bundles in *Arabidopsis*, both in drought- and non-stressed environments. Phloem companion cells exhibit elevated activity of ABA biosynthesizing enzymes. Guard cells have the ability to synthesize ABA on their own, which allows them to react quickly to changes in their surroundings and support the maintenance of water status homeostasis. The exact mechanisms behind the possible relationship between stomatal control and ABA production in vascular and guard cells are still unknown.

Comparatively speaking, ABA catabolism plays a simpler role than biosynthesis. A variety of methyl groups can hydroxylate ABA. Of the hydroxylated products, 8′-hydroxy ABA can subsequently be converted to phaseic acid (PA) and subsequently dihydrophaseic acid (DPA). With no ABA-like action, DPA is the final by-product of ABA catabolism (Yang et al. 2023). In barley under drought stress, there is an up-regulation of *HvABA8′OH-1*, which codes for ABA8′OH (cytochrome P450 belonging to the CYP707 family) and breaks down active ABA into 8′hydroxy-ABA (Collin et al. 2020). This up-regulation is essential to quickly reduce excess ABA in response to drought stress and regulate ABA concentration. Exogenous ABA treatment and dehydration both increased the mRNA levels of CYP707As (ABA 8-hydroxylase) in *Arabidopsis* (Khan et al. 2020). Apart from the processes of ABA biosynthesis and catabolism, ABA de/conjugation has been recognized as a swift method for releasing active ABA in reaction to external stressors (Chen et al. 2020a). Inactivated ABA gathers in the apoplastic space or vacuoles as ABA glucosyl ester (ABA-GE) catalyzed by ABA glucosyltransferase through conjugation reaction. Lower ABA levels and an ABA-deficient phenotype were found in analyses of the *Arabidopsis atbg1* knockout mutant (Jiao et al. 2020). Conversely, ABA accumulation brought on by the drought was more noticeable in over-expression lines. Subsequent research revealed that *BG1* expression in wild-type *Arabidopsis* was substantially higher under water stress than it was under normal circumstances, highlighting the significance of ABA de/conjugation pathways in controlling ABA levels in response to stress.

7.3 ROLE OF STRESS HORMONES IN DROUGHT STRESS

7.3.1 Role of Ethylene under Drought Stress

Under drought stress, plants undergo an array of detrimental effects such as diminished cell growth, stomatal closure, and reduced water potential ultimately reducing photosynthesis and yield or causing death under severe stress (Zia et al. 2021). The most negative impact undergone by plant due to water scarcity is oxidative stress. Drought stress leads to the formation of reactive oxygen species (ROS) such as superoxide anion (O_2^-), hydrogen peroxide (H_2O_2), and hydroxyl anion (OH^-) resulting in lipid peroxidation, membrane instability, and cell death (Zou et al. 2021). Under drought, ethylene depicts an antagonistic response where it enhances ROS in some species while decreasing ROS in others (Yu et al. 2017). Ethylene enabled the osmolyte-mediated scavenging of ROS in *Arabidopsis* under drought-stressed conditions (Table 7.1; Cui et al. 2015).

Exogenous ethylene treatment entails a positive effect on plant growth and mitigates drought stress. Moreover, the direct precursor of ethylene, ACC, has been thoroughly studied due to its involvement in a plant's resistance to drought. Plants facing drought-related challenges benefit greatly from ethylene signalling regulation of multiple physiological and metabolic processes, which includes preserving the balance between antioxidant defence and oxidative stress. Studies suggest the enhanced expression of genes involved in antioxidant metabolism under drought stress after ethylene treatment in Kentucky bluegrass (*Poa pratensis*; Zhang et al. 2018, 2021a). Tobacco (*Nicotiana tabacum*) and sugarcane (*Saccharum officinarum*) seedlings also depicted a positive response where ACC implementation decreased the ROS content via enhanced activity of peroxidase (POX) and superoxide dismutase (SOD) under drought stress (Xiao Le et al. 2014; Wang et al. 2016). This suggests that ethylene regulated the antioxidant enzymes under drought stress to maintain osmoticum of the plant. Furthermore, plants tend to accumulate osmolytes such as proline, mannitol, phenols, soluble sugars, alpha-tocopherol, and flavonoids under drought stress to counteract the negative impact of drought stress and safeguard the plant. Osmolytes tend to protect plants against stress via signalling stress defence mechanisms (Saha et al. 2022). Ethylene (1.0 mM)

TABLE 7.1
Ethylene-Mediated Changes in Physiological Traits under Drought Stress

Plant species	Source	Mode of application	Physiological changes	References
Poa pratensis (Kentucky bluegrass)	Ethephon; 150 mg L^{-1}	Seed priming	Enhanced antioxidant metabolism through increasing ascorbate peroxidase (APX), peroxidase (POD), and catalase (CAT) activities, along with proline content. Decreased ROS level and maintained high relative water content (RWC).	Zhang et al. (2018)
Nicotiana tabacum (Tobacco)	ACC; 0, 0.5, 5, and 100 μM	Through Murashige and Skoog (MS) medium	Lessened the leaf water loss (LWL). Enhanced antioxidant enzymes (POD and SOD). High accumulation of soluble sugar and proline contents.	Wang et al. (2016)
Saccharum officinarum (Sugarcane)	Ethephon; 300 mg L^{-1}	Foliar spray	Enhanced antioxidant activity	Xiao-Le et al. (2014)
Kentucky bluegrass	Ethephon; 200 mg L^{-1}	Foliar spray	Enhanced gene expression related to cuticular wax synthesis and ROS scavenging proteins.	Zhang et al. (2021a)
Zea mays (Maize)	Ethephon; 1.0 mM	Foliar spray	Increased cuticular wax synthesis gene expression such as *ZmERE*, *ZmGL1*, *ZmGL15*, *ZmFDH1*, and *ZmFAE1*. Reduction in LWL and ROS. Increase in proline, antioxidant enzymes, and root/shoot ratio.	Yu et al. (2017)
Poncirus trifolia (Trifoliate orange)	Ethephon; 1.0 mol L^{-1}	Seed priming	Enhanced root hairs.	Zhang et al. (2013b)
Triticum aestivum (Wheat)	ACC; 0.1 μM	Foliar spray	Enhanced shoot biomass, chlorophyll content, leaf area, and carbohydrate content.	Valluru et al. (2016)
Maize	Ethephon; 0, 0.56, and 0.84 kg ha^{-1}	Foliar spray	Enhanced grain yield.	Shekoofa and Emam (2007)

treatment enhanced the proline accumulation and cuticular wax under drought stress in maize to reduce leaf water loss and oxidative stress (Yu et al. 2017). Ethephon treatment enhances proline biosynthesis in plants under drought stress. Proline protects plants against drought stress by protecting cell membrane integrity (Hosseinifard et al. 2022). Soluble sugars such as glucose and sucrose regulate many key processes in plant growth and development. In addition to that, soluble sugars have shown promising effects in cell regulation, photosynthesis, and stress defence under drought stress. Enhanced soluble sugars accumulation improved tolerance to drought stress in plants (Sami et al. 2016; Ozturk et al. 2021). In *Arabidopsis* plants, wild plants with impaired enhanced soluble sugars and proline content were found more tolerant to drought stress as compared to ethylene-defective mutants (Cui et al. 2015). This further emphasizes the importance of ethylene in drought stress alleviation via enhanced soluble sugars and proline content. Root architecture significantly influences the plant's adaptability to confront drought stress. It mainly comprises structural features of root including surface area, root hairs, root length, and lateral roots which show flexibility under drought conditions and play a key role in developing drought-tolerant plants (Ranjan et al. 2022). Deep-rooted plants are more tolerant to drought-stressed conditions as they can take up water from deeper parts of the soil (Wasaya et al. 2018). In *Arabidopsis*, enhanced root length in *Col-0* seedlings conferred greater tolerance under drought-stressed conditions compared to ethylene mutants (Cui et al. 2015). Ethylene mediates roots water absorbance through expanding, which eventually causes plants to adjust and minimize loss of water though stomata. In trifoliate orange (*Poncirus trifolia*), ethylene application enhanced photosynthesis and root hair development, aiding in drought tolerance (Zhang et al. 2013b). In rice plants, enhanced resistance to drought stress was observed in plants with increased ethylene biosynthesis, which suggests the importance of ethylene in withstanding stress (Zhang et al. 2013a). In addition, ethylene regulates photosynthesis to optimize plant growth and development under stress conditions (Riyazuddin et al. 2020). In wheat (*Triticum aestivum*), exogenous ACC supplementation at lower concentrations enhanced photosynthesis and chlorophyll and carbohydrate content under drought stress (Valluru et al. 2016).

Ethylene interacts with other phytohormones to regulate plant adaptability in drought stress. Ethylene response factor EIN3 up-regulates root development induced by jasmonic acid (JA) under drought stress in *Arabidopsis* (Cui et al. 2015). ABA and ethylene interaction in stomatal responses impacts the plant's regulation in drought stress. Under drought stress, ABA mediates stomatal closure to minimize water loss while exogenous ethylene application inhibits ABA-induced stomatal closure in *Arabidopsis* (Tanaka et al. 2005). Ethylene regulates stomatal activity by reducing ROS produced due to ABA (Watkins et al. 2014). The ultimate focus of the plant is to maintain a higher yield while enduring drought stress. In maize (*Zea mays*), the exogenous treatment of ethephon has significantly enhanced yield under drought stress (Shekoofa and Emam 2007).

7.3.2 Role of ABA under Drought Stress

In response to drought stress, plants employ a sophisticated and finely tuned regulatory system based on ABA to regulate stomatal closure. One essential line of defence against water conservation is stomatal closure. The increase in ABA levels during drought triggers a series of actions that lead to the closure of stomatal pores, which is essential for minimizing water loss (Hasan et al. 2021). In this intricate mechanism, the ABA-induced cellular calcium (Ca^{2+}) signals are recognized by calcium B-like proteins (CBLs) and CBL-interacting proteins kinase (CIPKs; Liu et al. 2022b). The anion channels whose activity these proteins in turn control, changing the volume of guard cells, including slow anion channel-associated 1 (SLAC1) and R-type channels like QUAC1/ALMT12 (Malcheska et al. 2017). These channels constitute a complex network that reacts to ABA signalling, and their activation is carefully regulated by many kinases, including OST1 (SnRK2.6), CPKs/CDPKs, CBL5/CIPK11, and CBL1/9-CIPK23. To emphasize the complex nature of ABA-mediated stomatal control, OST1, a crucial kinase in ABA signalling, is also involved in stimulating the generation of ROS and suppressing the expression of *KAT1*, a potassium (K^+) channel (Qi et al. 2018). The intricate network of contacts that is essential for accurate stomatal responses is shown by the coordinated action of ABA, Ca^{2+}, ROS, and kinases on many additional channels in guard cells (Kim et al. 2010). Multiple kinases, including CPKs and CIPKs, which activate different channels in guard cells, are further regulated by the complex ABA signalling network. As a result of ABA-induced nicotinamide adenine dinucleotide phosphate hydrogen (NADPH) oxidase activation, ROS is produced. These ROS inhibit ABA insensitive 2 (ABI2), which in turn increases guard cell hydrogen peroxide resistant1 (GHR1) activity (Hua et al. 2012). Furthermore, nitric oxide (NO) generation is enhanced by ABA-induced ROS build-up via *NR1*, which contributes to the complex regulation of stomatal closure (Hancock et al. 2011). This complex ABA-mediated signalling cascade in guard cells is comprised of a web of interacting proteins, including ion channels, transcription factors, and kinases. By ensuring water conservation and strengthening defence mechanisms against potential pathogenic threats invading through stomata, plants are able to quickly and efficiently adjust to changing environmental conditions due to the dynamic and responsive character of this network. By combining inputs from several environmental stimuli, ABA functions as a central orchestrator, adjusting stomatal responses. Multiple levels of complexity are added to this regulatory network by the inclusion of ROS generation, NO synthesis, and Ca^{2+} signalling. Plants exhibit amazing environmental sophistication in their ability to navigate and adapt to dry conditions, because of the dynamic interaction between ABA and its downstream effectors.

In response to both internal and external stimuli, plants finely regulate the growth and morphogenesis of their organs. One such regulator that affects many aspects of plant development is ABA. Renowned for its function in seed germination and dormancy, ABA influences root growth in response to environmental stressors including drought and salinity. The negative regulator protein phosphatases type 2C (PP2C) and the cytosolic receptors pyrabactin resistance/PYR1-like/regulatory components of ABA receptors (PYR/PYL/RCAR) are involved in the ABA signalling pathway (Umezawa et al. 2010). The activation of transcription factors downstream is determined by the amount of ABA. The effects of ABA on primary root growth exhibit a delicate balance: high doses hinder, while low amounts promote the creation of new roots (Miao et al. 2021). In order to precisely control root architecture, ABA works in concert with other phytohormones such as auxin (Aux), ethylene, Ca^{2+}, and ROS. Gene expression linked to cell division, differentiation, and stem cell activity in the root meristem is influenced by the interaction of ABA with these signalling components. Notably, by lowering Aux levels and suppressing the expression of genes linked to Aux transport, ABA adversely affects root growth (Xie et al. 2021). Further connecting ABA signalling to the control of root development is the complex network of transcription factors, which includes auxin response factor 2 (ARF2) and PLETHORA (PLTs). The regulatory network is further complicated by the basic region/leucine zipper motif (bZIP) transcription factor Hypocotyl 5 (HY5), which functions as a molecular bridge between ethylene and ABA signalling. ABA also has an impact on root hair growth, negatively regulating it by up-regulating OBF binding protein (OBP4) (Shibata and Sugimoto 2019). An additional layer to the control of root growth is the complex role that ROS play in ABA signalling. The variability of ABA's effects on main and lateral root growth, as well as its complicated interactions with Aux, ethylene, and other signalling components, underscores the intricacy of plant signalling pathways mediated by ABA (Figure 7.1). Determining the precise processes behind these interactions and their unique roles in various plant species would require more investigation.

Gibberellins (GA) and ABA, two important phytohormones that induce and sustain seed dormancy, respectively, are involved in the complicated regulation of seed dormancy (Vishal and Kumar 2018). The main regulatory mechanism controlling seed germination and dormancy is the sensitive and delicate balance between the levels of ABA and GA. The molecular mechanisms of seed dormancy and germination have been clarified by genomic studies that have revealed important genes involved in the metabolism and signalling pathways of ABA and GA. NCED is the rate-limiting enzyme in the multistep process that makes up ABA production. Conversely, ABA 8′-hydroxylase (ABA8′OH), which hydroxylates ABA at the 8′ position, is encoded by the CYP707A gene, which controls catabolism (Millar et al. 2006). The amount of ABA in latent seeds is mostly determined by the ratio of NCED to CYP707A expression. Proteins such as PP2C, which function as negative regulators of ABA signalling and impact seed germination, can regulate the balance between GA and ABA metabolism. PP2C suppresses *OsbZIP10* in rice, increasing the expression of α-amylase and facilitating seed germination (Gong et al. 2022). Comprehending the molecular principles that underpin these processes offers significant insights into crop management, facilitating the formulation of tactics aimed at augmenting seed quality, consistent germination, and aggregate crop yield.

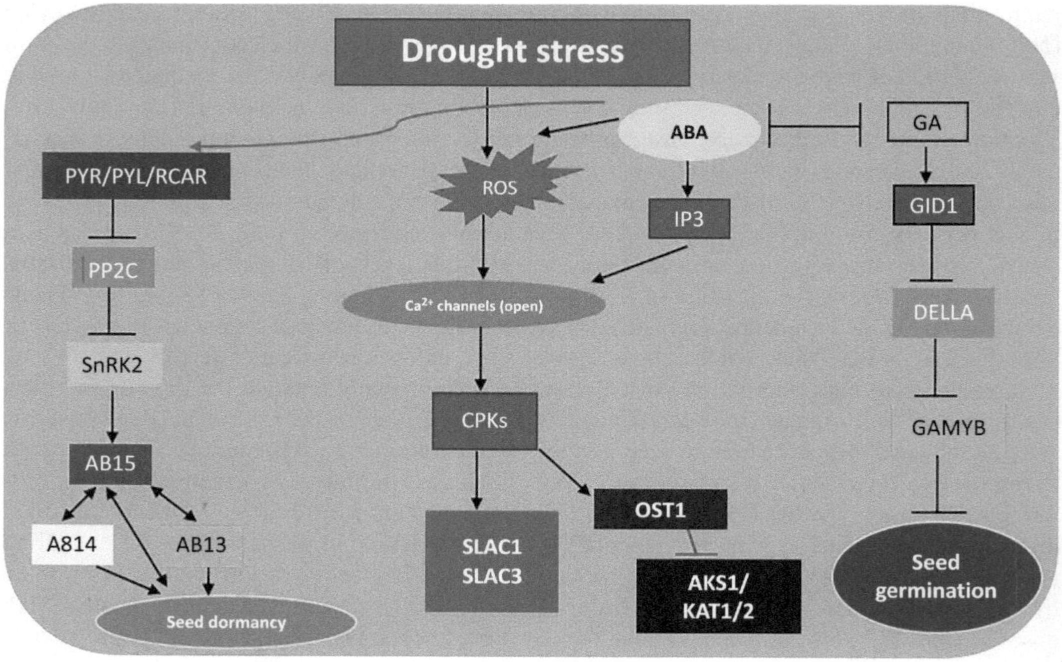

FIGURE 7.1 Role of ABA in drought stress. ABA, abscisic acid; Ca^{2+}, calcium ion; CPK, calcium-dependent protein kinases; GA, gibberellic acid; PP2C, protein phosphatases type 2C; PYR/RCAR/PYL, pyrabactin resistance/PYR1-like/regulatory components of ABA receptors; ROS, reactive oxygen species; SLAC, slow anion channel-associated 1; SnRK2, SNF1-related protein kinase 2.

According to early research, ABA produced under stress regulates the growth and development of shoots in response to water stress (Sharp 2002). Examples of reduced shoot growth were noted when soil dryness raised xylem ABA levels. Recent research has brought attention to the variations in the responses of roots and shoots to ABA. Studies revealed that the suppression of ethylene synthesis, a growth inhibitor under stress, may be part of ABA's function. Furthermore, stress-induced ABA can help maintain restricted root and shoot growth in situations of water scarcity, hence improving drought tolerance. Further research backed up the idea that ABA plays two different physiological roles. Because of its inhibitory function, ABA builds up dramatically in stressed plants, promoting their survival by preventing growth and stomatal opening. On the other hand, under more typical circumstances, ABA functions as a promoter at lower concentrations and is crucial for vegetative growth in a variety of organs, including primary root growth and post-germination seedling development (Rodríguez-Gacio et al. 2009). Furthermore, by catalyzing multiple steps from carotenoid-derived xanthoxin to abscisic aldehyde in the cytosol, short chain dehydrogenases (SDR1) aids in the production of ABA (Shahrajabian et al. 2021). Current approaches, in particular the use of ABA-deficient or ABA-insensitive mutants, have clarified that there are two main routes via which the signalling of water stress may be understood: the ABA-dependent and ABA-independent pathways for gene expression. The ABA-dependent pathway can be further divided into two pathways: the first requires the synthesis of new proteins, whilst the second does not. The ABA-responsive element known as ABRE (PyACGTGGC, where Py stands for a pyrimidine base, C or T) is found in the promoter domain of ABA-responsive genes in the event that new protein synthesis is not required. ABRE can trigger ABA-mediated gene expression through binding to members of the homologous bZIP family of transcription factors, including EmBP-1 (Gao et al. 2022). The dehydration-responsive rd29B gene in *Arabidopsis* which possesses two crucial ABREs in its promoter region, is one example of a gene that is triggered by ABA (Wu et al. 2023). Through phosphorylation by an ABA-activated protein kinase, two bZIP transcription factors, AREB1 and AREB2, most likely orchestrate the ABA-induced production of rd29B (Chen et al. 2021b). De novo protein synthesis is a precondition for the pathway that needs new protein synthesis to be activated by ABA. ABREs are absent from genes that go down this path, and their ABA-responsive regions interact with transcription factors belonging to the myelecytomatosis oncogenes (MYC) family (Zhang et al. 2023b). Before any ABA-initiated genes can be activated, the genes encoding the transcription factors of the MYC family must be activated and their transcription factors synthesized (Singh et al. 2021).

7.4 SIGNAL TRANSDUCTION OF STRESS HORMONES UNDER DROUGHT STRESS

7.4.1 Signal Transduction of Ethylene under Drought Stress

Ethylene undergoes a complicated array of signalling cascade to regulate plant growth. Ethylene signalling has been widely studied in *Arabidopsis* plants (Wang et al. 2024). Nonetheless, related signalling pathways can be found in variety of plants, thus knowledge gained from ethylene signalling can be applied to other plants. Several ethylene receptor isoforms are present in plants. In *Arabidopsis*, five receptor isoforms have been identified, that is ethylene response 1 (ETR1), ETR2, ethylene response sensor 1 (ERS1), ERS2, and EIN4 which plays a crucial role in ethylene signalling. Ethylene receptors show homology to bacterial receptors that signal via histidine autophosphorylation. Ethylene receptors are mainly present on the endoplasmic reticulum (ER) membrane (Monthony et al. 2024). A key feature of ethylene signalling is that it contains both positive and negative regulators where one protein triggers the responses while others stifle those (Zhao et al. 2021). In the absence of ethylene, a Raf-like Ser/Thr protein kinase, constitutive triple response1 (CTR1), is activated. CTR1 in turn phosphorylates the C-terminal of EIN2 (EIN2-CEND) thereby deactivating EIN2. In the absence of EIN2, EBF1 and EBF2, two F-box proteins, degrade ethylene insensitive3/ethylene insensitive3-like1 (EIN3 and EIL1), which regulates the main ethylene downstream signalling, thereby halting the signalling cascade (Khan et al. 2023).

However, in the presence of ethylene, CTR1 gets deactivated which protects EIN2 from phosphorylation. EIN2 further ascends to the nucleus and activates EIN3 and EIL1, the major regulators responsible for downstream ethylene signalling, which activates ethylene response factors (ERFs). Also, EIN2 stops the translation of EBF1 and EBF2, thereby protecting EIN2 from deactivation (Chien et al. 2024).

Ethylene detection and signalling triggers an array of physiological responses in plants. In addition to that, ethylene imparts stress tolerance which increases plant survival under challenging circumstances (Figure 7.2). In wild pear (*Pyrus betulaefolia*) over-expression of *PbERF3* restored the redox balance and enhanced plant's drought tolerance. On the other hand, knockdown of *PbERF3* rendered vulnerability to drought conditions (Zhang et al. 2024). In rice plants, genetic analysis revealed the up-regulation of APETALA2/ethylene responsive factor, *OsERF*103, in response to drought stress. Furthermore, subsequent studies indicated *OsERF*103's association with stress-responsive NAC1 (SNAC1), which positively modulates drought stress tolerance and maintains expression of many defence genes against drought stress (Yang et al. 2024). In soybean, over-expression of *GmAP2/ERF*144 showed enhanced drought tolerance in respect to wild plants owing to decreased ROS and enhanced leaf water content (Wang et al. 2022). Another AP2/ERF, *ZmEREBP60*, was up-regulated under drought stress conditions in maize. *ZmEREBP60* over-expression enhanced drought stress tolerance and reduced ROS content. Furthermore, transcriptome analysis revealed that *ZmEREBP60* was involved in up-regulation of genes responsible for ROS scavenging and stress defence (Qi et al. 2023). Over-expression of *JrERF2–2* enhances drought tolerance in plants via enhancing the stress defence mechanism, proline, and decreasing oxidative stress conditions. WRKY transcription factors are known to play an important role in the

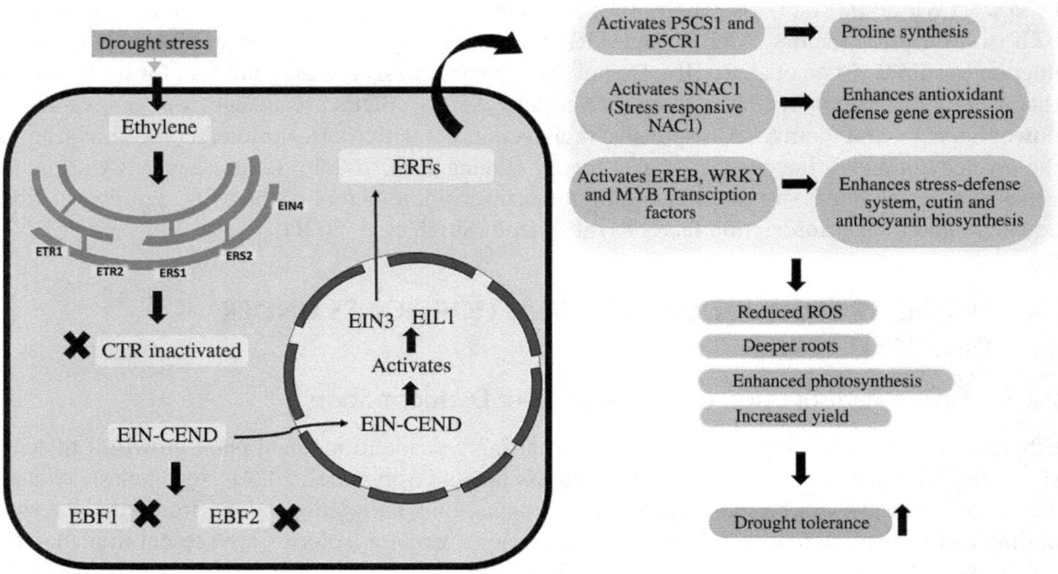

FIGURE 7.2 Signal transduction of ethylene under drought stress. CTR, constitutive triple response; EBF, EIN3-binding F-box; EIN/EIL, ethylene insensitive; EREB, ethylene responsive element binding protein; ERF, ethylene response factors; ERS, ethylene response sensor; ETR, ethylene receptor proteins; MYB, myeloblastosis; P5CS1, delta1-pyrroline-5-carboxylate synthase; P5CR1, pyrroline-5-carboxylate reductase 1; SNAC, stress-responsive NAC1.

stress tolerance mechanism. Studies suggest *JrERF2–2* interacts with *JrWRKY7* to promote drought tolerance (Yang et al. 2021).

Secondary metabolites aid in safeguarding the plant against drought stress (Yadav et al. 2021). Cuticular wax imparts a major role as a barrier against water losses through stomata, thereby playing a major role in adaptability under drought stress (Guo et al. 2016). In maize, *ZmEREB46* enhanced drought tolerance via increased wax deposition. Furthermore, *ZmEREB46* binds to *ZmCER2*, *ZmCER3.2*, and *ZmKCS1* responsible for the biosynthesis of cutin, thereby proving the role of *ZmEREB46* in drought stress tolerance (Yang et al. 2022). Anthocyanin is linked with the scavenging of ROS under stress conditions. Under stressful conditions, plants tend to modify light perception through anthocyanins to redirect metabolism of stress-related molecules from growth-related photosynthates (Cirillo et al. 2021). It has been discovered that the ERF protein *MdERF38* in apples (*Malus domestica*) is implicated in drought resistance. Furthermore, it was discovered that *MdMYB1*, a positive regulator of anthocyanin production, interacted with MdERF38 aiding in drought stress tolerance (An et al. 2020). Over-expression of *TaERF87* in wheat enhances resistance of plants to drought stress via stimulating the expression of *TaP5CS1* and *TaP5CR1*, the proline biosynthesis genes (Du et al. 2023). In transgenic rice, *JERF1* over-expression enhanced drought tolerance via regulating *OsP5CS* which encodes the major proline biosynthesis enzyme, deltal-pyrroline-5-carboxylate synthetase (Zhang et al. 2010). Another AP2/ERF transcription factor, AP2/ERF-N22 derived from the drought-tolerant rice variety *Oryza sativa* L. cv. N-22, was over-expressed in *Arabidopsis* when the plant was under drought stress. Through increased proline, wax, and decreased ROS levels, transgenic *Arabidopsis* demonstrated improved drought tolerance (Mawlong et al. 2015). In tomato (*Solanum lycopersicum*), enhanced *SlERF5* expression was observed under drought-stressed conditions. Furthermore, over-expression of the gene resulted in greater tolerance, emphasizing the role of *SlERF5* in enhanced drought tolerance (Pan et al. 2012).

Root architecture is significantly modulated by ethylene-induced ERFs to reduce the impact of drought stress (Ranjan et al. 2022). Transgenic rice seedlings with over-expressed *JeERF1* exhibited longer roots in comparison to wild-type plants and thus resulted in better drought stress tolerance (Zhang et al. 2010). A multitude of physiological and morphological parameters influence yield, which is the total amount of photosynthates allocated from source to sink tissues. ERFs also affect yield by influencing the photosynthesis. In transgenic *Trifolium alexandrium*, *HARDY* gene belonging to the AP2/ERF superfamily was over-expressed. It was found that over-expression of *HARDY* significantly induced drought tolerance by improving photosynthesis and reduced Na^+ accumulation (Abogadallah et al. 2011). Similarly, over-expression of ERF-194 in poplar and higher yield rice (HYR) encoding an APETALA2/ethylene responsive factor (AP2/ERF) in rice enhanced drought stress tolerance by enhancing antioxidant enzymes and chlorophyll, sugar, and starch (Ambavaram et al. 2014; Huan et al. 2023). An AP2/ERF protein, *TSRF1* found in tomato was found to enhance drought resistance in tobacco plants. Over-expression of *TSRF1* in rice plants also reduced drought stress via enhanced expression of photosynthesis-related genes. It was found that over-expressed *TSRF1* significantly enhanced the expression of *OsrbcS*, a gene encoding ribulose 1, 5-bisphosphate carboxylase (Rubisco) small subunit as well as *OsPLAS*, which encodes plastocyanin (Quan et al. 2010). Another ERF gene, *OsERF101*, enhanced drought stress resistance in rice via increased pollen fertility and seed germination (Jin et al. 2018). These results suggest that under drought stress, ERFs play a key role in influencing photosynthetic performance, stress-defence mechanisms, and secondary metabolite production. However, more research is needed to understand how ERFs function in these situations.

7.4.2 Signal Transduction of ABA under Drought Stress

A complicated and tightly controlled process known as signal transduction of ABA during drought stress entails a number of biochemical events that provide the plant the ability to adapt to water constraint. Receptors, protein kinases, and certain transcription factors are some of the elements that make up this complex signalling network, which works together to coordinate adaptive responses at the cellular and whole-plant levels (Figure 7.3). As a major regulator, ABA in plants controls gene expression, stomatal closure, and overall stress tolerance by starting a series of reactions. When some receptors sense elevated ABA levels, the ABA signalling pathway gets started.

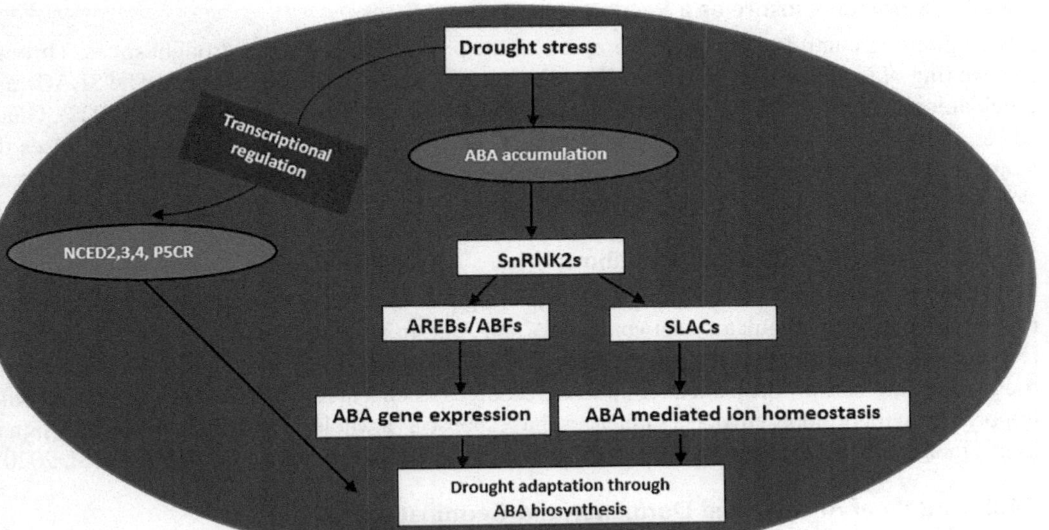

FIGURE 7.3 Signal transduction of ABA under drought stress. ABA, abscisic acid; NCED, 9-cis epoxycarotenoid dioxygenase; SnRNK2, SNF1-related protein kinase 2; P5CR, pyrroline-5-carboxylate; AREB, ABA-responsive element-binding protein/ABRE-binding factor; SLAC, slow anion channel-associated 1.

These receptors, which are a part of the PYR/PYL/RCAR family, are found in wheat and other plants (Zhou et al. 2019; Nian et al. 2021). The receptors experience conformational changes upon binding ABA, which enable them to interact with and inhibit PP2C (Lozano-Juste et al. 2023). By dephosphorylating and deactivating downstream components, specifically sucrose non-fermenting 1-related protein kinases 2 (SnRK2s), these PP2Cs function as negative regulators of the system (Mao et al. 2020).

7.4.2.1 Activation of SnRK2s

SnRK2 homologs, such as TaSnRK2.4 and TaSnRK2.8, are found in wheat and are essential for ABA signalling (Zhang et al. 2023a). Through dephosphorylation, PP2Cs keep SnRK2s inactive when ABA is not present. On the other hand, ABA perception causes PP2Cs and SnRK2s to no longer connect, which activates these protein kinases. SnRK2s initiate adaptive responses to drought stress by phosphorylating downstream targets once they are activated. This process propagates the ABA signal (Soma et al. 2020).

7.4.2.2 Role of Transcription Factors in ABA-Mediated Responses

The regulation of gene expression by certain transcription factors is a noteworthy consequence of ABA signalling. ABA-activated SnRK2s activate essential transcription factors belonging to the ABA-responsive element-binding protein/ABRE-binding factor (AREB/ABF) family (Saddhe et al. 2017). These transcription factors bind to target gene promoters containing ABA-responsive elements (ABREs), which starts the transcription of the target gene. A number of physiological functions that improve the plant's resistance to drought are facilitated by the active genes. Other wheat transcription factors, such as those in the AREB/ABF family, are also essential for responding to drought (Singh and Laxmi 2015). TaERA1, TaWRKY2/19, and TaNAC2a have been found to be transcription factors linked to phenotypes related to drought (Liu et al. 2020). These elements probably have a role in the general transcriptional control of genes related to stress tolerance. The intricate interplay of multiple transcription factors underscores the complexity of ABA-mediated gene expression in wheat.

7.4.2.3 Stomatal Closure as a Response to Drought Stress

ABA signalling controls stomatal closure, one of the primary responses to drought stress. Through the opening of ion channels in guard cells, ABA causes stomatal closure in wheat. The SLAC1 and quick anion channel 1 (QUAC1) are important channels in this process (Sasaki et al. 2022). Guard cell's ion fluxes are modulated by these channels, which alters the turgor pressure and causes the stomata to close. One of the most important adaptations to water shortage, this physiological reaction helps limit water loss through transpiration.

7.4.2.4 Crosstalk with Other Phytohormones

The complex interactions between the ABA signalling pathway and other phytohormones contribute to the intricacy of the regulatory network. ABA and Aux, in particular, interact to adjust the plant's response to drought (Singh and Roychoudhury 2023). Aux affects the growth and architecture of the roots, which affects how the plant distributes its resources when it lacks water (Hu et al. 2021). Another important phytohormone, ethylene, influences the response as a whole, emphasizing how ABA signalling is integrated in coordinating many hormonal pathways during drought stress (Bhatt et al. 2020).

7.4.2.5 Role of ABA in Seed Dormancy and Germination

ABA is essential for controlling seed dormancy and germination in addition to its role in drought responses (Ali et al. 2022). The ratio of ABA to GA is a key factor in determining the dormant state of seeds in wheat, rice, and other plants (Yu et al. 2020; Bahrabadi et al. 2022; Liao et al. 2023). GA encourages germination whereas ABA suppresses it in order to keep seeds dormant (Chen et al. 2020a). The complex regulation of GA and ABA levels depends on the expression of certain genes, including those that code for the enzymes involved in the metabolism of GA and ABA.

7.4.2.6 Transcriptional Regulation of ABA Biosynthesis and Catabolism

The content of ABA in plant tissues is determined by the strictly controlled mechanisms of ABA production and catabolism. The key factor influencing ABA levels in wheat is the genes that encode the enzymes involved in ABA production, specifically 9-cis-epoxycarotenoid deoxygenase (NCED; Lang et al. 2021). Furthermore, the *CYP707A* gene encodes ABA 8′-hydroxylase (ABA8OH), which catabolizes ABA (Fidler et al. 2022). These genes' dynamic regulation affects how much ABA is present in response to environmental stimuli, such as drought stress.

7.4.2.7 Recent Advances in Understanding ABA Signalling

The complex signalling of water stress in wheat has been better understood because of recent methods, such as the use of ABA-deficient and ABA-insensitive mutants. These investigations have demonstrated that the two main gene expression pathways – ABA-dependent and ABA-independent – can be used to explain the signalling of water stress in wheat. Our knowledge of the molecular mechanisms underlying ABA-mediated responses to drought stress in wheat has increased as a result of the identification and characterization of certain components in these pathways.

To sum up, the process of ABA signal transduction in wheat under drought stress is intricate and multifaceted, comprising a complex network of transcription factors, protein kinases, and receptors. To regulate gene expression, stomatal closure, and general stress tolerance, important elements like *TaSnRK2.4*, *TaSnRK2.8*, and several transcription factors – including those outside of the AREB/ABF family – are involved (Ahmed et al. 2023). The interplay between ABA signalling and other phytohormones, like Aux and ethylene, contributes to the intricate structure of the regulatory network by integrating many hormonal pathways during drought stress. Gaining an understanding of these molecular mechanisms is essential to the development of wheat varieties that can withstand stress through focused breeding techniques, ultimately enhancing global food security in the face of climate change.

7.5 CONCLUSION AND FUTURE PROSPECTS

Drought stress endangers the food security of millions of people by negatively affecting the plant's growth and physiology. Due to the changing climate and global warming drought stress is rising by the day. Such situations place us in dire need for remedies to regulate plant production even under such worsening situations. Stress hormones, ABA, and ethylene are known to increase under drought stress conditions and regulate stress defence mechanisms to reduce damage. Biosynthesis of ABA and ethylene was up-regulated under stress conditions. Further, exogenous application of ABA and ethylene proved to be beneficial in sustaining plants encountering drought stress. ABA and ethylene triggers an intricate network of signal transduction pathways to mediate stress. This chapter elucidates strategies by which ABA and ethylene could be enhanced in plants by providing them exogenously or inducing trans-genetic mutations in plants to reduce drought stress.

7.6 AUTHORSHIP CONTRIBUTION

Conceptualization, M.I.R.K.; Writing – Original draft preparation, M.M., S.K., S.P., and M.I.R.K.; Writing – Review and editing, M.M., S.K., and M.I.R.K. All authors have read and agreed to the published version of the book chapter.

7.7 DECLARATION OF COMPETING INTEREST

The authors declare that they have no known competing financial interests or personal relationships that could have appeared to influence the work reported in this book chapter.

REFERENCES

Ablazov, Abdugaffor, Jianing Mi, Muhammad Jamil, Kun-Peng Jia, Jian You Wang, Qitong Feng, and Salim Al-Babili. "The apocarotenoid zaxinone is a positive regulator of strigolactone and abscisic acid biosynthesis in Arabidopsis roots." *Frontiers in Plant Science* 11 (2020): 518781.

Abogadallah, Gaber M., Reham M. Nada, Robert Malinowski, and Paul Quick. "Overexpression of HARDY, an AP2/ERF gene from Arabidopsis, improves drought and salt tolerance by reducing transpiration and sodium uptake in transgenic *Trifolium alexandrinum* L." *Planta* 233 (2011): 1265–1276.

Ahmadizadeh, Mostafa, Jen-Tsung Chen, Soosan Hasanzadeh, Sunny Ahmar, and Parviz Heidari. "Insights into the genes involved in the ethylene biosynthesis pathway in Arabidopsis thaliana and *Oryza sativa*." *Journal of Genetic Engineering and Biotechnology* 18, no. 1 (2020): 62.

Ahmed, Borhan, Fakhrul Hasan, Anika Tabassum, Rasel Ahmed, Rajnee Hassan, Md Ruhul Amin, and Mobashwer Alam. "Genome-wide investigation of SnRK2 gene family in two jute species: *Corchorus olitorius* and *Corchorus capsularis*." *Journal of Genetic Engineering and Biotechnology* 21, no. 1 (2023): 5.

Albacete, Alfonso A., Cristina Martínez-Andújar, and Francisco Pérez-Alfocea. "Hormonal and metabolic regulation of source–sink relations under salinity and drought: from plant survival to crop yield stability." *Biotechnology Advances* 32, no. 1 (2014): 12–30.

Ali, Faiza, Ghulam Qanmber, Fuguang Li, and Zhi Wang. "Updated role of ABA in seed maturation, dormancy, and germination." *Journal of Advanced Research* 35 (2022): 199–214.

Ali, Rohaya, Sumaya Hassan, Durdana Shah, Nasreena Sajjad, and Eijaz Ahmed Bhat. "Role of polyamines in mitigating abiotic stress." In *Protective chemical agents in the amelioration of plant abiotic stress: Biochemical and molecular perspectives*, pp. 291–305. Wiley, 2020.

Ambavaram, Madana M. R., Supratim Basu, Arjun Krishnan, Venkategowda Ramegowda, Utlwang Batlang, Lutfor Rahman, Niranjan Baisakh, and Andy Pereira. "Coordinated regulation of photosynthesis in rice increases yield and tolerance to environmental stress." *Nature Communications* 5, no. 1 (2014): 5302.

An, Jian-Ping, Xiao-Wei Zhang, Si-Qi Bi, Chun-Xiang You, Xiao-Fei Wang, and Yu-Jin Hao. "The ERF transcription factor MdERF38 promotes drought stress-induced anthocyanin biosynthesis in apple." *The Plant Journal* 101, no. 3 (2020): 573–589.

Arraes, Fabricio Barbosa Monteiro, Magda Aparecida Beneventi, Maria Eugenia Lisei de Sa, Joaquin Felipe Roca Paixao, Erika Valeria Saliba Albuquerque, Silvana Regina Rockenbach Marin, Eduardo Purgatto, Alexandre Lima Nepomuceno, and Maria Fatima Grossi-de-Sa. "Implications of ethylene biosynthesis and signaling in soybean drought stress tolerance." *BMC Plant Biology* 15 (2015): 1–20.

Ashrafi, Mohsen, Mohammad-Reza Azimi-Moqadam, Ehsan Mohseni Fard, Farid Shekari, Hossein Jafary, Parviz Moradi, Mariachiara Pucci, Giulia Abate, and Andrea Mastinu. "Physiological and molecular aspects of two Thymus species differently sensitive to drought stress." *BioTech* 11, no. 2 (2022): 8.

Baharudin, Nur Farhana, and Nurul Izzati Osman. "Plant development, stress responses, and secondary metabolism under ethylene regulation." *Plant Stress* 7 (2023): 100146.

Bahrabadi, Elham, Reza Tavakkol Afshari, M. Nassiri Mahallati, and S. M. Seyyedi. "Abscisic, gibberellic, and salicylic acids effects on germination indices of corn under salinity and drought stresses." *Journal of Crop Improvement* 36, no. 1 (2022): 73–89.

Bhatt, Deepesh, Manoj Nath, Mayank Sharma, Megha D. Bhatt, Deepak Singh Bisht, and Naresh V. Butani. "Role of growth regulators and phytohormones in overcoming environmental stress." In *Protective chemical agents in the amelioration of plant abiotic stress: Biochemical and molecular perspectives*, pp. 254–279. Wiley, 2020.

Cao, Xiaochuang, Longlong Wu, Meiyan Wu, Chunquan Zhu, Qianyu Jin, and Junhua Zhang. "Abscisic acid mediated proline biosynthesis and antioxidant ability in roots of two different rice genotypes under hypoxic stress." *BMC Plant Biology* 20 (2020): 1–14.

Chaudhry, Smita, and Gagan Preet Singh Sidhu. "Climate change regulated abiotic stress mechanisms in plants: A comprehensive review." *Plant Cell Reports* 41, no. 1 (2022): 1–31.

Chen, Hao, David A. Bullock Jr, Jose M. Alonso, and Anna N. Stepanova. "To fight or to grow: The balancing role of ethylene in plant abiotic stress responses." *Plants* 11, no. 1 (2021a): 33.

Chen, Huhui, Jiuxiao Ruan, Pu Chu, Wei Fu, Zhenwei Liang, Yin Li, Jianhua Tong et al. "AtPER1 enhances primary seed dormancy and reduces seed germination by suppressing the ABA catabolism and GA biosynthesis in Arabidopsis seeds." *The Plant Journal* 101, no. 2 (2020a): 310–323.

Chen, Kong, Guo-Jun Li, Ray A. Bressan, Chun-Peng Song, Jian-Kang Zhu, and Yang Zhao. "Abscisic acid dynamics, signaling, and functions in plants." *Journal of Integrative Plant Biology* 62, no. 1 (2020b): 25–54.

Chen, Xuexue, Yanglin Ding, Yongqing Yang, Chunpeng Song, Baoshan Wang, Shuhua Yang, Yan Guo, and Zhizhong Gong. "Protein kinases in plant responses to drought, salt, and cold stress." *Journal of Integrative Plant Biology* 63, no. 1 (2021b): 53–78.

Chien, Yuan-Chi, and Gyeong Mee Yoon. "Subcellular dynamics of ethylene signaling drive plant plasticity to growth and stress: Spatiotemporal control of ethylene signaling in Arabidopsis." *BioEssays* (2024): 2400043.

Cirillo, Valerio, Vincenzo D'Amelia, Marco Esposito, Chiara Amitrano, Petronia Carillo, Domenico Carputo, and Albino Maggio. "Anthocyanins are key regulators of drought stress tolerance in tobacco." *Biology* 10, no. 2 (2021): 139.

Collin, Anna, Agata Daszkowska-Golec, Marzena Kurowska, and Iwona Szarejko. "Barley ABI5 (Abscisic Acid INSENSITIVE 5) is involved in abscisic acid-dependent drought response." *Frontiers in Plant Science* 11 (2020): 557684.

Cui, Mengying, Yingchao Lin, Yuangang Zu, Thomas Efferth, Dewen Li, and Zhonghua Tang. "Ethylene increases accumulation of compatible solutes and decreases oxidative stress to improve plant tolerance to water stress in Arabidopsis." *Journal of Plant Biology* 58 (2015): 193–201.

Czékus, Zalán, Gabriella Szalai, Irma Tari, M. Iqbal R. Khan, and Péter Poór. "Role of ethylene in ER stress and the unfolded protein response in tomato (*Solanum lycopersicum* L.) plants." *Plant Physiology and Biochemistry* 181 (2022): 1–11.

Dalal, Monika, Sarika Sahu, Sneha Tiwari, Atmakuri R. Rao, and Kishor Gaikwad. "Transcriptome analysis reveals interplay between hormones, ROS metabolism and cell wall biosynthesis for drought-induced root growth in wheat." *Plant Physiology and Biochemistry* 130 (2018): 482–492.

Du, Hao, Nai Wu, Fei Cui, Lei You, Xianghua Li, and Lizhong Xiong. "A homolog of ETHYLENE OVERPRODUCER, O s ETOL 1, differentially modulates drought and submergence tolerance in rice." *The Plant Journal* 78, no. 5 (2014): 834–849.

Du, Linying, Xueling Huang, Li Ding, Zhongxue Wang, Dongling Tang, Bin Chen, Lanjiya Ao, Yuling Liu, Zhensheng Kang, and Hude Mao. "TaERF87 and TaAKS1 synergistically regulate TaP5CS1/TaP5CR1-mediated proline biosynthesis to enhance drought tolerance in wheat." *New Phytologist* 237, no. 1 (2023): 232–250.

Fidler, Justyna, Jakub Graska, Marta Gietler, Małgorzata Nykiel, Beata Prabucka, Anna Rybarczyk-Płońska, Ewa Muszyńska, Iwona Morkunas, and Mateusz Labudda. "PYR/PYL/RCAR receptors play a vital role in the abscisic-acid-dependent responses of plants to external or internal stimuli." *Cells* 11, no. 8 (2022): 1352.

Gao, Weiwei, Mingkang Li, Songguang Yang, Chunzhi Gao, Yan Su, Xuan Zeng, Zhengli Jiao, Weijuan Xu, Mingyong Zhang, and Kuaifei Xia. "miR2105 and the kinase OsSAPK10 co-regulate OsbZIP86 to mediate drought-induced ABA biosynthesis in rice." *Plant Physiology* 189, no. 2 (2022): 889–905.

Gautam, Harsha, Sheen Khan, Nidhi, Adriano Sofo, and Nafees A. Khan. "Appraisal of the role of gaseous signaling molecules in thermo-tolerance mechanisms in plants." *Plants* 13, no. 6 (2024): 791.

Gong, Diankai, Fei He, Jingyan Liu, Cheng Zhang, Yanrong Wang, Shujun Tian, Chi Sun, and Xue Zhang. "Understanding of hormonal regulation in rice seed germination." *Life* 12, no. 7 (2022): 1021.

Guo, Jun, Wen Xu, Xiaocong Yu, Hao Shen, Haosheng Li, Dungong Cheng, Aifeng Liu et al. "Cuticular wax accumulation is associated with drought tolerance in wheat near-isogenic lines." *Frontiers in Plant Science* 7 (2016): 1809.

Habben, Jeffrey E., Xiaoming Bao, Nicholas J. Bate, Jason L. DeBruin, Dennis Dolan, Darren Hasegawa, Timothy G. Helentjaris et al. "Transgenic alteration of ethylene biosynthesis increases grain yield in maize under field drought-stress conditions." *Plant Biotechnology Journal* 12, no. 6 (2014): 685–693.

Hancock, J. T., S. J. Neill, and I. D. Wilson. "Nitric oxide and ABA in the control of plant function." *Plant Science* 181, no. 5 (2011): 555–559.

Hasan, Md Mahadi, Lei Gong, Zheng-Fei Nie, Feng-Ping Li, Golam Jalal Ahammed, and Xiang-Wen Fang. "ABA-induced stomatal movements in vascular plants during dehydration and rehydration." *Environmental and Experimental Botany* 186 (2021): 104436.

Hosseinifard, Marjanossadat, Szymon Stefaniak, Majid Ghorbani Javid, Elias Soltani, Łukasz Wojtyla, and Małgorzata Garnczarska. "Contribution of exogenous proline to abiotic stresses tolerance in plants: A review." *International Journal of Molecular Sciences* 23, no. 9 (2022): 5186.

Hsu, Po-Kai, Guillaume Dubeaux, Yohei Takahashi, and Julian I. Schroeder. "Signaling mechanisms in abscisic acid-mediated stomatal closure." *The Plant Journal* 105, no. 2 (2021): 307–321.

Hu, Qi-Qi, Jian-Qin Shu, Wen-Min Li, and Guang-Zhi Wang. "Role of auxin and nitrate signaling in the development of root system architecture." *Frontiers in Plant Science* 12 (2021): 690363.

Hua, Deping, Cun Wang, Junna He, Hui Liao, Ying Duan, Ziqiang Zhu, Yan Guo, Zhizhong Chen, and Zhizhong Gong. "A plasma membrane receptor kinase, GHR1, mediates abscisic acid-and hydrogen peroxide-regulated stomatal movement in Arabidopsis." *The Plant Cell* 24, no. 6 (2012): 2546–2561.

Huan, Xuhui, Xingqi Wang, Shengqiang Zou, Kai Zhao, Youzhi Han, and Shengji Wang. "Transcription factor ERF194 modulates the stress-related physiology to enhance drought tolerance of poplar." *International Journal of Molecular Sciences* 24, no. 1 (2023): 788.

Iqbal, Noushina, Nafees A. Khan, Antonio Ferrante, Alice Trivellini, Alessandra Francini, and M. I. R. Khan. "Ethylene role in plant growth, development and senescence: interaction with other phytohormones." *Frontiers in Plant Science* 8 (2017): 235913.

Jia, Kun-Peng, Jianing Mi, Shawkat Ali, Hajime Ohyanagi, Juan C. Moreno, AbdugafforAblazov, Aparna Balakrishna et al. "An alternative, zeaxanthin epoxidase-independent abscisic acid biosynthetic pathway in plants." *Molecular Plant* 15, no. 1 (2022): 151–166.

Jiao, Qingsong, Tianshu Chen, Guanting Niu, Huchen Zhang, Chang Fang Zhou, and Zhi Hong. "N-glycosylation is involved in stomatal development by modulating the release of active abscisic acid and auxin in Arabidopsis." *Journal of Experimental Botany* 71, no. 19 (2020): 5865–5879.

Jin, Yue, Weiyang Pan, Xiufang Zheng, Xuan Cheng, Mengmeng Liu, Hong Ma, and Xiaochun Ge. "OsERF101, an ERF family transcription factor, regulates drought stress response in reproductive tissues." *Plant Molecular Biology* 98 (2018): 51–65.

Khan, Irfan Ullah, Akhtar Ali, Haris Ali Khan, Dongwon Baek, Junghoon Park, Chae Jin Lim, Shah Zareen et al. "PWR/HDA9/ABI4 complex epigenetically regulates ABA dependent drought stress tolerance in Arabidopsis." *Frontiers in Plant Science* 11 (2020): 623.

Khan, M. Iqbal R., Sudhakar R. Palakolanu, Priyanka Chopra, Ashish B. Rajurkar, Ravi Gupta, Noushina Iqbal, and Chirag Maheshwari. "Improving drought tolerance in rice: Ensuring food security through multi-dimensional approaches." *Physiologia Plantarum* 172, no. 2 (2021): 645–668.

Khan, Naeem, Asghari Bano, and M. D. Ali Babar. "Metabolic and physiological changes induced by plant growth regulators and plant growth promoting rhizobacteria and their impact on drought tolerance in *Cicer arietinum* L." *PloS One* 14, no. 3 (2019): e0213040.

Khan, Nafees A., Mohd Asgher, Tasir S. Per, Asim Masood, Mehar Fatma, and M. Iqbal R. Khan. "Ethylene potentiates sulfur-mediated reversal of cadmium inhibited photosynthetic responses in mustard." *Frontiers in Plant Science* 7 (2016): 1628.

Khan, Sheen, Zebus Sehar, Mohammed Albaqami, and Nafees A. Khan. "Ethylene crosstalk with isoprenoid-derived signaling molecules in the context of salinity tolerance." *Environmental and Experimental Botany* 212 (2023): 105379.

Kim, Tae-Houn, Maik Böhmer, Honghong Hu, Noriyuki Nishimura, and Julian I. Schroeder. "Guard cell signal transduction network: Advances in understanding abscisic acid, CO2, and Ca2+signaling." *Annual Review of Plant Biology* 61 (2010): 561–591.

Lang, Jing, Yuxin Fu, Yong Zhou, Mengping Cheng, Min Deng, Maolian Li, Tingting Zhu et al. "Myb10-D confers PHS-3D resistance to pre-harvest sprouting by regulating NCED in ABA biosynthesis pathway of wheat." *New Phytologist* 230, no. 5 (2021): 1940–1952.

Li, Li, Jiayi Xing, Haiyang Ma, Fulai Liu, and Yaosheng Wang. "In situ determination of guard cell ion flux underpins the mechanism of ABA-mediated stomatal closure in barley plants exposed to PEG-induced drought stress." *Environmental and Experimental Botany* 187 (2021): 104468.

Liang, Shanshan, Wei Xiong, Cuicui Yin, Xiaodong Xie, Ya-jun Jin, Siju Zhang, Bo Yang, Guoyou Ye, Shouyi Chen, and Wei-jiang Luan. "Overexpression of OsARD1 improves submergence, drought, and salt tolerances of seedling through the enhancement of ethylene synthesis in rice." *Frontiers in Plant Science* 10 (2019): 1088.

Liao, Zhigang, Yunchao Zhang, Qing Yu, Weicong Fang, Meiyao Chen, Tianfei Li, Yi Liu et al. "Coordination of growth and drought responses by GA-ABA signaling in rice." *New Phytologist* 240, no. 3 (2023): 1149–1161.

Liu, Daofeng, Xiaojing Liu, Yonglu Meng, Cuihui Sun, Hongshu Tang, Yudong Jiang, Muhammad Ali Khan, Jingqi Xue, Nan Ma, and Junping Gao. "An organ-specific role for ethylene in rose petal expansion during dehydration and rehydration." *Journal of Experimental Botany* 64, no. 8 (2013): 2333–2344.

Liu, Hao, Ying Yang, Dandan Liu, Xiaoyu Wang, and Linsheng Zhang. "Transcription factor TabHLH49 positively regulates dehydrin WZY2 gene expression and enhances drought stress tolerance in wheat." *BMC Plant Biology* 20 (2020): 1–10.

Liu, Hui, Songbo Song, Hui Zhang, Yanhua Li, Liangjie Niu, Jinghua Zhang, and Wei Wang. "Signaling transduction of ABA, ROS, and Ca2+ in plant stomatal closure in response to drought." *International Journal of Molecular Sciences* 23, no. 23 (2022a): 14824.

Liu, Jie, Shenglan Li, Xin Yang, Zhenhua Wei, and Fulai Liu. "Effects of soil drought and vapour pressure deficit (VPD) on water use efficiency of tomato plants with contrasting endogenous ABA levels." *Scientia Horticulturae* 295 (2022b): 110797.

Lozano-Juste, Jorge, Lourdes Infantes, Irene Garcia-Maquilon, Rafael Ruiz-Partida, Ebe Merilo, Juan Luis Benavente, Adrian Velazquez-Campoy et al. "Structure-guided engineering of a receptor-agonist pair for inducible activation of the ABA adaptive response to drought." *Science Advances* 9, no. 10 (2023): eade9948.

Malcheska, Frosina, Altaf Ahmad, Sundas Batool, Heike M. Müller, Jutta Ludwig-Müller, Jürgen Kreuzwieser, Dörte Randewig et al. "Drought-enhanced xylem sap sulfate closes stomata by affecting ALMT12 and guard cell ABA synthesis." *Plant Physiology* 174, no. 2 (2017): 798–814.

Man, Da, Yong-Xia Bao, Lie-Bao Han, and Xunzhong Zhang. "Drought tolerance associated with proline and hormone metabolism in two tall fescue cultivars." *Hort Science* 46, no. 7 (2011): 1027–1032.

Manavalan, Lakshmi Praba, and Henry T. Nguyen. "Drought tolerance in crops: Physiology to genomics." *Plant Stress Physiology* 1 (2017): 1–23.

Mao, Xinguo, Yuying Li, Shoaib Ur Rehman, Lili Miao, Yanfei Zhang, Xin Chen, Chunmei Yu, Jingyi Wang, Chaonan Li, and Ruilian Jing. "The sucrose non-fermenting 1-related protein kinase 2 (SnRK2) genes are multifaceted players in plant growth, development and response to environmental stimuli." *Plant and Cell Physiology* 61, no. 2 (2020): 225–242.

Mawlong, Ibandalin, Kishwar Ali, R. Srinivasan, R. D. Rai, and Aruna Tyagi. "Functional validation of a drought-responsive AP2/ERF family transcription factor-encoding gene from rice in Arabidopsis." *Molecular Breeding* 35 (2015): 1–14.

Mercado-Reyes, Joel A., Talitha Soares Pereira, Anju Manandhar, Ian M. Rimer, and Scott AM McAdam. "Extreme drought can deactivate ABA biosynthesis in embolism-resistant species." *Plant, Cell & Environment* 47, no. 2 (2024): 497–510.

Miao, Rui, Wei Yuan, Yue Wang, Irene Garcia-Maquilon, Xiaolin Dang, Ying Li, Jianhua Zhang, Yiyong Zhu, Pedro L. Rodriguez, and Weifeng Xu. "Low ABA concentration promotes root growth and hydrotropism through relief of ABA INSENSITIVE 1-mediated inhibition of plasma membrane H+-ATPase 2." *Science Advances* 7, no. 12 (2021): eabd4113.

Millar, Anthony A., John V. Jacobsen, John J. Ross, Chris A. Helliwell, Andrew T. Poole, Graham Scofield, James B. Reid, and Frank Gubler. "Seed dormancy and ABA metabolism in Arabidopsis and barley: the role of ABA 8′-hydroxylase." *The Plant Journal* 45, no. 6 (2006): 942–954.

Monthony, Adrian S., Maxime de Ronne, and Davoud Torkamaneh. "Exploring ethylene-related genes in Cannabis sativa: Implications for sexual plasticity." *Plant Reproduction* (2024): 1–19.

Mottaleb, Khondoker Abdul, Gideon Kruseman, Aymen Frija, Kai Sonder, and Santiago Lopez-Ridaura. "Projecting wheat demand in China and India for 2030 and 2050: Implications for food security." *Frontiers in Nutrition* 9 (2023): 1077443.

Mukarram, Mohammad, Sadaf Choudhary, Daniel Kurjak, Anja Petek, and M. Masroor A. Khan. "Drought: Sensing, signalling, effects and tolerance in higher plants." *Physiologia Plantarum* 172, no. 2 (2021): 1291–1300.

Naing, Aung Htay, Jova Riza Campol, Hyunhee Kang, Junping Xu, Mi Young Chung, and Chang Kil Kim. "Role of ethylene biosynthesis genes in the regulation of salt stress and drought stress tolerance in petunia." *Frontiers in Plant Science* 13 (2022): 844449.

Nazir, Faroza, Poor Peter, Ravi Gupta, Sarika Kumari, Kashif Nawaz, and M. Iqbal R. Khan. "Plant hormone ethylene: A leading edge in conferring drought stress tolerance." *Physiologia Plantarum* 176, no. 1 (2024): e14151.

Nian, Lili, Xiaoning Zhang, Xianfeng Yi, Xuelu Liu, Noor ul Ain, Yingbo Yang, Xiaodan Li, Fasih Ullah Haider, and Xiaolin Zhu. "Genome-wide identification of ABA receptor PYL/RCAR gene family and their response to cold stress in *Medicago sativa* L." *Physiology and Molecular Biology of Plants* 27 (2021): 1979–1995.

Ojuederie, Omena Bernard, Oluwaseyi Samuel Olanrewaju, and Olubukola Oluranti Babalola. "Plant growth promoting rhizobacterial mitigation of drought stress in crop plants: implications for sustainable agriculture." *Agronomy* 9, no. 11 (2019): 712.

Ozturk, Munir, Bengu Turkyilmaz Unal, Pedro García-Caparrós, Anum Khursheed, Alvina Gul, and Mirza Hasanuzzaman. "Osmoregulation and its actions during the drought stress in plants." *Physiologia Plantarum* 172, no. 2 (2021): 1321–1335.

Pan, Yu, Graham B. Seymour, Chungui Lu, Zongli Hu, Xuqing Chen, and Guoping Chen. "An ethylene response factor (ERF5) promoting adaptation to drought and salt tolerance in tomato." *Plant Cell Reports* 31 (2012): 349–360.

Park, Chanung, Han Yong Lee, and Gyeong Mee Yoon. "The regulation of ACC synthase protein turnover: a rapid route for modulating plant development and stress responses." *Current Opinion in Plant Biology* 63 (2021): 102046.

Parwez, Rukhsar, Tariq Aftab, Sarvajeet Singh Gill, and M. Naeem. "Abscisic acid signaling and crosstalk with phytohormones in regulation of environmental stress responses." *Environmental and Experimental Botany* 199 (2022): 104885.

Pattyn, Jolien, John Vaughan-Hirsch, and Bram Van de Poel. "The regulation of ethylene biosynthesis: a complex multilevel control circuitry." *New Phytologist* 229, no. 2 (2021): 770–782.

Qi, Huanhuan, Kun Liang, Yinggen Ke, Jing Wang, Pingfang Yang, Feng Yu, and Fazhan Qiu. "Advances of Apetala2/ethylene response factors in regulating development and stress response in maize." *International Journal of Molecular Sciences* 24, no. 6 (2023): 5416.

Qi, Junsheng, Chun-Peng Song, Baoshan Wang, Jianmin Zhou, Jaakko Kangasjärvi, Jian-Kang Zhu, and Zhizhong Gong. "Reactive oxygen species signaling and stomatal movement in plant responses to drought stress and pathogen attack." *Journal of Integrative Plant Biology* 60, no. 9 (2018): 805–826.

Quan, Ruidang, Shoujing Hu, Zhili Zhang, Haiwen Zhang, Zhijin Zhang, and Rongfeng Huang. "Overexpression of an ERF transcription factor TSRF1 improves rice drought tolerance." *Plant Biotechnology Journal* 8, no. 4 (2010): 476–488.

Ranjan, Alok, Ragini Sinha, Sneh L. Singla-Pareek, Ashwani Pareek, and Anil Kumar Singh. "Shaping the root system architecture in plants for adaptation to drought stress." *Physiologia Plantarum* 174, no. 2 (2022): e13651.

Rehman, Bushra, Javeria Javed, Mamoona Rauf, Sumera Afzal Khan, Muhammad Arif, Muhammad Hamayun, Humaira Gul et al. "ACC deaminase-producing endophytic fungal consortia promotes drought stress tolerance in *M. oleifera* by mitigating ethylene and H2O2." *Frontiers in Plant Science* 13 (2022): 967672.

Riyazuddin, Riyazuddin, Radhika Verma, Kalpita Singh, Nisha Nisha, Monika Keisham, Kaushal Kumar Bhati, Sun Tae Kim, and Ravi Gupta. "Ethylene: A master regulator of salinity stress tolerance in plants." *Biomolecules* 10, no. 6 (2020): 959.

Rodrigues, Maria A., Ricardo E. Bianchetti, and Luciano Freschi. "Shedding light on ethylene metabolism in higher plants." *Frontiers in Plant Science* 5 (2014): 665.

Rodríguez-Gacio, María del Carmen, Miguel A. Matilla-Vázquez, and Angel J. Matilla. "Seed dormancy and ABA signaling: The breakthrough goes on." *Plant Signaling & Behavior* 4, no. 11 (2009): 1035–1048.

Saddhe, Ankush Ashok, Kumar Kundan, and Dwivedi Padmanabh. "Mechanism of ABA signaling in response to abiotic stress in plants." In *Mechanism of plant hormone signaling under stress*, Vol. 1, pp. 173–195. Wiley, 2017.

Saha, Debanjana, Prince Choyal, Udit Nandan Mishra, Prajjal Dey, Bandana Bose, M. D. Prathibha, Narendra Kumar Gupta et al. "Drought stress responses and inducing tolerance by seed priming approach in plants." *Plant Stress* 4 (2022): 100066.

Salvi, Prafull, Mrinalini Manna, Harmeet Kaur, Tanika Thakur, Nishu Gandass, Deepesh Bhatt, and Mehanathan Muthamilarasan. "Phytohormone signaling and crosstalk in regulating drought stress response in plants." *Plant Cell Reports* 40 (2021): 1305–1329.

Sami, Fareen, Mohammad Yusuf, Mohammad Faizan, Ahmad Faraz, and Shamsul Hayat. "Role of sugars under abiotic stress." *Plant Physiology and Biochemistry* 109 (2016): 54–61.

Sasaki, Takayuki, Michiyo Ariyoshi, Yoko Yamamoto, and Izumi C. Mori. "Functional roles of ALMT-type anion channels in malate-induced stomatal closure in tomato and Arabidopsis." *Plant, Cell & Environment* 45, no. 8 (2022): 2337–2350.

Sehar, Zebus, Noushina Iqbal, M. Iqbal R. Khan, Asim Masood, Md Tabish Rehman, Afzal Hussain, Mohamed F. AlAjmi, Altaf Ahmad, and Nafees A. Khan. "Ethylene reduces glucose sensitivity and reverses photosynthetic repression through optimization of glutathione production in salt-stressed wheat (*Triticum aestivum* L.)." *Scientific Reports* 11, no. 1 (2021): 12650.

Shahrajabian, Mohamad Hesam, Wenli Sun, and Qi Cheng. "Molecular breeding and the impacts of some important genes families on agronomic traits, a review." *Genetic Resources and Crop Evolution* 68, no. 5 (2021): 1709–1730.

Sharp, R. E. "Interaction with ethylene: Changing views on the role of abscisic acid in root and shoot growth responses to water stress." *Plant, Cell & Environment* 25, no. 2 (2002): 211–222.

Shekoofa, A., and Y. Emam. "Maize (*Zea mays* L.) Growth and yield response to Ethephon application under water stress conditions." *Iran Agricultural Research* 24, no. 1.2 (2007): 39–52.

Shibata, Michitaro, and Keiko Sugimoto. "A gene regulatory network for root hair development." *Journal of Plant Research* 132, no. 3 (2019): 301–309.

Shivhare, Radha, Mehar H. Asif, and Charu Lata. "Comparative transcriptome analysis reveals the genes and pathways involved in terminal drought tolerance in pearl millet." *Plant Molecular Biology* 103, no. 6 (2020): 639–652.

Singh, Ankur, and Aryadeep Roychoudhury. "Abscisic acid in plants under abiotic stress: Crosstalk with major phytohormones." *Plant Cell Reports* 42, no. 6 (2023): 961–974.

Singh, D., and A. Laxmi. "Transcriptional regulation of drought response: a tortuous network of transcriptional factors." *Frontiers in Plant Science* 6 (2015): 165462.

Singh, Priyanka, Yamshi Arif, Andrzej Bajguz, and Shamsul Hayat. "The role of quercetin in plants." *Plant Physiology and Biochemistry* 166 (2021): 10–19.

Singh, Rajnish Prakash, Ying Ma, and Afreen Shadan. "Perspective of ACC-deaminase producing bacteria in stress agriculture." *Journal of Biotechnology* 352 (2022): 36–46.

Soma, Fumiyuki, Fuminori Takahashi, Takamasa Suzuki, Kazuo Shinozaki, and Kazuko Yamaguchi-Shinozaki. "Plant Raf-like kinases regulate the mRNA population upstream of ABA-unresponsive SnRK2 kinases under drought stress." *Nature Communications* 11, no. 1 (2020): 1373.

Tanaka, Yoko, Toshio Sano, Masanori Tamaoki, Nobuyoshi Nakajima, Noriaki Kondo, and Seiichiro Hasezawa. "Ethylene inhibits abscisic acid-induced stomatal closure in Arabidopsis." *Plant Physiology* 138, no. 4 (2005): 2337–2343.

Ullah, Abid, Hakim Manghwar, Muhammad Shaban, Aamir Hamid Khan, Adnan Akbar, Usman Ali, Ehsan Ali, and Shah Fahad. "Phytohormones enhanced drought tolerance in plants: A coping strategy." *Environmental Science and Pollution Research* 25 (2018): 33103–33118.

Umezawa, Taishi, Kazuo Nakashima, Takuya Miyakawa, Takashi Kuromori, Masaru Tanokura, Kazuo Shinozaki, and Kazuko Yamaguchi-Shinozaki. "Molecular basis of the core regulatory network in ABA responses: Sensing, signaling and transport." *Plant and Cell Physiology* 51, no. 11 (2010): 1821–1839.

Valluru, Ravi, William J. Davies, Matthew P. Reynolds, and Ian C. Dodd. "Foliar abscisic acid-to-ethylene accumulation and response regulate shoot growth sensitivity to mild drought in wheat." *Frontiers in Plant Science* 7 (2016): 461.

Van Dijk, Michiel, Tom Morley, Marie Luise Rau, and Yashar Saghai. "A meta-analysis of projected global food demand and population at risk of hunger for the period 2010–2050." *Nature Food* 2, no. 7 (2021): 494–501.

Vishal, Bhushan, and Prakash P. Kumar. "Regulation of seed germination and abiotic stresses by gibberellins and abscisic acid." *Frontiers in Plant Science* 9 (2018): 368905.

Wahid, Iram, Sarika Kumari, Rafiq Ahmad, Sofi J. Hussain, Saud Alamri, Manzer H. Siddiqui, and M. Iqbal R. Khan. "Silver nanoparticle regulates salt tolerance in wheat through changes in ABA concentration, ion homeostasis, and defense systems." *Biomolecules* 10, no. 11 (2020): 1506.

Wang, Haitang, Danqing Ni, Jiacheng Shen, Sushuang Deng, Huidong Xuan, Congcong Wang, Jianyu Xu et al. "Genome-wide identification of the AP2/ERF gene family and functional analysis of GmAP2/ERF144 for drought tolerance in soybean." *Frontiers in Plant Science* 13 (2022): 848766.

Wang, Honglin, Feifei Wang, Fangfang Zheng, Lijuan Wang, Hiaxia Pei, and Chun-Hai Dong. "Ethylene-insensitive mutants of *Nicotiana tabacum* exhibit drought stress resistance." *Plant Growth Regulation* 79 (2016): 107–117.

Wang, Kai, Hongqing Guo, and Yanhai Yin. "AP2/ERF transcription factors and their functions in arabidopsis responses to abiotic stresses." *Environmental and Experimental Botany* (2024): 105763.

Wang, Xiao Le, Li Ming, Cai Li Qun, Fang Wei Kuan, Liang Zhao Xu, Tan Fang, He Shan Shan et al. "Effect of ethephon and 1-MCP under water stress on antioxidase activities of sugarcane leaves at seedling stage." *Journal of Southern Agriculture* 45 no. 9 (2014): 1558–1565.

Wasaya, Allah, Xiying Zhang, Qin Fang, and Zongzheng Yan. "Root phenotyping for drought tolerance: A review." *Agronomy* 8, no. 11 (2018): 241.

Watkins, Justin M., Paul J. Hechler, and Gloria K. Muday. "Ethylene-induced flavonol accumulation in guard cells suppresses reactive oxygen species and moderates stomatal aperture." *Plant Physiology* 164, no. 4 (2014): 1707–1717.

Wu, Guofan, Nongfu Tian, Fawen She, Aohua Cao, Wangze Wu, Sheng Zheng, and Ning Yang. "Characteristics analysis of early responsive to dehydration genes in *Arabidopsis thaliana* (AtERD)." *Plant Signaling & Behavior* 18, no. 1 (2023): 2105021.

Xie, Qijun, Jemaa Essemine, Xiaochen Pang, Haiying Chen, Jing Jin, and Weiming Cai. "Abscisic acid regulates the root growth trajectory by reducing auxin transporter PIN2 protein levels in *Arabidopsis thaliana*." *Frontiers in Plant Science* 12 (2021): 632676.

Xue, Dawei, Xiaoqin Zhang, Xueli Lu, Guang Chen, and Zhong-Hua Chen. "Molecular and evolutionary mechanisms of cuticular wax for plant drought tolerance." *Frontiers in Plant Science* 8 (2017): 258957.

Yadav, Bindu, Abhimanyu Jogawat, Md Samiur Rahman, and Om Prakash Narayan. "Secondary metabolites in the drought stress tolerance of crop plants: A review." *Gene Reports* 23 (2021): 101040.

Yang, Guiyan, Shaobing Peng, Tianyu Wang, Xiangqian Gao, Dapei Li, Mengge Li, Shuwen Chen, and Zhenggang Xu. "Walnut ethylene response factor JrERF2–2 interact with JrWRKY7 to regulate the GSTs in plant drought tolerance." *Ecotoxicology and Environmental Safety* 228 (2021): 112945.

Yang, Lijia, Liang Xu, Jiazhuo Guo, Aipeng Li, Haoyue Qi, Jiaxuan Wang, and Shiyong Song. "SNAC1-OsERF103-OsSDG705 module mediates drought response in rice." *New Phytologist* 241, no. 6 (2024): 2480–2494.

Yang, Ruizhen, Ziyi Yang, Meng Xing, Yexing Jing, Yunwei Zhang, Kewei Zhang, Yun Zhou, Huixian Zhao, Weihua Qiao, and Jiaqiang Sun. "TaBZR1 enhances wheat salt tolerance via promoting ABA biosynthesis and ROS scavenging." *Journal of Genetics and Genomics* 50, no. 11 (2023): 861–871.

Yang, Yue, Jianxin Shi, Limei Chen, Wenhan Xiao, and Jingjuan Yu. "ZmEREB46, a maize ortholog of Arabidopsis WAX INDUCER1/SHINE1, is involved in the biosynthesis of leaf epicuticular very-long-chain waxes and drought tolerance." *Plant Science* 321 (2022): 111256.

Yu, Haiyue, Yushi Zhang, Yan Xie, Yubin Wang, Liusheng Duan, Mingcai Zhang, and Zhaohu Li. "Ethephon improved drought tolerance in maize seedlings by modulating cuticular wax biosynthesis and membrane stability." *Journal of Plant Physiology* 214 (2017): 123–133.

Yu, Xiaofen, Jiapeng Han, Li Li, Qian Zhang, Guangxiao Yang, and Guangyuan He. "Wheat PP2C-a10 regulates seed germination and drought tolerance in transgenic Arabidopsis." *Plant Cell Reports* 39 (2020): 635–651.

Zhang, Dejian, Renxue Xia, Xiu Cao, Bo Shu, and Chunli Chen. "Root hair development of *Poncirus trifoliata* grown in different growth cultures and treated with 3-indolebutyric acid and ethephon." *Scientia Horticulturae* 160 (2013b): 389–397.

Zhang, Feng, Zhijian Pan, Chenyang Han, Huizhen Dong, Likun Lin, Qinghai Qiao, Keke Zhao et al. "*Pyrus betulaefolia* ERF3 interacts with HsfC1a to coordinately regulate aquaporin PIP1; 4 and NCED4 for drought tolerance." *Horticulture Research* (2024): uhae090.

Zhang, Haiwen, Jianfei Zhang, Ruidang Quan, Xiaowu Pan, Liyun Wan, and Rongfeng Huang. "EAR motif mutation of rice OsERF3 alters the regulation of ethylene biosynthesis and drought tolerance." *Planta* 237 (2013a): 1443–1451.

Zhang, Jiahang, Yanan Gao, Lixin Xu, and Liebao Han. "Transcriptome analysis of *Kentucky bluegrass* subject to drought and ethephon treatment." *Plos One* 16, no. 12 (2021a): e0261472.

Zhang, Na, Lu Han, Lixin Xu, and Xunzhong Zhang. "Ethephon seed treatment impacts on drought tolerance of *kentucky bluegrass* seedlings." *Hort Technology* 28, no. 3 (2018): 319–326.

Zhang, Yanfei, Jingyi Wang, Yuying Li, Zihui Zhang, Lili Yang, Min Wang, Yining Zhang et al. "Wheat TaSnRK2. 10 phosphorylates TaERD15 and TaENO1 and confers drought tolerance when overexpressed in rice." *Plant Physiology* 191, no. 2 (2023a): 1344–1364.

Zhang, Yuqin, Himabindu Vasuki Kilambi, Jie Liu, Hamutal Bar, Shani Lazary, Aiman Egbaria, Dagmar Ripper et al. "ABA homeostasis and long-distance translocation are redundantly regulated by ABCG ABA importers." *Science Advances* 7, no. 43 (2021b): eabf6069.

Zhang, Zhijin, Fang Li, Dingjun Li, Haiwen Zhang, and Rongfeng Huang. "Expression of ethylene response factor JERF1 in rice improves tolerance to drought." *Planta* 232 (2010): 765–774.

Zhang, Zijie, Jie Fang, Lei Zhang, Huiyin Jin, and Shengzuo Fang. "Genome-wide identification of bHLH transcription factors and their response to salt stress in *Cyclocaryapaliurus*." *Frontiers in Plant Science* 14 (2023b): 1117246.

Zhao, He, Cui-Cui Yin, Biao Ma, Shou-Yi Chen, and Jin-Song Zhang. "Ethylene signaling in rice and Arabidopsis: New regulators and mechanisms." *Journal of Integrative Plant Biology* 63, no. 1 (2021): 102–125.

Zhou, Yuyi, Rui He, Yuling Guo, Keke Liu, Guanmin Huang, Chuanxi Peng, Yiguo Liu, Mingcai Zhang, Zhaohu Li, and Liusheng Duan. "A novel ABA functional analogue B2 enhances drought tolerance in wheat." *Scientific Reports* 9, no. 1 (2019): 2887.

Zia, Rabisa, Muhammad Shoib Nawaz, Muhammad Jawad Siddique, Sughra Hakim, and Asma Imran. "Plant survival under drought stress: Implications, adaptive responses, and integrated rhizosphere management strategy for stress mitigation." *Microbiological Research* 242 (2021): 126626.

Zou, Y.-N., Q.-S. Wu, and K. Kuča. "Unravelling the role of arbuscular mycorrhizal fungi in mitigating the oxidative burst of plants under drought stress." *Plant Biology* 23 (2021): 50–57.

8 Assessing the Biofortification Potential of Nanofertilizers in Enhancing Nutrient Use Efficiency and Crop Yield under Abiotic Stress

Urooj Hassan Bhat, Moin Uddin, Sangram Singh, Sarika Singh, Aman Sobia Chishti, and M. Masroor A. Khan

8.1 INTRODUCTION

In the present era, ensuring food security encounters a significant hurdle due to the ongoing rise in the global population and the dramatic shifts in climate conditions. Agriculture plays a crucial role in addressing this issue by producing nutrient-rich crops (Shalaby et al. 2022). However, various stresses often hinder crop productivity, potentially resulting in substantial losses ranging from 50% to 70% for abiotic stresses and 40% to 60% for biotic stresses. Prevalent abiotic stresses encompass salinity, drought, flooding, and heat stress, while biotic stresses primarily involve bacteria, viruses, and fungi (Tiwari et al. 2020). These stresses can be ameliorated in cultivated plants by employing a diverse array of stress-alleviating agents. These include compounds such as nitric oxide, pyraclostrobin, rhizobia, and plant-growth-promoting rhizobacteria (PGPR), as well as melatonin (Seth et al. 2023; Tyagi et al. 2023), strigolactones, phytohormones, and essential nutrients like silicon (Khan et al. 2021) and selenium (Nazir et al. 2023). Their nanoscale counterparts also hold potential in this regard (Hossain et al. 2021). Biofortification, or 'biological fortification,' which involves cultivating nutrient-rich crops, is a cost-effective approach to improve accessibility to essential nutrients (Khan et al. 2021). Essentially, two primary categories of biofortification, genetic and agronomic biofortification, are employed to augment the nutritional characteristics of agricultural produce (Stein et al. 2007). Agronomic biofortification predominantly involves the administration of nutrient-enriched fertilizers via either soil incorporation or foliar application (Athar et al. 2020). The utilization of macronutrients, namely nitrogen (N), phosphorus (P), and potassium (K), as well as micronutrients like copper (Cu), zinc (Zn), iron (Fe), and manganese (Mn) in the agronomic biofortification approach has successfully augmented crop yields and played a pivotal role in averting widespread famine among human populations (Khan et al. 2021). Nanotechnology is a new and promising technology that can be used in agriculture to strengthen crops and overcome the limitations of traditional methods of fortifying crops with genetic and conventional methods (Thakur et al. 2018). Nanomaterials, particularly in the configuration of nanofertilizers (NFs), exhibit the potential to ameliorate various stress factors while concurrently elevating crop yield through the augmentation of nitrogen metabolism, seedling growth, carbohydrate and protein synthesis, photosynthesis, and the heightened translocation of nutrients from the root system to the leaves (Zulfiqar et al. 2019). Furthermore, a reduced quantity of NF application can proficiently elevate nutrient content with minimized ecological impact, attributed to its precisely directed and gradual release mechanism (Al-Mamun et al. 2021). Nanofertilizers are synthesized or modified versions of traditional fertilizers created using nanotechnology. These nanoparticles, ranging

DOI: 10.1201/9781032630380-8

in size from 1 to 100 nm exhibit distinct physical and chemical properties compared to their larger counterparts. Their increased surface area relative to volume and nano dimensions enhances absorption and availability, promoting healthier crop growth by boosting photosynthate and biomass production through improved uptake via soil or leaves (Kumar et al. 2021). Therefore, this chapter focuses on the potential of NFs in biofortification, specifically enhancing nutrient use efficiency and crop yield under abiotic stress conditions.

8.2 ADVANTAGES OF NANOBIOFORTIFICATION OVER AGRONOMIC BIOFORTIFICATION

Agricultural biofortification stands as a successful approach for increasing the mineral content of cereal crops on a wide scale (Khan et al. 2021). However, due to either poor soil quality or an ineffective drainage system, agronomic biofortification often fails to produce the desired results, leading to significant fertilizer waste (Yashveer et al. 2014). Previous reports have indicated losses of approximately 40%–70% for applied N, 80%–90% for P, and 50%–90% for K fertilizers (Zulfiqar et al. 2019). The repeated use of chemical fertilizers across multiple agricultural seasons results in increased production costs for farmers. Furthermore, their continued use contributes to decreased soil fertility and increased salinity, ultimately leading to lower agricultural yields in the future. Additionally, the uncontrolled release of nutrients from chemical fertilizers further diminishes the nutritional value of crop grains.

Agronomic biofortification presents significant environmental challenges due to the extensive use of chemical fertilizers, leading to the accumulation of hazardous by-products such as nitrates and heavy metals in both soil and water (Khan et al. 2021). The long-term consequences of chemical fertilizer usage include disruptions to the soil's nutrient balance, degradation of soil fertility, and damage to its structure (Elemike et al. 2019). Therefore, it is imperative to develop successful strategies for increasing crop yield and quality without increasing the annual usage of chemical fertilizers (Khan et al. 2021).

Numerous challenges associated with agronomic biofortification could potentially find solutions through the application of nanotechnology (Khan et al. 2021). Nanofertilizers based on nanoparticles (NPs) possess distinctive qualities that set them apart from synthetic fertilizers (El-Ramady et al. 2021; Kumari et al. 2022). Due to their diminutive particle dimensions and notable reactivity, nanoparticles (NPs) offer expansive surface areas, minimal wastage, and the capacity to ingress stomata upon foliar application (Seleiman et al. 2020). Both intelligent fertilizers and environmentally friendly technologies fall under the purview of nanotechnology. Its main objective is the efficient preservation of nutrients, with N being of particular focus. Furthermore, the application of nanoscale particles improves the carbon balance in agriculture, promotes plant growth, and may improve the utilization of both micro- and macronutrients in plants, thereby reducing the per-acre requirement for chemical fertilizers. These advancements collectively address significant environmental challenges (El-Ramady et al. 2021). Nanotechnology not only enhances understanding of diverse physiological processes in plants but also contributes to disease management and the mitigation of abiotic stress factors, ultimately leading to increased crop yields and improved nutritional content (Khan et al. 2021).

8.3 PENETRATION AND TRANSLOCATION OF NANOFERTILIZERS

Foliar fertilizers encounter various structural impediments due to their salt-based nature, comprising cations and anions, which may face challenges in permeating the interior cellular tissue of plants. This hindrance is primarily attributed to the restricted pore dimensions of the cell wall, falling within the range of 5 to 20 nm (Schwab et al. 2015). The utilization of NFs shows promise in facilitating efficient nutrient translocation to the specific plant regions of interest (Deepa et al. 2015). Nanoparticles administered through foliar application undergo transport from the application site to heterotrophic cells, primarily conveyed through the phloem vessels. This transport is likely facilitated by passage through plasmodesmata, which typically have a diameter of approximately 40 nm (Etxeberria et al. 2016). The internalization of NPs within plant cells occurs through their interaction with carrier proteins and involves aquaporins, ion channels, and endocytosis (Rico et al. 2011). Nanoparticles may also enter the plant by creating complexes with membrane transport proteins (Kurepa et al. 2010). Corredor et al. (2009) demonstrated that NPs can ingress the plant system and translocate through the vascular network. Nanomaterials can ingress through leaf stomata from the atmosphere and disperse throughout various plant structures (Figure 8.1), as evidenced by Wang et al. (2013) in watermelon plants and Hong et al. (2014) in cucumber leaves. Additionally, Deepa et al. (2015) reported the transportation of calcium oxide nanoparticles (n-CaO) via the phloem tissue in groundnut plants.

8.4 BIOFORTIFICATION THROUGH NANOTECHNOLOGY

The growing global population and declining quality of farmland have spurred interest in harnessing nanotechnology for agricultural purposes. Fertilizers are used to address nutrient deficiencies, enhancing nutrient levels in plants as needed. Nanomaterials function as NFs by either serving as nutrients themselves or acting as carriers for nutrients (Sultana et al. 2021). Nanotechnology presents potential solutions to various limitations of agronomic biofortification. The advantageous characteristics of nanomaterials, including their significant sorption capacity, controlled and gradual release at specific locations, and high surface-to-volume ratio, make them suitable for producing NFs (Feregrino-Perez et al. 2018). Encapsulating nutrients using nanomaterials enhances nutrient absorption in plants by facilitating the controlled release of NPs and enabling easy penetration

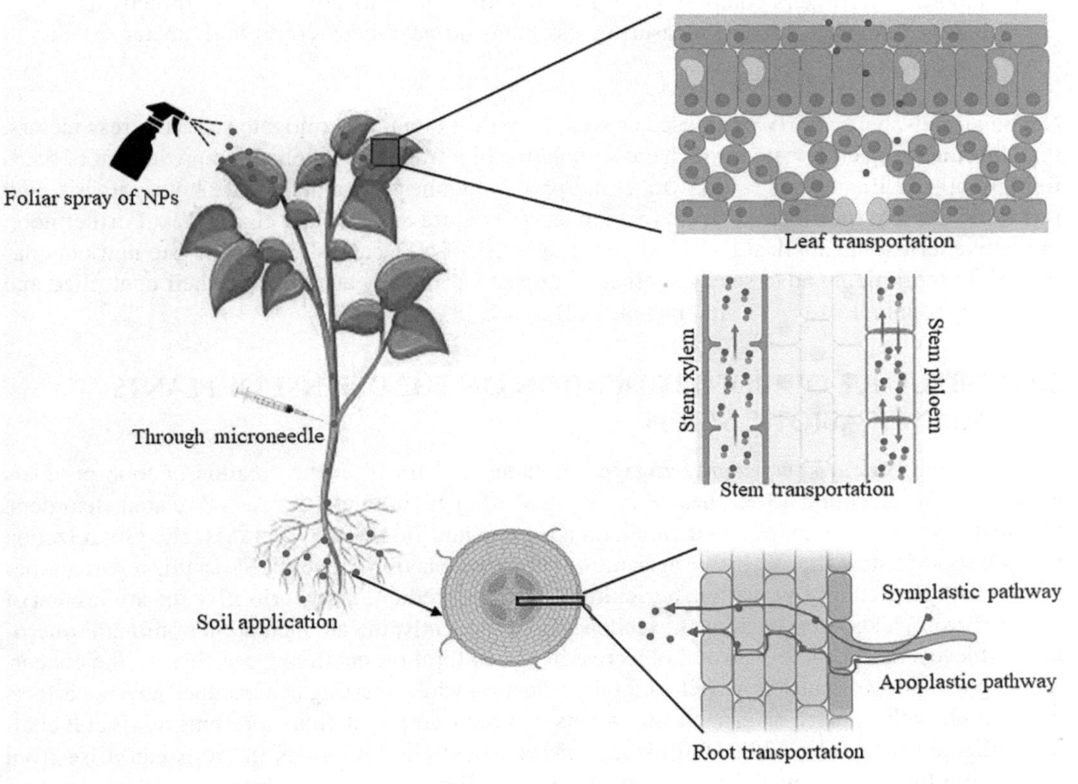

FIGURE 8.1 Translocation of nanofertilizers inside the plant.

through biological barriers into the plant vascular system (De La Torre-Roche et al. 2020). The consistent and prolonged nutrient delivery achieved through NFs promotes superior crop growth compared to traditional fertilizers. Since NFs are applied in minimal quantities, they also prevent soil from overloading with chemical fertilizer by-products, thus reducing environmental risks (Khan et al. 2021). Unlike chemical fertilizers, NFs can be custom-manufactured and tailored to meet the specific nutritional requirements of crops and soil conditions through biosensors (Kah et al. 2018). Additionally, NFs offer advantages over chemical fertilizers, such as enhanced bioavailability of minerals in plants owing to their smaller size, higher reactivity, and increased surface area (Khan et al. 2021). The efficacy of NFs exhibits variability contingent upon their constituent elements, size, chemical attributes, and, notably, the specific crop to which they are applied (Thakur et al. 2018). Nanofertilizers are defined as compounds characterized by exceedingly diminutive dimensions, typically equal to or less than one nanometer in size, encompassing a spectrum of macro- and microelements, including N, P, K, magnesium (Mg), calcium (Ca), sulfur (S), iron (Fe), zinc (Zn), copper (Cu), manganese (Mn), boron (B), nickel (Ni), molybdenum (Mo), and their corresponding compounds, such as cerium oxide (CeO_2), titanium oxide (TiO_2), silver (Ag), gold (Au), zinc oxide (ZnO), iron oxide (FeO), carbon nanotubes, aluminum oxide (Al_2O_3), and so forth (Mejias et al. 2021). Biofortification programs have successfully implemented three distinct categories of nanofertilizers:

i. Nanoscale encapsulation fertilizers involve conventional fertilizers encased within nanoparticles or embedded within nanoporous materials like zeolites and nanoclays (Tarafder et al. 2020).
ii. Nanoscale additive fertilizers, wherein conventional fertilizers are augmented with NPs containing specific nutrients (Khan et al. 2021).
iii. Nanoscale fertilizers composed of NPs containing nutrients are employed directly as fertilizers, with each particle measuring less than 100 nanometers (nm) in diameter (Mejias et al. 2021).

Nanomaterials, particularly when used as NFs, have the potential to mitigate various stress factors, thereby enhancing crop yield through the stimulation of nitrogen metabolism, improvement of seedling growth, facilitation of carbohydrate and protein synthesis, promotion of photosynthesis, and increased nutrient translocation from roots to leaves (Figure 8.2; Zulfiqar et al. 2019). Furthermore, in relative terms, the application of a reduced quantity of NFs can efficiently elevate nutrient content while minimizing adverse environmental impacts, primarily attributed to their controlled and targeted gradual release (Al-Mamun et al. 2021).

8.5 INFLUENCE OF BIOFORTIFICATION ON THE DEFENSE OF PLANTS AGAINST ABIOTIC STRESS

Global researchers are persistently engaged in endeavors aimed at the creation of food products enriched with selenium (Se), zinc (Zn), and iron (Fe) to mitigate deficiency-related disorders. Optimal human health relies significantly on adequate nutrition, as the World Health Organization (WHO) stipulated, with sustainable agricultural practices playing a pivotal role in this regard (Athar et al. 2020). Regrettably, prevailing agricultural systems predominantly prioritize the attainment of elevated crop yields over nutritional excellence. Consequently, the augmentation of mineral micronutrient levels has become a pivotal objective in agricultural production. Nevertheless, the concurrent augmentation of nutrient-enriched food production while averting conspicuous adverse effects on plant physiology, such as growth limitations and reduced productivity, presents a difficult challenge (Szerement et al. 2022). Diminished micronutrient concentrations in crops can arise from various limiting factors, including inadequate levels or limited bio-accessibility of critical elements within the soil, suboptimal abiotic circumstances encompassing extreme temperature fluctuations,

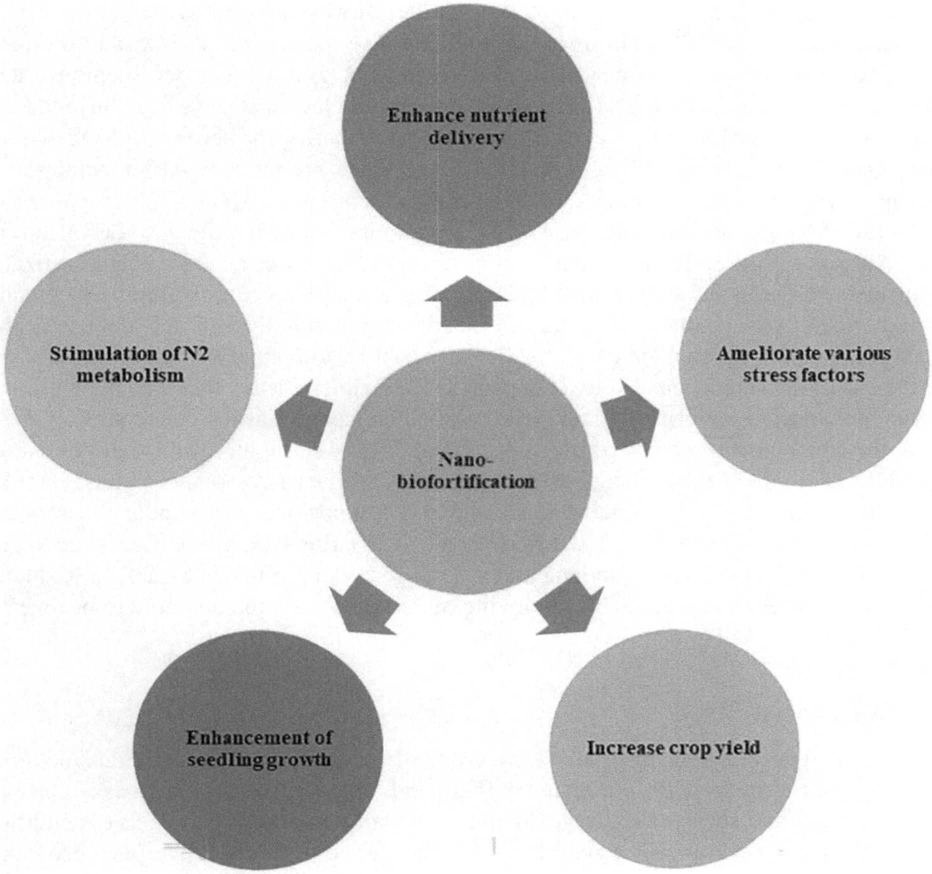

FIGURE 8.2 Various applications of nanobiofortification.

pH imbalances, water scarcity, and anaerobic conditions, as well as the coexistence of other elemental species, both micro and macro, alongside heavy metals (Manojlović et al. 2019). Approximately 50% of soils used for cultivating cereals were estimated to exhibit a deficiency in Zn. Fe insufficiency is predominantly observed in calcareous soils (Jalal et al. 2020). Micronutrient deficiencies are prevalent in regions characterized by a humid temperate and tropical climate, where substantial leaching occurs due to heightened precipitation. Another contributing factor is the cultivation of plant species with a limited capacity to accumulate adequate micronutrients in their consumable plant parts (Manojlović et al. 2019). Biofortification represents a viable strategy for augmenting the micronutrient content within crops (Huang et al. 2020). Studies have demonstrated that the cultivation of biofortified crops leads to elevated micronutrient consumption and substantially impacts human health (Bouis and Saltzman 2017). Moreover, numerous reports substantiate that biofortification, in addition to enhancing the micronutrient content of plants, exerts a notable influence on the biosynthesis of other compounds possessing nutritional attributes (Newman et al. 2021).

Nanoparticles can directly modify plant and seed metabolism and disrupt the production of hormones, ultimately enhancing the plants' resilience to environmental challenges (Salam et al. 2022; Wahid et al. 2022; Khatoon et al. 2024). An elevation in the generation of reactive oxygen species (ROS), a key player in several metabolic processes, is accompanied by heightened mobilization of storage proteins and an upsurge in phytohormone levels (Chandrasekaran et al. 2020). Furthermore, NPs can boost the water absorption capacity of seeds, leading to increased enzyme activity (Amna et al. 2022). Abiotic stresses, such as salinity, extreme temperature variations, heavy metal exposure, and drought, result in the excessive generation of ROS and the provocation of oxidative stress within plant systems (Szerement et al. 2022). It has been demonstrated that biofortification involving the application of various types and formulations of Zn, Se, and Fe fertilizers can mitigate oxidative stress-induced damage by augmenting the activity of ROS-scavenging enzymes, including superoxide dismutase (SOD), ascorbate peroxidase (APX), catalase (CAT), glutathione peroxidase (GPX), monodehydroascorbate reductase (MDHAR), dehydroascorbate reductase (DHAR), glutathione reductase (GR), glutathione sulfur-transferase (GST), and peroxiredoxin (PRX), within distinct cellular compartments (Noreen et al. 2021). The utilization of micronutrients has been shown to boost the levels of non-enzymatic antioxidants like glutathione (GSH), ascorbate (ASA), carotenoids, and proline, which are equally vital for preserving the balance of ROS in plants (Szerement et al. 2022). Furthermore, to uphold the equilibrium of ROS in plants, the use of micronutrients led to a noteworthy reduction in heavy metal content within plant tissues and enhanced the morphological growth indicators, as well as the concentration of Zn, Se, and Fe in the edible portions of crops (Szerement et al. 2022). Drought, salinity, elevated temperatures, and the presence of toxic metals are recognized as primary stressors for plants, significantly impacting food security by influencing the scope and composition of secondary metabolic processes in crop species (Salem et al. 2021; Nazir et al., 2023). Enhancing crop resilience to particular abiotic stressors constitutes a fundamental goal for breeders (Khan et al. 2017). Reducing the effects of abiotic stressors on crops by employing metallic NPs has the potential to be an effective approach (Khan et al. 2017).

8.5.1 Heavy Metal Stress

Over an extended period, soils have soils have gradually accumulated heavy metals including Fe, Mn, Cu, Ni, Co, cadmium (Cd), Zn, mercury (Hg), and arsenic (As), primarily as a consequence of industrial waste and sewage discharge. While several of these metals serve as essential micronutrients vital for various physiological processes in plants, their excessive presence can exert adverse aftermaths, directly impacting plant growth, metabolism, physiology, and senescence (Ghori et al. 2019). Experts have confirmed that using NPs to mitigate the different impacts of heavy metals on crops is a successful strategy (Rizwan et al. 2019). In their study, Huang et al. (2018) reported that Fe_3O_4NPs effectively mitigated the absorption of As by rice plants when exposed to low As concentrations. According to the findings presented by Mohammadi et al. (2018), the utilization of FeNPs can lead to a reduction in the uptake of chromium (Cr) in both the root and shoot systems of sunflower plants. In the context of Chen et al.'s (2018) study, the application of silicon nanoparticles (SiNPs) via foliar spraying on the rice variety Xiangzaoxian resulted in a diminished accumulation of Cd in both the grain and the root tissues (Chen et al. 2018). Furthermore, when ZnONPs were used to prime wheat seeds, the amount of absorbed Cd in the roots, shoots, and grains of wheat grown in high-concentration Cd soil was greatly decreased (Rizwan et al. 2019).

8.5.2 Drought Stress

Drought stress is a pervasive factor across diverse ecological settings, significantly impeding plant biomass production, quality, and energy conversion. It manifests as a critical environmental stressor arising from fluctuations in temperature and light intensity, coupled with inadequate precipitation (Seleiman et al. 2020; Khan et al. 2021). Drought, often described as 'water deficit,' 'water shortage,' or 'inadequate available water,' induces morphological, physiological, and biochemical changes that ultimately diminish agricultural output (Kumar et al. 2018). Plants under drought stress have significant physiological changes, such as variations in stomatal closure, reduced relative water content, photosynthetic activity, and transpiration, as well as increased electrolyte leakage (Ansari et al. 2019). The plant undergoes a biochemical compositional change in response to drought stress. This includes a decrease in chlorophyll levels, an accumulation of proline and other compatible osmolytes, an increase in the production of ROS like superoxide and hydrogen peroxide, an increase in lipid peroxidation, and the activation of antioxidant defense mechanisms (Shivakrishna et al. 2018). Drought-stressed plants typically exhibit a significant increase in abscisic acid (ABA) accumulation, thereby activating water-stress-responsive genes like *MYC*, *MYB*, *ASPG1*, and *PYL8* (Kandhol et al. 2022). Alabdallah et al. (2021) have documented that NPs induce various morphological, physiological, and biochemical changes in plants, enhancing their resilience to drought stress. Additionally, during the wheat flowering stage, Ghassemi and Farahvash (2018) found that the application of ZnNPs to the leaves positively influenced plant height. Additionally, this treatment increased relative water content, wheat yield, and yield-related characteristics. The application of selenium oxide (SiO_2) and titanium dioxide nanoparticles (TiO_2 NPs) to cotton foliage has the potential to enhance cotton traits and increase production under drought stress conditions, as demonstrated in the study by Shallan et al. (2016).

8.5.3 Temperature Stress

The earth's rising temperatures due to global warming can have detrimental effects on ecosystems, leading to decreased crop yields when temperatures exceed the optimal range (Khalid et al. 2022). Extreme temperatures can adversely affect the physiological and molecular processes of plants (Guha et al. 2022). Nanoparticles play a crucial role in helping plants to cope with abiotic stresses, particularly temperature stress (Dimkpa et al. 2017). The application of NPs in various concentrations lead to enhanced plant growth and hydration, mitigating the impacts of heat stress (Ali et al. 2021). When NPs are applied at low concentrations, plants display antioxidative properties, whereas they experience oxidative stress at high NP concentrations. In response to heat stress, plants synthesize molecular chaperones and heat shock proteins (HSPs; Khalid et al. 2020). Heat shock proteins play a crucial role in stabilizing other proteins during challenging conditions, aiding in heat stress resistance. Multiwall carbon nanotubes have been found to up-regulate the gene expression of heat shock proteins like HSP90, thereby contributing to their synthesis (Shafqat et al. 2021). Additionally, CeO_2NPs applied to maize lead to an excessive generation of H_2O_2 and an up-regulation of HSP70, potentially enhancing heat stress tolerance (Wang et al. 2022). Nanoparticles regulate stomatal opening, mitigating the impact of heat stress, as demonstrated by studies such as that of Wang et al. (2022). Furthermore, when wheat seedlings were treated with ZnO and TiO_2 NPs under heat stress, improvements in root and shoot parameters were observed, attributed to the strengthening of the antioxidant defense system and promotion of membrane stability (Thakur et al. 2021).ZnO NPs have shown the capacity to control osmotic potential and reduce thylakoid damage, ultimately leading to increased plant productivity.

Furthermore, the use of ZnONPs has been linked to enhanced cold tolerance in rice, with its application leading to a reduction in oxidative stress, improved photosynthetic activity, and increased root and shoot length as well as dry mass (Song et al. 2021). The use of NPs (such as nSiO_2, nSe, nZnO, and GNRs) on sugarcane effectively alleviated the effects of cold stress. This mitigation was achieved by elevating chlorophyll content, which, in turn, enhanced the photosynthetic rate and counteracted the adverse influences on gas exchange parameters and oxidative activity (Elsheery et al. 2020).

8.5.4 Salinity Stress

On a global scale, soil salinity poses a significant threat to key food crops, threatening their growth by hampering water absorption capabilities and impeding overall plant development

(Khalid et al. 2022). Salinity stress commonly impacts plants by reducing soil osmotic potential, causing nutritional imbalances, and intensifying ionic toxicity, which is commonly known as salt stress (Ahmed et al. 2016). As noted by Khalid et al. (2020) and Ahmed et al. (2021), the reactions of plants to salt stress are notably complex, involving processes such as osmotic regulation, ion compartmentalization or exclusion, uptake of toxic ions, generation of ROS, and electron transport within the context of photosynthesis. In their study, Babaei et al. (2017) discovered that the application of (Zn, Fe) oxide NPs to wheat plants subjected to salt stress resulted in an enhanced grain yield. Alharby et al. (2017) mitigated the detrimental impacts of salinity by employing ZnONPs in the callus culture of five tomato varieties exposed to sodium chloride (NaCl) stress. According to the research by Pérez-Labrada et al. (2019), applying CuNPs to the leaves of tomato plants improved their growth performance and helped balance the Na^+/K^+ ratio following exposure to salt stress. Proteomic analysis of tomato plants under salinity stress revealed that silicon influenced genes associated with light-harvesting complexes, cytochrome b6f (Cytb6f), and ATP-synthesizing complexes (Khalid et al. 2020). Multiwalled carbon nanotubes have been observed to modify the gene expression associated with both the antioxidant system and the salt overly sensitive 1 (SOS1) pathway in rapeseed plants when they are subjected to salinity stress (Zhao et al. 2019). In a study by Rossi et al. (2016), it was asserted that while salinity levels remained unaltered, the introduction of CeO_2NPs into *Brassica napus* enhanced their ability to cope with salt stress.

8.6 CONCLUSION AND FUTURE PROSPECTS

In conclusion, the challenges posed by a growing global population, environmental stress factors, and the COVID-19 pandemic underscore the need for innovative solutions in agriculture and nutrition. Biofortification and nanotechnology represent promising avenues to address these issues, although they are still in their early stages of development. Biofortification can enhance the nutritional content of food crops while making them more resilient to environmental stressors. On the other hand, nanotechnology offers the potential for controlled nutrient delivery but requires further research to mitigate any adverse effects on crops. The ongoing exploration of biologically fortified foods to boost human immunity in the face of pandemics presents an important step towards a healthier and more resilient future. Looking ahead, the prospects for these technologies are encouraging. Continued research and development in biofortification and nanotechnology have the potential to revolutionize agriculture, making it more sustainable and capable of meeting the nutritional needs of a growing population. Governments should play a proactive role in promoting these innovations and providing support to ensure equitable access, thereby mitigating biases against biofortified staple foods. Furthermore, the valuable lessons learned from global crises like COVID-19 and climate change should guide our investment in sustainable solutions for the future of humanity. By harnessing the potential of these technologies and building on our collective experiences, we can pave the way for a more resilient, nourished, and healthier world.

8.7 AUTHORSHIP CONTRIBUTION

Conceptualization, U.H.B.; Writing – Original draft preparation, U.H.B., Moin Uddin, S.S., S. Singh, A.S.C., and M.M.A.K.; Writing – Review and editing, U.H.B. All authors have read and agreed to the published version of the book chapter.

8.8 DECLARATION OF COMPETING INTEREST

The authors declare that they have no known competing financial interests or personal relationships that could have appeared to influence the work reported in this book chapter.

REFERENCES

Ahmed, Muhammad Zaheer, Bilquees Gul, M. Ajmal Khan, and Kazuo N. Watanabe. "Characterization and function of sodium exchanger genes in *Aeluropus lagopoides* under NaCl stress." In *Halophytes for food security in dry lands*, pp. 1–16. Academic Press, 2016.

Ahmed, Talaat, Ahmed Abou Elezz, and Muhammad Fasih Khalid. "Hydropriming with moringa leaf extract mitigates salt stress in wheat seedlings." *Agriculture* 11, no. 12 (2021): 1254.

Alabdallah, Nadiyah M., Md Mahadi Hasan, Inès Hammami, Azzah Ibrahim Alghamdi, Dikhnah Alshehri, and Hanan Ali Alatawi. "Green synthesized metal oxide nanoparticles mediate growth regulation and physiology of crop plants under drought stress." *Plants* 10, no. 8 (2021): 1730.

Alharby, Hesham F., Ehab M. R. Metwali, Michael P. Fuller, and Amal Y. Aldhebiani. "Impact of application of zinc oxide nanoparticles on callus induction, plant regeneration, element content and antioxidant enzyme activity in tomato (*Solanum lycopersicum* Mill) under salt stress." *Arhiv Za Bioloske Nauke* 68, no. 4, (2017): 723–735.

Ali, Shahid, Asif Mehmood, and Naeem Khan. "Uptake, translocation, and consequences of nanomaterials on plant growth and stress adaptation." *Journal of Nanomaterials* 2021, no. 1 (2021): 6677616.

Al-Mamun, Md Rashid, Md Rafiul Hasan, Md Sohel Ahommed, Md Sadek Bacchu, Md Romzan Ali, and Md Zaved Hossain Khan. "Nanofertilizers towards sustainable agriculture and environment." *Environmental Technology & Innovation* 23 (2021): 101658.

Amna, Baber Ali, Muhammad Atif Azeem, Ayesha Qayyum, Ghazala Mustafa, Muhammad Ansar Ahmad, Muhammad Tariq Javed, and Hassan Javed Chaudhary. "Bio-fabricated silver nanoparticles: a sustainable approach for augmentation of plant growth and pathogen control." In *Sustainable agriculture reviews 53: nanoparticles: a new tool to enhance stress tolerance*, pp. 345–371. Cham: Springer International Publishing, 2022.

Ansari, Waquar Akhter, Neelam Atri, Maneesh Pandey, Anil Kumar Singh, Bijendra Singh, and Sudhakar Pandey. "Influence of drought stress on morphological, physiological and biochemical attributes of plants: a review." *Biosciences Biotechnology Research Asia* 16, no. 4 (2019): 697–709.

Athar, Tabinda, Mohd Kamran Khan, Anamika Pandey, Fatma Gökmen Yilmaz, Mehmet Hamurcu, Erdogan Esref Hakki, and Sait Gezgin. "Biofortification and the involved modern approaches." *Journal of Elementology* 25, no. 2 (2020).

Babaei, Khadijeh, Raouf Seyed Sharifi, Alireza Pirzad, and Razieh Khalilzadeh. "Effects of bio fertilizer and nano Zn-Fe oxide on physiological traits, antioxidant enzymes activity and yield of wheat (*Triticum aestivum* L.) under salinity stress." *Journal of Plant Interactions* 12, no. 1 (2017): 381–389.

Bouis, Howarth E., and Amy Saltzman. "Improving nutrition through biofortification: a review of evidence from HarvestPlus, 2003 through 2016." *Global Food Security* 12 (2017): 49–58.

Chandrasekaran, Umashankar, Xiaofeng Luo, Qichao Wang, and Kai Shu. "Are there unidentified factors involved in the germination of nanoprimed seeds?" *Frontiers in Plant Science* 11 (2020): 832.

Chen, Rui, Changbo Zhang, Yanling Zhao, Yongchun Huang, and Zhongqi Liu. "Foliar application with nano-silicon reduced cadmium accumulation in grains by inhibiting cadmium translocation in rice plants." *Environmental Science and Pollution Research* 25 (2018): 2361–2368.

Corredor, Eduardo, Pilar S. Testillano, María-José Coronado, Pablo González-Melendi, Rodrigo Fernández-Pacheco, Clara Marquina, M. Ricardo Ibarra et al. "Nanoparticle penetration and transport in living pumpkin plants: in situ subcellular identification." *BMC Plant Biology* 9 (2009): 1–11.

Deepa, Manchala, Palagiri Sudhakar, Kandula Venkata Nagamadhuri, Kota Balakrishna Reddy, Thimmavajjula Giridhara Krishna, and Tollamadugu Naga Venkata Krishna Vara Prasad. "First evidence on phloem transport of nanoscale calcium oxide in groundnut using solution culture technique." *Applied Nanoscience* 5 (2015): 545–551.

De La Torre-Roche, Roberto, Jesus Cantu, Carlos Tamez, Nubia Zuverza-Mena, Helmi Hamdi, Ishaq O. Adisa, Wade Elmer, Jorge Gardea-Torresdey, and Jason C. White. "Seed biofortification by engineered nanomaterials: a pathway to alleviate malnutrition?" *Journal of Agricultural and Food Chemistry* 68, no. 44 (2020): 12189–12202.

Dimkpa, Christian O., and Prem S. Bindraban. "Nanofertilizers: new products for the industry?" *Journal of Agricultural and Food Chemistry* 66, no. 26 (2017): 6462–6473.

Elemike, Elias E., Ifeyinwa Monica Uzoh, Damian C. Onwudiwe, and Olubukola Oluranti Babalola. "The role of nanotechnology in the fortification of plant nutrients and improvement of crop production." *Applied Sciences* 9, no. 3 (2019): 499.

El-Ramady, Hassan, Neama Abdalla, Heba Elbasiouny, Fathy Elbehiry, Tamer Elsakhawy, Alaa El-Dein Omara, Megahed Amer et al. "Nano-biofortification of different crops to immune against COVID-19: a review." *Ecotoxicology and Environmental Safety* 222 (2021): 112500.

Elsheery, Nabil I., Mohamed N. Helaly, Hanan M. El-Hoseiny, and Shamel M. Alam-Eldein. "Zinc oxide and silicone nanoparticles to improve the resistance mechanism and annual productivity of salt-stressed mango trees." *Agronomy* 10, no. 4 (2020): 558.

Etxeberria, Ed, Pedro Gonzalez, Priyanka Bhattacharya, Parvesh Sharma, and Pu Chun Ke. "Determining the size exclusion for nanoparticles in citrus leaves." *HortScience* 51, no. 6 (2016): 732–737.

Feregrino-Perez, Ana A., Ernesto Magaña-López, C. Guzmán, and K. Esquivel. "A general overview of the benefits and possible negative effects of the nanotechnology in horticulture." *Scientia Horticulturae* 238 (2018): 126–137.

Ghassemi, Azizollah, and Farhad Farahvash. "Effect of nano-zinc foliar application on wheat under drought stress." *Fresenius Environmental Bulletin* 27, no. 7 (2018): 5022–5026.

Ghori, N.-H., T. Ghori, M. Q. Hayat, S. R. Imadi, A. Gul, V. Altay, and M. Ozturk. "Heavy metal stress and responses in plants." *International Journal of Environmental Science and Technology* 16 (2019): 1807–1828.

Guha, Anirban, Talent Vharachumu, Muhammad F. Khalid, Mark Keeley, Thomas J. Avenson, and Christopher Vincent. "Short-term warming does not affect intrinsic thermotolerance but induces strong sustaining photoprotection in tropical evergreen citrus genotypes." *Plant, Cell & Environment* 45, no. 1 (2022): 105–120.

Hong, Jie, Jose R. Peralta-Videa, Cyren Rico, Shivendra Sahi, Marian N. Viveros, Jane Bartonjo, Lijuan Zhao, and Jorge L. Gardea-Torresdey. "Evidence of translocation and physiological impacts of foliar applied CeO2 nanoparticles on cucumber (*Cucumis sativus*) plants." *Environmental Science & Technology* 48, no. 8 (2014): 4376–4385.

Hossain, Akbar, Ali Raza, Sagar Maitra, Md Asaduzzaman, Md Rafiqul Islam, Md Jamil Hossain, Ayman EL Sabagh et al. "Strigolactones: a novel carotenoid-derived phytohormone–biosynthesis, transporters, signalling, and mechanisms in abiotic stress." In *Plant growth regulators: signalling under stress conditions*, pp. 275–303. Springer, 2021.

Huang, Qing, Qing Liu, Lina Lin, Fang-Jun Li, Yifan Han, and Zheng-Guo Song. "Reduction of arsenic toxicity in two rice cultivar seedlings by different nanoparticles." *Ecotoxicology and Environmental Safety* 159 (2018): 261–271.

Huang, Sheng, Peitong Wang, Naoki Yamaji, and Jian Feng Ma. "Plant nutrition for human nutrition: hints from rice research and future perspectives." *Molecular Plant* 13, no. 6 (2020): 825–835.

Jalal, Arshad, Shahen Shah, M. Carvalho Minhoto Teixeira Filho, Ayesha Khan, Tariq Shah, Muhammad Ilyas, and Poliana Aparecida Leonel Rosa. "Agro-biofortification of zinc and iron in wheat grains." *Gesunde Pflanzen* 72, no. 3 (2020): 227–236.

Kah, Melanie, Rai Singh Kookana, Alexander Gogos, and Thomas Daniel Bucheli. "A critical evaluation of nanopesticides and nanofertilizers against their conventional analogues." *Nature Nanotechnology* 13, no. 8 (2018): 677–684.

Kandhol, Nidhi, Mukesh Jain, and Durgesh Kumar Tripathi. "Nanoparticles as potential hallmarks of drought stress tolerance in plants." *Physiologia Plantarum* 174, no. 2 (2022): e13665.

Khalid, Muhammad Fasih, Sajjad Hussain, Muhammad Akbar Anjum, Shakeel Ahmad, Muhammad Arif Ali, Shaghef Ejaz, and Raphael Morillon. "Better salinity tolerance in tetraploid vs diploid volkamer lemon seedlings is associated with robust antioxidant and osmotic adjustment mechanisms." *Journal of Plant Physiology* 244 (2020): 153071.

Khalid, M. F., R. Iqbal Khan, M. Z. Jawaid, W. Shafqat, S. Hussain, T. Ahmed, M. Rizwan, S. Ercisli, O. L. Pop, and R. Alina Marc. "Nanoparticles: the plant saviour under abiotic stresses." *Nanomaterials* 12, no. 21 (2022): 3915.

Khan, M. Iqbal R., Farha Ashfaque, Himanshu Chhillar, Mohammad Irfan, and Nafees A. Khan. "The intricacy of silicon, plant growth regulators and other signaling molecules for abiotic stress tolerance: an entrancing crosstalk between stress alleviators." *Plant Physiology and Biochemistry* 162 (2021): 36–47.

Khan, M. N., M. Mobin, Z. K. Abbas, K. A. AlMutairi, and Z. H. Siddiqui. "Role of nanomaterials in plants under challenging environments." *Plant Physiology and Biochemistry* 110 (2017): 194–209. https://doi.org/10.1016/j.plaphy. 2016.05.038.

Khatoon, Sayeda, Moksh Mahajan, Sarika Kumari, Noushina Iqbal, Iram Wahid, and M. Iqbal R. Khan. "Green-synthesized gold nanoparticles induce adaptation in photosynthetic responses, sugar and nitrogen metabolism, and seed yield of salt-stressed mustard plants." *Clean Technologies and Environmental Policy* (2024): 1–17.

Kumar, S., S. Sachdeva, K. V. Bhat, and S. Vats. "Plant responses to drought stress: physiological, biochemical and molecular basis." In *Biotic and abiotic stress tolerance in plants*, pp. 1–25. Singapore: Springer, 2018. https://doi.org/10.1007/978-981-10-9029-5_1.

Kumar, Yogendra, K. N. Tiwari, T. Singh, and R. Raliya. "Nanofertilizers and their role in sustainable agriculture." *Annals of Plant and Soil Research* 23, no. 3 (2021): 238–255.

Kumari, Sarika, Risheek Rahul Khanna, Faroza Nazir, Mohammed Albaqami, Himanshu Chhillar, Iram Wahid, and M. Iqbal R. Khan. "Bio-synthesized nanoparticles in developing plant abiotic stress resilience: a new boon for sustainable approach." *International Journal of Molecular Sciences* 23, no. 8 (2022): 4452.

Kurepa, J., T. Paunesku, S. Vogt, H. Arora, B. M. Rabatic, J. Lu, M. B. Wanzer, G. E. Woloschak, and J. A. Smalle. "Uptake and distribution of ultrasmall anatase TiO2 Alizarin red S nanoconjugates in *Arabidopsis thaliana*." *Nano Letter* 10 (2010): 2296–2302.

Manojlović, M. S., Z. Lončarić, R. R. Cabilovski, B. Popović, K. Karalić, V. Ivezić, and B. R. Singh. "Biofortification of wheat cultivars with selenium." *Acta Agriculturae Scandinavica, Section B – Soil & Plant Science* 69, no. 8 (2019): 715–724.

Mejias, J. H., F. Salazar, L. Pérez Amaro, S. Hube, M. Rodriguez, and M. Alfaro. "Nanofertilizers: a cutting-edge approach to increase nitrogen use efficiency in grasslands." *Frontiers in Environmental Science* 9 (2021): 52.

Mohammadi, H., M. Hatami, K. Feghezadeh, and M. Ghorbanpour. "Mitigating effect of nano-zerovalent iron, iron sulfate and EDTA against oxidative stress induced by chromium in *Helianthus annuus* L." *Acta Physiologiae Plantarum* 40, no. 4 (2018): 69. https://doi.org/10.1007/s11738-018-2647-2.

Nazir, Faroza, Sarika Kumari, Moksh Mahajan, and M. Iqbal R. Khan. "An explicit story of plant abiotic stress resilience: overtone of selenium, plant hormones and other signaling molecules." *Plant and Soil* 486, no. 1 (2023): 135–163.

Newman, R. G., Y. Moon, C. E. Sams, J. C. Tou, and N. L. Waterland. "Biofortification of sodium selenate improves dietary mineral contents and antioxidant capacity of culinary herb microgreens." *Frontiers in Plant Science* 12 (2021): 716437.

Noreen, S., M. Sultan, M. S. Akhter, K. H. Shah, U. Ummara, H. Manzoor, and P. Ahmad. "Foliar fertigation of ascorbic acid and zinc improves growth, antioxidant enzyme activity and harvest index in barley (*Hordeum vulgare* L.) grown under salt stress." *Plant Physiology and Biochemistry* 158 (2021): 244–254.

Pérez-Labrada, F., E. R. López-Vargas, H. Ortega-Ortiz, G. Cadenas-Pliego, A. Benavides-Mendoza, and A. Juárez-Maldonado. "Responses of tomato plants under saline stress to foliar application of copper nanoparticles." *Plants* 8, no. 6 (2019): 151.

Rico, C. M., S. Majumdar, M. Duarte-Gardea, J. R. Peralta-Videa, and J. L. Gardea-Torresdey. "Interaction of nanoparticles with edible plants and their possible implications in the food chain." *Journal of Agriculture and Food Chemistry* 59, no. 8 (2011): 3485–3498.

Rizwan, M., S. Ali, B. Ali, M. Adrees, M. Arshad, A. Hussain, and A. A. Waris. "Zinc and iron oxide nanoparticles improved the plant growth and reduced the oxidative stress and cadmium concentration in wheat." *Chemosphere* 214 (2019): 269–277. https://doi.org/10.1016/j.chemosphere.2018.09.120.

Rossi, L., W. Zhang, L. Lombardini, and X. Ma. "The impact of cerium oxide nanoparticles on the salt stress responses of *Brassica napus* L." *Environmental Pollution* 219 (2016): 28–36. https://doi.org/10.1016/j.envpol.2016.09.060.

Salam, A., M. S. Afridi, M. A. Javed, A. Saleem, A. Hafeez, A. R. Khan, and Y. Gan. "Nano-priming against abiotic stress: a way forward towards sustainable agriculture." *Sustainability* 14, no. 22 (2022): 14880.

Salem, K. F., M. T. Alloosh, M. M. Saleh, L. M. Alnaddaf, A. K. Almuhammady, and J. M. Al-Khayri. "Utilization of nanofertilizers in crop tolerance to abiotic stress." In *Nanobiotechnology: mitigation of abiotic stress in plants*, pp. 261–289. Springer, 2021.

Schwab, F., G. Zhai, M. Kern, A. Turner, J. L. Schnoor, and M. R. Wiesner. "Barriers, pathways and processes for uptake, translocation and accumulation of nanomaterials in plants critical review." *Nanotoxicology* 10 (2015): 257–278.

Seleiman, M. F., K. F. Almutairi, M. Alotaibi, A. Shami, B. A. Alhammad, and M. L. Battaglia. "Nano fertilization as an emerging fertilization technique: why can modern agriculture benefit from its use?" *Plants* 10 (2020): 2.

Seth, Tanashvi, Sejal Asija, Sayeda Khatoon, Noushina Iqbal, Shahid Umar, and M. Iqbal R. Khan. "A new perspective of melatonin in stress tolerance through regulation of nutrients." *South African Journal of Botany* 163 (2023): 316–329.

Shafqat, W., M. J. Jaskani, R. Maqbool, W. S. Chattha, Z. Ali, S. A. Naqvi, and C. I. Vincent. "Heat shock protein and aquaporin expression enhance water conserving behavior of citrus under water deficits and high temperature conditions." *Environmental and Experimental Botany* 181 (2021): 104270.

Shalaby, T. A., Y. Bayoumi, Y. Eid, H. Elbasiouny, F. Elbehiry, J. Prokisch, and W. Ling. "Can nanofertilizers mitigate multiple environmental stresses for higher crop productivity?" *Sustainability* 14, no. 6 (2022): 3480.

Shallan, M. A., H. Hassan, A. A. Namich, and A. A. Ibrahim. "Biochemical and physiological effects of TiO2 and SiO2 nanoparticles on cotton plant under drought stress." *Research Journal of Pharmaceutical, Biological and Chemical Sciences* 7, no. 4 (2016): 1540–1551.

Shivakrishna, P., K. A. Reddy, and D. M. Rao. "Effect of PEG-6000 imposed drought stress on RNA content, relative water content (RWC), and chlorophyll content in peanut leaves and roots." *Saudi Journal of Biological Sciences* 25, no. 2 (2018): 285–289.

Song, Y., M. Jiang, H. Zhang, and R. Li. Zinc oxide nanoparticles alleviate chilling stress in rice (*Oryza Sativa* L.) by regulating antioxidative system and chilling response transcription factors. *Molecules* 26, no. 8 (2021): 2196.

Stein, A. J., P. Nestel, J. V. Meenakshi, M. Qaim, H. P. S. Sachdev, and Z. A. Bhutta. "Plant breeding to control zinc deficiency in India: how cost-effective is biofortification?" *Public Health Nutrition* 10, no. 5 (2007): 492–501.

Sultana, Alviya, Neetu Talreja, Divya Chauhan, and Mohammad Ashfaq. "Nanotechnology-based biofortification: a plant–soil interaction modulator/enhancer." In *Frontiers in plant-soil interaction*, pp. 83–105. Academic Press, 2021.

Szerement, J., A. Szatanik-Kloc, J. Mokrzycki, and M. Mierzwa-Hersztek. "Agronomic biofortification with Se, Zn, and Fe: an effective strategy to enhance crop nutritional quality and stress defense – a review." *Journal of Soil Science and Plant Nutrition* 22, no. 1 (2022): 1129–1159.

Tarafder, C., M. Daizy, M. M. Alam, M. R. Ali, M. J. Islam, R. Islam, and M. Z. H. Khan. "Formulation of a hybrid nanofertilizer for slow and sustainable release of micronutrients." *ACS Omega* 5, no. 37 (2020): 23960–23966.

Thakur, S., B. Asthir, G. Kaur, A. Kalia, and A. Sharma. "Zinc oxide and titanium dioxide nanoparticles influence heat stress tolerance mediated by antioxidant defense system in wheat." *Cereal Research Communications* (2021): 1–12.

Thakur, S., S. Thakur, and R. Kumar. "Bio-nanotechnology and its role in agriculture and food industry." *Journal of Molecular and Genetic Medicine* 12, no. 324 (2018): 1747–1862.

Tiwari, R. K., M. K. Lal, K. C. Naga, R. Kumar, K. N. Chourasia, S. Subhash, and S. Sharma. "Emerging roles of melatonin in mitigating abiotic and biotic stresses of horticultural crops." *Scientia Horticulturae* 272 (2020): 109592.

Tyagi, Komal, V. Prathap, Prachi Tyagi, Arti Kumari, Rakesh Pandey, Nand Lal Meena, M. Iqbal R. Khan, Aruna Tyagi, and Chirag Maheshwari. "Seed priming with melatonin induces rhizogenesis and modulates physio-biochemical traits in high-yielding rice (*Oryza sativa* L.) genotypes." *South African Journal of Botany* 163 (2023): 191–200.

Wahid, Iram, Pratibha Rani, Sarika Kumari, Rafiq Ahmad, Sofi J. Hussain, Saud Alamri, Nirmalya Tripathy, and M. Iqbal R. Khan. "Biosynthesized gold nanoparticles maintained nitrogen metabolism, nitric oxide synthesis, ions balance, and stabilizes the defense systems to improve salt stress tolerance in wheat." *Chemosphere* 287 (2022): 132142.

Wang, S., Y. Fu, S. Zheng, Y. Xu, and Y. Sun. "Phytotoxicity and accumulation of copper-based nanoparticles in *brassica* under cadmium stress." *Nanomaterials* 12, no. 9 (2022): 1497.

Wang, W. N., J. C. Tarafdar, and P. Biswas. "Nanoparticle synthesis and delivery by an aerosol route for watermelon plant foliar uptake." *Journal of Nanoparticle Research* 15 (2013). https://doi.org/10.1007/s11051-013-1417-8.

Yashveer, S., V. Singh, V. Kaswan, A. Kaushik, and J. Tokas. "Green biotechnology, nanotechnology and biofortification: perspectives on novel environment-friendly crop improvement strategies." *Biotechnology and Genetic Engineering Reviews* 30, no. 2 (2014): 113–126.

Zhao, G., Y. Zhao, W. Lou, J. Su, S. Wei, X. Yang, and W. Shen. "Nitrate reductase-dependent nitric oxide is crucial for multi-walled carbon nanotube-induced plant tolerance against salinity." *Nanoscale* 11, no. 21 (2019): 10511–10523.

Zulfiqar, F., M. Navarro, M. Ashraf, N. A. Akram, and S. Munné-Bosch. "Nanofertilizer use for sustainable agriculture: advantages and limitations." *Plant Science* 289 (2019): 110270.

9 Role of Biostimulants in Improving Plant Growth under Abiotic Stresses

Muskan Parveen, Noor Fatima, Rizwan A. Ansari, Irshad Mahmood, and Rose Rizvi

9.1 INTRODUCTION

Climate is the most important aspect that has a major impact on agricultural yield and production, influencing global food security and the sustainability of crop production systems (Swaminathan and Kesavan 2012; Khan et al. 2023). Several prevalent factors affecting plant growth and productivity in the present era are extreme climate conditions such as desiccation, floods, salinity, torridity, and extreme high and low temperatures, which cause huge crop losses worldwide (Singh and Thakur 2018; Nazir et al. 2023). Biostimulants are incorporated into production systems with the ultimate goal of altering plant physiological processes to enhance productivity as a means to prevent these losses (Yakhin et al. 2017). A biostimulant is a naturally occurring chemical or combination of several organic compounds that can stimulate plant growth and development under different environmental challenges (Bhupenchandra et al. 2020; Rouphael and Colla 2020). In 2012, the industry originally proposed a definition of biostimulants, which stated, "plant biostimulants are substances and/or microbes that, when applied to plants or the rhizosphere, accelerate natural processes that promote nutrient efficiency, nutrient uptake, increase tolerance to abiotic stress, and improve crop quality. Since biostimulants don't directly target pests, they are exempt from the framework of pesticides." The definition proposed by du Jardin (2015), "Biostimulant is any substance or microbe administered to plants to boost nutrition efficiency, increase their ability to withstand abiotic stress, and improve crop quality attributes, irrespective of their nutritional value," represents the definite and most succinct definition of biostimulants. The application of biostimulants is growing and spreading incredibly in the current situation regarding food safety and ecological sustainability. According to Jing et al. (2022), applying biostimulants to grain crops, legumes, fruits, and vegetables has augmented yield by 17.9%.

Apart from raising the production standard, biostimulants have also shown the ability to reduce greenhouse gas emissions by decreasing the need for fertilizer in farming. Biostimulants are separated into two main categories: microbial biostimulants (plant growth-promoting rhizobacteria, arbuscular mycorrhizal fungi, and *Trichoderma* spp.) and non-microbial biostimulants (chitosan, fulvic acids, humic acids, silicon, phosphites, and seaweed extracts) (Colla and Rouphael 2015). Due to their high solubilization and proliferation capability in the rhizospheric soil, biostimulants could overcome the yield barrier of horticultural and field crops caused by environmental stressors. Additionally, they may assist with attaining the highest possible production while utilizing the fewest synthetic pesticides and fertilizers. The application of biostimulants includes foliar fertilization, or direct application to soil; these methods improve the growth, development, and quality of the crops (Van Oosten et al. 2017). Biostimulant formulations may be added directly to the soil as a pre-sowing treatment to promote synergistic crop growth, either by themselves or in combination with micronutrients. Biostimulants could lessen the harmful effects of chemical application by reducing the environmental risks to plants.

DOI: 10.1201/9781032630380-9

9.2 NON-MICROBIAL BIOSTIMULANTS

They can be classified further as organic and inorganic biostimulants.

9.2.1 Organic Biostimulants

9.2.1.1 Humic Acid and Fulvic Acid

These include humic compounds derived from organic matter decomposing aerobically. They are composed of various organic matter, polypeptides, lipids, sugars, hydrocarbons, and aromatic chemical substances (Shah et al. 2018; Lipczynska Kochany 2018; Guo et al. 2019). Humic substances contain an organic functional group that contains oxygen, and this includes phenolics and carboxyls. These substances can be extracted from a wide range of sources and are effective against both biotic and abiotic stresses, promoting plant development and growth. They are beneficial for both agricultural and horticultural crops, augmenting organic carbon in the soil, which controls physio-biochemical processes and the expression of genes linked to plant adaptation strategies. Therefore, they have the potential to revolutionize environmentally friendly agricultural systems.

9.2.1.2 Protein Hydrolysates and Nitrogen-Containing Compounds

Biostimulants derived from hydrolyzed proteins consist of a combination of oligopeptides, polypeptides, and amino acids obtained through biochemical or enzyme-mediated hydrolysis of animal tissues or plant remnants (Caruso et al. 2019). They induce the defense responses in a plant under various adverse growth conditions and thus aid in improving growth and yield-promoting capability. They enhance the gaseous exchange in the leaves by synthesizing and protecting plant pigments, and they also contribute to boosting the water utilization efficiency of the entire plant system (Colla et al. 2017; Rakkammal et al. 2022). Additionally, they possess hormone-like properties that modulate carbon and nitrogen metabolism. They alter the plant's growth and reproduction and help enhance root length and root surface area for better absorption of nutrients, thus maintaining the C:N ratio.

9.2.1.3 Seaweed Extract (SWE)

It is one of the most widely applied biostimulants, derived from various species of seaweed. Commercial seaweeds contain 30%–40% polysaccharides (Battacharyya et al. 2015). The extract can be obtained either from brown seaweeds, for example, *Fucus* spp., *Sargassum hildebrandtii*, *Ascophyllum nodosum*, *Laminaria setchellii*, *Turbinaria* spp., *Ecklonia arborea*, *Macrocystis pyrifera*, *Durvillaea antarctica*, and *Sargassum horridum*, or from red seaweeds, for example, *Acanthophora spicifera*, *Gracilaria parvispora*, *Gelidium robustum*, *Gracilaria edulis*, and *Kappaphycus alvarezii* (Singh et al. 2018; Sharma et al. 2019; Ali et al. 2021). SWE may influence the physio-chemical properties of the soil by modifying the nutritional content, promoting the proliferation of advantageous microorganisms, and assisting the soil in retaining water. Seaweed extracts are abundant in secondary metabolites, which act as signaling molecules in several stress-responsive pathways, thus alleviating nutritional and environmental stresses on the plants. Seaweed products have been shown to promote increasing rates of germination and also help to increase seedling vigor by enhancing the size and density of roots. The seaweed extracts are also useful in protecting seedlings from transplantation shock in various plants, for example, tomatoes, cabbage, and marigolds.

9.2.1.4 Inorganic Biostimulants

They include different metals such as aluminum, sodium, selenium, silicon, cobalt, and so on.

9.3 MICROBIAL BIOSTIMULANTS

Different types of microorganisms, including plant growth-promoting rhizobacteria or microbes, comprise microbial biostimulants, which help stimulate the uptake of various nutrients, and the production of secondary metabolites, siderophores, organic acids, and hormones. They also aid in N_2 fixation. They include many species of *Trichoderma* and arbuscular mycorrhizal fungi. Many microorganisms present in the soil, such as *Pseudomonas* spp., *Bacillus* spp., *Promicromonospora* spp., *Arbuscular micorrhizae, Paraburkholderia phytophirmans*, and *Azospirillum* spp., among others, respond to different abiotic stress conditions (Kang et al. 2012; Duc et al. 2018; Esmaeel et al. 2018; Crovadore et al. 2020; Del Carmen Orozco-Mosqueda et al. 2020; Oliveira et al. 2020). Endophytic microbial biostimulants release some secondary metabolites, which help scavenge the reactive oxygen species and thus help overcome stress.

9.4 RESPONSES OF PLANTS TO ABIOTIC STRESS CONDITION

Abiotic stress is defined as an ambient factor that lowers growth, development, and yields below optimal levels. Agriculture systems exposed to unfavorable conditions have been greatly affected by climate change. According to projections, the increasing extreme weather conditions (heat, salinity, and drought) brought on by climatic changes have a detrimental effect on entire agricultural production systems, making a significant difference in both the quality and the quantity of plant production (Buono 2021). Due to ongoing climatic changes and the deterioration of the environment brought on by human activities, abiotic stress has grown to be a serious threat to food security.

One sustainable way to secure food security is through the use of biostimulants, which may increase plant resistance to challenges associated with climate change (Calvo et al. 2014; Yakhin et al. 2017). The negative impact of rising temperatures on agricultural plants may result in heat-induced cell damage, disturbances in protein synthesis, and disruptions in the functioning of many important enzymes (Devi et al. 2017). Plants are impacted by drought in both their morphological and their physiological aspects. These could impact availability, absorption, and nutrient translocation (Rouphael et al. 2012; Devi et al. 2017), as well as cause the buildup of toxic reactive oxygen species and the release of ethylene (Ali et al. 2014). Furthermore, damage caused by drought stress might worsen if there is a buildup of salt in the soil during dispersed and low-intensity rainfall. Plants undergo osmotic stress at high salinity levels, which has a detrimental impact on their development, metabolism, and nutritional composition (Bulgari et al. 2019a). In this regard, by inducing various processes, such as molecular change and physiological, biochemical, and anatomical modulations, biostimulants could significantly reduce the negative consequences of stress on plants (Figure 9.1). Biostimulants also activate the innate immune response of plants, specifically cellular hypersensitivity, lignin synthesis, and callose deposition in response to biotic stress. In crops, biostimulants such as seaweed extracts, humic acids, and rhizobacteria increase the transcript of genes related to plant defense (Agarwal et al. 2016; Pereira et al. 2021). Biostimulants maintain the excessive production of reactive oxygen species prompted by unfavorable environmental conditions by speeding up antioxidant machinery (Hasanuzzaman et al. 2021; Rakkammal et al. 2022).

9.5 ROLE OF BIOSTIMULANTS IN PROMOTING PLANT GROWTH

9.5.1 Acquisition and Mobilization of Nutrients

Humic substances (HSs), which are rich in carbon, are excellent sources of nutrients. The vermicompost-derived humic acid shows a positive effect on various nutrient acquisitions and metabolisms and also affects membrane fluidity and the pyrophosphatase enzyme (H^+-PPase) activity. It increases the activation of the plasma membrane H^+-ATPase enzyme, which further facilitates

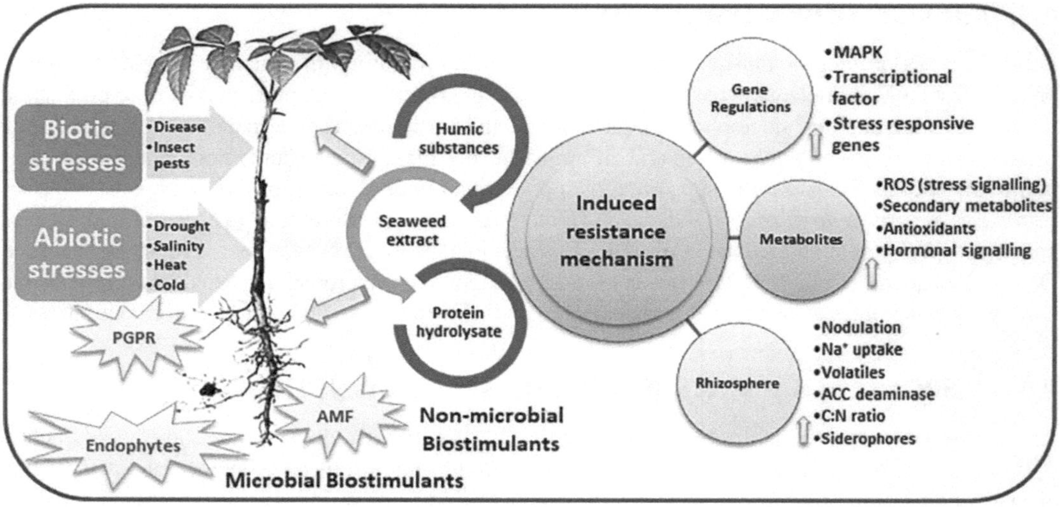

FIGURE 9.1 The role of biostimulants in alleviating the effects of stresses on plants by various mechanisms (physiological, biochemical, anatomical, and molecular alterations).

Source: Bhupenchandra et al. (2022)

the uptake of N, P, and K and amplifies the uptake of other essential plant nutrients, including S, Ca, Mg, Mn, and Fe (Morozesk et al. 2017). HSs augment the solubility of micronutrients in soil water by forming a metal-humic complex (Chen et al. 2004). According to Halpern et al. (2015), humic substances increase root surface area and secondary root hairs, altering the morphology of the roots and activating H^+-ATPase and NO_3-assimilating enzymes. Therefore, by modifying the soil properties and the plant responses, HSs contribute to enhanced nutrient assimilation. Liquefied humic substances modify the metabolism of nitrogen by upregulating genes associated with nitrate acquisition, including *ZmNRT2.1*, *ZmNRT2.2* (encode high-affinity nitrate transporters), *ZmMHA2* (code plasma membrane H^+-proton pump), *ZmNADPH*: NR, *ZmNADH*: NR (code nitrate reduction), and *ZmNiR* (code nitrite reduction) (Vujinović et al. 2020). Leonardite-derived HA (30%–80%) increases the transcription of genes encoding H^+-ATPase (*CsHa1* and *CsHa2*), Fe (III) chelate-reductase (CsFRO1), and Fe (II) transporter (*CsIRT1*). This, in turn, causes increased levels of iron buildup in the aerial portion of the plants (Aguirre et al. 2009). Low-temperature stress causes nutrient deficit conditions, which might be remedied by employing seaweed extracts (Van Oosten et al. 2017). SWE is abundant in micro- and macronutrients, vitamins, amino acids, and plant growth regulators (auxins, cytokinins, and abscisic acid). These substances can significantly alter cellular metabolism, which in turn affects crop yield and growth (Khan et al. 2009). The demonstration of the synergetic interaction between many chemical constituents of seaweed extracts and agricultural output has been found through the up- and downregulation of several genes involved in primary metabolism, plant development, modulation of phytohormone signaling, and antioxidant activities (Kumar et al. 2020; Trivedi et al. 2021). Because they are rich in carbon, HSs provide food for many plant growth-promoting microorganisms in the soil. Nitrogen assimilation has been augmented by the synergistic interaction between PGPM and other crops, including legumes (Sanginga 2003). Additionally, PGPM has been shown to enhance the physicochemical properties of soil, alter the root structure for nutrient translocation (Lopez-Bucio et al. 2007; Halpern et al. 2015), and increase the solubilization of phosphorous and iron. PGPMs such as *Brevi bacillus*, *Bacillus*, and *Rhizobium* spp. enhance the chlorophyll amount by either decreasing the rate of degradation or boosting the rate of synthesis of pigments. This raises the photosynthetic rate and, in turn, boosts agricultural production (Khan et al. 2009). Moreover, *Pseudomonas* spp. boost antioxidant activities, which may assist in controlling desiccation conditions (Heidari and Golpayegani 2012). By enhancing root biomass, promoting K^+ uptake, and lowering Na^+/K^+, microalgae-cyanobacteria-based biostimulants enhance the total nutritional value and assist in maintaining ion equilibrium under salt stress conditions (Mutale-Joan et al. 2021). Microbial biostimulants (MBSts), such as *Trichoderma koningii* and arbuscular mycorrhizal fungi (AMF), boosted the uptake of P, Mg, Fe, Mn, and Zn multiple times, as well as several phenolics, thereby enhancing crop production (Saia et al. 2019). Therefore, biostimulants have the potential to ameliorate agricultural plants' total nutritional content.

9.5.2 Biostimulants and Abiotic Stress Tolerance

9.5.2.1 Drought Stress Tolerance

Drought is the foremost environmental stress that affects plants' normal functioning, ultimately reducing growth, yield, and quality of the crop plants. It has been observed that the application of biostimulants (microbial and non-microbial) profoundly impact plant growth and development, nutrient (micro and macro) absorption, and translocation in different crops, eventually leading to enhanced biomass production and output (Rouphael and Colla 2020). Simultaneously, they also promote endurance to different environmental challenges in agricultural and horticultural crops. Biostimulants like humic substances, seaweed extracts, protein hydrolysates, amino acids, and many advantageous microorganisms on vegetable plants mitigate the utmost frequent abiotic stresses, that is, drought stress conditions (Bulgari et al. 2019b).

Under drought conditions, foliar application of pollen grain extract (PGE) boosted plant development, antioxidant enzymatic activity, and ethereal oil production of basil (*Ocimum basilicum*). It also decreased electrolyte leakage and increased relative water content (Taha et al. 2020). Pure organic substances may be utilized as natural biostimulants to help plants cope with various stresses (Garcia-Garcia et al. 2020). According to La et al. (2020), glutamate enhanced the production of proline biosynthetic genes and the concentrations of compatible solutes in canola, improving its resistance to drought. A foliar spray of proline boosted gross growth and reduced the harmful effects of drought in water-deficit maize (Ali et al. 2008). Proline functions as an osmolyte and helps tobacco and barley withstand drought stress (Hoque et al. 2007; Per et al. 2017; Abdelaal et al. 2020). Hanif et al. (2020) reported that proline foliar sprays enhanced the antioxidant enzymatic activity, boosting the plants' resistance to high temperatures and drought. Besides proline, another osmolyte, glycine betaine (GB), mitigated the negative impacts of salt and drought stress by root and foliar application in many crops, such as corn, red beans, faba bean, barley, and lettuce (Gadallah 1999; Shams et al. 2016; Wang et al. 2019; Sofy et al. 2020). Aminobutyric acid (GABA) enhanced the ability to withstand stress (Bown and Shelp 2016; Khan et al. 2021; Khanna et al. 2021) via improvements in secondary metabolites, osmoregulation, and energy production in carpet bentgrass (Li et al. 2017). GABA also promotes drought tolerance in maize by activating ROS scavenging machinery as well as producing abscisic acid (ABA) via the jasmonic acid pathway (Shaw et al. 2016). Alginate oligosaccharides (AOS) are the natural polysaccharides derived from brown algae that have been shown to upregulate desiccation-resistant genes and regulate ABA-dependent signaling pathways, thereby improving endurance to PEG-prompted dry conditions in tomato, wheat, and cucumber (Liu et al. 2009, 2013; Li et al. 2018). Utilizing BACSTIMR 100, a microbial BSt composition of five *Bacilli* strains, improved the plants' ability to withstand drought by changing key metabolic pathways involved in mechanisms such as increased energy production, osmoregulation, redox homeostasis, cell wall strengthening, and remodeling of membrane (Nephali et al. 2021). Plants' resistance to salt and drought stress can be increased through exogenously applied cytokinins, zeatin (cis and trans), and plant extract (Schafer et al. 2015). According to Alharby et al. (2020), pre-treating wheat seeds with organic BSt (produced from maize grain) and zeatin improved hormone contents, polyamine gene expression, and ability to withstand salt and drought stress.

Recent molecular research elucidates the cellular and gene-level mechanisms activated by specific BSts (Baltazar et al. 2021). The molecular basis of the morphological changes caused due to applying BSts has been clarified by a variety of techniques, including *in vitro* bio-assay, multi-trait high-throughput screening, high-throughput phenotyping, micro-phenotyping, and microarray techniques (Smith et al. 2021; Geelen and Xu 2020), the Illumina IG sequencing technique, and more lately, omic technologies such as proteomics, metabolomics, and transcriptomics (Xu et al. 2020). Recently, the effectiveness of a novel biostimulant (EnNuVi® ALPAN®) was evaluated against drought conditions on tomato plants, and the impact of molecular basis and gene expression was clarified (Hamedeh et al. 2022). ALPAN® elevates the expression of many genes participating in stomatal closure, cell homeostasis, translocation, source translocation, and carbohydrate metabolism (source-to-sink). To summarize, BSts promote stress tolerance through two distinct mechanisms: the first involves morpho-physiological adaptation to establish an avoidance mechanism, and the second involves gene regulation, biochemical, and biomolecular science to develop a tolerance mechanism inside the plant.

9.5.2.2 Salinity Stress Tolerance

Salt stress poses a significant constraint on plant productivity, global food security, and ecological sustainability in the modern era (Kumari et al. 2021; Mukhopadhyay et al. 2021; Ondrasek et al. 2022). It occurs due to factors such as evaporation induced by high temperatures, reduced precipitation associated with degraded water quality, and groundwater contamination with salt (Akbari et al. 2020). Salt intrusion into soil leads to coastal flooding caused by sea level rise, particularly affecting dry and semi-arid areas and coastal regions, thus negatively impacting soil quality (Shrivastava and Kumar 2015). Salts in soil water hinder plant growth by reducing water absorption capacity, thereby slowing growth rates. Additionally, excessive salt absorption by plants through the transpiration stream can damage cells in transpiring leaves, leading to significant growth reductions. The soluble form of excess salt in the soil disrupts physiological processes in plants and inhibits their growth, resulting in salt stress (Akbari et al. 2020).

The utilization of environmentally friendly, non-toxic organic techniques like BSts may enhance crop development, growth, and resilience to stresses like salt stress (Ondrasek et al. 2022). By modulating physiological processes, BSts reduce the effects of salt stress on a variety of plant species, enhancing their growth and production. By upregulating the stress tolerance gene, BSts from various origins increase the plant's resistance to salt (Campobenedetto et al. 2021; Islam et al. 2021). According to Tognetti et al. (2010), the biosynthesis of indole butyric acid (IBA) is potentially enhanced by the gene UGT74E2, which stimulates the buildup of enzyme UDP-glucosyltranferase that exhibits glycosylation activity in certain stress circumstances such as high salt condition. BSt represents one of the innovative and promising methods for modulating physiological functions in numerous plant species.

Plants develop many alterations to their observed biochemical and physiological systems to live in high salinity soil concentrations; Table 9.1 details these modifications. Table 9.2 lists the many BSTs used to reduce salt stress.

9.5.2.3 High-Temperature Stress Tolerance

Rising temperatures lead to alterations in growth periods, crop plant distribution, and production levels. Elevated temperatures have been found to harm the cell membrane and certain proteins, deactivate key enzymes, impair the synthesis of macromolecules, and limit the cell division process. Increasing temperature also alters plants' physiology by raising transpiration and respiration rates and modifying photosynthetic distribution (Munns 2002; Malhotra 2017). At high temperatures, Rubisco enzyme affinity for CO_2 reduces while its affinity for O_2 rises (Sangiorgio et al. 2020). As temperature rises, carbon dioxide becomes less soluble than oxygen, which lowers the chloroplast's carbon dioxide concentration (Ku and Edwards 1977). In addition, the closing of stomata occurs in response to increasing temperatures to stop or reduce evapotranspiration loss. As stomata close, the concentration of carbon dioxide rapidly decreases while the oxygen concentration rises, constraining photosynthesis and increasing photorespiration (Bhattacharya 2019). BSts, including MBSt, may have a significant stimulatory impact on reducing plant responses to heat

TABLE 9.1
Physio-Biochemical Modulations by Biostimulants in Plants

Particular	Action/Role	References
Ionic compartmentalization and homeostasis		
Na^+/Cl^- ratio	Plant ionic homeostasis is maintained by regular regulation of Na^+ entry into the cytoplasm of the cell through the Na^+/H^+ antiporter via vacuolar H^+ ATPase (V-ATPase) and vacuolar pyrophosphatase (V-PPase) and its compartmentalization into the vacuole.	Mansour (2022)
	Na^+ and Cl^- ion conc. inside the cell maintains the plant's growth.	Assaha et al. (2017); Hasegawa et al. (2000)
Osmotic adjustment		
Compatible solute	Increased synthesis of soluble sugar, glycine betaine, and polyols maintain the cell's necessary osmotic potential.	Kerepesi and Galiba (2000)
	Maintains the cell's pH and detoxifies free radicals and hazardous chemicals produced during the oxidation process.	Mansour (2000); Slama et al. (2015)
Proline	Proline acts as reserve organic nitrogen to recover from the stress; pyrroline carboxylic acid synthetase and reductase are essential enzymes in the biosynthesis of proline from the major precursor glutamate during stress conditions.	Qamar et al. (2015)
	Enhancing the enzymatic activity that produces proline in antioxidant defense system.	Hoque et al. (2007); Meena et al. (2019)
Antioxidants		
Ascorbic acid	Heat dissipation, ROS scavenging, growth-regulating phytohormone production, stimulation of cell signaling, and cell division.	Gallie (2013); Massange-Sánchez et al. (2022); Smirnoff (2000)
Carotenoids	Heat dissipation, defense of photosynthetic apparatus, hormone biosynthesis regulation.	Jin et al. (2015); Ruiz-Sola and Rodríguez-Concepción (2012)
Flavonoids	ROS-producing enzymes inhibition.	Agati et al. (2012); Mierziak et al. (2014)
Glutathione	Expression of stress-responsive gene, ROS scavenging, xenobiotics detoxification.	Mullineaux and Rausch (2005); Massange-Sánchez et al. (2021); Noctor et al. (2012)
Tocopherol	Decreased lipid peroxidation, cell signaling, growth regulation phytohormone regulation.	Krieger-Liszkay and Trebst (2006); Maeda et al. (2006); Munné-Bosch and Alegre (2002)

stress. The same advantage was noticed in plants inhabited by microorganisms, including *Bacillus*, *Pseudomonas*, *Septoglomus constrictum*, and *Septoglomus deserticola* (Duc et al. 2018). *P. fluorescens* and *P. aeruginosa* are found in SoilPro® (Liventia, TX, USA), which enhances soil quality and tolerance to high-temperature stress (Sangiorgio et al. 2020). By utilizing microorganisms that restrict the formation of ethylene, the detrimental impacts of heat stress can be mitigated (Glick 2014). *Paraburkholderia phytophirmans* strain PsJN exhibits ACC deaminase activity in potatoes and tomatoes (Bensalim et al. 1998; Esmaeel et al. 2018) to reduce and promote tolerance against heat stress. This activity was also observed in *Pseudomonas fluorescens* YsS6, *Pseudomonas migulae* 8R6, *Bacillus subtilis* BERA7, and *Leclerciaa decarboxylata* MOI (Del Carmen Orozco-Mosqueda et al. 2020). Khan et al. (2020) investigated the impact of thermotolerant SA1, an isolate

TABLE 9.2
Role of Biostimulants in Mitigating Salt Stress in Crops

Types of BS	Crop	Mechanisms	References
Algal extract	Soybean (*Glycine max*)	Increased seed yield	Islam et al. (2021)
Arbuscular mycorrhizal fungi (AMF)	Tomato (*Lycopersicon esculentum*)	Negation of salinity-induced oxidative stress.	Abdel Latef and Chaoxing (2011)
Humic acid extracts	Common bean (*Phaseolus vulgaris*), Rice (*Oryza sativa*)	Activation of antioxidant enzyme function, enhancement of ROS-scavenging enzymes, detoxification of synthesized harmful oxygen free radicals, increased endogenous proline levels, and minimized membrane leakage in plants manifested to desiccation and salt stress.	Aydin et al. (2012); García et al. (2012)
Plant growth-promoting rhizobacteria (PGPR)	Wheat (*Triticum aestivum*)	Improved growth and yield.	Upadhyay and Singh (2015)
Retrosal® (Valagro S.p.A)	Lettuce (*Lactuca sativa*)	Maintain the level of salinity (NaCl) exposed to the water, greater proline accumulation, and lower ABA content.	Bulgari et al. (2019b)

of *Bacillus cereus*, and HA on tomato seedlings. Under heat-stress conditions, they found that the simultaneous administration of SA1+HA greatly increased chlorophyll fluorescence and biomass of tomato plants. Additionally, the levels of salicylic acid and abscisic acid decreased. Antioxidant enzymes like APX, SOD, and GSH were observed to be more active in the plants treated with SA1+HA mixture. *Ascophyllum nodosum* (SWE) causes the overexpression of various transcription factors and heat shock protein families involved in heat stress. Under BSt therapy, AtHSP17 (HSP family) is highly expressed, especially AtHPS17.4 and AtHPS17.6A (Cocetta et al. 2022). The carbohydrate-rich fraction of *A. nodosum* plays a critical role in heat tolerance in tomatoes, as it overexpresses HSP70.9 and HSP101.1 in flowers and fruits (Carmody et al. 2020). Therefore, BSts help plants withstand extreme temperatures by lowering photorespiration, upregulating HSPs' antioxidant enzymes, and taking part in the transduction of stress signals and the stress response.

9.5.2.4 Low-Temperature Stress Tolerance

All crop plants, including cereals, fruit trees, vegetables, and woody crops, grow and develop at optimum temperatures. Each crop species has a limited ability to withstand low temperatures under stress. Extreme cold and rapid drops in temperature can be fatal to crops due to the production of small ice crystals inside and outside of plant cells. This may limit the plant's ability to respire, photosynthesize, or transport vital nutrients and water, all of which could impact on growth, development, and yield of the plant. Microbial BSts can assist plants in resisting the effects of cold temperatures by generating growth-related hormones or increasing the quantity of ethylene (Sangiorgio et al. 2020). Application of psychrotolerant (cold-tolerant) microorganisms to the soil can provide resistance to freezing. *Burkholderia phytofirman* is a PGPR that may colonize a variety of plant species. It is a psychrotolerant soil bacterium. According to Van Oosten et al. (2017), *B. phytofirman* enhanced the ability of *Vitis vinifera* to cope with the cold by upregulating genes linked to stress and efficiently scavenging reactive oxygen species (ROS). Inoculated plants exhibited quicker recovery from chilling stress and returned to normal metabolic levels faster compared to their untreated counterparts. Using morphometrical techniques, Gaveliene et al. (2018) investigated the impact of biostimulants rich in macro- and microelements as well as free amino acids on chilling resistance in winter wheat and winter rapeseeds under regulated

TABLE 9.3
Different Biostimulants Used in Mitigating Temperature Stress

Biostimulants	Crop	Mode of application	Role	References
High-temperature stress				
Boosten and Megafol extracted from seaweed, Sabion extracted from collagen (animal)	Tomato	Foliar	Through stabilizing chlorophyll pigments, CO_2 assimilation enhanced root growth aid in nutrient uptake.	Niu et al. (2022)
Humic acid along with *Bacillus cereus*	Tomato	Soil application	Decreased ABA and higher SA on contrary to non-stressed plants, improved PSII stability under heat stress.	Khan et al. (2020)
KIEM® (2% organic nitrogen, 2% molybdenum, and 21% organic carbon)	Soybean	Seed priming	Increased germination rate, upregulation of DNA repair genes, defense against enzymatic degradation and methyl transferases.	Campobenedetto et al. (2021)
Protein hydrolysate (CycoFlow)	Tomato	Soil application	Increased antioxidant conc., decreased H_2O_2 buildup and MDA content.	Francesca et al. (2022)
Seaweed extract	Wheat	Seed pre-treatment and fertigation	Enhanced antioxidant activity and stability of chlorophyll.	Jócsák et al. (2022)
Jasmonic acid (JA) Salicylic acid (SA) Nitric oxide (NO) as sodium nitroprusside (SNP)	Kimchi cabbage (*Brassica rapa* L. ssp. *pekinensis*)	Foliar	Decreased the transpiration up to 50% compared to only heat-treated plants; biostimulant able to maintain higher rate of photosynthesis and yield under heat stress.	Lee et al. (2019)
Low-temperature stress				
Asahi SL (synthetic biostimulants)	*Coriandrum sativum*	Foliar	Increased transpiration rate, maximum stomatal conductance, quantum yield of PSI.	Pokluda et al. (2016)
Ruter AA, Terra Sorb, and Razormin	*Brassica napus* *Triticum aestivum*	Foliar spray	High accumulation of free amino acids (most probably proline and glutamine).	Gaveliene et al. (2018)
Smoke-water and KAR_1	*Ceratotheca triloba*	Seed priming	Improved germination by reducing detrimental effects of low temp.	Masondo et al. (2018)

cold circumstances. They discovered that the BSts applied in both natural and controlled cold stress conditions enhanced the capacity to withstand freezing temperatures in both crops. An enzyme, 1-aminocyclopropane-1-carboxylic acid (ACC) deaminase-secreting, psychrophilic bacteria (*Pseudomonas fluorescens*, *P. chlororaphis P. fragi*, *P. proteolytica*, and *B. frigoritolerans*), were found to inoculate beans, resulting in decreased ROS generation and increased lipid frost resilience (Tiryaki et al. 2019). Table 9.3 summarizes numerous biostimulants, their administration methods, and the mechanisms responsible for reducing high- and low-temperature stresses.

9.6 CONCLUSION AND FUTURE PROSPECTS

Biostimulants have demonstrated great potential to mitigate the negative consequences of abiotic stressors induced by climatic changes while simultaneously maintaining crop productivity, quality, and yield, with the ultimate goal of meeting the demands for nutrition and food security. BSts exhibit the ability to enhance sustainability and resilience in agriculture. According to an overview of studies, several types of BSts are known to possess beneficial effects against abiotic stresses. Such natural agri-inputs offer an alternative to establishing ecologically sustainable and environmentally friendly farming methods, thus serving as a potential substitute for chemical protectants. Nonetheless, it is important and necessary to comprehend and raise awareness about the fundamental concepts of these organic matters among farmers. Due to their variable effectiveness in different plants across varying environmental conditions, evaluating and verifying BSts in particular areas, such as their effectiveness and safety, remains an ongoing challenge and concern. Further formulation of BSt substances and targeted crop-based research and development are critical considering the current shifting climatic conditions.

9.7 AUTHORSHIP CONTRIBUTION

Conceptualization, R.R.; Writing – Original draft preparation, M.P., N.F., R.A.A., I.M., and R.R.; Writing – Review and editing, M.P., N.F., R.A.A., I.M., and R.R. All authors have read and agreed to the published version of the book chapter.

9.8 DECLARATION OF COMPETING INTEREST

The authors declare that they have no known competing financial interests or personal relationships that could have appeared to influence the work reported in this book chapter.

REFERENCES

Abdel Latef, A.A.H., and Chaoxing, H. "Arbuscular Mycorrhizal Influence on Growth, Photosynthetic Pigments, Osmotic Adjustment and Oxidative Stress in Tomato Plants Subjected to Low Temperature Stress." *Acta Physiol. Plant* 33 (2011): 1217–1225.

Abdelaal, K.A.A., Attia, K.A., Alamery, S.F., et al. "Exogenous Application of Proline and Salicylic Acid Can Mitigate the Injurious Impacts of Drought Stress on Barley Plants Associated with Physiological and Histological Characters." *Sustainability* 12, no. 5 (2020): 1–15.

Agarwal, P., Patel, K., Das, A.K., et al. "Insights into the Role of Seaweed *Kappaphycus alvarezii* sap Towards Phytohormone Signalling and Regulating Defence Responsive Genes in *Lycopersicon esculentum*." *J. Appl. Phycol.* 28 (2016): 2529–2537.

Agati, G., Azzarello, E., Pollastri, S., et al. "Flavonoids as Antioxidants in Plants: Location and Functional Significance." *Plant Sci.* 196 (2012): 67–76.

Aguirre, E., Leménager, D., Bacaicoa, E., et al. "The Root Application of a Purified Leonardite Humic Acid Modifies the Transcriptional Regulation of the Main Physiological Root Responses to Fe Deficiency in Fe-Sufficient Cucumber Plants." *Plant Physiol. Biochem.* 47, no. 3 (2009): 215–223.

Akbari, M., Najafi Alamdarlo, H., and Mosavi, S.H. "The Effects of Climate Change and Groundwater Salinity on Farmers' Income Risk." *Ecol. Ind.* 110 (2020): 1–8.

Alharby, H.F., Alzahrani, Y.M., and Rady, M.M. "Seeds Pretreatment with Zeatins or Maizegrain-Derived Organic Biostimulant Improved Hormonal Contents, Polyamine Gene Expression, and Salinity and Drought Tolerance of Wheat." *Int. J. Agric. Biol.* 24 (2020): 714–724.

Ali, S., Charles, T.C., and Glick, B.R. "Amelioration of High Salinity Stress Damage by Plant Growth-Promoting Bacterial Endophytes that Contains ACC Deaminase." *Plant Physiol. Biochem.* 80 (2014): 160–167.

Ali, A., Ali, Z., Quraishi, U.M., Kazi, A.G., Malik, R.N., Sher, H., and Mujeeb-Kazi, A. "Integrating Physiological and Genetic Approaches for Improving Drought Tolerance in Crops," in *Emerging Technologies and Management of Crop Stress Tolerance* (Academic Press, 2014), 315–345.

Ali, O., Ramsubhag, A., and Jayaraman, J. "Biostimulant Properties of Seaweed Extracts in Plants: Implications Towards Sustainable Crop Production." *Plants* 10 (2021): 531.

Ali, Q., Muhammad, A., Muhammad, S., et al. "Ameliorating Effect of Foliar Applied Proline on Nutrient Uptake in Water Stressed Maize (*Zea mays* L.) Plants." *Pakistan J. Bot.* 40 (2008): 211–219.

Assaha, D.V.M., Ueda, A., Saneoka, H., et al. "The Role of Na+ and K+ Transporters in Salt Stress Adaptation in Glycophytes." *Front Physiol.* 8 (2017): 1–19.

Aydin, A., Kant, C., and Turan, M. "Humic Acid Application Alleviates Salinity Stress of Bean (*Phaseolus vulgaris* L.) Plants Decreasing Membrane Leakage." *Afric. J. Agri. Res.* 7 (2012): 1073–1086.

Baltazar, M., Correia, S., Guinan, K.J., et al. "Recent Advances in the Molecular Effects of Biostimulants in Plants: An Overview." *Biomolecules* 11 (2021): 1–27.

Battacharyya, D., Babgohari, M. Z., Rathor, P., et al. "Seaweed Extracts as Biostimulants in Horticulture." *Sci. Hortic.* 196 (2015): 39–48.

Bensalim, S., Nowak, J., and Asiedu, S.K. "A Plant Growth Promoting Rhizobacterium and Temperature Effects on Performance of 18 Clones of Potato." *Am. J. Potato Res.* 75 (1998): 145–152.

Bhattacharya, A. "Effect of High Temperature on Carbohydrate Metabolism in Plants," in *Effect of High Temperature on Crop Productivity and Metabolism of Macromolecules*. Ed. A. Bhattacharya (Cambridge, MA: Academic Press, 2019), 115–216.

Bhupenchandra, I., Devi, S.H., Basumatary, A., et al. "Biostimulants: Potential and Prospects in Agriculture." *Int. Res. J. Pure Appl. Chem.* 21 (2020): 20–35.

Bhupenchandra, I., Chongtham, S.K., Devi, E.L., et al. "Role of Biostimulant in Mitigating the Effects of Climate Change on Crop Performance." *Front. Plant Sci.* 13 (2022): 1–19.

Bown, A.W., and Shelp, B. J. "Plant GABA: Not Just a Metabolite." *Trends Plant Sci.* 21 (2016): 811–813.

Bulgari, R., Franzoni, G., and Ferrante, A. "Biostimulants Application in Horticultural Crops Under Abiotic Stress Conditions." *Agronomy* 9, no. 6 (2019a): 1–30.

Bulgari, R., Trivellini, A., and Ferrante, A. "Effects of Two Doses of Organic Extract-Based Biostimulant on Greenhouse Lettuce Grown Under Increasing NaCl Concentrations." *Front. Plant Sci.* 9 (2019b): 1–14.

Buono, D.D. "Can Biostimulants Be Used to Mitigate the Effect of Anthropogenic Climate Change on Agriculture? It Is Time to Respond." *Sci. Total Environ.* 751 (2021): 1–12.

Calvo, P., Nelson, L., and Kloepper, J.W. "Agricultural Uses of Plant Biostimulants." *Plant Soil* 383 (2014): 3–41.

Campobenedetto, C., Mannino, G., Beekwilder, J., et al. "The Application of a Biostimulant Based on Tannins Affects Root Architecture and Improves Tolerance to Salinity in Tomato Plants." *Sci. Rep* 11, no. 1 (2021): 1–15.

Carmody, N., Goñi, O., Łangowski, Ł., et al. "*Ascophyllum nodosum* Extract Biostimulant Processing and its Impact on Enhancing Heat Stress Tolerance During Tomato Fruit Set." *Front. Plant Sci.* 11 (2020): 1–14.

Caruso, G., De Pascale, S., Cozzolino, E., et al. "Protein Hydrolysate or Plant Extract-Based Biostimulants Enhanced Yield and Quality Performances of Greenhouse Perennial Wall Rocket Grown in Different Seasons." *Plants* 8 (2019): 208.

Chen, Y., Clapp, C.E., and Magen, H. "Mechanisms of Plant Growth Stimulation by Humic Substances: The Role of Organo-Iron Complexes." *Soil Sci. Plant Nutr.* 50 (2004): 1089–1095.

Cocetta, G., Landoni, M., Pilu, R., et al. "Priming Treatments with Biostimulants to Cope the Short-Term Heat Stress Response: A Transcriptomic Profile Evaluation." *Plants* 11, no. 9 (2022): 1–20.

Colla, G., Hoagland, L., Ruzzi, M., et al. "Biostimulant Action of Protein Hydrolysates: Unraveling Their Effects on Plant Physiology and Microbiome." *Front. Plant Sci.* 8 (2017): 1–14.

Colla, G., and Rouphael, Y. "Biostimulants in Horticulture." *Sci. Hortic.* 196 (2015): 1–2.

Crovadore, J., Cochard, B., Grizard, D., et al. "Draft Genome Sequence of *Bacillus licheniformis* Strain UASWS1606, a Plant Biostimulant for Agriculture." *Microbiol. Resour. Announc.* 9 (2020): e00740–e00720.

Del Carmen Orozco-Mosqueda, M., Glick, B. R., and Santoyo, G. "ACC Deaminase in Plant Growth–Promoting Bacteria (PGPB): An Efficient Mechanism to Counter Salt Stress in Crops." *Microbiol. Res.* 235 (2020): 126439.

Devi, E.L., Kumar, S., Singh, T. B., et al. "Adaptation Strategies and Defence Mechanisms of Plants During Environmental Stress," in *Medicinal Plants and Environmental Challenges*. Eds. M. Ghorbanpour and A. Varma (Cham, Switzerland: Springer International Publishing AG, 2017).

Duc, N.H., Csintalan, Z., and Posta, K. "Arbuscular Mycorrhizal Fungi Mitigate Negative Effects of Combined Drought and Heat Stress on Tomato Plants." *Plant Physiol. Biochem.* 132 (2018): 297–307.

du Jardin, P. "Plant Biostimulants: Definition, Concept, Main Categories and Regulation." *Science Horticulture* 30, no. 196 (2015): 3–14.

Esmaeel, Q., Miotto, L., Rondeau, M., et al. "*Paraburkholderia phytofirmans* PsJN-Plants Interaction: From Perception to the Induced Mechanisms." *Front. Microbiol.* 9 (2018): 1–14.

Francesca, S., Najai, S., Zhou, R., et al. "Phenotyping to Dissect the Biostimulant Action of a Protein Hydrolysate in Tomato Plants Under Combined Abiotic Stress." *Plant Physiol. Biochem.* 179 (2022): 32–43.

Gadallah, M.A.A. "Effects of Proline and Glycinebetaine on *Vicia faba* Responses to Salt Stress." *Biol. Plant* 42 (1999): 249–257.

Gallie, D.R. "L-Ascorbic Acid: A Multifunctional Molecule Supporting Plant Growth and Development." *Scientifica* (2013): 1–24.

García, A.C., Santos, L.A., Izquierdo, F.G., et al. "Vermicompost Humic Acids as an Ecological Pathway to Protect Rice Plant Against Oxidative Stress." *Eco. Eng.* 47 (2012): 203–208.

Garcia-Garcia, A.L., Garcia-Machado, F.J., Borges, A.A., et al. "Pure Organic Active Compounds Against Abiotic Stress: A Biostimulant Overview." *Front. Plant Sci.* 11 (2020): 1–17.

Gaveliene, V., Pakalniskyte, L., Novickiene, L., et al. "Effect of Biostimulants on Cold Resistance and Productivity Formation in Winter Rapeseed and Winter Wheat." *Irish J. Agric. Food Res.* 57, no. 1 (2018): 71–83.

Geelen, D., and Xu, L. "The Chemical Biology of Plant Biostimulants," in *Wiley Series in Renewable Resource* (Chichester, UK: John Wiley & Sons, 2020).

Glick, B.R. "Bacteria with ACC Deaminase can Promote Plant Growth and Help to Feed the World." *Microbiol. Res.* 169 (2014): 30–39.

Guo, X.X., Liu, H. T., and Wu, S.B. "Humic Substances Developed During Organic Waste Composting: Formation Mechanisms, Structural Properties, and Agronomic Functions." *Sci. Total Environ.* 662 (2019): 501–510.

Halpern, M., Bar Tal, A., Ofek, M., et al. "The Use of Biostimulants for Enhancing Nutrient Uptake." *Adv. Agron.* 130 (2015): 141–174.

Hamedeh, H., Antoni, S., Cocciaglia, L., et al. "Molecular and Physiological Effects of Magnesium-Polyphenolic Compound as Biostimulant in Drought Stress Mitigation in Tomato." *Plants* 11, no. 5 (2022): 586.

Hanif, S., Saleem, M.F., Sarwar, M., et al. "Biochemically Triggered Heat and Drought Stress Tolerance in Rice by Proline Application." *J. Plant Growth Regul.* 40, no. 20 (2020): 1–8.

Hasanuzzaman, M., Parvin, K., Bardhan, K., et al. "Biostimulants for the Regulation of Reactive Oxygen Species Metabolism in Plants Under Abiotic Stress." *Cells* 10, no. 10 (2021): 1–29.

Hasegawa, P.M., Bressan, R.A., Zhu, J.K., et al. "Plant Cellular and Molecular Responses to High Salt." *Annu. Rev. Plant Physiol. Plant Mol. Biol.* 51 (2000): 463–499.

Heidari, M., and Golpayegani, A. "Effects of Water Stress and Inoculation with Plant Growth-Promoting Rhizobacteria (PGPR) on Antioxidant Status and Photosynthetic Pigments in Basil (*Ocimumbasilicum* L.)." *J. Saudi Soc Agric. Sci.* 11 (2012): 57–61.

Hoque, M.A., Banu, M.N.A., Okuma, E., et al. "Exogenous Proline and Glycinebetaine Increase NaCl-Induced Ascorbate-Glutathione Cycle Enzyme Activities, and Proline Improves Salt Tolerance More than Glycinebetaine in Tobacco Bright Yellow-2 Suspension Cultured Cells." *J. Plant Physiol.* 164 (2007): 1457–1468.

Islam, M.T., Ckurshumova, W., Fefer, M., et al. "A Plant Based Modified Biostimulant (*Copper chlorophyllin*), Mediates Defense Response in *Arabidopsis thaliana* Under Salinity Stress." *Plants* 10, no. 4 (2021): 1–18.

Jin, C., Ji, J., Zhao, Q., et al. "Characterization of Lycopene β-cyclase Gene from Lyciumchinense Conferring Salt Tolerance by Increasing Carotenoids Synthesis and Oxidative Stress Resistance in Tobacco." *Mol Breed.* 35, no. 12 (2015): 1–18.

Jing, L., Thijs, V.G., and Danny, G. "A Meta-Analysis of Biostimulants Yield Effectiveness in Field Trials." *Front. Plant Sci.* 14 (2022): 1–13.

Jócsák, I., Gyalog, H., Hoffmann, R., et al. "*In-Vivo* Biophoton Emission, Physiological and Oxidative Responses of Biostimulant-Treated Winter Wheat (*Triticum aestivum* L.) as Seed Priming Possibility, for Heat Stress Alleviation." *Plants* 11, no. 5 (2022): 1–13.

Kang, S.M., Khan, A.L., Hamayun, M., et al. "Gibberellin-Producing *Promicromonospora* Sp. SE188 Improves *Solanum lycopersicum* Plant Growth and Influences Endogenous Plant Hormones." *J. Microbiol.* 50 (2012): 902–909.

Kerepesi, I., and Galiba, G. "Osmotic and Salt Stress-Induced Alteration in Soluble Carbohydrate Content in Wheat Seedlings." *Crop Sci.* 40, no. 2 (2000): 482–487.

Khan, M. Iqbal R., Faroza Nazir, Chirag Maheshwari, et al. "Mineral Nutrients in Plants Under Changing Environments: A Road to Future Food and Nutrition Security." *The Plant Genome* 16, no. 4 (2023): e20362.

Khan, M. Iqbal R., Syed Uzma Jalil, Priyanka Chopra, et al. "Role of GABA in Plant Growth, Development and Senescence." *Plant Gene* 26 (2021): 100283.

Khan, M.A., Asaf, S., Khan, A.L., et al. "Extending Thermotolerance to Tomato Seedlings by Inoculation with SA1 Isolate of *Bacillus cereus* and Comparison with Exogenous Humic Acid Application." *Plos One* 15, no. 4 (2020): 1–25.

Khan, W., Rayirath, U.P., Subramanian, S., et al. "Seaweed Extracts as Biostimulants of Plant Growth and Development." *J. Plant Growth Regul.* 28 (2009): 386–399.

Khanna, Risheek Rahul, Badar Jahan, Noushina Iqbal, et al. "GABA Reverses Salt-Inhibited Photosynthetic and Growth Responses Through its Influence on NO-Mediated Nitrogen-Sulfur Assimilation and Antioxidant System in Wheat." *Journal of Biotechnology* 325 (2021): 73–82.

Krieger-Liszkay, A., and Trebst, A. "Tocopherol Is the Scavenger of Singlet Oxygen Produced by the Triplet States of Chlorophyll in the PSII Reaction Centre." *J. Exp. Bot.* 57, no. 8 (2006): 1677–1684.

Ku, S.B., and Edwards, G.E. "Oxygen Inhibition of Photosynthesis: I. Temperature Dependence and Relation to O2/CO2 Solubility Ratio." *Plant Physiol.* 59 (1977): 986–990.

Kumar, R., Trivedi, K., Anand, K.G.V., et al. "Science Behind Biostimulant Action of Seaweed Extract on Growth and Crop Yield: Insights into Transcriptional Changes in Roots of Maize Treated with *Kappaphycus alvarezii* Seaweed Extract Under Soil Moisture Stressed Conditions." *J. Appl. Phycol.* 32 (2020): 599–613.

Kumari, Sarika, Himanshu Chhillar, Priyanka Chopra, et al. "Potassium: A Track to Develop Salinity Tolerant Plants." *Plant Physiology and Biochemistry* 167 (2021): 1011–1023.

La, V.H., Lee, B.R., Islam, M.T., et al. "Characterization of Glutamate-Mediated Hormonal Regulatory Pathway of the Drought Responses in Relation to Proline Metabolism in *Brassica napus* L." *Plants* 9 (2020): 2–18.

Lee, H.J., Lee, J.H., Lee, S.G., et al. "Foliar Application of Biostimulants Affects Physiological Responses and Improves Heat Stress Tolerance in Kimchi Cabbage." *Hortic. Environ. Biotechnol.* 60, no. 6 (2019): 841–851.

Li, J., Wang, X., Lin, X., et al. "Alginate-Derived Oligosaccharides Promote Water Stress Tolerance in Cucumber (*Cucumis sativus* L.)." *Plant Physiol. Biochem.* 130 (2018): 80–88.

Li, Z., Yu, J., Peng, Y., and Huang, B. "Metabolic Pathways Regulated by Abscisic Acid, Salicylic Acid and γ-Aminobutyric Acid in Association with Improved Drought Tolerance in Creeping Bentgrass (Agrostis stolonifera)." *Physiologia Plantarum* 159, no. 1 (2017): 42–58.

Lipczynska Kochany, E. "Humic Substances, Their Microbial Interactions and Effects on Biological Transformations of Organic Pollutants in Water and Soil: A Review." *Chemosphere* 202 (2018): 420–437.

Liu, H., Zhang, Y.H., Yin, H., et al. "Alginate Oligosaccharides Enhanced *Triticum aestivum* L. Tolerance to Drought Stress." *Plant Physiol. Biochem.* 62 (2013): 33–40.

Liu, R., Jiang, X., Guan, H., et al. "Promotive Effects of Alginate-Derived Oligosaccharides on the Inducing Drought Resistance of Tomato." *J. Ocean Univ. China* 8 (2009): 303–311.

Lopez-Bucio, J., Campos-Cuevas, J.C., Hernandez-Calderon, E., et al. "*Bacillus megaterium* Rhizobacteria Promote Growth and Alter Root-System Architecture Through an Auxin- and Ethylene-Independent Signalling Mechanism in *Arabidopsis thaliana*." *Mol. Plant Microbe Interact* 20, no. 2 (2007): 207–217.

Maeda, H., Song, W., Sage, T.L., et al. "Tocopherols Play a Crucial Role in Low-Temperature Adaptation and Phloem Loading in Arabidopsis." *Plant Cell.* 18, no. 10 (2006): 2710–2732.

Malhotra, S. "Horticultural Crops and Climate Change: A Review." *Indian J. Agric. Sci.* 87 (2017): 12–22.

Mansour, M.M.F. "Nitrogen Containing Compounds and Adaptation of Plants to Salinity Stress." *Biol. Plant.* 43 (2000): 491–500.

Mansour, M.M.F. "Role of Vacuolar Membrane Transport Systems in Plant Salinity Tolerance." *J. Plant Growth Regul.* (2022): 1–18.

Masondo, N.A., Kulkarni, M.G., Finnie, J.F., et al. "Influence of Biostimulants-Seed-Priming on *Ceratotheca triloba* Germination and Seedling Growth Under Low Temperatures, Low Osmotic Potential and Salinity Stress." *Ecotoxicol. Environ. Saf.* 147 (2018): 43–48.

Massange-Sánchez, J.A., Sánchez-Hernández, C.V., Hernández-Herrera, R.M., et al. "The Biochemical Mechanisms of Salt Tolerance in Plants," in *Plant Stress Physiology – Perspective in Agriculture*. Eds. M. Hasanuzzaman and K. Nahar (BoD–Books on Demand, 2022).

Meena, M., Divyanshu, K., Kumar, S., et al. "Regulation of L-proline Biosynthesis, Signal Transduction, Transport, Accumulation and its Vital Role in Plants During Variable Environmental Conditions." *Heliyon* 5, no. 12 (2019): 1–20.

Mierziak, J., Kostyn, K., and Kulma, A. "Flavonoids as Important Molecules of Plant Interactions with the Environment." *Molecules* 19, no. 10 (2014): 16240–16265.

Morozesk, M., Bonomo, M.M., da Costa Souza, I., Rocha, L.D., Duarte, I. D., Martins, I.O., . . ., and Matsumoto, S.T. "Effects of Humic Acids from Landfill Leachate on Plants: An Integrated Approach Using Chemical, Biochemical and Cytogenetic Analysis." *Chemosphere* 184 (2017): 309–317.

Mukhopadhyay, R., Sarkar, B., Jat, H.S., et al. "Soil Salinity Under Climate Change: Challenges for Sustainable Agriculture and Food Security." *J. Environ. Manage.* 280 (2021): 1–14.

Mullineaux, P.M., and Rausch, T. "Glutathione, Photosynthesis and the Redox Regulation of Stress-Responsive Gene Expression." *Photosynth Res.* 86, no. 3 (2005): 459–474.

Munné-Bosch, S., and Alegre, L. "The Function of Tocopherols and Tocotrienols in Plants." *CRC Crit. Rev. Plant Sci.* 21, no. 1 (2002): 31–57.

Munns, R. "Comparative Physiology of Salt and Water Stress." *Plant Cell Environ.* 25 (2002): 239–250.

Mutale-Joan, C., Rachidi, F., Mohamed, H.A., et al. "Microalgae-Cyanobacteria-Based Biostimulant Effect on Salinity Tolerance Mechanisms, Nutrient Uptake, and Tomato Plant Growth Under Salt Stress." *J. Appl. Phycol.* 33, no. 6 (2021): 3779–3795.

Nazir, Faroza, Sarika Kumari, Moksh Mahajan, et al. "An explicit story of plant abiotic stress resilience: Overtone of selenium, plant hormones and other signaling molecules." *Plant and Soil* 486, no. 1 (2023): 135–163.

Nephali, L., Moodley, V., Piater, L., et al. "A Metabolomic Landscape of Maize Plants Treated with a Microbial Biostimulant Under Well-Watered and Drought Conditions." *Front. Plant Sci.* 12 (2021): 1–15.

Niu, C., Wang, G., Sui, J., et al. "Biostimulants Alleviate Temperature Stress in Tomato Seedlings." *Sci. Hortic.* 293 (2022).

Noctor, G., Mhamdi, A., Chaouch, S., et al. "Glutathione in Plants: An Integrated Overview." *Plant Cell Environ.* 35, no. 2 (2012): 454–484.

Oliveira, C.E.D.S., Zoz, T., Vendruscolo, E P., et al. "Does *Azospirillumbrasilense* and Biostimulant Improve the Initial Growth of Rice Sown at Greater Depths?" *J. Crop Sci. Biotechnol.* 23 (2020): 461–468.

Ondrasek, G., Rathod, S., Manohara, K.K., et al. "Salt Stress in Plants and Mitigation Approaches." *Plants* 11, no. 6 (2022): 717.

Per, Tasir S., Nafees A. Khan, Palakolanu Sudhakar Reddy, et al. "Approaches in Modulating Proline Metabolism in Plants for Salt and Drought Stress Tolerance: Phytohormones, Mineral Nutrients and Transgenics." *Plant Physiology and Biochemistry* 115 (2017): 126–140.

Pereira, R.V., Filgueiras, C.C., Dória, J., et al. "The Effects of Biostimulants on Induced Plant Defense." *Front. Agron.* 3 (2021): 1–9.

Pokluda, R., Sękara, A., Jezdinský, A., et al. "The Physiological Status and Stress Biomarker Concentration of *Coriandrum sativum* L. Plants Subjected to Chilling Are Modified by Biostimulant Application." *Biol. Agric. Hortic.* 32, no. 4 (2016): 258–268.

Qamar, A., Mysore, K.S., and Senthil-Kumar, M. "Role of Proline and Pyrroline-5-Carboxylate Metabolism in Plant Defense Against Invading Pathogens." *Front. Plant Sci.* (2015): 1–9.

Rakkammal, K., Maharajan, T., Ceasar, S.A., et al. "Biostimulants and Their Role in Improving Plant Growth Under Drought and Salinity." *Cereal Res. Commun.* 51 (2022): 61–74.

Rouphael, Y., Cardarelli, M., Schwarz, D., et al. "'Effects of Drought on Nutrient Uptake and Assimilation in Vegetable Crops' in Plant Responses to Drought Stress." *R. Aroca* (2012): 171–195.

Rouphael, Y., and Colla, G. "Editorial: Biostimulants in Agriculture." *Front. Plant Sci.* 11 (2020): 1–7.

Ruiz-Sola, M. Á., and Rodríguez-Concepción M. "Carotenoid Biosynthesis in Arabidopsis: A Colorful Pathway." *The Arabidopsis Book* 10 (2012): 1–28.

Saia, S., Colla, G., Raimondi, G., et al. "An Endophytic Fungi-Based Biostimulant Modulated Lettuce Yield, Physiological and Functional Quality Responses to Both Moderate and Severe Water Limitation." *Sci. Hortic.* 256 (2019): 1–10.

Sanginga, N. "Role of Biological Nitrogen Fixation in Legume Based Cropping Systems; a Case Study of West Africa Farming Systems." *Plant Soil* 252 (2003): 25–39.

Sangiorgio, D., Cellini, A., Donati, I., et al. "Facing Climate Change: Application of Microbial Biostimulants to Mitigate Stress in Horticultural Crops." *Agronomy* 10, no. 6 (2020): 1–24.

Schafer, M., Brutting, C., Meza-Canales, I.D., et al. "The Role of Cis-Zeatin-Type Cytokinins in Plant Growth Regulation and Mediating Responses to Environmental Interactions – a Review." *J. Exp. Bot.* 66 (2015): 4873–4884.

Shah, Z.H., Rehman, H.M., Akhtar, T., et al. "Humic Substances: Determining Potential Molecular Regulatory Processes in Plants." *Front. Plant Sci.* (2018): 9.

Shams, M., Yildirim, E., Ekinci, M., et al. "Exogenously Applied Glycinebetaine Regulates Some Chemical Characteristics and Antioxidative Defence Systems in Lettuce Under Salt Stress." *Hortic. Environ. Biotechnol.* 57 (2016): 225–231.

Sharma, S., Chen, C., Khatri, K., et al. "*Gracilaria dura* Extract Confers Drought Tolerance in Wheat by Modulating Abscisic Acid Homeostasis." *Plant Physiol. Biochem.* 136 (2019): 143–154.

Shaw, A.K., Bhardwaj, P.K., Ghosh, S., et al. "B-aminobutyric Acid Mediated Drought Stress Alleviation in Maize (*Zea mays* L.)." *Environ. Sci. Pollut. Res.* 23 (2016): 2437–2453.

Shrivastava, P., and Kumar, R. "Soil Salinity: A Serious Environmental Issue and Plant Growth Promoting Bacteria as One of the Tools for its Alleviation." *Saudi Journal of Biological Sciences* 22, no. 2 (2015): 123–131.

Singh, I., Gopalakrishnan, V. A. K., Solomon, S., et al. "Can We Not Mitigate Climate Change Using Seaweed Based Biostimulant: A Case Study with Sugarcane Cultivation in India." *J. Clean. Prod.* 204 (2018): 992–1003.

Singh, J., and Thakur, J.K. "Photosynthesis and Abiotic Stress in Plants," in *Biotic and Aboitic Stress Tolerance in Plants* (Vats, 2018), 27–46.

Slama, I., Abdelly, C., Bouchereau, A., et al. "Diversity, Distribution and Roles of Osmoprotective Compounds Accumulated in Halophytes Under Abiotic Stress." *Ann. Bot.* 115, no. 3 (2015): 433–447.

Smirnoff, N. "Ascorbate Biosynthesis and Function in Photoprotection." *Phil. Trans. R. Soc. Lond B.* 355, no. 1402 (2000): 1455–1464.

Smith, D.T., Potgieter, A.B., and Chapman, S.C. "Scaling Up High-Throughput Phenotyping for Abiotic Stress Selection in the Field." *Appl. Genet.* 134 (2021): 1845–1866.

Sofy, M.R., Elhawat, N., and Alshaal, T. "*Glycine betaine* Counters Salinity Stress by Maintaining High K+/Na+ Ratio and Antioxidant Defense Via Limiting Na+ Uptake in Common Bean (*Phaseolus vulgaris* L.)." *Ecotoxicol. Environ. Saf.* 200, no. 30 (2020): 1–12.

Swaminathan, M.S., and Kesavan, P.C. "Agricultural Research in an Era of Climate Change." *Agri. Res.* 1, no. 1 (2012): 3–11.

Taha, R.S., Alharby, H.F., Bamagoos, A.A., et al. "Elevating Tolerance of Drought Stress in Ocimumbasilicum Using Pollengrains Extract: A Natural Biostimulant by Regulation of Plant Performance and Antioxidant Defense System." *S. Afr. J. Bot.* 128 (2020): 42–53.

Tiryaki, D., Aydın, I., and Atici, O. "Psychrotolerant Bacteria Isolated From the Leaf Apoplast of Cold-Adapted Wild Plants Improve the Cold Resistance of Bean (*Phaseolus vulgaris* L.) Under Low Temperature." *Cryobiology* 86 (2019): 111–119.

Tognetti, V.B., Van Aken, O., Morreel, K., et al. "Perturbation in Indole-3-Butyric Acid Homeostasis by the UDP-Glucosyltransferase UTG74E2 Modulates Arabidopsis Architecture and Water Stress Tolerance." *Plant Cell* 22 (2010): 2660–2679.

Trivedi, K., Gopalakrishnan, V.A.K., Kumar, R., et al. "Transcriptional Analysis of Maize Leaf Tissue Treated with Seaweed Extract Under Drought Stress." *Front. Sustain. Food Syst.* 5 (2021): 1–14.

Upadhyay, S.K., and Singh, D.P. "Effect of Salt-Tolerant Plant Growth-Promoting Rhizobacteria on Wheat Plants and Soil Health in a Saline Environment." *Plant Biol.* 17, no. 1 (2015): 288–293.

Van Oosten, M.J., Pepe, O., De Pascale, S., et al. "The Role of Biostimulants and Bioeffectors as Alleviators of Abiotic Stress in Crop Plants." *Chem. Biol. Technol. Agric.* 4 (2017): 2–12.

Vujinović, T., Zanin, L., Venuti, S., et al. "Biostimulant Action of Dissolved Humic Substances From a Conventionally and an Organically Managed Soil on Nitrate Acquisition in Maize Plants." *Front. Plant Sci.* 10 (2020): 1–14.

Wang, W., Paschalidis, K., Feng, J.C., et al." Polyamine Catabolism in Plants: A Universal Process with Diverse Functions." *Front. Plant Sci.* 10 (2019): 1–13.

Xu, L., Trinh, H.K., and Geelen, D. "Biostimulant Mode of Action: Impact of PBs on Molecular Level," in *The Chemical Biology of Plant Biostimulants* (Hoboken: John Wiley & Sons, 2020), 245–259.

Yakhin, O.I., Lubyanov, A.A., Yakhin, I.A., et al. "Biostimulants in Plant Science: A Global Perspective." *Fron. Plant Sci.* 7 (2017): 1–32.

10 Harnessing CRISPR/Cas9 System to Enhance Abiotic Stress Tolerance in Plants

Mohd Anas, Faroza Nazir, and Qazi Fariduddin

10.1 INTRODUCTION

Clustered regularly interspaced short palindromic repeats/CRISPR-associated proteins (CRISPR/Cas9) was discovered initially in bacteria and archaea. CRISPR was created in 2013 as a gene editing method. CRISPR is used by scientists in gene editing and transcriptional control in animals and plants. This technique is a game changer in plant biology since it is very efficient, the CRISPR/Cas9 method is continually optimized, and its use has grown substantially (Liu et al., 2017). CRISPR is used to change any genomic sequence in plants or other organisms to obtain the desired features, with the only known constraint being the existence of the protospacer adjacent motif (PAM) sequence (Wang et al., 2018). Instead of CRISPR, certain other tools for genome editing are routinely utilized, such as zinc finger nucleases (ZFNs) and transcriptional activator-like effector nucleases (TALENs). They offer a mechanism for introducing targeted mutations, insertions/deletions (indels), and precise sequence changes in a wide range of species using tailored nucleases (Jain and Kumar, 2015). However, when compared to other genome editing methods such as ZFNs and TALENs, CRISPR technology has been found to be much simpler, less expensive, more precise, and highly effective, even at multiplex genome editing (Wang et al., 2018). CRISPR/Cas9 applications in plant genome editing are one of the most rapidly advancing technologies in the biosciences, and for the advancement of non-transgenic genome edited crop plants with changing environmental conditions, CRISPR/Cas9 is becoming a user-friendly tool, ensuring future food security (Haque et al., 2018). The CRISPR/Cas9 system is derived from bacteria and archaea's adaptive immune systems, which identify and digest intrusive DNA from bacteriophages and plasmids (Fineran and Charpentier, 2012). In bacteria, the CRISPR/Cas9 system comprises three components: a Cas9 nuclease, CRISPR RNA (crRNA), and trans-activating crRNA (trRNA). The crRNA and trRNA create a complex by base pairing, which then directs Cas9 to target sites that are complementary to the crRNA's 20nt sequence. Cas9, RuVc, and HNH are DNA endonucleases with two nuclease domains that introduce double-stranded breaks (DSBs) in target DNA (Mojica et al., 2009). The Cas9 nuclease from the type-II CRISPR/Cas9 system of *Streptococcus pyogenes* is the most extensively used system for genome editing, which requires any nucleotide and guanine (NGG) PAM sequence of DNA targeting (Gasiunas et al., 2012). Exploitation of this system for biotechnological breakthroughs in plants contains Cas9 endonuclease and the other synthetic single guide RNA (sgRNA), with the letter generally under the oversight of a U6 or U3 small nuclear RNA promoter (Cong et al., 2013). Cas9 is an RNA-guided endonuclease whose specificity is determined by the single guide RNA (sgRNA) guide sequence (20nt). The Cas9 nuclease causes DSBs at the target location, which may be repaired by the biological system by non-homologous end joining (NHEJ) or homologous recombination (HR), resulting in gene disruption via indels or gene addition/correction, respectively (Mushtaq et al., 2018). When a DNA repair template is provided exogenously, NHEJ can cause modest insertions or deletions at the repair junction, whereas homology-directed repair (HDR) induces precise sequence modifications, including targeted sequence correction along with DNA fragment insertion

 DOI: 10.1201/9781032630380-10

and exchange. The technique has been effectively tested in staple crops such as maize, wheat, rice, and soybean. (Cai et al., 2015; Du et al., 2016). Plant researchers and crop breeders are primarily interested in knowing the genetic basis of plant responses to environmental stresses. In recent decades, numerous genes and regulatory networks implicated in plant responses to stresses such as biotic and abiotic stresses have been discovered (Nasr Esfahani et al., 2017).

10.1.1 How Does the CRISPR/Cas9 System Work?

Cas9 and sgRNA are the two components of the CRISPR/Cas9 system that breaks foreign DNA into fragments. Cas9 is a DNA endonuclease derived from a variety of bacteria, including *Streptococcus pyogenes* (Geng et al., 2016). *Streptococcus thermophilus* was discovered in 2015 by Steinert et al. (2015), *Laterosporus brevibacillus* by Karvelis et al. (2015), and *Streptococcus aureus* by Ran et al. (2015). Cas9 isolation is commonly performed from *Streptococcus pyogenes*. Cas9 contains two domains: the Ruvc-like domain and the HNH domain. The HNH domain cleaves the complementary strand of crRNA, while the Ruvc-like domain splits the opposite strand of the double-stranded DNA (dsDNA). It is synthesized RNA with a length of roughly 100nt . The 20nt sequence on its 5' end serves as a guide sequence for recognizing the target sequence, which is frequently the consensus NGG (N, any nucleotide; G, guanine). The loop structure at the 3' end of the sgRNA can attach to the target sequence and form a complex with Cas9, which generates a DSB at this point by cleaving the dsDNA. When a DSB is created, the NHEJ or HDR DNA repair pathways are activated. In most cases, a DSB is repaired by NHEJ, a straightforward technique that produces mismatch and indels of genes, resulting in gene knockout. HDR causes selective replacement of genes or insertion of foreign DNA knock-ins in the presence of an oligo template (Jacobs et al., 2015). By utilizing these mechanisms, CRISPR/Cas9 can efficiently modify the genomes of a wide range of organisms, including plants, animals, and humans (Liu et al., 2017).

10.2 CRISPR/CAS9 FOR MITIGATING THE IMPACT OF ABIOTIC STRESSES

10.2.1 Salinity Stress

To fulfil rising global food demand, the UN estimates that agricultural output needs to be boosted by 70%–100% by the year 2050. In the meanwhile, increased agricultural cultivation has resulted in decreased fertility and increased salinization of soil, both of which are unfavourable for crop development and cultivation. Salinization of the soil happens when soluble salts accumulate in the crop root zone, preventing roots from absorbing water. More than half of agricultural areas may become significantly salinized by 2050 (Jha et al., 2019; Kumari et al., 2021; Nazir et al., 2024a). Salt stress produces a variety of physiological and morphological changes in plants due to changes in signalling pathways and gene expression (Prusty et al., 2018; Nazir et al., 2023a). The primary detrimental effects of salt stress include necrosis, early mortality of aged plants, and severe ion stoppage in cells (Julkowska and Testerink, 2015). Salinity also increases the number of reactive oxygen species (ROS), which harms plant cellular and metabolic activities (Sahin et al., 2018; Kumari et al., 2023, 2024). ROS damages DNA and proteins, as well as causing lipid peroxidation and membrane disintegration (El Ghazali, 2020). Many genes have been found and characterized to boost plant salt tolerance utilizing CRISPR/Cas9-based genome editing. In *A. thaliana* plants, using CRISPR/Cas9 to knock out the *AtWRKY3* and *AtWRKY4* genes resulted in considerable increased expression of genes under salt and methyl jasmonate stressors. These double mutant plants were susceptible to salt and methyl jasmonate, with increased ion leakage and decreased antioxidant activities such as peroxidase (POD), catalase (CAT), and superoxide dismutase (SOD) (Chen et al., 2021). According to Kim et al. (2021) CRISPR-generated mutants of the acquired osmotolerance (*AtACQOS*) gene also conferred salt stress tolerance in *Arabidopsis*. CRISPR/Cas9-based knockout mutants of ABA-induced transcription repressors (*AITRs*) genes were used to confer salt stress tolerance in soybean

(*Glycine max*) plants (Wang et al., 2021). These mutant plants were more ABA sensitive than wild type (WT) plants and had longer shoots and roots. *GmDrb2a* and *GmDrb2b* knockout mutants in soybean also displayed increased salinity stress tolerance (Curtin et al., 2018). CRISPR/Cas9-mediated editing of *OsRAV2* gene expression in rice was stimulated by the GT-1 element's regulatory function and resulted in salt stress tolerance (Duan et al., 2016). CRISPR mutants lacking *SnRK2* and the osmotic stress/ABA-activated protein kinase genes *SAPK-1* and *SAPK-2* displayed salt tolerance in rice (Lou et al., 2017). According to Zheng et al. (2021) *TaHAG1* CRISPR/Cas9 knockout mutants in wheat plants displayed increased salt tolerance. Alam et al. (2022) generated gene-edited rice mutant plants by targeting the *OsbHLH024* gene to examine its role in salt stress situations. Violaxanthin de-epoxidase (VDE) is required for ABA synthesis, plant development, and stress responses. In rice, the functional advantages of *OsVDE* in salt tolerance have been demonstrated. Under seedling stage salt stress, overexpressed transgenic rice with *OsVDE* loci had higher ABA levels, stomatal closure percentage, and survival rate than the natural type. Many plant transcription factor family genes are involved in the salt stress response. *OsNAC041*, an NAC transcription factor coding gene, was found to be helpful in seed germination under salt stress. The CRISPR/Cas9 mutant *OsNAC041* was shown to be more salt sensitive than wild plants (Bo et al., 2019). By regulating ethylene synthesis in the root meristem, the transcription factor *OsDOF15* positively influences primary root elongation. *OsDOF15* deficiency impeded primary root expansion and root meristem cell proliferation (Qin et al., 2019). The DELLA protein SLENDER RICE1 (*SLR1*) and *OsPQT3* are two examples of negative regulator editing for salt stress tolerance. According to Mo et al. (2020), under salt stress, *OsPQT3* knockout mutants demonstrated enhanced oxidative and salt stress resistance due to increased expression of *OsGPX1*, *OsAPX1*, and *OsSOD1*, as well as *SLR1* loss of function, and mesocotyl and root development was accelerated, especially in the dark and under salt stress. Despite the fact that there are few papers on the potential consequences of the CRISPR/Cas9 approach for improving salt tolerance, there is no question that the CRISPR/Cas9 system is a promising technique for increasing salt tolerance in a wide range of crops.

10.2.2 Drought Stress

Drought is threatening agricultural output in many parts of the world, and it is being worsened by climate change effects such as variations in precipitation and rising temperatures. Drought tolerance is a complex quantitative trait associated with a number of biological and physiological processes (Per et al., 2017; Bhat et al., 2020; Khan et al., 2021a; Nazir et al., 2024b). Drought-responsive gene and transcription factor overexpression improves drought tolerance in plants (Kumar et al., 2020). Drought-sensitive (S) gene expression makes plants more vulnerable to drought through hormonal instability, reduced antioxidant activity, and enhanced reactive oxygen species production. Many efforts were made to modify these genes and incorporate other genes using transgenic techniques. The genetically modified crops were not marketed due to the negligible influence of transgene(s) on drought tolerance and rigorous regulatory restrictions for the discharge of GM crops into the environment. Following the discovery of genome editing technologies, trials are being planned to change genes implicated in drought tolerance pathways in order to increase public acceptability of genome-edited crops (Li et al., 2022). Many studies have shown that CRISPR can impart drought tolerance in plants. Downregulation of *DERF1*, *PMS3*, *MSH1*, *MYB5*, and *SPP* regulatory genes, for example, has been demonstrated to improve drought tolerance in rice (Zhang et al., 2014). Drought resistance was established by a mutation produced in the *Arabidopsis OST2* structural gene using CRISPR/Cas9 (Osakabe et al., 2016). Another study found that CRISPR/Cas9-mediated deletion of the *miR169a* gene in *Arabidopsis* improved drought tolerance significantly (Zhao et al., 2016). Similarly, CRISPR/Cas9 activation of the vacuolar H^+-pyrophosphate (*AVP1*) regulatory gene improved *Arabidopsis* drought tolerance (Park et al., 2017). Similarly, CRISPR/Cas9 activation of the abscisic acid-responsive element binding gene (*AREB1*) increased drought tolerance in *Arabidopsis* (Roca Paixão et al., 2019). Drought resistance was obtained in

Arabidopsis by employing CRISPR/Cas9 to silence the trehalase (*TRE1*) gene (Nuñez-Muñoz et al., 2021). Furthermore, modifying the *STL1* structural gene improved drought tolerance in *A. thaliana*. CRISPR/Cas9 editing of the *ARGOS8* gene, which is a negative regulator of ethylene response in maize, increased drought tolerance (Shi et al., 2017).

Furthermore, lowering the expression of the abscisic acid hydroxylase 2 (*abh2*) gene improved maize drought tolerance (Liu et al., 2020b). CRISPR/Cas9 was utilized to modify the *A6.RGA* gene in rapeseed, resulting in a significant increase in drought tolerance (Wu et al., 2020). The deletion of the *SRL1*, *SRL2*, and *ERA1* genes in rice using CRISPR/Cas9 improves drought tolerance (Ogata et al., 2020). Multiple genes in tomato plants were also altered using the CRISPR/Cas9 assay to improve drought tolerance. For example, the gibberellin-insensitive dwarf1 (*GID1*) gene (Illouz-Eliaz et al., 2020) and the *LBD40* gene (Liu et al., 2020a) were altered. The *SAL1* gene, a negative regulator of drought tolerance in wheat, was altered using a multiplex CRISPR/Cas9 test, which improved drought tolerance at the seedling stage (Abdallah et al., 2022). According to He et al. (2020), drought tolerance in cotton can be increased using CRISPR/Cas9 genome editing of the *HB12* gene.

10.2.3 Heavy Metal Stress

According to Shahid et al. (2015) heavy metals (As, Ni, Mn, Co, Cu, Zn, etc.) have accumulated in soils as a result of various anthropogenic activities and can have a detrimental influence on plant development by disrupting cellular membranes, photosynthetic ability, and cellular respiration, ultimately limiting crop productivity. Furthermore, heavy metals generate hydrogen peroxide (H_2O_2), free radicals such hydroxyl radicals ($^{\bullet}OH$), and superoxide radicals ($O_2{}^{-\bullet}$), all of which contribute to oxidative stress (Ghori et al., 2019; Khan et al., 2021b). Also, metal ion build-up in plants disrupts cellular ionic equilibrium. As a result, plants have evolved detoxifying systems to limit the harmful effects and build-up of heavy metals. According to Sruthi et al. (2017) controlled removal of harmful ions from roots, metal absorption, effective neutralization of metal ions in the protoplast, and appropriation or translocation to distant organs are examples of such systems. Several genes drive these pathways to improve resistance to heavy metal stress (Hasanuzzaman et al., 2013). For example, the g-glutamyl cyclotransferase loss-of-function mutant showed protective properties against heavy metal toxicity, implying that functional loss mutants of *OXP1* and g-glutamyl cyclotransferase show heavy metal and xenobiotic detoxification through increased glutathione (GSH) accumulation (Paulose et al., 2013). As a consequence, creating CRISPR/Cas9-mediated mutants might help plants withstand heavy metal stress. Baeg et al. (2021) recently generated *oxp1*/CRISPR mutant *Arabidopsis* plants that were resistant to cadmium (Cd), suggesting that mutant plants have greater heavy metal detoxifying capacity than WT Col0 plants. As a consequence, this study established a method for imparting resistance to xenobiotics and heavy metals in plants via indel mutations using a gene-editing methodology. The transporters *OsNrampl*, *OsNramp5*, and *OsCdl* help rice roots absorb Cd from the soil. According to Chen et al. (2019) *OsHMA3* is engaged in Cd sequestration within the root vacuole and adversely affects xylem loading, whereas *OsLCT1* is involved in Cd transport to the grains.

Using genome editing to manipulate the expression of these transporter genes has shown some effect in decreasing Cd in grain crops. CRISPR/Cas9-based mutations of the *OsNramp5* and *OsLCT1* genes allowed Cd levels in rice to be allowed (Lu et al., 2019). Similarly, in rice, *OsARM1* regulates the expression of associated transporter genes. It is located in the phloem of the vascular bundle in the basal and upper nodes. *OSARM1* CRISPR knockout improves tolerance, while overexpression enhances susceptibility to arsenic (As) (Baeg et al., 2021). Researchers used the CRISPR-Cas9 technology to develop transgenic plants lacking *OsHAK1* activity to limit radioactive caesium (Cs) uptake by rice plants in Fukushima soil contaminated with 137 Cs^+. Nieves-Cordones et al. (2017) state that plants lacking *OsHAK1* showed much lower levels of 137 Cs^+ in their roots, as well as reduced radioactive Cs components. Another example of applying CRISPR-Cas9-based editing in rice is to determine the function of a possible target, *OsPRX2*, for better potassium (K^+) deficit tolerance. In a K^+ limiting environment, *OsPRX2* has been reported to suppress ROS generation. Overexpression of *OsPRX2* induces stomatal closure and increased tolerance to K^+ deprivation. Simultaneously, knocking down *OsPRX2* causes substantial deficits in leaf phenotypic and stomatal opening under K^+-deficiency tolerance (Miao et al., 2017).

10.2.4 High-Temperature Stress

According to Tubiello and Fischer (2007), high-temperature stress is one of the key abiotic stressors that have become a severe problem in agricultural productivity in many regions of the world, leading to global warming. High heat stress has emerged as a significant issue that is causing massive yield losses and is predicted to worsen in the future (Khan et al., 2013; Harvey, 2018; Poór et al., 2022). Heat stress has a devastating effect on plants at all phases of development, from germination to harvesting (Ejaz et al., 2022). It induces a number of molecular, biochemical, and physiological changes that can have a negative impact on plant development and production, as well as visible indications such as leaf burn and discoloration (Hasanuzzaman et al., 2013; Nazir et al., 2023b). Heat stress mostly influences reproductive development, as temperatures of 30°C can cause pollen shedding, poor pollen viability, poor germination, and pollen growth (Fahad et al., 2016). According to Jha et al. (2017), heat stress causes plants to activate complex molecular networks that include heat stress-responsive gene expression, signal transduction, and metabolite synthesis. As functional and structural genomics approaches in plants have evolved, several heat stress–associated genes have been found and described in order to improve heat resistance using modern biotechnological technologies. As per the studies of Awasthi et al. (2015), heat shock proteins (HSPs) and heat shock transcription factors (HSFs) are important components that work through the heat stress-response signal transduction pathway, which is associated with ROS build-up. As a result, heat stress tolerance can be increased by improving plants' ability to detoxify ROS components (Parmar et al., 2017). CRISPR/Cas9 genome editing is a novel precision genome editing method for studying heat stress–related biological pathways and increasing crop heat tolerance (Duan et al., 2016). Slagamous-like 6 (*SLAGL6*) genome editing using CRISPR/Cas9 has recently been created for heat tolerance in tomatoes, enhancing fruit setting under heat stress conditions (Klap et al., 2017). The *SlMAPK3* gene in tomatoes belongs to the mitogen-activated protein kinase family and is involved in a number of environmental stress responses. CRISPR/Cas9-mediated genome editing produced *slmapk3* mutants with increased heat tolerance relative to WT plants, implicating a role as a negative regulator of thermo-tolerance (Yu et al., 2019). *BRZ1* is a component of heat tolerance and positively controls ROS generation in the apoplastic area of tomatoes. CRISPR/Cas9-based *bzr1* mutants showed reduced H_2O_2 production in apoplast and heat tolerance by reduced respiratory burst oxidase homolog 1 (*RBOH1*) (Yin et al., 2018). Qiu et al. (2018) discovered that CRISPR/Cas9-mediated heat stress–sensitive albino 1 (*HSA1*) mutant tomato plants are more sensitive to heat stress than wild-type plants. According to Li et al. (2017) in maize, CRISPR mutants of the thermo-sensitive genic male sterile 5 (*TMS5*) gene improved thermo-sensitive male-sterile plants. Knockouts of *NCED4*, a critical regulating enzyme in the synthesis of ABA, were utilized to increase the temperature of lettuce seed germination. As a result, *LsNCED4* mutations may have economic value in high-temperature industrial environments (Bertier et al., 2018).

10.2.5 Low-Temperature Stress

Climate change has emerged as the primary danger to global food security in recent years. Cold/low-temperature stress is classified as freezing stress (0°C) and chilling stress (0°C–15°C), both of which have a negative impact on crop development and yield (Shi et al., 2018). According to Jha et al. (2017), low temperature is a major abiotic stressor that has a negative impact on plant development and production. Low temperatures can impair crop species, affecting their growth, output, and survival (Sanghera et al., 2011). Rice editing through CRISPR/Cas9 is being used to increase plant resilience to cold by deleting particular negative regulator transcription factors. *OsMYB30* is a transcription factor that attaches to the b-amylase promoter and reduces cold tolerance. According to Lv et al. (2017), when exposed to frigid temperatures, *OsMYB30* binds to *OsJAZ9* and suppresses b-amylase gene expression, changing starch breakdown and maltose accumulation and perhaps leading to enhanced cold sensitivity. Zeng et al. (2020) state that CRISPR/Cas9 genome editing that targeted three genes, *OsPIN5b*, *GS3*, and *OsMYB30*, led to improved yield and resilience to cold stress. The rice annexin genes *OsAnn3* and *OsAnn5* act as positive regulators of cold stress resistance. Cold sensitivity was caused by the deletion of *OsAnn3* and *OsAnn5* (Shen et al., 2017). Furthermore, by regulating antioxidants and maintaining signalling pathways, *OsPRP1* improves rice cold tolerance. *OsPRP1* deletion increased cold sensitivity in rice, and mutant lines accumulated lower antioxidant activity of the enzyme as well as lower quantities of proline, chlorophyll, ABA, and ascorbic acid (AsA) content as compared to WT. Because tomato plants are subject to chilling stress, the tomatoes they produce are more prone to cold stress. CRISPR/Cas9 *cbf1* mutants have been demonstrated to protect tomato plants against chilling/cold damage and minimize leakage of electrolytes (Li et al., 2018). Cucumbers overexpressing the *CsWRKY6* gene demonstrated increased resilience to cold stress in addition to susceptibility to ABA and proline build-up (Zhang et al., 2016). *Brassica napus* sequencing of RNA showed many genes from the *WRKY* family that play essential roles in cold response (Ke et al., 2020). At low temperatures (4°C), *SlNPR1* expression and protein levels were found to be increased in tomatoes. Shu et al. (2023) state that CRISPR/Cas9-mediated genome editing of *SlNPR1* induced chilling injury symptoms in tomato plants, which were supported by an increase in oxidative stress markers, as well as a decrease in the amino acid proline content, antioxidant enzymatic activities, and soluble protein content.

10.2.6 Flooding Stress

Flooding and waterlogging may both cause severe harm to crops and soil, but their effects on plant growth and development differ (Jackson and Colmer, 2005; Khan et al., 2020; Mondal et al., 2020). While certain plants may be evolved to handle periodic floods, most plants are extremely

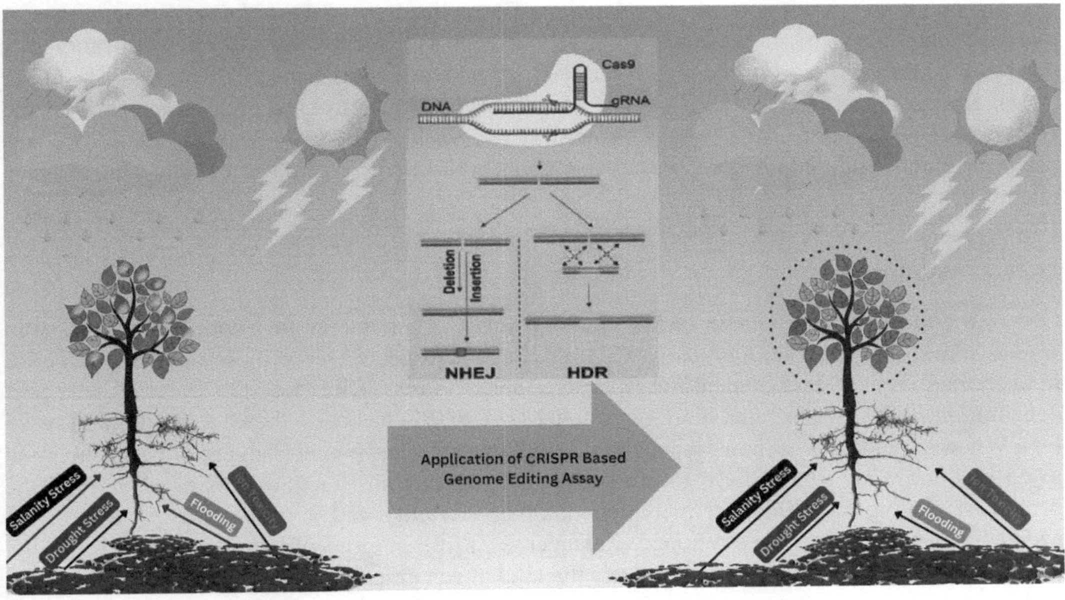

FIGURE 10.1 Illustrating CRISPR-based genome editing for enhancing resilience to abiotic stresses.

susceptible to waterlogging and can suffer from stunted growth, root damage, and even death (Pan et al., 2021). Xiao et al. (2021) employed the CRISPR/Cas9 technology to alter two paralog genes (*GmpPLA-II* and *GmpPLA-IIz*) to investigate their involvement in modifying soybean response to flood stress. There are 112 *GmPLAs* in the soybean genome, including 78 *PLA1*, 29 patatin-like *PLAs*, and five secretory *sPLA2*. *PLAs* in the soybean genome have been extended because of occurrences of segmental and tandem duplication.

Some CRISPR/Cas9 mutant strains performed better under flooding conditions. The soybean mutants would be useful genetic resources for unravelling molecular pathways governing soybean response to flood stress and other abiotic stressors (Xiao et al., 2021). Ye et al. (2018) conducted an intriguing investigation that confirmed CRISPR-mediated knockout of the gene *OsCBL10* is embryo-lethal, whereas spontaneous polymorphisms in the gene's promoter were connected with flooding tolerance. Because drought and flood may coexist in plants and perhaps share mechanisms for regulation, it is crucial to test previously identified drought-tolerant mutants for flood tolerance. This might be an effective method for accelerating the discovery of genes that confer this trait.

10.3 THE CRISPR/CAS9 SYSTEM FOR ENGINEERING ABIOTIC STRESS TOLERANCE IN CROP PLANTS

The CRISPR/Cas9 genome editing technique has been used effectively in *Arabidopsis* (Miki et al., 2018), rice (Wang et al., 2017b), wheat (Shan et al., 2014), tomato (Brooks et al., 2014), and maize (Zong et al., 2017). Notably, abiotic stress-associated genes that disrupt cellular and molecular function continue to be a biological stigma that can be readily targeted using the CRISPR/Cas9 genome editing technology.

10.3.1 *Arabidopsis*

It is an array model in understanding its molecular exercise and abiotic stress. CRISPR/Cas9 system has been broadly employed for knockout mutant studies in *Arabidopsis*. To produce the mutant lines by the system of CRISPR/Cas9, *CBF1*, *CBF2*, and *CBF3* have all been employed which are cold-inducible C-repeats/DRE-binding factors (Cho et al., 2017). *OST2*, a novel allele, has been produced which shows a proton pump for the acceleration of osmotic stress in *Arabidopsis*. The presence of bi-allelic mutations in germ lines was demonstrated using a CRISPR/Cas9 system modification. So, the altered stress response of the *OST2* gene has been identified as a novel allele (Osakabe et al., 2016). To maximize antioxidant properties to cope with abiotic stress, ectopic articulation of the *UGT79B2*/*UGT79B3* has been created. According to Li et al. (2017), glycosyltransferase enzymes perform an unusual role in *Arabidopsis* by producing anthocyanin colours that rely on abiotic stress tolerance.

10.3.2 Wheat

One of the most difficult jobs in tackling abiotic stress is genome editing with CRISPR/Cas9 in wheat. Wheat genome alteration using CRISPR/Cas9 is projected to be more open and effective in encouraging plant development in response to abiotic stress. Kim et al. (2018) state that wheat dehydration-responsive element binding protein2 (*TaDREB2*) has been shown to be drought resistant. Wheat genome editing can produce stable transformants and passable changes from generation to generation. Many studies have shown substantial explanations for transients in wheat utilizing protoplast testing, claiming that increased mutant variability and accountability in a selective Mendelian technique may be developed (Zhang et al., 2016). In terms of wheat genome editing, the powerful tool Cas9 endonuclease speculates the known genomics sequences to improve qualitative and quantitative wheat characteristics for the worldwide availability of food.

10.3.3 Tomato

Abiotic stress is one of the most important environmental factors limiting tomato crop productivity. Wang et al. (2017a) state that the ablation of *SlMAPK3* lowered the expression of *SlLOX*, *SlGST*, and *SlDREB*, resulting in resistance to salt stress. As a result, it has a protective role in tomato crops, calibrating cellular damage (Wang et al., 2017a). According to Li et al. (2018), the C-repeat binding factor *SlCBF1*, a chilling-sensitive gene responding to cold stress, has been altered using the CRISPR/Cas9 system. Furthermore, the *slcbf1* mutant boosted the plant's H_2O_2 and indole acetic acid (IAA) levels. As a result, we now have a better knowledge of *CBF1* and how it might be used to treat cold stress in tomato crops. According to various experts, using genetic engineering to increase the expression of the *SLMAPK3* gene in tomato plants may result in increased drought tolerance and higher yield (Klap et al., 2017). The tomato *SLAGL6* gene was knocked down by CRISPR/Cas9, allowing mutant tomato plants to produce parthenocarpic fruits under heat stress. Klap et al. (2017) state that the pollen viability is unaltered and the seedless fruits have normal weight and form, making *SLAGL6* a promising candidate gene for facultative parthenocarpy.

10.3.4 Rice

CRISPR/Cas9 is employed for producing knockdown mutants for the japonica rice annexin gene (*Osann3*), a Ca^{2+}-dependent phospholipid binding protein, leading to a large increase in relative electrical conductivity and a significant decrease in the rate of survival of T1 mutant lines in cold stress as compared to WT plants. This shows that *OsANN3* plays an important role in rice cold resistance, and hence this gene might be a good target for developing cold-resistant rice (Shen et al., 2017). The CRISPR/Cas9 system is used to develop mutant rice with a loss of function in the stress/ABA-activated protein kinase 2 (*SAPK2*) genes, which is an important regulator of ABA signalling. The researchers employed nickase CRISPR/Cas9 (nCas9) or nuclease-deficient Cas9 (dCas9) connected to Petromyzon marinus cytidine deaminase (*PmCDA1*) to generate point mutations in rice plants, resulting in herbicide-resistant mutant lines.

10.3.5 Maize

In maize, the physiological network and genetic analysis play a more important role in abiotic stress adaptation. It has been discovered that *WRKY* genes in maize respond to abiotic stress. Approximately 119 *WRKY* genes have been identified to trigger various stress responses, mainly in drought. In the future, *ZmWRKY40*, a putative transcription factor gene, might be altered using the CRISPR/Cas9 toolbox to better understand maize functional genomics (Wang et al., 2018). Shi et al. (2017) state that CRISPR/Cas9 could be used to develop plants that express the maize auxin-regulated gene that regulates organ size (*ARGOS8*) constitutively through substituting the natural promoter. *ARGOS8* transcript levels were significantly higher in *ARGOS8-v1* and *ARGOS8-v2* lines compared to WT plants. According to Shi et al. (2017), variants of the *ARGOS8* gene showed a considerable increase in grain output under field drought circumstances, with no yield penalty detected under normal growth conditions.

10.3.6 Other Crops

There are some other crops like cotton (*Gossypium* sp.) that have successfully accommodated CRISPR/Cas9 technology. Chen et al. (2017) have stated that it effectively discovers the mutational exercise in the marked genes. To fight with the drought stress, CRISPR editing has been found recently on two new genes, *GhRDLI* and *GhPINI-3*, in cotton (Haque et al., 2018). A numerous abiotic stress such as osmosis, salt, drought, and cold gene *MeKUPs* in cassava crop were accommodate by employing CRISPR/Cas9 system (Ou et al., 2018). In bananas, genetic alteration is also being done *MaAPSI* and *MaAPL3* in chilling and salt stress (Miao et al., 2017). In sugarcane, the gene from the family of non-specific lipid transfer protein (NsLTP) regulates during cold, and PEG stress is also being edited by employing the CRISPR/Cas9 system (Chen et al., 2017). To improve salt tolerance, the beta 1,3-glucanase gene isolated from sugarcane was edited using CRISPR/Cas9 genome editing technology (Su et al., 2016).

10.4 CONCLUSION AND FUTURE PROSPECTS

In this chapter, we have explored the enormous potential of the CRISPR/Cas9 genome editing technique for improving the tolerance of agricultural plants to abiotic stresses. Abiotic stressors such as salinity, drought, heavy metal toxicity, extreme temperatures, and flooding persistently threaten global agricultural output and food security. CRISPR/Cas9 has emerged as a flexible and accurate technology for enhancing crops, offering promising solutions to these challenges. We have presented an in-depth analysis of the CRISPR/Cas9 system and its main components and mechanisms, demonstrating how it directs Cas9 to target sites and causes DNA double-stranded breaks, therefore aiding repair processes. CRISPR/Cas9 outperforms other genome editing technologies such as zinc finger nucleases and transcriptional activator-like effector nucleases in terms of simplicity, cost-effectiveness, precision, and efficiency, making it the preferred choice for improving crop stress tolerance. Throughout this chapter, we have explored effective CRISPR/Cas9 applications in various agricultural species, including *Arabidopsis*, wheat, tomato, rice, maize, soybean, and others. As a result of these applications, crops with enhanced salt tolerance, drought resilience, and other stress-related characteristics have been developed. While research on flood tolerance is still in its early stages, CRISPR/Cas9 shows promise for addressing this abiotic stress as well. Global collaboration among scientists, biotechnologists, and politicians is essential to fully realize the promise of CRISPR/Cas9 in crop development. Sharing information and resources can expedite development and address global food security concerns more effectively. Ultimately, the CRISPR/Cas9 system has opened new avenues for crop development in the face of abiotic stressors. This technology holds the key to producing more robust crop types, ensuring food security, and mitigating the impact of environmental uncertainties through ongoing research and innovation.

10.5 AUTHORSHIP CONTRIBUTION

Conceptualization, F.N. and Q.F.; Writing – Original draft preparation, M.A., F.N., and Q.F.; Writing – Review and editing, F.N. and Q.F. All authors have read and agreed to the published version of the book chapter.

10.6 DECLARATION OF COMPETING INTEREST

The authors declare that they have no known competing financial interests or personal relationships that could have appeared to influence the work reported in this book chapter.

REFERENCES

Abdallah, N. A., Elsharawy, H., Abulela, H. A., Thilmony, R., Abdelhadi, A. A., & Elarabi, N. I. (2022). Multiplex CRISPR/Cas9-mediated genome editing to address drought tolerance in wheat. *GM Crops & Food*, 1–17.

Alam, M. S., Kong, J., Tao, R., Ahmed, T., Alamin, M., Alotaibi, S. S., . . . Xu, J. H. (2022). CRISPR/Cas9 mediated knockout of the OsbHLH024 transcription factor improves salt stress resistance in rice (*Oryza sativa* L.). *Plants*, *11*(9), 1184.

Awasthi, R., Bhandari, K., & Nayyar, H. (2015). Temperature stress and redox homeostasis in agricultural crops. *Frontiers in Environmental Science*, *3*, 11.

Baeg, G. J., Kim, S. H., Choi, D. M., Tripathi, S., Han, Y. J., & Kim, J. I. (2021). CRISPR/Cas9-mediated mutation of 5-oxoprolinase gene confers resistance to sulfonamide compounds in Arabidopsis. *Plant Biotechnology Reports*, *15*, 753–764.

Bertier, L. D., Ron, M., Huo, H., Bradford, K. J., Britt, A. B., & Michelmore, R. W. (2018). High-resolution analysis of the efficiency, heritability, and editing outcomes of CRISPR/Cas9-induced modifications of NCED4 in lettuce (*Lactuca sativa*). *G3: Genes, Genomes, Genetics*, *8*(5), 1513–1521.

Bhat, J. A., Deshmukh, R., Zhao, T., Patil, G., Deokar, A., Shinde, S., & Chaudhary, J. (2020). Harnessing high-throughput phenotyping and genotyping for enhanced drought tolerance in crop plants. *Journal of Biotechnology*, *324*, 248–260.

Bo, W., Zhaohui, Z., Huanhuan, Z., Xia, W., Binglin, L. I. U., Lijia, Y., . . . Yong, Z. (2019). Targeted mutagenesis of NAC transcription factor gene, OsNAC041, leading to salt sensitivity in rice. *Rice Science*, *26*(2), 98–108.

Brooks, C., Nekrasov, V., Lippman, Z. B., & Van Eck, J. (2014). Efficient gene editing in tomato in the first generation using the clustered regularly interspaced short palindromic repeats/CRISPR-associated9 system. *Plant Physiology*, *166*(3), 1292–1297.

Cai, Y., Chen, L., Liu, X., Sun, S., Wu, C., Jiang, B., . . . Hou, W. (2015). CRISPR/Cas9-mediated genome editing in soybean hairy roots. *PLoS One*, *10*(8), e0136064.

Chen, S., Zhang, N., Zhou, G., Hussain, S., Ahmed, S., Tian, H., & Wang, S. (2021). Knockout of the entire family of AITR genes in Arabidopsis leads to enhanced drought and salinity tolerance without fitness costs. *BMC Plant Biology*, *21*, 1–15.

Chen, J., Zou, W., Meng, L., Fan, X., Xu, G., & Ye, G. (2019). Advances in the uptake and transport mechanisms and QTLs mapping of cadmium in rice. *International Journal of Molecular Sciences*, *20*(14), 3417.

Chen, X., Lu, X., Shu, N., Wang, S., Wang, J., Wang, D., . . . Ye, W. (2017). Targeted mutagenesis in cotton (*Gossypium hirsutum* L.) using the CRISPR/Cas9 system. *Scientific Reports*, *7*(1), 44304.

Cho, S., Yu, S. I., Park, J., Mao, Y., Zhu, J. K., Yun, D. J., & Lee, B. H. (2017). Accession-dependent CBF gene deletion by CRISPR/Cas system in Arabidopsis. *Frontiers in Plant Science*, *8*, 1910.

Cong, L., Ran, F. A., Cox, D., Lin, S., Barretto, R., Habib, N., . . . Zhang, F. (2013). Multiplex genome engineering using CRISPR/Cas systems. *Science*, *339*(6121), 819–823.

Curtin, S. J., Xiong, Y., Michno, J. M., Campbell, B. W., Stec, A. O., Čermák, T., . . . Stupar, R. M. (2018). CRISPR/Cas9 and TALENs generate heritable mutations for genes involved in small RNA processing of glycine max and medicagotruncatula. *Plant Biotechnology Journal*, *16*(6), 1125–1137.

Du, H., Zeng, X., Zhao, M., Cui, X., Wang, Q., Yang, H., . . . Yu, D. (2016). Efficient targeted mutagenesis in soybean by TALENs and CRISPR/Cas9. *Journal of Biotechnology*, *217*, 90–97.

Duan, Y. B., Li, J., Qin, R. Y., Xu, R. F., Li, H., Yang, Y. C., . . . Yang, J. B. (2016). Identification of a regulatory element responsible for salt induction of rice OsRAV2 through ex situ and in situ promoter analysis. *Plant Molecular Biology*, *90*, 49–62.

Ejaz, M., Abbas, G., Fatima, Z., Iqbal, P., Raza, M. A., Kheir, A. M., . . . Ahmad, S. (2022). Modelling climate uncertainty and adaptations for soybean-based cropping system. *International Journal of Plant Production*, *16*(2), 235–250.

El Ghazali, G. E. (2020). SuaedavermiculataForssk. ex JF Gmel.: Structural characteristics and adaptations to salinity and drought: A review. *International Journal of Science*, *9*(02), 28–33.

Fahad, S., Hussain, S., Saud, S., Khan, F., Hassan, S., Amanullah, . . . Huang, J. (2016). Exogenously applied plant growth regulators affect heat-stressed rice pollens. *Journal of Agronomy and Crop Science*, *202*(2), 139–150.

Fineran, P. C., & Charpentier, E. (2012). Memory of viral infections by CRISPR-Cas adaptive immune systems: Acquisition of new information. *Virology*, *434*(2), 202–209.

Gasiunas, G., Barrangou, R., Horvath, P., & Siksnys, V. (2012). Cas9–crRNA ribonucleoprotein complex mediates specific DNA cleavage for adaptive immunity in bacteria. *Proceedings of the National Academy of Sciences*, *109*(39), E2579–E2586.

Geng, Y., Deng, Z., & Sun, Y. (2016). An insight into the protospacer adjacent motif of *Streptococcus pyogenes* Cas9 with artificially stimulated RNA-guided-Cas9 DNA cleavage flexibility. *RSC Advances*, *6*(40), 33514–33522.

Ghori, N. H., Ghori, T., Hayat, M. Q., Imadi, S. R., Gul, A., Altay, V., & Ozturk, M. (2019). Heavy metal stress and responses in plants. *International Journal of Environmental Science and Technology*, *16*, 1807–1828.

Haque, E., Taniguchi, H., Hassan, M. M., Bhowmik, P., Karim, M. R., Śmiech, M., . . . Islam, T. (2018). Application of CRISPR/Cas9 genome editing technology for the improvement of crops cultivated in tropical climates: Recent progress, prospects, and challenges. *Frontiers in Plant Science*, *9*, 617.

Harvey, L. D. (2018). *Global Warming*. Routledge.

Hasanuzzaman, M., Nahar, K., Alam, M. M., Roychowdhury, R., & Fujita, M. (2013). Physiological, biochemical, and molecular mechanisms of heat stress tolerance in plants. *International Journal of Molecular Sciences*, *14*(5), 9643–9684.

He, X., Luo, X., Wang, T., Liu, S., Zhang, X., & Zhu, L. (2020). GhHB12 negatively regulates abiotic stress tolerance in Arabidopsis and cotton. *Environmental and Experimental Botany*, *176*, 104087.

Illouz-Eliaz, N., Nissan, I., Nir, I., Ramon, U., Shohat, H., & Weiss, D. (2020). Mutations in the tomato gibberellin receptors suppress xylem proliferation and reduce water loss under water-deficit conditions. *Journal of Experimental Botany*, *71*(12), 3603–3612.

Jackson, M. B., & Colmer, T. (2005). Response and adaptation by plants to flooding stress. *Annals of Botany*, *96*(4), 501–505.

Jacobs, T. B., LaFayette, P. R., Schmitz, R. J., & Parrott, W. A. (2015). Targeted genome modifications in soybean with CRISPR/Cas9. *BMC Biotechnology*, *15*, 1–10.

Jain, M., & Kumar, V. (2015). The CRISPR-Cas system for plant genome editing: Advances and opportunities. *Journal of Experimental Botany*, *66*(1), 47–57.

Jha, U. C., Bohra, A., Jha, R., & Parida, S. K. (2019). Salinity stress response and 'omics' approaches for improving salinity stress tolerance in major grain legumes. *Plant Cell Reports*, *38*, 255–277.

Jha, U. C., Bohra, A., Parida, S. K., & Jha, R. (2017). Integrated "omics" approaches to sustain global productivity of major grain legumes under heat stress. *Plant Breeding*, *136*(4), 437–459.

Julkowska, M. M., & Testerink, C. (2015). Tuning plant signaling and growth to survive salt. *Trends in Plant Science*, *20*(9), 586–594.

Karvelis, T., Gasiunas, G., Young, J., Bigelyte, G., Silanskas, A., Cigan, M., & Siksnys, V. (2015). Rapid characterization of CRISPR-Cas9 protospacer adjacent motif sequence elements. *Genome Biology*, *16*, 1–13.

Ke, L., Lei, W., Yang, W., Wang, J., Gao, J., Cheng, J., . . . Yu, D. (2020). Genome-wide identification of cold responsive transcription factors in *Brassica napus* L. *BMC Plant Biology*, *20*(1), 1–13.

Khan, M. I. R., Iqbal, N., Masood, A., Per, T. S., & Khan, N. A. (2013). Salicylic acid alleviates adverse effects of heat stress on photosynthesis through changes in proline production and ethylene formation. *Plant Signaling & Behavior*, *8*(11), e26374.

Khan, M. I. R., Jahan, B., AlAjmi, M. F., Rehman, M. T., Iqbal, N., Irfan, M., Sehar, Z., & Khan, N. A. (2021b). Crosstalk of plant growth regulators protects photosynthetic performance from arsenic damage by modulating defense systems in rice. *Ecotoxicology and Environmental Safety*, *222*, 112535.

Khan, M. I. R., Palakolanu, S. R., Chopra, P., Rajurkar, A. B., Gupta, R., Iqbal, N., & Maheshwari, C. (2021a). Improving drought tolerance in rice: Ensuring food security through multi-dimensional approaches. *Physiologia Plantarum*, *172*(2), 645–668.

Khan, M. I. R., Trivellini, A., Chhillar, H., Chopra, P., Ferrante, A., Khan, N. A., & Ismail, A. M. (2020). The significance and functions of ethylene in flooding stress tolerance in plants. *Environmental and Experimental Botany*, *179*, 104188.

Kim, D., Alptekin, B., & Budak, H. (2018). CRISPR/Cas9 genome editing in wheat. *Functional & Integrative Genomics*, *18*, 31–41.

Kim, S. T., Choi, M., Bae, S. J., & Kim, J. S. (2021). The functional association of ACQOS/VICTR with salt stress resistance in *Arabidopsis thaliana* was confirmed by CRISPR-Mediated mutagenesis. *International Journal of Molecular Sciences*, *22*(21), 11389.

Klap, C., Yeshayahou, E., Bolger, A. M., Arazi, T., Gupta, S. K., Shabtai, S., . . . Barg, R. (2017). Tomato facultative parthenocarpy results from Sl AGAMOUS-LIKE 6 loss of function. *Plant Biotechnology Journal*, *15*(5), 634–647.

Kumar, S., Ayachit, G., & Sahkumoo, L. (2020). Screening of mungbean for drought tolerance and transcriptome profiling between drought-tolerant and susceptible genotype in response to drought stress. *Plant Physiology and Biochemistry*, *157*, 229–238.

Kumari, S., Chhillar, H., Chopra, P., Khanna, R. R., & Khan, M. I. R. (2021). Potassium: A track to develop salinity tolerant plants. *Plant Physiology and Biochemistry*, *167*, 1011–1023.

Kumari, S., Nazir, F., Jain, K., & Khan, M. I. R. (2023). GABA and potassium modulatesdefence systems, assimilation of nitrogen and carbon, and yield traits under salt stress in wheat. *Journal of Plant Growth Regulation*, *42*(10), 6721–6740.

Kumari, S., Nazir, F., Singh, A., Haroon, H., Khan, N. R., Sahoo, R. N., Albaqami, M., Siddique, K. H. M., & Khan, M. I. R. (2024). γ-Aminobutyric acid (GABA) strengthened nutrient accumulation, defense metabolism, growth and yield traits against salt and endoplasmic reticulum stress conditions in wheat plants. *Plant and Soil*, 1–21.

Li, R., Zhang, L., Wang, L., Chen, L., Zhao, R., Sheng, J., & Shen, L. (2018). Reduction of tomato-plant chilling tolerance by CRISPR–Cas9-mediated SlCBF1 mutagenesis. *Journal of Agricultural and Food Chemistry*, *66*(34), 9042–9051.

Li, X., Xie, Y., Zhu, Q., & Liu, Y. G. (2017). Targeted genome editing in genes and cis-regulatory regions improves qualitative and quantitative traits in crops. *Molecular Plant*, *10*(11), 1368–1370.

Li, X., Xu, S., Fuhrmann-Aoyagi, M. B., Yuan, S., Iwama, T., Kobayashi, M., & Miura, K. (2022). CRISPR/Cas9 technique for temperature, drought, and salinity stress responses. *Current Issues in Molecular Biology*, *44*(6), 2664–2682.

Liu, H., Ding, Y., Zhou, Y., Jin, W., Xie, K., & Chen, L. L. (2017). CRISPR-P 2.0: an improved CRISPR-Cas9 tool for genome editing in plants. *Molecular Plant*, *10*(3), 530–532.

Liu, L., Zhang, J., Xu, J., Li, Y., Guo, L., Wang, Z., . . . Zhang, N. (2020a). CRISPR/Cas9 targeted mutagenesis of SlLBD40, a lateral organ boundaries domain transcription factor, enhances drought tolerance in tomato. *Plant Science*, *301*, 110683.

Liu, S., Li, C., Wang, H., Wang, S., Yang, S., Liu, X., . . . Qin, F. (2020b). Mapping regulatory variants controlling gene expression in drought response and tolerance in maize. *Genome Biology*, *21*, 1–22.

Lou, D., Wang, H., Liang, G., & Yu, D. (2017). OsSAPK2 confers abscisic acid sensitivity and tolerance to drought stress in rice. *Frontiers in Plant Science*, *8*, 993.

Lu, C., Zhang, L., Tang, Z., Huang, X. Y., Ma, J. F., & Zhao, F. J. (2019). Producing cadmium-free Indica rice by overexpressing OsHMA3. *Environment International*, *126*, 619–626.

Lv, Y., Yang, M., Hu, D., Yang, Z., Ma, S., Li, X., & Xiong, L. (2017). The OsMYB30 transcription factor suppresses cold tolerance by interacting with a JAZ protein and suppressing β-amylase expression. *Plant Physiology*, *173*(2), 1475–1491.

Miao, H., Sun, P., Liu, Q., Liu, J., Xu, B., & Jin, Z. (2017). The AGPase family proteins in banana: genome-wide identification, phylogeny, and expression analyses reveal their involvement in the development, ripening, and abiotic/biotic stress responses. *International Journal of Molecular Sciences*, *18*(8), 1581.

Miki, D., Zhang, W., Zeng, W., Feng, Z., & Zhu, J. K. (2018). CRISPR/Cas9-mediated gene targeting in Arabidopsis using sequential transformation. *Nature Communications*, *9*(1), 1967.

Mo, W., Tang, W., Du, Y., Jing, Y., Bu, Q., & Lin, R. (2020). PHYTOCHROME-INTERACTING FACTOR-LIKE14 and SLENDER RICE1 interaction controls seedling growth under salt stress. *Plant Physiology*, *184*(1), 506–517.

Mojica, F. J., Díez-Villaseñor, C., García-Martínez, J., & Almendros, C. (2009). Short motif sequences determine the targets of the prokaryotic CRISPR defence system. *Microbiology*, *155*(3), 733–740.

Mondal, S., Khan, M. I. R., Entila, F., Dixit, S., Cruz, P. C. S., Ali, M. P., Pittendrigh, B., Septiningsih, E. M., & Ismail, A. M. (2020). Responses of AG1 and AG2 QTL introgression lines and seed pre-treatment on growth and physiological processes during anaerobic germination of rice under flooding. *Scientific Reports*, *10*(1), 10214.

Mushtaq, M., Bhat, J. A., Mir, Z. A., Sakina, A., Ali, S., Singh, A. K., . . . Bhat, R. (2018). CRISPR/Cas approach: A new way of looking at plant-abiotic interactions. *Journal of Plant Physiology*, *224*, 156–162.

Nasr Esfahani, M., Inoue, K., Chu, H. D., Nguyen, K. H., Van Ha, C., Watanabe, Y., . . . Tran, L. S. P. (2017). Comparative transcriptome analysis of nodules of two Mesorhizobium–chickpea associations with differential symbiotic efficiency under phosphate deficiency. *The Plant Journal*, *91*(5), 911–926.

Nazir, F., Jahan, B., Kumari, S., Iqbal, N., Albaqami, M., Sofo, A., & Khan, M. I. R. (2023b). Brassinosteroid modulates ethylene synthesis and antioxidant metabolism to protect rice (*Oryza sativa*) against heat stress-induced inhibition of source–sink capacity and photosynthetic and growth attributes. *Journal of Plant Physiology*, *289*, 154096.

Nazir, F., Mahajan, M., Khatoon, S., Albaqami, M., Ashfaque, F., Chhillar, H., Chopra, P., & Khan, M. I. R. (2023a). Sustaining nitrogen dynamics: A critical aspect for improving salt tolerance in plants. *Frontiers in Plant Science*, *14*, 1087946.

Nazir, F., Khan, L., Umar, S., & Khan, M. I. R. (2024b). Omics-based strategies for improving salt tolerance in rice. In *Current Omics Advancement in Plant Abiotic Stress Biology* (pp. 63–70). Academic Press.

Nazir, F., Peter, P., Gupta, R., Kumari, S., Nawaz, K., & Khan, M. I. R. (2024a). Plant hormone ethylene: A leading edge in conferring drought stress tolerance. *Physiologia Plantarum*, *176*(1), e14151.

Nieves-Cordones, M., Mohamed, S., Tanoi, K., Kobayashi, N. I., Takagi, K., Vernet, A., . . . Véry, A. A. (2017). Production of low-Cs+ rice plants by inactivation of the K+ transporter Os HAK 1 with the CRISPR-Cas system. *The Plant Journal*, *92*(1), 43–56.

Nuñez-Muñoz, L., Vargas-Hernández, B., Hinojosa-Moya, J., Ruiz-Medrano, R., & Xoconostle-Cázares, B. (2021). Plant drought tolerance provided through genome editing of the trehalase gene. *Plant Signaling & Behavior*, *16*(4), 1877005.

Ogata, T., Ishizaki, T., Fujita, M., & Fujita, Y. (2020). CRISPR/Cas9-targeted mutagenesis of OsERA1 confers enhanced responses to abscisic acid and drought stress and increased primary root growth under non-stressed conditions in rice. *PloS One*, *15*(12), e0243376.

Osakabe, Y., Watanabe, T., Sugano, S. S., Ueta, R., Ishihara, R., Shinozaki, K., & Osakabe, K. (2016). Optimization of CRISPR/Cas9 genome editing to modify abiotic stress responses in plants. *Scientific Reports*, *6*(1), 26685.

Ou, W., Mao, X., Huang, C., Tie, W., Yan, Y., Ding, Z., . . . Hu, W. (2018). Genome-wide identification and expression analysis of the KUP family under abiotic stress in cassava (*Manihot esculenta* Crantz). *Frontiers in Physiology*, *9*, 17.

Pan, J., Sharif, R., Xu, X., & Chen, X. (2021). Mechanisms of waterlogging tolerance in plants: Research progress and prospects. *Frontiers in Plant Science*, *11*, 627331.

Park, J. J., Dempewolf, E., Zhang, W., & Wang, Z. Y. (2017). RNA-guided transcriptional activation via CRISPR/dCas9 mimics overexpression phenotypes in Arabidopsis. *PLoS One*, *12*(6), e0179410.

Parmar, N., Singh, K. H., Sharma, D., Singh, L., Kumar, P., Nanjundan, J., . . . Thakur, A. K. (2017). Genetic engineering strategies for biotic and abiotic stress tolerance and quality enhancement in horticultural crops: A comprehensive review. *3 Biotech*, *7*, 1–35.

Paulose, B., Chhikara, S., Coomey, J., Jung, H. I., Vatamaniuk, O., & Dhankher, O. P. (2013). A γ-glutamyl cyclotransferase protects Arabidopsis plants from heavy metal toxicity by recycling glutamate to maintain glutathione homeostasis. *The Plant Cell*, *25*(11), 4580–4595.

Per, T. S., Khan, N. A., Reddy, P. S., Masood, A., Hasanuzzaman, M., Khan, M. I. R., & Anjum, N. A. (2017). Approaches in modulating proline metabolism in plants for salt and drought stress tolerance: Phytohormones, mineral nutrients and transgenics. *Plant Physiology and Biochemistry*, *115*, 126–140.

Poór, P., Nawaz, K., Gupta, R., Ashfaque, F., & Khan, M. I. R. (2022). Ethylene involvement in the regulation of heat stress tolerance in plants. *Plant Cell Reports*, *41*(3), 675–698.

Prusty, M. R., Kim, S. R., Vinarao, R., Entila, F., Egdane, J., Diaz, M. G., & Jena, K. K. (2018). Newly identified wild rice accessions conferring high salt tolerance might use a tissue tolerance mechanism in leaf. *Frontiers in Plant Science*, *9*, 417.

Qin, H., Wang, J., Chen, X., Wang, F., Peng, P., Zhou, Y., . . . Huang, R. (2019). Rice Os DOF 15 contributes to ethylene-inhibited primary root elongation under salt stress. *New Phytologist*, *223*(2), 798–813.

Qiu, Z., Kang, S., He, L., Zhao, J., Zhang, S., Hu, J., . . . Zhu, L. (2018). The newly identified heat-stress sensitive albino 1 gene affects chloroplast development in rice. *Plant Science*, *267*, 168–179.

Ran, F. A., Cong, L., Yan, W. X., Scott, D. A., Gootenberg, J. S., Kriz, A. J., . . . Zhang, F. (2015). In vivo genome editing using Staphylococcus aureus Cas9. *Nature*, *520*(7546), 186–191.

Roca Paixão, J. F., Gillet, F. X., Ribeiro, T. P., Bournaud, C., Lourenço-Tessutti, I. T., Noriega, D. D., . . . Grossi-de-Sa, M. F. (2019). Improved drought stress tolerance in Arabidopsis by CRISPR/dCas9 fusion with a Histone AcetylTransferase. *Scientific Reports*, *9*(1), 8080.

Sahin, U., Ekinci, M., Ors, S., Turan, M., Yildiz, S., & Yildirim, E. (2018). Effects of individual and combined effects of salinity and drought on physiological, nutritional and biochemical properties of cabbage (*Brassica oleracea* var. capitata). *Scientia Horticulturae*, *240*, 196–204.

Sanghera, G. S., Wani, S. H., Hussain, W., & Singh, N. B. (2011). Engineering cold stress tolerance in crop plants. *Current Genomics*, *12*(1), 30.

Shahid, M., Khalid, S., Abbas, G., Shahid, N., Nadeem, M., Sabir, M., . . . Dumat, C. (2015). Heavy metal stress and crop productivity. In *Crop Production and Global Environmental Issues* (pp. 1–25). Springer Nature.

Shan, Q., Wang, Y., Li, J., & Gao, C. (2014). Genome editing in rice and wheat using the CRISPR/Cas system. *Nature Protocols*, *9*(10), 2395–2410.

Shen, C., Que, Z., Xia, Y., Tang, N., Li, D., He, R., & Cao, M. (2017). Knock out of the annexin gene OsAnn3 via CRISPR/Cas9-mediated genome editing decreased cold tolerance in rice. *Journal of Plant Biology*, *60*, 539–547.

Shi, J., Gao, H., Wang, H., Lafitte, H. R., Archibald, R. L., Yang, M., . . . Habben, J. E. (2017). ARGOS 8 variants generated by CRISPR-Cas9 improve maize grain yield under field drought stress conditions. *Plant Biotechnology Journal*, *15*(2), 207–216.

Shi, Y., Ding, Y., & Yang, S. (2018). Molecular regulation of CBF signaling in cold acclimation. *Trends in Plant Science*, *23*(7), 623–637.

Shu, P., Li, Y., Xiang, L., Sheng, J., & Shen, L. (2023). SlNPR1 modulates chilling stress resistance in tomato plant by alleviating oxidative damage and affecting the synthesis of ferulic acid. *Scientia Horticulturae*, *307*, 111486.

Sruthi, P., Shackira, A. M., & Puthur, J. T. (2017). Heavy metal detoxification mechanisms in halophytes: An overview. *Wetlands Ecology and Management*, *25*, 129–148.

Steinert, J., Schiml, S., Fauser, F., & Puchta, H. (2015). Highly efficient heritable plant genome engineering using Cas9 orthologues from *Streptococcus thermophilus* and *Staphylococcus aureus*. *The Plant Journal*, *84*(6), 1295–1305.

Su, Y., Wang, Z., Liu, F., Li, Z., Peng, Q., Guo, J., . . . Que, Y. (2016). Isolation and characterization of ScGluD 2, a new sugarcane beta-1, 3-glucanase D family gene induced by *Sporisorium scitamineum*, ABA, H_2O_2, NaCl, and $CdCl_2$ stresses. *Frontiers in Plant Science*, *7*, 1348.

Tubiello, F. N., & Fischer, G. (2007). Reducing climate change impacts on agriculture: Global and regional effects of mitigation, 2000–2080. *Technological Forecasting and Social Change*, *74*(7), 1030–1056.

Wang, L., Chen, L., Li, R., Zhao, R., Yang, M., Sheng, J., & Shen, L. (2017a). Reduced drought tolerance by CRISPR/Cas9-mediated SlMAPK3 mutagenesis in tomato plants. *Journal of Agricultural and Food Chemistry*, *65*(39), 8674–8682.

Wang, M., Wang, S., Liang, Z., Shi, W., Gao, C., & Xia, G. (2018). From genetic stock to genome editing: Gene exploitation in wheat. *Trends in Biotechnology*, *36*(2), 160–172.

Wang, T., Xun, H., Wang, W., Ding, X., Tian, H., Hussain, S., . . . Wang, S. (2021). Mutation of GmAITR genes by CRISPR/Cas9 genome editing results in enhanced salinity stress tolerance in soybean. *Frontiers in Plant Science*, *12*, 779598.

Wang, Y., Geng, L., Yuan, M., Wei, J., Jin, C., Li, M., . . . Li, X. (2017b). Deletion of a target gene in Indica rice via CRISPR/Cas9. *Plant Cell Reports*, *36*, 1333–1343.

Wu, J., Yan, G., Duan, Z., Wang, Z., Kang, C., Guo, L., . . . Dai, C. (2020). Roles of the *Brassica napus* DELLA protein BnaA6. RGA, in modulating drought tolerance by interacting with the ABA signaling component BnaA10. ABF2. *Frontiers in Plant Science*, *11*, 577.

Xiao, Y., Karikari, B., Wang, L., Chang, F., & Zhao, T. (2021). Structure characterization and potential role of soybean phospholipases A multigene family in response to multiple abiotic stress uncovered by CRISPR/Cas9 technology. *Environmental and Experimental Botany*, *188*, 104521.

Ye, N. H., Wang, F. Z., Shi, L., Chen, M. X., Cao, Y. Y., Zhu, F. Y., . . . Liu, Y. G. (2018). Natural variation in the promoter of rice calcineurin B-like protein10 (Os CBL 10) affects flooding tolerance during seed germination among rice subspecies. *The Plant Journal*, *94*(4), 612–625.

Yin, Y., Qin, K., Song, X., Zhang, Q., Zhou, Y., Xia, X., & Yu, J. (2018). BZR1 transcription factor regulates heat stress tolerance through FERONIA receptor-like kinase-mediated reactive oxygen species signaling in tomato. *Plant and Cell Physiology*, *59*(11), 2239–2254.

Yu, W., Wang, L., Zhao, R., Sheng, J., Zhang, S., Li, R., & Shen, L. (2019). Knockout of SlMAPK3 enhances tolerance to heat stress involving ROS homeostasis in tomato plants. *BMC Plant Biology*, *19*(1), 1–13.

Zeng, Y., Wen, J., Zhao, W., Wang, Q., & Huang, W. (2020). Rational improvement of rice yield and cold tolerance by editing the three genes OsPIN5b, GS3, and OsMYB30 with the CRISPR–Cas9 system. *Frontiers in Plant Science*, *10*, 1663.

Zhang, H., Zhang, J., Wei, P., Zhang, B., Gou, F., Feng, Z., . . . Zhu, J. K. (2014). The CRISPR/C as9 system produces specific and homozygous targeted gene editing in rice in one generation. *Plant Biotechnology Journal*, *12*(6), 797–807.

Zhang, Y., Liang, Z., Zong, Y., Wang, Y., Liu, J., Chen, K., . . . Gao, C. (2016). Efficient and transgene-free genome editing in wheat through transient expression of CRISPR/Cas9 DNA or RNA. *Nature Communications*, *7*(1), 12617.

Zhao, Y., Zhang, C., Liu, W., Gao, W., Liu, C., Song, G., . . . Xie, C. (2016). An alternative strategy for targeted gene replacement in plants using a dual-sgRNA/Cas9 design. *Scientific Reports*, *6*(1), 23890.

Zheng, M., Lin, J., Liu, X., Chu, W., Li, J., Gao, Y., . . . Hu, Z. (2021). Histone acetyltransferase TaHAG1 acts as a crucial regulator to strengthen salt tolerance of hexaploid wheat. *Plant Physiology*, *186*(4), 1951–1969.

Zong, Y., Wang, Y., Li, C., Zhang, R., Chen, K., Ran, Y., . . . Gao, C. (2017). Precise base editing in rice, wheat and maize with a Cas9-cytidine deaminase fusion. *Nature Biotechnology*, *35*(5), 438–440.

11 Metabolomics as a Prospective Tool for Optimizing Plant Health in Response to Abiotic Stresses

Pravneet Kaur, Faroza Nazir, and M. Iqbal R. Khan

11.1 INTRODUCTION

The advent of new technologies has transformed various facets of plant biology research, opening up new pathways to understanding the functional properties of cellular components (Mehta et al., 2019; Iqbal et al., 2021; Gupta et al., 2022; Nazir et al., 2024). Among the many approaches, metabolomics stands out as a fundamental branch of systems biology and a powerful tool for investigating how organisms respond to external factors at the metabolite level (Khan et al., 2017; Choudhury et al., 2021). Metabolomics is crucial for decoding the chemical signals that guide plant growth and development. In the context of climate change, abiotic stresses such as drought, floods, high temperatures, and heavy metal toxicity (stemming from human activities) pose a significant global threat to agriculture, with the frequency of such events expected to rise considerably in the near future (Chaudhry and Sidhu, 2022). These stresses are the primary causes of reduced crop yield and quality, potentially endangering food security. Economic losses due to abiotic stresses are estimated to be around USD 14–19 million annually worldwide (Rawat et al., 2021; Khan et al., 2023). Therefore, understanding the components involved in plants' responses, tolerance, and adaptability to these stresses is imperative.

In the current context, the use of metabolomics to understand plant responses to abiotic stresses is significantly propelled by advancements in molecular genetics and the enhancement and diversification of various analytical tools. These developments have enabled the identification of numerous metabolites of substantial biological importance. Techniques such as gas chromatography–mass spectrometry (GC-MS), liquid chromatography–mass spectrometry (LC-MS), and nuclear magnetic resonance spectroscopy (NMR) are commonly employed in plant metabolomics research (Hayashi et al., 2021; Kar et al., 2022; Tansawat et al., 2023). These methods are often used in combination because they are largely complementary, with each method preferentially covering different types of metabolites. Following this, studies utilizing metabolomics to dissect plant responses to abiotic stresses will be examined. Additionally, we will analyze existing data to elucidate both general and stress-specific responses. Finally, we will provide an updated overview of the metabolomic signatures of plants in response to major abiotic stresses.

11.2 UNLOCKING THE PLANT METABOLOME: INNOVATIVE TECHNIQUES USED FOR UNRAVELING THE SECRETS OF PLANT METABOLISM

11.2.1 Gas Chromatography–Mass Spectrometry (GC-MS)

Owing to its ability to comprehensively analyze metabolites by facilitating their separation, consequent identification, and quantification via gas chromatography and mass spectroscopic detection

DOI: 10.1201/9781032630380-11

respectively, GC-MS contributes significantly in understanding cellular processes and disease states on the basis of metabolic profiles. Multiple studies evince the increasing deployment of GC-MS metabolomics in diverse studies as described further. A study by Yuliana et al. (2020) was the first to employ GC-MS metabolomics for identification of compounds with cytotoxic activity against WiDr colon cancer cells via analysis of phytochemical profiles of Indonesian white, red, and black rice brans, along with Japanese white rice brans. Using orthogonal projection to the latent structure (OPLS) analysis, protocatechuic acid (PA) was identified as the discriminating factor specifically present in black rice brans, which exhibited notable cytotoxicity against WiDr colon cancer cells, with an IC50 value of 8.53 ± 0.26 μM. This study succeeded in providing valuable insights into the cytotoxic potential of PA in black rice brans against WiDr colon cancer cells, highlighting its impact on distinct cellular pathways (Yuliana et al., 2020). Another study deployed static headspace gas chromatography–mass spectroscopy (SHS-GC-MS) for the identification of key volatile aroma compounds and their biosynthetic pathways in Thai native-colored rice varieties also known as jasmine rice (Tansawat et al., 2023). In this study, analysis of 23 rice varieties including aromatic white, aromatic black, and non-aromatic red was carried out, in which 48 compounds were detected using GC-MS. 2-methylpropanal emerged as the most distinctive volatile compound in black- and red-colored rice, 2-methoxy phenol was primarily identified in both aromatic and non-aromatic black rice, while 3-methylbutanal was majorly found in aromatic black rice. Moreover, this study aided in highlighting the nutritional benefits of colored rice as a sustainable and healthy plant-based food source. The untargeted metabolomics approach identified key compounds and pathways, contributing to understanding the unique aroma and flavor profiles of Thai native-colored rice (Tansawat et al., 2023). Further, GC-MS has also been utilized in another investigation to analyze 60 rice seeds from two main origins in Heilongjiang Province (Feng et al., 2019). This method succeeded in identifying specific metabolites distinguishing the origins. A total of 173 peaks were detected, with 54 identified metabolites, including amino acids and sugars. Nine unique metabolites were found in Wuchang, eight in Jiansanjiang. Ten differentially significant metabolites indicated origin-specific information in rice. The proposed GC-MS method proved to be useful for metabolomic research on rice seeds as it provides insights into origin-related metabolite variations and its impacts on rice's terminal metabolic pathways. The approach can possibly be extended to analyze metabolites in similar samples by adjusting experimental conditions (Feng et al., 2019). Additionally, GC-MS has been employed along with liquid chromatography–mass spectrometry (LC-MS) to facilitate the examination of physiological and chemical changes during germination (Kim et al., 2020). In this study, the germination of unpolished brown rice (BR) for up to 48 hours to produce germinated brown rice (GBR) was studied. Twenty-five metabolites were identified, which were found to be associated with carbohydrates and lipid metabolism. The study proposed a metabolic pathway suggesting that carbohydrate metabolism, citric acid cycle, and lipid metabolism were the key processes impacted during germination. The detailed insights provided by GC-MS regarding the intricacies of metabolic changes occurring in the aforementioned pathways played a crucial role in advancing our comprehension of the nutritional changes occurring in germinated brown rice (Kim et al., 2020). Furthermore, GC-MS has also been harnessed by researchers to genetically analyze metabolites and isozymes in diverse rice cultivars (Ranjitha et al., 2019). Five isozymes, inclusive of peroxidase, catalase, malate dehydrogenase, and alcohol dehydrogenase (ADH) were studied to characterize 30 rice cultivars and two hybrids, namely KRH-2 and KRH-4 and their parental lines. ADH proved to be useful in detecting specific cultivars. GC-MS facilitated the identification of 66 metabolites in the rice samples and played a key role in showcasing significant variations in metabolite profiles among cultivars, thereby demonstrating its effectiveness. Overall, the optimized GC-MS method proved efficient in identifying plausible biomarkers, thereby laying the foundation for future research to enhance rice grain quality and nutritional value through genetic diversity.

11.2.2 Liquid Chromatography–Mass Spectrometry (LC-MS)

LC-MS is deployed in plant metabolomics, sometimes alongside GC-MS to complement its capabilities of analysis and quantification of a wide range of metabolites to offer a detailed view of a plant's metabolic pathways and provide insights into plant responses to environmental factors, contributing to advancements in plant science and crop improvement. While GC-MS is proficient in analyzing volatile and thermally stable compounds, LC-MS excels in handling polar and non-volatile molecules. Hence, the utilization of LC-MS has not only broadened the spectrum of detectable metabolites but has also enhanced the analytical depth and accuracy, thereby contributing to our understanding of plant biochemistry. Experiments during different growth stages are proposed to enhance understanding of iron homeostasis and identify marker metabolites (Kar et al., 2022). Additionally, it has also been integrated in a study to distinguish between rice of varied geographical origins based on their metabolic profiles which was hypothesized to be influenced by factors such as soil, climate, and water source in the planting area. Rice samples from Heilongjiang and Jiangsu Provinces, including Daohuaxiang (DH), Huaidao No. 5 (HD), Songjing (SJ), and Changlixiang (CL), were studied via LC-MS-based non-targeted metabolomics technology to trace the differences in metabolites and pathways of these rice samples. It led to the identification of 328 differential metabolites, highlighting significant distinctions in rice metabolites from different origins (Li et al., 2022). Another study investigating the impacts of heavy metal stress, specifically cadmium (Cd) toxicity, on metabolism of two indica rice varieties, WZ11 and WC01, employed a comparable approach utilizing LC-MS as a central tool. It led to the identification and quantification of a varied range of metabolites encompassing carbohydrates, amino acids, organic acids, and nucleotides. In addition, activation of certain organic acids associated with key metabolic pathways, such as alpha-linolenic acid metabolism and jasmonic acid production, were activated. These findings serve as valuable insights for facilitating selection of rice varieties with low cadmium accumulation and deepen our understanding of the metabolic responses to cadmium stress in rice (Zeng et al., 2021). Further, Pérez-Cova et al. (2022) also applied LC-MS to analyze rice lipidome and metabolome for gauging the impact of arsenic exposure on rice crops, specifically in the early stages of growth owing to their vulnerability to arsenic, especially under flooding conditions. A total of 100 lipids and 40 metabolites were found to be considerably altered due to arsenic exposure, chiefly in glycerophospholipids, glycerolipids, and amino acid–associated pathways.

11.2.3 Nuclear Magnetic Resonance (NMR) Spectroscopy

Leveraging NMR spectroscopy in plant metabolomics facilitates non-destructive, high-sensitivity identification, quantification, and structural elucidation of metabolites, thereby enabling comprehensive metabolite profiling and aiding in plant physiology and responses. Widespread utilization of NMR in various studies is a testament to its remarkable capabilities and signifies its indispensable role in advancing our understanding of plant physiology and responses. A study by Saeed et al. (2022) utilized NMR metabolite profiling for a rather peculiar purpose of distinguishing and predicting the geographical origin of white and brown rice samples from Korea and China. This study involved the investigation of 44 white and 20 brown rice samples collected from different regions of Korea and China using one-dimensional nuclear magnetic resonance (1H-NMR) spectroscopy. It led to the identification of 24 metabolites inclusive of amino acids (4-aminobutyrate, alanine, asparagine, aspartate, glutamate, glutamine, glycine, isoleucine, leucine, methionine, threonine, valine), organic acids (malate, fumarate, succinate, acetate, glycolate), sugars (glucose, maltose, sucrose), alcohol (ethanol), and others (pyruvate, threonate, choline). The metabolite profiles were influenced by environmental factors such as rainfall and temperature in the regions where the rice samples were collected. This was followed by processing of data so collected via several data normalization and scaling methods that in turn aided in determining the best-suited preprocessing method for each type of rice. The outcomes demonstrated the potential application of machine learning based on NMR data for distinguishing the geographical origins of rice. The identification of metabolites provided valuable information for facilitating differentiation of rice samples from Korea and China with high accuracy. Another study conducted by Elshamey et al. (2021) leveraged NMR metabolomics to investigate the physiological and amino acid responses of different rice varieties to water deficit conditions, specifically deficit irrigation (DI). Via this study the researchers sought to identify DI-tolerant rice cultivars, compare genetic differences in DI tolerance among seven genotypes, evaluate water productivity under DI, and analyze metabolite responses to be able to recommend promising rice genotypes for higher yields and quality under water deficit conditions. Via deployment of NMR-metabolomic analysis, alteration in metabolic profiles of full irrigation (FI) and DI conditions were identified. Changes were particularly noticed in amino acids, sugars, and organic compounds. Metabolites recognized under DI entailed sucrose, threonine, glutamate, and allantoin, which were indicative of the metabolic adaptations and responses of rice plants to water deficit stress. This study provided insights into the metabolic pathways and compounds closely associated with DI tolerance in rice. Further, utilization of NMR particularly, 1H-NMR, was done by another study for the classification of Japanese sake based on its metabolic profile (Hayashi et al., 2021). Factors like rice polishing ratio and the addition of alcohol heavily impacted the traditional methods of sake classification. However, the presence of dominant ethanol signals in 1H-NMR spectra posed challenges for metabolome analysis in sake classification. To overcome this issue, the researchers deployed a technique to selectively suppress ethanol signals in the 1H-NMR spectra, thus allowing the detection of trace components. The study identified several metabolites via NMR analysis which were associated with the fermentation process and composition of sake, despite the challenges posed by the dominant ethanol signals. Key metabolites recognized encompassed organic acids, amino acids, and sugars which perform crucial roles in the brewing and flavor profile of sake. In addition, in another study conducted by Du et al. (2019), the chemical composition of various rice cultivars with a particular focus of pigmented rice varieties was investigated using NMR spectroscopy, specifically 1H-NMR. Additionally, with the assistance of decision tree analysis, this study discerned metabolic profiles of each rice type, further shedding light upon their nutritional disparities and potential health benefits. Using NMR, researchers recognized crucial metabolites inclusive of sugars such as sucrose and fructose, as well as non-sugars like γ-amino butyric acid (GABA) and citric acid. This research depicted the utility of NMR spectroscopy in elucidating the chemical composition of agricultural products and its implications for food science and nutrition. Furthermore, a study by Ma et al. (2021) leveraged NMR spectroscopy to unveil the metabolic alterations induced by salinity stress and subsequent recovery in rice plants. Via metabolomic analysis, metabolites such as glutamine, asparagine, and malonic acid were associated with salinity stress resilience and recovery. These metabolites were found to be upregulated in recovered-tolerant genotype, thereby indicating their potential role in regulation of essential processes like panicle branching and spikelet formation critical for plant survival under salt stress conditions. This study furnished valuable insights into the mechanisms governing salt stress management at the cellular level, thereby furthering the development of novel strategies to enhance salt tolerance in agriculture.

11.3 DECIPHERING THE METABOLOMICS SIGNATURES OF PLANTS UNDER ABIOTIC STRESSES

Table 11.1 furnishes crucial insights into plant metabolomics under diverse abiotic stress conditions from various studies. It unveils the potential strategies that plants deploy for withstanding such adverse conditions. By leveraging advanced metabolomics techniques as mentioned later, researchers have attempted to reveal a myriad of metabolites pivotal for stress response encompassing amino acids, sugars, organic acids, and secondary metabolites like alkaloids and flavonoids. These metabolites are known to play crucial roles in energy metabolism, osmotic regulation, detoxification,

TABLE 11.1
Deciphering the Metabolomics Signatures of Plants under Abiotic Stresses

S. No.	Abiotic stress	Technique deployed	Metabolites identified	Role of metabolites identified	Reference
1.	Partial submergence stress	Short-term experiment: gas chromatography–mass spectrometry (GC-MS); Long-term experiment: identified a comprehensive set of metabolites using GC, CE, and LC-MS	Short-term experiment: identified 91 known metabolite peaks and 28 mass spectral tags (MSTs) using GC-MS; raffinose, galactinol, caffeate, quinate conjugates, asparagine, arginine, homoserine, methionine, histidine, trehalose, glycolysis metabolites, branched chain amino acids, lysine, threonine, proline, phenylalanine, tryptophan, IAA-Leu, IAA-Asp	GC-MS detected metabolites involved in providing energy and building blocks for elongation under PS stress.	Fukushima et al. (2020)
2.	Cadmium (Cd) Stress	GC-MS	Amino acids: glutamic acid, glycine, aspartic acid, tyrosine; sugars: sucrose, maltose, fructose, glucose; Organic acids: citric acid, malic acid, succinic acid; Others: glutathione, phytochelatins (PCs)	Amino acids play roles in stress response, signaling, and osmotic regulation. Sugars serve as energy sources and signaling molecules. Organic acids are involved in the tricarboxylic acid (TCA) cycle, central to cellular respiration. Glutathione is an antioxidant involved in detoxification processes and scavenging of ROS. Phytochelatins (PCs) are involved in chelation and detoxification of heavy metals like Cd.	Wang et al. (2021)
3.	Exogenous GSH (Reduced glutathione) treatment under Cd stress conditions	GC-MS	D-glucoheptanose, sorbitol, altrose, hydroxyhippuric acid, sucrose, D-heptanose, sophorose, 3-phosphoglycerate, malic acid, maleic acid, allo-inositol, prunin degr. prod. 1; Fatty acids: stearic acid, palmitic acid, linoleic acid, linoleic acid methyl ester; Others: guanine, aspartic acid, flavanone	-D-glucoheptanose: increased in both roots and leaves after GSH treatment. Sorbitol, altrose, hydroxyhippuric acid: decreased in leaves after GSH treatment. Sucrose, D-heptanose, sophorose, 3-phosphoglycerate, malic acid, maleic acid, allo-inositol, prunin degr. prod. 1: increased in leaves after GSH treatment. Stearic acid, palmitic acid, linoleic acid, linoleic acid methyl ester: increased in roots after GSH treatment. Guanine, aspartic acid, flavanone: identified metabolites with roles in stress response, energy metabolism, and detoxification processes.	Wang et al. (2021)

TABLE 11.1 (*Continued*)
Deciphering the Metabolomics Signatures of Plants under Abiotic Stresses

S. No.	Abiotic stress	Technique deployed	Metabolites identified	Role of metabolites identified	Reference
4.	Cadmium (Cd) stress	LC-QTOF/MS (Liquid chromatography coupled with quadrupole time-of-flight mass spectrometry)		Alkaloids, flavonoids, and other identified metabolites may act as antioxidants, regulate cell growth, provide resistance to biotic and abiotic stress, and play roles in signaling pathways and stress responses. They contribute to the plant's defense mechanisms against Cd-induced oxidative damage and disruption of lipid metabolism, and may promote cell wall synthesis and provide energy sources for cellular responses to stress.	Lan et al. (2023)
5.	Arsenic (As) stress	Ultra performance liquid chromatography–triple time-of-flight mass spectrometry (UPLC-Triple TOF-MS)	Dermatan L-iduronate, gentioflavine, N,N'-Bis(gamma-glutamyl)cysteine, 6-(carboxymethoxy)-3,4,5-trihydroxyoxane-2-carboxylic acid, arginine, proline, phenylalanine, tyrosine, tryptophan, phospholipids, carboxylic acids, monosaccharides	Dermatan L-iduronate is a glycosaminoglycan component involved in extracellular matrix formation and signaling. Gentioflavine may act as an antioxidant, protecting against oxidative stress induced by arsenic. N,N'-Bis(gamma-glutamyl)cysteine is involved in glutathione metabolism, which plays a crucial role in detoxification. 6-(carboxymethoxy)-3,4,5-trihydroxyoxane-2-carboxylic acid may have antioxidant properties, helping to mitigate oxidative stress. Arginine and proline act as osmoprotectant, helping to maintain cellular integrity and cope with osmotic stress. Phenylalanine, tyrosine, and tryptophan are precursors for various metabolites involved in plant growth and stress response. Phospholipids are essential components of cell membranes and play a role in signaling and stress response. Carboxylic acids are involved in various metabolic pathways and may be affected by stress-induced changes. Monosaccharides are essential for energy production and signaling in plants, and their levels can indicate stress responses.	Ma et al. (2023)

(*Continued*)

TABLE 11.1 (*Continued*)
Deciphering the Metabolomics Signatures of Plants under Abiotic Stresses

S. No.	Abiotic stress	Technique deployed	Metabolites identified	Role of metabolites identified	Reference
6.	Cadmium (Cd) toxicity	Mass spectrometry-based metabolomics	D-glucose, D-xylose, myoinositol, trehalose, D-mannitol, 1-pyrroline-4-hydroxy-2-carboxylate, O-acetylserine, L-DOPA, L-threonine, L-proline, L-cysteine, L-phenylalanine, L-tyrosine, citric acid, 13(S)-hydroperoxy-9(Z),11(E),15(Z)-octadecatrienoic acid (13(S)-HPOT), (9R, 10R, 12Z)-9,10-dihydroxyoctadec-12-enoic acid (9, 10-DHOME), palmitic acid, stearic acid, 2-isopropylmalic acid, uridine, guanosine	Carbohydrates: energy source, osmolyte, stabilizer; Amino acids: stress response, protein synthesis; Organic Acids: Energy metabolism, stress response, jasmonic acid production.	Zeng et al. (2021)
7.	Salt stress	LC-MS, GC-MS	Organic acids: malate, citrate, succinate, shikimate, quinate, malonate, pipecolate, isocitrate, fumarate, 2-keto gluconic acid; Sugars: raffinose, sucrose, mannitol, trehalose, fructose, glucose, erythritol, ribose, rhamnose, fucose, galactitol, melibiose, xylose, galactose, turanose, beta-gentibiose, mannose, fructose-6-phosphate, glucose-6-phosphate, galactinol, xylitol	These metabolites act as indicators of salt stress response and are involved in osmotic regulation, energy storage, and stress tolerance mechanisms. They play roles in maintaining osmotic balance, protecting cellular structures, and regulating metabolic processes under salt stress conditions. Mannitol and trehalose, in particular, are highlighted as potential markers for salt stress tolerance in rice roots.	Chang et al. (2019)

and signaling pathways, thus depicting the dynamic metabolic adaptations that plants undergo. This understanding will not only enhance our grasp of plant stress physiology but also offer avenues for developing resilient crop varieties, thereby addressing global challenges like climate change and food security through sustainable agricultural practices.

11.4 CONCLUSION AND FUTURE PROSPECTS

The successful implementation of metabolomics techniques has enriched our knowledge of the biochemical landscape of plants; significant progress has been made in the elucidation of the metabolic state of plants under specified abiotic stress conditions. These techniques have allowed an in-depth evaluation of the plant's cellular homeostasis, metabolic reprogramming, and the adaptive mechanisms plants use in response to environmental stimuli. In future metabolic markers identifications can enhance the plant metabolic pathways detection. It can improve the capability of a breeder in order to plan and develop superior plants that adequately fulfill modern era needs. It can be associated with speed breeding and other approaches for rapid research purpose.

11.5 AUTHORSHIP CONTRIBUTION

Conceptualization, M.I.R.K.; Writing – Original draft preparation, P.K., F.N., and M.I.R.K.; Writing – Review and editing, P.K., F.N., and M.I.R.K. All authors have read and agreed to the published version of the book chapter.

11.6 DECLARATION OF COMPETING INTEREST

The authors declare that they have no known competing financial interests or personal relationships that could have appeared to influence the work reported in this book chapter.

REFERENCES

Chang, Jing, Bo Eng Cheong, Siria Natera, and Ute Roessner. "Morphological and metabolic responses to salt stress of rice (*Oryza sativa* L.) cultivars which differ in salinity tolerance." *Plant Physiology and Biochemistry* 144 (2019): 427–435.

Chaudhry, Smita, and Gagan Preet Singh Sidhu. "Climate change regulated abiotic stress mechanisms in plants: A comprehensive review." *Plant Cell Reports* 41, no. 1 (2022): 1–31.

Choudhury, Shuvasish, Parul Sharma, Debojyoti Moulick, and Muhammed Khairujjaman Mazumder. "Unrevealing metabolomics for abiotic stress adaptation and tolerance in plants." *Journal of Crop Science and Biotechnology* 24, no. 5 (2021): 479–493.

Du, Hongying, Yinqiang Huo, Huili Liu, Ghulam Mustafa Kamal, Jiaren Yang, Yongchao Zeng, Siming Zhao, and Youming Liu. "Fast nutritional characterization of different pigmented rice grains using a combination of NMR and decision tree analysis." *CyTA-Journal of Food* 17, no. 1 (2019): 128–136.

ElShamey, Essam A. Z., Esmat F. Ali, Mahmoud E. Selim, Mahmoud A. A. ElSayed, Mukhtar Ahmed, Fahad M. Alotaibi, Mohamed M. Kamara, Mohammad S. AL-Harbi, and Ahmed M. S. Kheir. "Water deficit induced physiological and amino acid responses in some rice varieties using NMR-metabolic analysis." *Agronomy Journal* 113, no. 6 (2021): 4690–4704.

Feng, Yuchao, TianXin Fu, Liyuan Zhang, Changyuan Wang, and Dongjie Zhang. "Research on differential metabolites in distinction of rice (*Oryza sativa* L.) origin based on GC-MS." *Journal of Chemistry* 2019 (2019).

Fukushima, Atsushi, Takeshi Kuroha, Keisuke Nagai, Yoko Hattori, Makoto Kobayashi, Tomoko Nishizawa, Mikiko Kojima et al. "Metabolite and phytohormone profiling illustrates metabolic reprogramming as an escape strategy of deepwater rice during partially submerged stress." *Metabolites* 10, no. 2 (2020): 68.

Gupta, R., M. I. R. Khan, J. L. Gonzalez Hernandez, W. Wang, and L. V. Bindschedler. Improvement of rice through "-omics" approaches. *Frontiers in Plant Science* 13 (2022): 926976.

Hayashi, Yuji, Takanori Komatsu, Kazuhiro Iwashita, and Eiichiro Fukusaki. "1H-NMR metabolomics-based classification of Japanese sake and comparative metabolome analysis by gas chromatography–mass spectrometry." *Journal of Bioscience and Bioengineering* 131, no. 5 (2021): 557–564.

Iqbal, Zahra, Mohammed Shariq Iqbal, M. Iqbal R. Khan, and Mohammad Israil Ansari. "Toward integrated multi-omics intervention: Rice trait improvement and stress management." *Frontiers in Plant Science* 12 (2021): 741419.

Kar, Saradia, Raj Kishan Agrahari, Emiko Yanase, Yuriko Kobayashi, Hiroyuki Koyama, and Sanjib Kumar Panda. "Liquid chromatography-mass spectrometry (LC-MS) based metabolomic fingerprinting in contrasting rice varieties for iron (Fe) excess." *Plant Stress* 4 (2022): 100078.

Khan, M. Iqbal R., Mohammad Irfan, and Ravi Gupta. "Improving crop nutritional security for sustainable agriculture in the era of climate change." *Frontiers in Plant Science* 14 (2023): 1292264.

Khan, Naeem, Shahid Ali, Muhammad Adnan Shahid, and Ardashir Kharabian-Masouleh. "Advances in detection of stress tolerance in plants through metabolomics approaches." *Plant Omics* 10, no. 3 (2017): 153–163.

Kim, Hoon, Oui-Woung Kim, Jae-Hwan Ahn, Bo-Min Kim, Juhong Oh, and Hyun-Jin Kim. "Metabolomic analysis of germinated brown rice at different germination stages." *Foods* 9, no. 8 (2020): 1130.

Lan, Zhenni, Qing He, Mingxia Zhang, Huahong Liu, Liusen Fang, and Jinfang Nie. "Assessing the effects of cadmium stress on the growth, physiological characteristics, and metabolic profiling of rice (*Oryza sativa* L.) using HPLC-QTOF/MS." *Chemosensors* 11, no. 11 (2023): 558.

Li, Zhanming, Mengmeng Tan, Huxue Deng, Xu Yang, Yue Yu, Dongren Zhou, and Hao Dong. "Geographical origin differentiation of rice by LC–MS-based non-targeted metabolomics." *Foods* 11, no. 21 (2022): 3318.

Ma, Li, Jin Zeng, Rui qi Zhang, Lin Wang, Fawen Zhang, Xuejin Zhao, Yuan Yuan, and Limei Li. "Integrated transcriptomic and metabolomic analysis the variation of rice cultivars response to arsenite stress." *Environmental Technology & Innovation* 31 (2023): 103207.

Ma, Nyuk Ling, Su Datt Lam, Wan Afifudeen Che Lah, Aziz Ahmad, Jörg Rinklebe, Christian Sonne, and Wanxi Peng. "Integration of environmental metabolomics and physiological approach for evaluation of saline pollution to rice plant." *Environmental Pollution* 286 (2021): 117214.

Mehta, Sahil, Donald James, and M. K. Reddy. "Omics technologies for abiotic stress tolerance in plants: current status and prospects." In *Recent Approaches in Omics for Plant Resilience to Climate Change*, pp. 1–34. Springer, 2019.

Nazir, Faroza, Laiba Khan, Shahid Umar, and M. Iqbal R. Khan. "Omics-based strategies for improving salt tolerance in rice." In *Current Omics Advancement in Plant Abiotic Stress Biology*, pp. 63–70. Academic Press, 2024.

Nazir, Faroza, Sarika Kumari, Moksh Mahajan, and M. Iqbal R. Khan. "An explicit story of plant abiotic stress resilience: Overtone of selenium, plant hormones and other signaling molecules." *Plant and Soil* 486, no. 1 (2023): 135–163.

Pérez-Cova, Miriam, Romà Tauler, and Joaquim Jaumot. "Adverse effects of arsenic uptake in rice metabolome and lipidome revealed by untargeted liquid chromatography coupled to mass spectrometry (LC-MS) and regions of interest multivariate curve resolution." *Separations* 9, no. 3 (2022): 79.

Ranjitha, Hurugalavadi Puttaswamaiah, Rame Gowda, Nagarajappa Nethra, Narayanappa Amruta, and Hemanth Kumar Kandikattu. "Biochemical and metabolomics on rice cultivars." *Rice Science* 26, no. 3 (2019): 189–194.

Rawat, P., D. Shankhdhar, and S. C. Shankhdhar. "Abiotic stress: Its outcome and tolerance in plants." In *Microbes and Signaling Biomolecules Against Plant Stress: Strategies of Plant-Microbe Relationships for Better Survival*, pp. 79–106. Springer Nature, 2021.

Saeed, Maham, Jung-Seop Kim, Seok-Young Kim, Ji Eun Ryu, Ju Hee Ko, Syed Farhan Alam Zaidi, Jeong-Ah Seo, Young-Suk Kim, Do Yup Lee, and Hyung-Kyoon Choi. "Differentiation of geographical origin of white and brown rice samples using NMR spectroscopy coupled with machine learning techniques." *Metabolites* 12, no. 11 (2022): 1012.

Tansawat, Rossarin, Supawat Jindawatt, Paweena Ekkaphan, Siriphat Ruengphayak, Apichart Vanavichit, Nitima Suttipanta, Sornkanok Vimolmangkang, and Wanchai De-Eknamkul. "Metabolomics approach to identify key volatile aromas in Thai colored rice cultivars." *Frontiers in Plant Science* 14 (2023): 973217.

Wang, Runfeng, Kaina Lin, Huabin Chen, Zhenyu Qi, Bohan Liu, Fangbin Cao, Hao Chen, and Feibo Wu. "Metabolome analysis revealed the mechanism of exogenous glutathione to alleviate cadmium stress in maize (*Zea mays* L.) seedlings." *Plants* 10, no. 1 (2021): 105.

Yuliana, N. D., M. Z. Tuarita, A. Khatib, F. Laila, and S. Sukarno. "GC–MS metabolomics revealed protocatechuic acid as a cytotoxic and apoptosis-inducing compound from black rice brans." *Food Science and Biotechnology* 29 (2020): 825–835.

Zeng, Ting, Baohua Fang, Fenglin Huang, Li Dai, Zhi Tang, Jinglin Tian, Guodong Cao et al. "Mass spectrometry-based metabolomics investigation on two different indica rice grains (*Oryza sativa* L.) under cadmium stress." *Food Chemistry* 343 (2021): 128472.

12 Revisiting the Role of MicroRNAs in the Battle against Abiotic Stress

Shalini Dhiman, Raman Tikoria, Mohd Ali, Deepak Kumar, Roohi Sharma, Prakriti, Vikram, Isha Madaan, Rajesh Kumar Singh, Puja Ohri, Geetika Sirhindi, and Renu Bhardwaj

12.1 INTRODUCTION

According to Shriram et al. (2016), plants have developed microRNA (miRNA) regulatory networks throughout evolution to better adapt to various environmental stresses. Growing research specifies that miRNAs play a significant role in regulating a number of plant activities, including abiotic and biotic stress responses, growth and development, and both (Singroha et al., 2021). The miRNA mediated abiotic stress responses involve various molecular, biochemical, and physiological mechanisms (Singroha et al., 2021). When the plant body is under stress, miRNAs are produced in higher amounts. These miRNAs can lower the amount of target transcripts, which may function as negative controllers of stress tolerance. However, other miRNAs have their levels drop during stress, boosting the expression of their target mRNAs, further aiding in adaptation responses (Sunkar et al., 2012; Singroha et al., 2021). Therefore, it is absolutely necessary to have a solid understanding of miRNA control in plants under stress. This chapter begins with an exploration of the crucial role of miRNAs in plant adaptation to a various form of environmental abiotic stressors. These minor, non-coding RNA molecules are at the forefront of research in plant biology, especially in the context of abiotic stressors like water scarcity, saltiness, and temperature extremes.

In recent years, miRNAs have emerged as a significant area of intensive research. Flowering, morphogenesis, signal transduction, and gene feedback control (Meng et al., 2010; Pervaiz et al., 2021) are just a few examples of the many cellular and metabolic processes that microRNAs are involved in during abiotic stress. A study of the functions of conserved miRNAs showed that they are connected to many biological and developmental processes (Pervaiz et al., 2021; Nazir et al., 2024). They regulate a broad range of metabolic activities, such as the establishment of meristem boundaries, organ separation, auxin signaling, transition from the juvenile to adult development stages, and stress tolerance (Pervaiz et al., 2021). To better improve plant responses to abiotic stressors, it may be useful to gain a deeper comprehension of the role of microRNAs.

This chapter provides a foundational overview of miRNAs, detailing their nature, synthesis process, and pivotal role in gene regulation, while also touching upon their evolutionary significance. We then navigate through the historical evolution of research in this field, highlighting key milestones that have shaped our current understanding of miRNAs in stress response. The focus shifts to examining the specific roles played by various miRNAs in responding to different abiotic stresses, unraveling their complex molecular functions and interactions within plant cells. This chapter not only aims to deepen our comprehension of plant stress mechanisms but also explores the potential of miRNAs in advancing agricultural biotechnology, particularly in developing resilient crop varieties.

DOI: 10.1201/9781032630380-12

12.2 BRIEF OVERVIEW OF MICRORNAs (MiRNAs)

MicroRNAs, small non-coding RNA molecules transcribed from miRNA genes, typically range from 20 to 24 nucleotides in length and serve crucial functions in regulating gene expression at the post-transcriptional level (Zhang et al., 2006; Laubinger et al., 2008). They engage in diverse biological processes, encompassing developmental events, metabolic activities, responses to pathogens, and reactions to abiotic stressors (Min Yang and Chen, 2013; Yang et al., 2013; Raza et al., 2023).

The synthesis of miRNAs initiates within the nucleus and is transcribed from DNA by either RNA polymerase II (in animals) or RNA polymerase III (in plants; Bartel, 2004; Yang et al., 2013). Primary miRNAs (pri-miRNAs) are the initial transcripts of miRNAs and are often hundreds to thousands of nucleotides in length with a hairpin structure. The pri-miRNAs are further transformed by enzymes like Drosha and DGCR8 in animals or Dicer in plants and animals into precursor miRNAs (pre-miRNAs), around 70 nucleotides long, which are then exported to the cytoplasm for further processing and eventually give rise to mature miRNAs. Further, it is also possible for a mature miRNA strand to join the RNA-induced silence complex (RISC). This guides the complex to specific mRNA molecules to help control how genes work (O'Brien et al., 2018; Treiber et al., 2019).

The miRNAs have an important function in post-transcriptional gene expression regulation (Floris et al., 2009). In the light of abiotic stress, miRNAs have the capacity to exhibits action such as:

a. *Gene repression*: MiRNAs have the ability to interact with the 3' untranslated region (UTR) of target messenger RNAs (mRNAs) leading to the suppression of gene expression. The process of binding can lead to several outcomes, including degradation of mRNA, inhibition of translation, and sequestration in P-bodies (O'Brien et al., 2018).
b. *Fine-tune responses*: MiRNAs modulate the stress-related responses through their ability to target genes involved in stress signaling, stress tolerance, and responsiveness to environmental signals (Pagano et al., 2021; Iqbal et al., 2022).

The extent to which the sequences are complementary between a microRNA and the messenger RNA it targets plays a vital function in governing the accuracy of gene regulation facilitated by miRNAs (Patanun et al., 2013). This level of similarity in genetic sequences is a pivotal factor that determines how effectively miRNAs are able to alter gene expression. MiRNAs have the capability to control the expression of numerous genes implicated in stress responses (Wilczynska et al., 2015). In the context of abiotic stress reactions, miRNAs play a pivotal role in the evolutionary development of organisms by the following mechanisms:

a. *Conservation*: Many stress-responsive miRNAs exhibit substantial conservation across species, signifying their ancient origins and essential role in stress adaptation. This shared evolutionary history underscores their vital contribution to stress response mechanisms and survival strategies across diverse organisms, emphasizing their fundamental importance in biological functions and overall survival (Zhang et al., 2006; Li et al., 2022).
b. *Diversification*: While the majority of miRNAs are conserved, there exists a subset that has undergone evolutionary changes with the explicit purpose of regulating genes associated with stress responses in the face of environmental difficulties. The presence of species- or stress-specific miRNAs plays a noteworthy role in facilitating the adaptation and diversification of plant and animal species (Zhang, 2015).
c. *Role in stress adaptation*: MiRNAs are known to have significant implications in the modulation of stress-responsive genes, encompassing those implicated in stress signal perception, detoxification, and protective mechanisms. Organisms have an important role in facilitating the adaptation process to many abiotic stresses, including but not limited to drought, temperature extremes, and salinity (Zhang et al., 2022a).

d. *Plasticity in stress response*: MiRNAs confer a level of plasticity to stress responses, enabling organisms to finely regulate their responses to dynamic environmental stimuli. They assist in maintaining a delicate equilibrium between the promotion of development and stress reaction (Pan, 2013). In conclusion, miRNAs play essential functions in fine-tuning stress responses, aiding survival, and fostering the evolutionary diversification of species despite their dynamic and ever-changing roles in stress adaptation.

12.3 HISTORICAL PERSPECTIVE ON MiRNAs AND ABIOTIC STRESS

Caenorhabditis elegans was the first organism in which miRNAs were described (Lee et al., 1993). In the regulatory pathway controlling the timing of larval growth and development in the nematode *C. elegans*, lin-4 is a crucial regulatory protein. Ten years later, plant miRNAs were found. Various groups discovered plant miRNAs in 2002 by cloning tiny RNAs in the *Arabidopsis* plant (Llave et al., 2002; Park et al., 2002). There are 731 microRNA genes that have been discovered in numerous plants, with 117 coming from *Arabidopsis*, 178 from rice (*Oryza sativa*), 97 from maize, and the remaining genes coming from *Medicago truncatula*, *Populus trichocarpa*, *Saccharum officinarum*, *Sorghum bicolor*, and *Glycine max*, respectively (Yang et al., 2007). The following characteristics of miRNAs are obvious: miRNAs are non-coding, single-stranded RNAs that range in length from 20 to 24 nucleotides (Reinhart et al., 2002). Every miRNA precursor has a stem loop hairpin structure that has been accurately predicted, and this fold-back hairpin shape has little free energy (Bonnet et al., 2004). MiRNA's synthesis from miRNA genes is similarly a sequential process in *Arabidopsis*. A primary miRNA, often a lengthy sequence of greater than several hundred nucleotides, is the first product of the transcription of a miRNA gene. The Pol II enzymes in this process are in charge (Lee et al., 2004). The plant hormone abscisic acid (ABA) controls numerous critical physiological and developmental processes in plants, including seed maturation, germination, the production of seed storage proteins and lipids, stomatal closure, pathogen reactional response, and tolerance introduction (Shinozaki and Yamaguchi-Shinozaki, 2000). Antioxidant systems naturally present in plants comprise a range of enzyme scavengers that prevent the buildup of reactive oxygen species (ROS) caused by stress (Mittler, 2002). The initial line of defense against extremely harmful superoxide radicals is provided by superoxide dismutases (SODs). By converting O_2^- to H_2O_2, SOD helps to scavenge ROS (Fridovich, 1995). SODs are divided into three types based on the metal cofactor used, namely iron SOD (FeSOD), manganese SOD (MnSOD), and copper-zinc SOD (Cu/Zn-SOD; Fridovich, 1995). Gel blot analysis of RNA was carried out on seedlings treated with cold, desiccation, or NaCl to examine if any miRNAs are controlled by abiotic stressors.

Results demonstrated that exposure to cold, dehydration, and NaCl treatments significantly upregulate miR393 expression. While miR319c seems to be elevated by cold but not lack of moisture, miR389a is downregulated in all forms of stress treatments involving NaCl. On the other hand miR397b and miR402 are marginally upregulated by all stress treatments (Sunkar and Zhu, 2004). Lu extracted small RNAs from *Populus trichocarpa* stems' developing xylem and cloned 22 miRNAs to determine the expression of particular miRNAs. They compared the transcript levels of ptr-miRNA in xylem tissues under tension and compression stress to those in xylem tissues that were developing normally. By applying tension or compression pressures, ptr-miR156, 162, 164, 475, 480, and 481 expressions is similarly inhibited. Ptr-miR408 transcript levels are increased in tissues under tension and compression stress. According to these findings, one of the most important defense mechanisms for physical and mechanical fitness is the ability of plant miRNAs to be controlled by mechanical stress (Lu et al., 2005). MicroRNAs are important regulators of the differential expression of genes that affect plant development and abiotic stress responses. Small RNAs called miRNAs, which are single-stranded and 20–24 nucleotides elongated and do not code for proteins, are crucial for post-transcriptional gene silencing in a variety of organisms (Rogans and Rey, 2016).

Phytohormone signaling, growth, development, organogenesis, and adaptive responses to biotic and abiotic challenges are all controlled by miRNAs in plants (Njaci et al., 2018). Studies have confirmed that miRNAs play a noteworthy role in plants' ability to tolerate drought by influencing the cleavage of numerous drought-responsive genes, which in turn prevents the translation of those genes (Nadarajah and Kumar, 2019). The various plant species in which drought-responsive miRNAs have been discovered and characterized are sugarcane (Gentile et al., 2015), soybean (Zheng et al., 2016), *Arabidopsis* (Liu et al., 2008), wheat (Sun et al., 2014), sorghum (Zhang et al., 2011), tomato (Zhou et al., 2020), maize (Ding et al., 2009), and rice (Nadarajah and Kumar, 2019). Using high-resolution sequencing of minor endogenous RNA libraries from the leaf, stem, callus, and male and female floral tissues, cassava miRNAs were discovered. Several of these miRNAs targeted gene families that were linked to transcription factor (TF) families and drought tolerance. They comprised Mes-miR166j targeted basic-leucine zipper TF, Mes-miR319f targeted MYB1, Mes-miR156k targeted Squamosa promoter-binding protein-like, Mes-miR169a targeted transcription-repair coupling factor, MiR477 targeted sequence-peculiar to DNA binding transcription factor, and Mes-miR319f targeted Teosinte branched1/cycloidea and PCF TF (Khatabi et al., 2016). The miR156 was discovered to be a stomatal movement mediator, controlling stomatal activity and water relations. To increase drought tolerance in tomatoes, miR156 interacted with the ABA-dependent strigolactone signaling pathways (Visentin et al., 2020). A genome-scale systematic investigation of miRNAs in Euphorbiaceae was undertaken using computational prediction and experimentation based on drought. They discovered that 48 miRNAs in two kinds of cassava that underwent drought stress treatment were either up- or downregulated. Squamosa promoter-binding protein miR156, LIGULELESS1 protein miR157, r2r3-myb transcription factor miR159, and auxin response miR160 were some of them. In castor beans, miRNA target pairs, miR166 (DNA binding), miR164 (NAC domain-containing protein), miR162 (dicer-1), and acetoin catabolism (protein X), miR167, miR168, miR171 (DELLA protein GAIP-B), miR395 (WRKY), miR396 (heat shock protein binding protein), miR397 (laccase, putative), and miR403 (Argonaute 2) were co-expressed during both seedling development and in response to stress conditions (Zeng et al., 2010). Differential expression of the cassava miR164 (target MesNAC) and miR167 (targets MesARF6 and MesARF8) was detected using RT-qPCR and was linked to alterations in leaf shape, stomatal closing, and relative water content when exposed to a water shortage (Phookaew et al., 2014). In response to drought stress in cassava, long noncoding RNAs (lncRNAs) may act via miRNAs. In cassava under drought stress, for example, they discovered 11 scarcity-specific differentially expressed lncRNAs serving as probable target mimics of 24 miRNAs, among which miR156, miR164, miR169, miR172, and miR395 were all suggested to be important in abiotic stress reactional response (Ding et al., 2019). The characterization of microRNA gene targets that play a role in the control of post-transcriptional responses to abiotic stress using high-throughput sequencing and bioinformatics may be relevant for developing drought-resistant cassava. For the 60 conserved miRNA sequences, they identified 134 possible target genes for drought resistance. The discovered miRNAs and miRNA gene targets in this study may be involved in drought-induced posttranscriptional regulation and may be used to design drought-resistant cassava (Ballén-Taborda et al., 2013).

12.4 ROLE OF SPECIFIC MIRNAS IN ABIOTIC STRESS RESPONSE

MicroRNAs play a pivotal role in enhancing plants' tolerance to abiotic stress by orchestrating a sophisticated regulatory network that helps plants adapt to challenging environmental circumstances. Abiotic stressors, including nutrient shortage, drought, salt, and extremely high or low temperatures, can have detrimental effects on plant growth and productivity. MiRNAs have been discovered to play a crucial function in mediating defense against different kinds of abiotic stresses faced by a plant during its growth and development (Anwar and Kim, 2020). The mechanisms through which miRNAs confer abiotic stress tolerance vary with the type of stress as well as the plant. However, the most common process includes cleavage or post-transcriptional silencing of mRNA or inhibition of protein synthesis (Basso et al., 2019). MiRNAs act as crucial molecular switches that modulate gene expression in response to these stresses. When subjected to drought, for instance, specific miRNAs, like miR156, miR159, and miR160, target genes associated with root development, stress signaling, and hormonal responses. This regulation results in altered root architecture, improved water uptake, and enhanced abscisic acid signaling, which collectively contribute to drought tolerance. MiRNAs such as miR164 and miR166 target transcription factors like NAC and HD-ZIP, influencing organ development and growth (He et al., 2019).

MiR169 fine-tunes ABA signaling and stress-responsive gene expression, further improving plant resilience to water scarcity. Similarly, in response to salinity stress, miRNAs like miR397 and miR398 target genes related to ion homeostasis and antioxidant defense, while miR408 regulates copper homeostasis and cell wall integrity. MiRNAs are also implicated in the control of nutrient uptake and utilization under nutrient-deficient conditions, thus enhancing nutrient stress tolerance (Huang et al., 2018). In essence, miRNAs act as molecular regulators that optimize the plant's response to abiotic stressors, promoting survival and ensuring continued growth and reproduction. The dynamic interaction between target genes and miRNAs is a fundamental mechanism underlying the multifaceted adaptation of plants to various abiotic stresses, and the intricate miRNA-mediated regulatory pathways offer promising avenues for the improvement of stress-tolerant crops through genetic engineering and breeding strategies.

12.4.1 MiRNAs Associated with Drought Tolerance

MiRNAs have emerged as essential molecular players in the intricate web of gene regulation that enables plants to cope with the formidable challenge of drought stress. Drought is a pervasive abiotic stress that threatens agricultural productivity and food security worldwide. In response to water scarcity, plants must swiftly and precisely modulate their gene expression to adjust their physiological and developmental processes. MiRNAs, small non-coding RNA molecules typically 20–24 nucleotides in measurement, have been identified as key regulators in this process. MiRNAs operate by binding to messenger RNA molecules, effectively controlling the translation of these mRNAs into proteins. Within the framework of drought tolerance, definite miRNAs target genes participating in a multitude of plant developmental, growth, and stress response processes, orchestrating a finely tuned response to water deficit (Table 12.1; Singroha et al., 2021). One of the critical roles of miRNAs in drought tolerance is their involvement in root architecture and development. For instance, miR156, which targets the Squamosa promoter binding protein-like (SPL) family of transcription factors, has been shown to control plant growth and development (Zhang et al., 2018). By downregulating SPL genes, miR156 promotes changes in root architecture, leading to enhanced root growth and branching. This modified root system enables vegetation to penetrate a greater volume of soil, increasing their access to water resources (Zhou et al., 2019a). Additionally, miR160 and miR167, which target auxin response factors (ARFs), are intricate in modulating auxin signaling, a hormone critical for root development. By controlling ARF expression, these miRNAs influence root growth and architecture, which is crucial for efficient water uptake and drought tolerance. MiRNAs are also integral to the regulation of abscisic acid signaling, a key component of the plant's reaction to drought stress (He et al., 2020). MiR159, for instance, targets MYB transcription factors, including MYB33 and MYB101, which are involved in ABA signaling and stress reactions. By downregulating MYB genes, miR159 contributes to improved drought tolerance by fine-tuning the ABA-dependent stress response. Similarly, miR169 targets nuclear factor Y subunit A (NF-YA), a component of the NF-Y transcription factor complex involved in ABA signaling (Banerjee, 2020). This regulation by miR169 results in enhanced ABA signaling and stress-responsive gene expression, bolstering potential of the plant to withstand drought circumstances. In addition to these mechanisms, miRNAs influence plant reactional response to drought by regulating genes associated with the control of reactive oxygen species and antioxidant defense (Orek et al., 2023). Under drought conditions, the generation of ROS increases and leads to the oxidative damage in plant cells. MiR398, for instance, targets genes encoding copper superoxide dismutase (CSD), an enzyme involved in scavenging ROS (Singh et al., 2023). By modulating the expression of CSD

TABLE 12.1
Roles of Different miRNA in Drought Tolerance

miRNA	Target genes	Function in drought tolerance
miR156	Squamosa promoter binding protein-like (SPL) family	Regulates plant growth and development, promoting drought tolerance through altered root architecture and stress-responsive gene expression.
miR159	MYB transcription factors (e.g., MYB33, MYB101)	Regulates genes associated with abscisic acid signaling and stress response, contributing to improved drought tolerance.
miR160	Auxin response factors (ARFs)	Modulates auxin signaling and root development, enhancing drought tolerance by regulating root growth and architecture.
miR164	NAC transcription factors (e.g., NAC1, NAC2)	Regulates genes involved in root progression and abiotic stress responses, improving drought tolerance.
miR166	Homeobox (HD-ZIP) transcription factors	Regulates root and shoot development, enhancing drought tolerance through altered organ growth.
miR167	Auxin response factors (ARF6, ARF8)	Modulates auxin signaling and root development, contributing to improved drought tolerance.
miR169	Nuclear factor Y subunit A (NF-YA)	Affects gene expression via regulation of ABA signaling and stress responses, contributing to drought tolerance.
miR319	TCP transcription factors (e.g., TCP4)	Modulates leaf development, enhancing drought tolerance by controlling leaf morphology.
miR393	Transport inhibitor response 1 (TIR1)	Regulates auxin signaling and root development, influencing drought tolerance.
miR396	Growth-regulating factor (GRF) family	Regulates growth and development by targeting GRF transcription factors, impacting drought tolerance through organ development.
miR398	Copper superoxide dismutase (CSD)	Modulates the expression of CSD genes, impacting antioxidant defense mechanisms under drought stress.
miR408	Plantacyanin (PC) and laccase	Regulates copper homeostasis and lignin biosynthesis, which can contribute to drought tolerance through cell wall integrity.
miR472	WRKY transcription factors	Modulates the display of stress-responsive WRKY genes, impacting drought tolerance.
miR529	Vacuolar ATPase (V-ATPase)	Regulates proton transport and ion homeostasis, which can affect drought tolerance by regulating osmotic balance.
miR827	Nitrogen transporters (NRT1.1)	Regulates nitrogen uptake and utilization under drought conditions, impacting plant growth and stress tolerance.

genes, miR398 can enhance antioxidant defenses, reducing the detrimental effects of oxidative stress under drought conditions (Aravind et al., 2017; Ma and Hu, 2023).

Furthermore, miRNAs are involved in regulating genes related to drought-responsive transcription factors. MiR472 targets WRKY transcription factors, which play a vital role in plant stress responses. By influencing the expression of WRKY genes, miR472 can modulate the plant's ability to activate stress-responsive pathways, thereby improving drought tolerance (Zhang et al., 2022b). In summary, miRNAs are essential components of the plant's adaptive response to drought stress. They influence root architecture, hormonal signaling, and stress-responsive gene expression, all of which are critical aspects of drought tolerance. These small RNA molecules act as master regulators, fine-tuning the plant's gene expression to optimize its response to water scarcity. Understanding the intricate interplay between miRNAs and their target genes is crucial for unraveling the intricate mechanisms underlying drought tolerance in plants. This knowledge not only enhances our understanding of plant biology but also holds promise for the development of more resilient crop varieties through genetic engineering and breeding strategies, which is of paramount importance in the context of climate change and increasing water shortage.

12.4.2 MiRNAs Associated with Salt Tolerance

An excess of salts in the soil solution leads to salt stress in plants, ultimately resulting in detrimental effects on their development and growth. The negative impacts of salt stress include osmotic tension mediated by water stress; ion-mediated cytotoxicity, declined photosynthesis, oxidative stress, osmotic stress, nutritional disorders, reduced antioxidant capacity, and ultimately restrained crop production. Depending on their salt tolerance potential, plants can be categorized as glycophytes (salt-sensitive) or halophytes (salt-tolerant; Srivastava et al., 2022).

However, one concern is that many plants under salt stress have differential action of small RNAs, including changes in related genes and miRNA expression levels. Numerous miRNAs with varying levels of regulation have been discovered in various salt-stressed plants such as *Zea mays*, *Oryza sativa*, *Raphanus sativus*, *Arabidopsis thaliana*, *Triticum aestivum*, and *Cicer arietinum*. Using microarrays, it was observed that in salt-susceptible plants some miRNAs are upregulated while some others are downregulated in salt-tolerant plants (Ullah et al., 2022). A number of miRNAs have been discovered in many salt-tolerant plants. However, some common miRNAs in most plants include miR408, miR399, miR397, miR396, miR395, miR394, miR319, miR172, miR172, miR169, miR168, miR167, miR166, miR164, miR160, miR3159, and miR156. These miRNAs thus act on specific salt stress-related transcriptional factor genes and eventually regulate those genes to boost salt stress tolerance in plants (Gao et al., 2022). For example, in cotton plant, the miRNA ghr-miR414c inhibited the expression of gene *GhFSD1* which was an antioxidant gene. This resulted in salt hypersensitivity and improved salinity tolerance in plants (Wang et al., 2019a). And in maize plants, ZmmiR169q was responsive toward salt stress, and the main target of this miRNA was *ZmNF-YA8*, whose overexpression actually improves salt tolerance (Xing et al., 2022). In another study on rice plants, nine preserved and two novel differentially expressed miRNAs in root or shoot tissues were discovered with the help of northern blotting technique. The target transcription factors of many of these miRNAs were involved in salt tolerance, including ARF, MYB, HD-Zp III, SBP, TCP, NF-YA, and NAC. Additionally, the role of osa-miR12477 in regulating L-ascorbate oxidase (*LAO*) under salt stress was observed in this plant (Parmar et al., 2020). In a study on wheat and rice plants, a positive regulator of salt tolerance was found as miR172 . Its abundance was found to be higher during many developmental stages in salt treated plants. It suppresses a downstream target gene IDS (Cheng et al., 2021). In rice plants, another miRNA was also detected to be responding towards salt stress tolerance, in which miR168 was acting on target downregulating stress-responsive gene AGO1 and enhancing plant salt tolerance capacity (Wan et al., 2022). Usually, a number of salt stress-responsive genes and miRNAs play an important mechanistic role in mitigating saline stress in plants, as observed in *Vicia faba* plant, in which 693miRNAs in genotype ILB4347 (salt tolerant) and 527miRNAs in Hassawi-3 (salt sensitive) were found. Discovered in the salt-tolerant variety were 395 downregulated and 298 upregulated miRNAs and in the salt-sensitive variety 243 downregulated and 284 upregulated genes (Alzahrani et al., 2019). Different plants exhibit different expression patterns for the same miRNA during salt stress. For example, in *Arabidopsis thaliana*, *Nicotiana tabacum*, and *Oryza sativa* miR396 was downregulated, whereas in *Suaeda maritima*, *Panicum virgatum*, and *Saccharum* spp. it was upregulated under stress conditions (Gao et al., 2022). Similarly, in another study, the expression level of miR164 was upregulated under salt stress in chickpeas. In the guava plant, there was a downregulation of miR164 expression (Sharma et al., 2020b). Some other miRNAs involved in the salt stress mitigation process are discussed in Table 12.2.

TABLE 12.2
MiRNAs Involved in the Salt Stress Mitigation Process

miRNA	Host plant	Target gene	Response	Reference
miR164a	*Zea mays*	*GRMZM2G008819* and *GRMZM2G114850*	Better adaptation to salinity stress, increase yield, accelerate plant development	Shan et al. (2020)
miR156a	*Malus domestica*	*MdSPL13*	Upregulates gene promoter *MdWRKY100* to enhance salt tolerance in plant	Ma et al. (2021)
Osa-miR319b	*Panicum virgatum*	*PvPCF5*	Regulates ethylene synthesis positively and enhances salt tolerance	Liu et al. (2019b)
miR390	*Helianthus tuberosus*	*ARF3/4, TAS3*	Normalize plant growth and development timing as well as pattern, response to salt stress	Wen et al. (2020)
Osa-miR396a	*Agrostis stolonifera*	*GRF*	Controls leaf senescence, participates in signaling pathway to enhance salt tolerance	Yuan et al. (2019)
miR414c	*Gossypium*	*GhFSD1*	Regulates cellulose synthesis genes, improves salt tolerance	Wang et al. (2019b)
gma-miR5036	*Glycine max*	*NAC*	Involved in signal transduction, metabolic processes	Ning et al. (2019)
Ghr-miR393-Ghir_D11G006890	*Gossypium*	*Protein auxin signaling F-box 2*	Increases MDA content, improves leaf health, responds to salt stress	Zhan et al. (2021)
tae-miR9674a	*Triticum*	*Mta/sah nucleosidase (CA716329)*	Involved in cellular component constitution, modulates biochemical metabolism	Wang et al. (2023)
miR393a, miR162a	*Pennisetum glaucum*	*TIR1, auxin response factor (ARF)*	Positively regulate auxin signaling pathway, response to salinity stress	Shinde et al. (2020)
miR156, miRNA 390	*Cucumis melo*	*CmMATE genes*	Translocate secondary metabolites, enhance salt tolerance, perform various physiological functions	Shah et al. (2022)
miR393	*Arabidopsis*	*TIR1/AFB2*	Downregulates auxin signaling pathways, responds to salt stress	Denver and Ullah (2019)
miR156	*Tamarix chinensis*	*SPLs*	Responds to salt stress, regulates genes involved in reactive oxygen species, signal transduction	Wang et al. (2019b)
miR393, miR166, miR156, miR160 and *miR396*	*Hemerocallis fulva*	*SPL, TIR1, ARF, GRF4, CML*	Regulates plant hormones, responds to salt stress	Zhou et al. (2023)

12.4.3 MiRNAs Associated with Temperature Stress Tolerance

Temperature fluctuation brought about by a changing environment interrupts the normal progress and development of the plants. The main factor hindering plant growth is the creation of reactive oxygen species, which severely impairs cell functioning. Accumulation of ROS in plant cells damages several cell components and causes denaturation of cellular enzymes. To cope with such circumstances inducing stress in plants due to temperature variations, plants modify their gene expressions. In this respect, some miRNAs with temperature-responsive effects have been discovered in plants (Ullah et al., 2022). These miRNAs are either heat responsive or cold responsive. Usually, high temperatures reduce fertility in plants and affect their photosynthetic activities.

Similarly cold stress induces membrane damage and cell dehydration. Thus, temperature variation affects the normal functioning of all the biochemical and physiological processes of plants. To regulate the normal functioning of plants, the defense mechanism of plants activates several changes in metabolic pathways, physiological responses, and signal transduction pathways. Thus, these plants regulate various genes expressions via miRNAs.

A small alteration in miRNAs levels differentially expresses their content in different parts of the plants. These miRNAs either suppress or cleave the target gene for their regulation and alter various biological processes. Temperature stress induces oxidative damage that causes activation of antioxidant synthesizing genes which are also targeted by the temperature responsive miRNAs (Sihag et al., 2021; Zhang et al., 2023). For example, in a study on *Vitisvinifera* under high temperature, the miRNA Vvi-miR167b was overexpressed transiently to downregulate the expression of *vvARF8*, *vvARF6*, and *vvARF6-like* target genes (Zhang et al., 2023). Similarly, many miRNAs have been discovered to be functional against cold stress like in the *Astragalus membranaceus*a medicinal plant. Low temperature limits the growth of these plants and affects crop production; this may be due to accumulation of some active pharmacological ingredients. In this plant 27 miRNAs were discovered to be cold responsive, of which 17 were cold-repressed conserved miRNAs, four were cold-stress induced, and six were cold-induced non-conserved miRNAs. Such miRNAs direct the response of target genes toward cold tolerance and regulate the development, defense, hormone, secondary metabolism, and redox homeostasis of the plants (Abla et al., 2019). Sometimes even the same miRNAs are involved in both low-temperature and high-temperature stress conditions to enhance plant tolerance. In a study on the banana plant, the effect of temperature variations was observed on miRNAs of the plant. In the fruits of this plant, 93 heat-responsive and 22 cold-responsive genes were discovered, of which some miRNA families were responsive toward both stressors, and only a few were stress-specific ones. In cold stress, the number of both upregulating and downregulating miRNAs was the same, while under heat stress, fewer miRNAs were upregulated than downregulated. Out of overlapping miRNAs responsive in both stresses, the 10 miRNAs were opposite in their functions during both stresses. For example, mac-miR164/b was downregulated in fruits under cold stress but was upregulated in heat stress (Zhu et al., 2019). Some more miRNAs that enhance plant survival in fluctuating temperatures are discussed in Table 12.3.

TABLE 12.3
MiRNAs That Enhance Plant Survival in Fluctuating Temperatures

miRNA	Host plant	Temperature	Target gene	Response	Reference
miRNA159a, miRNA164f, miRNA160a	*Cynodondactylon*	High temperature	*GAMYB*, *NAC1*, *ARF17*	Regulate flower development and seed germination, activate plant growth, modulate heat tolerance, regulate auxin signaling	Amini et al. (2023)
miR156	*Medicago sativa*	High temperature	*SPL13*	Prevents antioxidant damage due to heat stress, enhances heat tolerance	Matthews et al. (2019)
miR319	*Solanum lycopersicum*	Low and high temperature	*MYB83* and *CBF1*, *HsfA1a*, *HsfA1b*, and *Hsp90*	Enhance stress tolerance and chlorophyll level, regulate MAD level and electrolyte leakage	Shi et al. (2019)
miR162, *miR403*	*Ipomoea batatas*	High and low temperature	*IbDCL1*, *IbZAT*, *IbMAA*	Respond to high and low temperatures	Yu et al. (2020)
osa-miR531b	*Oryza sativa*	High temperature	*HSP24*	Regulates auxin signaling pathway, enhances lateral root development, enhances stress tolerance	Kushawaha et al. (2021)

(*Continued*)

TABLE 12.3 (*Continued*)
MiRNAs That Enhance Plant Survival in Fluctuating Temperatures

miRNA	Host plant	Temperature	Target gene	Response	Reference
miR160, *miR167*, and *miR2949*	*Gossypium*	High temperature	*ARF17*, *ARF10*, *ARF8*, and *ARF6*	Involved in auxin signaling pathway, respond to heat stress	Chen et al. (2020)
mac-miR394a, *mac-miR169c*	*Musa paradisiaca*	Low and high temperature	*MaFAD2.3*, *MaFAD2.2*	Improve temperature stress tolerance, improve seed germination, high photosynthetic activity, enhance biotic stress resistance	Cheng et al. (2022)
miR528	*Musa paradisiaca*	Low temperature	*PPO*	Regulates peel browning in banana, involved in ROS generation and oxidative processes	Zhu et al. (2020)
MIR398B, *MIR399B*	*Arabidopsis thaliana*	Low temperature	*AT2G04070*, *AT5G20710*	Regulate plant fertility, affect pollen development, regulate plant growth, respond to fluctuating temperatures	Lin et al. (2020)
miR398	*Brassica napus*	Low temperature	*BnCAT1*, *BnCSD1*	Regulate hydrogen peroxide scavenging, improve defense system, respond to cold temperature	Taghvaei et al. (2022)
miR395	*Vitis*	Low temperature	*LAST3*, *ATP sulfurylas*, and *sat-1*	Involved in sulfur transport and metabolism	Ghabooli et al. (2019)
miR160, *miR319*, *miR395*, *miR528*	*Zea mays*	Low temperature	*MYBs*, *GAMYB*, *GRFs*, *TCPs*	Affect cell division, response to chill	Aydinoglu (2020)
MIR398b, *MIR398c*	*Arabidopsis thaliana*	High temperature	*CSDs*	Respond to cell death, infections, oxidative stress, heat stress	Li et al. (2020a)
miR966	*Triticum aestivum*	Low temperature	*mTERF15*	Involved in transcription initiation and chloroplast maturation, regulates growth of plants, responds to chill stress	Li et al. (2019)
Sl-miR164a/b-5p	*Solanum Lycopersicum*	Low temperature	*NAM3*	Regulates ethylene production in plants and cold stress	Dong et al. (2022b)
ath-miR858b, *cme-miR828*, *aly-miR858–3p*	*Cryptomeria fortunei*	Low temperature	*MYBs*	Regulates leaf coloration, secondary metabolites, and flavonoid biosynthesis	Zhang et al. (2021)

12.4.4 MiRNAs Associated with Heavy Metal Stress Tolerance

Worldwide, the pollution from heavy metals has increased due to rapid urbanization and industrialization. Heavy metals pose severe problems for both people and animals and also have a negative impact on soil quality and crop productivity (Asgher et al., 2018; Sharma et al., 2022; Alsafran et al., 2023; Nazir et al., 2023). Although heavy metals are readily accessible and difficult to degrade in the environment, plants can concentrate them, leading to their entrance into the food chain and posing substantial risks to human health (Perković et al., 2022). Diverse contaminants are currently causing detrimental effects on soil, people, plants, and the environment as a whole. The possible harmful metals from these pollutants have raised concerns due to their toxicity, increased mobility, and non-degradable nature (Gupta et al., 2023). The following possible hazardous metals pose severe concerns for agricultural productivity, food safety, food security, and human health: arsenic, chromium, cadmium, cobalt, lead, mercury, molybdenum, manganese, nickel, and silver (Haiying et al., 2022).

Due to their greater mobility, bioaccumulation behavior, and toxicity, the higher concentrations of heavy metals have a deleterious influence on crops (Dikilitas et al., 2019; Bharti and Sharma, 2022; Ali et al., 2023). A variety of plant processes, including nutritional homeostasis, photosynthesis, gas exchange characteristics, osmolytes buildup, secondary metabolites, and antioxidant activities are all impacted by heavy metals (Kumar and Ohri, 2022). Reactive oxygen species generation is another consequence of heavy metals that is considered significant (Yaashikaa et al., 2022). Cellular membranes, proteins, lipids, and different micro- and macromolecules are seriously harmed by heavy metals-induced reactive oxygen species (Ulhassan et al., 2022). Additionally, heavy metals also inhibit plant growth, photosynthetic pigments, and protein and free amino acid accumulation and increase malondialdehyde and hydrogen peroxide accumulation, all of which lower overall productivity (Noor et al., 2022). Current investigations (Table 12.4) have shown that miRNAs regulate several phytohormone signaling pathways, target a number of transcription factors, and regulate genes that participate in defense in response to a variety of biotic and abiotic stimuli (Singh et al., 2021; Chakraborty and Roychoudhury, 2023). According to Zhang and colleagues, plants with sophisticated levels of miR156 expression were more resilient to the effects of Cd stress and accumulated much less Cd in their branches. This is because miR156 affects Cd uptake/transport gene expression and reactive oxygen species levels to positively influence Cd stress tolerance (Zhang et al., 2020). Due to their altered states, lines overexpressing miR408 displayed a high degree of sensitivity to arsenite stress, whereas miR408 knockout lines displayed tolerance as a result of the controlled expression of genes engaged in the sulfur lessening pathway, which affected the buildup of sulphate and glutathione (Kumar et al., 2023b). MiR167a, novel_MiR15, and novel_MiR22 and their targets may be implicated in Cr transportation and chelation, according to research by Nie et al. (2021). Additionally, it was discovered that miR164, miR156a, miR396d, and novel_miR155 were involved in the detoxification of plant Cr. Further, 16 new and 54 well-known miRNAs displayed differential expression in response to Cr stress in *Raphanus sativus*. Under Cr stress, miR394, miR16, miR390, and miR172 have been elevated whereas miR408, miR398, miR156, miR160, miR169, miR159, miR168, miR319, miR397, and miR399 were downregulated (Liu et al., 2015). According to studies, miR156/157 , miR159 , and miR5293 may be crucial in controlling Cr homeostasis since they target several *SPLs* in radish, including *SPL15*, *SPL13*, *SPL9*, *SPL6*, and *SPL3*. Additionally, miR159, miR319 and miR858 may control the expression of numerous genes during Cr stress by targeting MYB3, 13, 101, 104, and 305. Therefore, miRNA-mediated regulation of Cr stress in radish is reduced by targeting different transcription factors (Liu et al., 2015). Additionally, the development of an*spl7* knockout mutant in *Arabidopsis thaliana* allowed researchers to detect the miRNAs that participate in the plant's reaction to the excess heavy metal elements Cu and Cd. Because Cd affects Cu homeostasis, transcription factor *SPL7* is a major regulator of Cu homeostasis. As a result, Cd-induced Cu deficit led to the opposing effects of miRNAs like miR857, miR397a and miR398b/c under Cu and Cd contacts (Gielen et al., 2016). Discovered by means of degradome sequencing and minor RNA high-throughput sequencing were 84 miRNAs and 802 target genes in *Brassica napus*. Amongst these, the target gene of miR167 was found to be the NRAMP metal transporter gene (Ding et al., 2018).

Further, 20 well-known miRNAs and 131 new miRNAs were found among the 151 differentially expressed miRNAs in Cd-stressed maize plants (L63, a tolerant genotype, and L42, a sensitive genotype). According to the findings, Cd upregulated 90 and downregulated 22 miRNAs in the Cd-tolerant genotype L63, while it upregulated 23 and downregulated 43 miRNAs in the Cd-sensitive genotype L42. Additionally, 26 miRNAs showed upregulation in L42 and either remained stable in L63 or experienced downregulation (Teng et al., 2023). Furthermore, 108 miRNAs were either unaffected in L63 or downregulated in L42. The target genes of these miRNAs

TABLE 12.4
Heavy Metal-Responsive miRNA in Different Plants

Heavy metal	miRNA	Target gene	Plant species	Response	References
Cr	miR444f and miR159c	*ABC transporter G29*	Maize	Accumulates more Cr in roots and impairs their growth	Adhikari et al. (2023)
Cd	miR168, miR390, and miR528	Genes involved in polyamines synthesis	Maize	Regulate polyamines synthesis in response to Cd stress	Hassani et al. (2023)
As	miR408	Plantacyanin (*ARPN*) and Laccase3 (*LAC3*)	*Arabidopsis*	MiR408 mutant exhibits tolerance to arsenite stress	Kumar et al. (2023b)
Cd	IamiR-4–3p from spinach	*GST3* and *AWPM19-like*	*Arabidopsis thaliana*	Lower the expression of *AWPM19-like* and *GST3*, which increases Cd absorption and translocation	Shen et al. (2022)
Cd	miR156	*Squamosa promoter binding protein-like (SPL) gene*	*Arabidopsis*	Increases the length of the primary roots, as well as increases biomass, level of chlorophyll, and enzyme activity	Zhang et al. (2022b)
Cr	miR156a, miR164, miR167a, miR396d, novel_miR15, novel_miR22, and novel_miR155	Sulfate transmembrane transporter, ABC transporter, Mg^{2+} transporter, and MT-like proteins and cytochrome P450 gene	*Miscanthus sinensis*	Exhibit involvement in Cr transportation, chelation, and detoxification	Nie et al. (2021)
Cd	miR396	*Growth regulating factor* (*GRF*)	*Arabidopsis thaliana*	Provides adoptive response to *Arabidopsis* against Cd	Pegler et al. (2021)
Ni	miR395, miR398, and miR838	*Cu-Zn/Superoxide fismutase sulphate transporter* and *sulphate adenyltransferase gene*	*Ricinus communis*	Increase expression of target genes to activate reactive oxygen species repression pathways	Çelik and Akdaş (2019)
Cd	miR398	*ACO1*, *CSD2*, *CSD1*, and *CCS* gene	*Brassica napus*	Involved in detoxification of superoxide radicals	Fu et al. (2019)
Mn	miR395	*ATP sulfurylase genes (APSs)* and *sulfate transporter SULTR2*	*Arabidopsis*	Regulate nutritional homeostasis and Mn toxicity response	Gong et al. (2019)
–	miR159, miR9664, and miR9774	*TaHMA2;1*	*Triticum aestivum*	Found to be involved in carbon fixation pathways, tyrosine and thymine metabolism, and ubiquitin-mediated proteolysis	Zhou et al. (2019b)
Cd	miR166	*Homeodomain containing protein4* (*OsHB4*) gene (*Os03g43930*)	Rice	Diminished translocation of Cd from roots and shoot and Cd buildup in grains	Ding et al. (2018)

TABLE 12.4 (*Continued*)
Heavy Metal-Responsive miRNA in Different Plants

Heavy metal	miRNA	Target gene	Plant species	Response	References
Cd	33 known and 44 novel miRNAs	460 putative target genes	*Brassica napus*	Involved in ion transport, transcription factor regulation	Jian et al. (2018)
Cd	miR268	Macrophage protein 3 (*NRAMP3*)	Rice	Negatively affects rice's ability to withstand Cd stress	Ding et al. (2017)
Cd	miR398	*CSD1* and *CSD2*	*Arabidopsis*	Involved in melatonin-mediated improvement of redox imbalance	Gu et al. (2017)
Cu	miR397, miR398, and miR408	*vi-LAC4*, *vvi-LAC17*, *vvi-PC*, and *vvi-LAC12*	Grapevine	Cu-responsive miRNAs were suppressed leading to upregulation of genes encoding Cu-containing proteins	Leng et al. (2017)
Zn	miR6189, miR7502, miR3437, miR1535, and miR3635	*HMA1*, *PPO*, *DHAR*, *ARF2*, and *KT*	*Solanum nigrum*	Involved in regulation of catabolism and metabolism of phenylpropanoid and lignin and programmed cell death in response to reactive oxygen species	Xie et al. (2017)

were the most highly abundant in peroxisomes, glutathione metabolism, ABC transporters, and the ubiquitin-protease system. Target genes implicated in the peroxisome pathway and GSH metabolism may be particularly important for L63's ability to tolerate Cd. Additionally, a number of ABC transporters that could be involved in Cd absorption and transport were found (Teng et al., 2023). In Cd-stressed wheat, miRNAs play a significant role by controlling targets like *TaHMA2;1*. Zhou et al. (2019b) discovered 1561 differentially expressed genes (DEGs) and a total of three known and 19 new differentially expressed miRNAs through a comparison of transcriptome and miRNA analysis.

12.4.5 Other Relevant Abiotic Stresses

Aside from the major abiotic factors or stressors mentioned earlier, plants are also impacted by a variety of abiotic and biotic stresses (Kaur et al., 2022; Kumar et al., 2023a; Sharma et al., 2023b). MiRNA has been stated to have a significant function in different stress tolerance, particularly nutrient stress (Table 12.5). Plants need nutrients to balance their growth, metabolism, and other developmental activities while maintaining the matrix of demand and supply. It should be noted that plants need the right combination of environmental factors in order to successfully develop, grow, and reproduce. Plants need at least 17 important nutrients during their existence, which are absorbed from the soil by their roots. These include copper, calcium, carbon, hydrogen, iron, molybdenum, magnesium, manganese, nitrogen, nickel, oxygen, potassium, phosphorus, sulphur, and zinc (Jamla et al., 2023). These elements, which are often absorbed in the form of compounds or salts, are essential for controlling and promoting the molecular, cellular, and physiological responses of plants. The elements are divided into macronutrients and micronutrients based on what the plant needs (Molina et al., 2019; Wulandhari et al., 2019; Tiwari et al., 2020). The main sources of nutrients available to plants are soil, air, and water. Storing organs such as vacuoles/cytosol, xylem tissues, and phloem tissues are part of a systemic signaling network that actively senses, absorbs, stores, and transports nutrients, facilitating their delivery to plants through proper compartmentalization of the root and branch systems (He et al., 2022; Jamla et al., 2023). However, plants may experience short-term or

TABLE 12.5
Nutrient and Hypoxia Stress-Responsive miRNA

Other stress	miRNA	Target gene	Plant species	Response	References
Hypoxia (waterlogging)	miR156, miR166, miR167, and miR399	Genes involved in the physiological and metabolic functions of cells	*Solanum melongena*	Affected the seedlings' capacity to withstand stress in various waterlogging situations	Jiang et al. (2023)
Phosphate deficiency	miR156, miR171, miR399, and miR2111	–	*Trifolium pratense*	Provided tolerance to phosphate limited conditions	Demirkol (2022)
Hypoxia (waterlogging)	miR528, miR408, miR393, miR167, miR164, and miR159	–	Maize	Involved in stress reactions and root formation	Liu et al. (2012)
Hypoxia (submergence)	miR172a	*ERTF RAP2–7 ISOFORM X2*	Maize	Shifted aerobic respiration to a less effective anaerobic fermentation	Azahar et al. (2020)
Hypoxia (submergence)	miR169c, miR167c, miR166b, and miR159a	–	Maize	Participated in development, growth, and the reaction to oxidative stress	Sepúlveda-García et al. (2020)
Iron deficiency	miR408	*LAC3*, *LAC12*, and *LAC13*	*Arabidopsis thaliana*	Altered hydrogen peroxide levels	Carrió-Seguí et al. (2019)
Boron toxicity	miR319 and miR171	Myeloblastosis (MYB) transcription factor gene, a cation transporting ATPase gene, and a SCARECROW-like protein gene	*Citrus grandis*	Are required for citrus to be resistant to long-term B poisoning.	Huang et al. (2019)
Nitrogen deficiency	osa-miR169	Genes encoding the nuclear transcription factor Y subunit alpha	Rice	–	Shin et al. (2018)
Phosphate deficiency	miR399	*AtUBC24* (*PHO2*)	*Arabidopsis*	Influenced phosphate uptake	Shukla et al. (2018)
Nitrogen deficiency	miR169o	*LOC_Os03g07880* and *LOC_Os03g48970 encoding* NF-YA1 and NF-YA4	Rice	Biomass accumulation and enhanced content of nitrate and amino acid	Yu et al. (2018)
Magnesium deficiency	miR158	SPFH/Band 7/PHB domain-containing membrane-associated protein family	*Citrus sinensis*	Adaptation to Mg deficiency by maintaining root growth, triggering DNA repair, inhibiting root respiration, and activating transport related genes	Liang et al. (2017)
Phosphorus deficiency	miR156, miR166, and miR171	Squamosa promoter-binding-like protein, PHB, PHV, and SCLs	Maize	Modified maize morphology	Li et al. (2016)

long-term shortages of one or more nutrients, which can impair their ability to grow and develop. Additionally, the production and accumulation of significant secondary metabolites depend on receiving the right amounts of nutrients at different phases of plant development (Li et al., 2020b; Pant et al., 2021). It has been discovered that in order for plants to survive, they have developed defense systems against stressful environments where nutrition supply is less than ideal (Francis et al., 2023). Additionally, the right ratios and adequate dosages are required to prevent nutritional deficiencies and toxicity. Therefore, in the affected plant varieties, nutritional balance needs to be closely monitored and controlled by various biochemical, physiological and molecular adaptations (Gong et al., 2020). Cellular component and cellular organelle interactions become vigorous under stress, including a wide range of cell-level and molecular-level processes. The involvement of nutritional stress-responsive small regulators, the miRNAs, in vegetation species has been strongly established by key findings from high-throughput sequencing techniques used in conjunction utilizing bioinformatics tools and databases (Jamla et al., 2023; Ma and Hu, 2023). For instance, under abiotic and Mn stress conditions, 37 miRNAs displayed differential expression. Eleven miRNAs were activated and 11 others were inhibited by Mn exposure. New Mn-responsive miRNAs, including miR1508, miR1515, miR1510/2110 and miR1532, have been characterized. Key targets of Mn-responsive miRNAs include receptor kinases, calcium-dependent protein kinases, and proteins with leucine-rich repeats (Valdés-López et al., 2010). Moreover, recent research finds that *Arabidopsis* miRNAs that are involved in cell development, nutritional balance, and ion transport respond differently to Mn stress conditions (Gong et al., 2019). Further, numerous earlier investigations have demonstrated that nitrogen triggers the synthesis of miRNA. MiR167 can inhibit root growth because it regulates how adventitious plants react to nitrogen and even the N-metabolizing enzymes that are created after nitrification and absorption (Yousuf et al., 2021). Nitrogen signals that are sent during nitrification and absorption activate miR393. In line with miR393- or afb3–1-mutant overexpression, nitrate does not have any effect on the formation of primary roots; however, in reaction to nitrate treatment, it does control horizontal root formation (Vega et al., 2019; Islam et al., 2022). Similarly, with *Arabidopsis* showing the most significant regulation for miR172 and miR319, Kayihan et al. (2019) demonstrated that the amount of microRNAs expressed for transcription factors associated to jasmonate and ethylene metabolism was dramatically elevated underneath moderate B poisoning but not severe B toxicity (Kayihan et al., 2019). Scientists have shown that poor dietary potassium utilization alters the expression of miRNA in cotton and wheat. The expression of 16 of the 20 miRNAs were changed as a result of K deficiency therapy. Through molecular pathways, wheat improves root development and nutrient uptake in response to K shortage. MiRNAs have a significant impact on root development in peanut plants and are crucial in K-deficient environments. It is hypothesized that miR156, miR390, miR160, miR164 and miR393 are elevated in response to a potassium shortage (Fontana et al., 2020; Thornburg et al., 2020). Another study reported that in *Panicum virgatum* the miR399 family (miR399-3, 4, and 6 subfamilies) significantly mediated stress responses by targeting aminocyclopropane-1-carboxylate synthase/ethylene-overproducer 1 (*ACS9/ETO3*) homolog, ubiquitin-conjugating24/phosphate overaccumulator2 (*UBC24/PHO2*), $Ca2^{+}/H^{+}$ (Ding et al., 2021). Further, Zeng et al. (2019) reported that miR528-5p, miR397b-5p, miR398a-5p, and miR171g-5p genes in *Oryza sativa* responded to Zn deficiency in a similar manner (Zeng et al., 2019). Different miRNA families were discovered in *Citrus sinensis* roots, together with their functions and objectives in response to magnesium insufficiency. Citrus exhibits Mg-shortage tolerance due to Mg-responsive miRNAs that were found to be participating in root signal transduction. Different expressions were produced by the miRNAs and their targets in response to magnesium insufficiency. The miRNAs miR158 and miR2919 were induced, which prevented root respiration; miR780, miR6190, miR1044, miR5261 and miR1151 were downregulated to recover the antioxidant system; miR6190, miR6485, miR1044, miR5029 and miR3437 control transporter-related proteins; miR1044, miR5261, miR1151 and miR5029 were downregulated to sustain ubiquitin.

Further, research study revealed that during submergence stress, certain miRNAs from the miR164 family, including miR164c-5p, miR164b-5p and miR164a were downregulated in maize seedlings (Azahar et al., 2020). Another study reported that target genes of hypoxia-responsive miRNAs showed that cellulose synthase, auxin, and mitochondrial ribosomal associated genes (*Csa2G315390*, *Csa6G141390*, *Csa4G053280*, and *Csa6G310480*) are likely involved in the enhancement of the cucumber seedlings' hypoxia tolerance with the injection of exogenous calcium (He et al., 2023). Similarly, miR398 was shown to be suppressed in adult *Arabidopsis* plants when submerged. Inhibitors of mitochondrial activity also inhibited miR398 expression, indicating that mitochondrial respiration is crucial in the control of miRNA production during submersion (Loreti et al., 2020). Hou et al. (2019) reported that 33 known miRNAs were weakly expressed, however only three miRNAs had sophisticated expression in wild tomato root treated with hypoxia than in untreated wild tomato root. In part through reactive oxygen species-mediated oxidative burst, increased UV-B radiation as a result of ozone-layer depletion adversely affects plant growth and development. Wang et al. (2013) identified a novel wheat Tae- miR6000 and demonstrated the expression diversity of this gene during UV-B treatments. Additionally, they discovered that UV-B responsive miRs include miR395, miR171, miR167a, miR164, miR159 and miR156 . At various times following UV-B exposure miR171, miR167a and miR159 were considerably elevated, but the other three miRNAs were downregulated (Liang et al., 2017).

12.5 MOLECULAR MECHANISMS OF MIRNAS IN ABIOTIC STRESS RESPONSE

Under natural conditions, plants are exposed to a variety of abiotic stressors that are mostly harmful to vegetation growth and development (Ali et al., 2023; Fadiji et al., 2023). Numerous physiological, molecular, and metabolic responses are significantly impacted by abiotic stressors (Ali et al., 2023; Yang et al., 2023). There has been significant development in understanding the complicated mechanisms of the stress response, especially in finding genes that are responsive to stress with the aid of biotechnological methods (Raza, 2021). Via base pairing with supplementary miRNA targets, such as transcription factors, microRNAs play an indispensable role in post-transcriptional regulation (Shaik et al., 2022) (Figure 12.1). Innovative strategies for enhancing plant responses to abiotic challenges may be developed with the aid of an understanding of the role of miRNAs in abiotic stresses (Noman et al., 2017; Ahmad et al., 2023). Under abiotic stressors, microRNAs participate in a numerous cellular and metabolic processes, including flowering, morphogenesis, signal transmission, and gene feedback control. MicroRNAs are a class of short, single-stranded RNAs that do not code for proteins and have an average length of 18 to 25 nucleotides (Shang et al., 2023). With unique stem-loop assemblies, microRNAs are created via antecedence. At different stages of plant development, miRNAs show a significant role in regulating gene expression in plants. For example, 959 founding members representing 178 miRNA families were discovered in *Arabidopsis* (*Arabidopsis thaliana*), rapeseed (*Brassica napus*), earth mosses (*Physcomitrella patens*), maize (*Zea mays*), black cottonwood (*Populus trichocarpa*), and barrel clover (*Medicago truncatula*; Islam et al., 2019). Negative gene expressions at the post-transcriptional level cause substantial damage to microRNAs during plant development, making them a prevalent molecular tool in modern biotechnology for investigating signal transduction, environmental extremes, stress responses, protein deprivation, biogenesis, and pathogen incursion (Haider et al., 2019). Over the past 10 years, researchers have focused on finding miRNAs that have a strong potential to increase resistance to abiotic stressors like cold, drought, and heavy metals. Various studies have discovered that plant cells' growth, transcription, protein breakdown, mechanisms for abiotic stress tolerance, and detoxification are all regulated by miRNAs (Chen and Yu, 2023; Zhang et al., 2022a).

Drought stress can alter development and growth of plants, uptake of minerals, photosynthesis, hormonal metabolic processes, protein metabolism, and other physiological processes in plants (Batool et al., 2022). Several studies have demonstrated that various types of stress can cause the miRNAs to function in various types of plants (Zhang, 2015). Through NF-Y, miR169 contributes

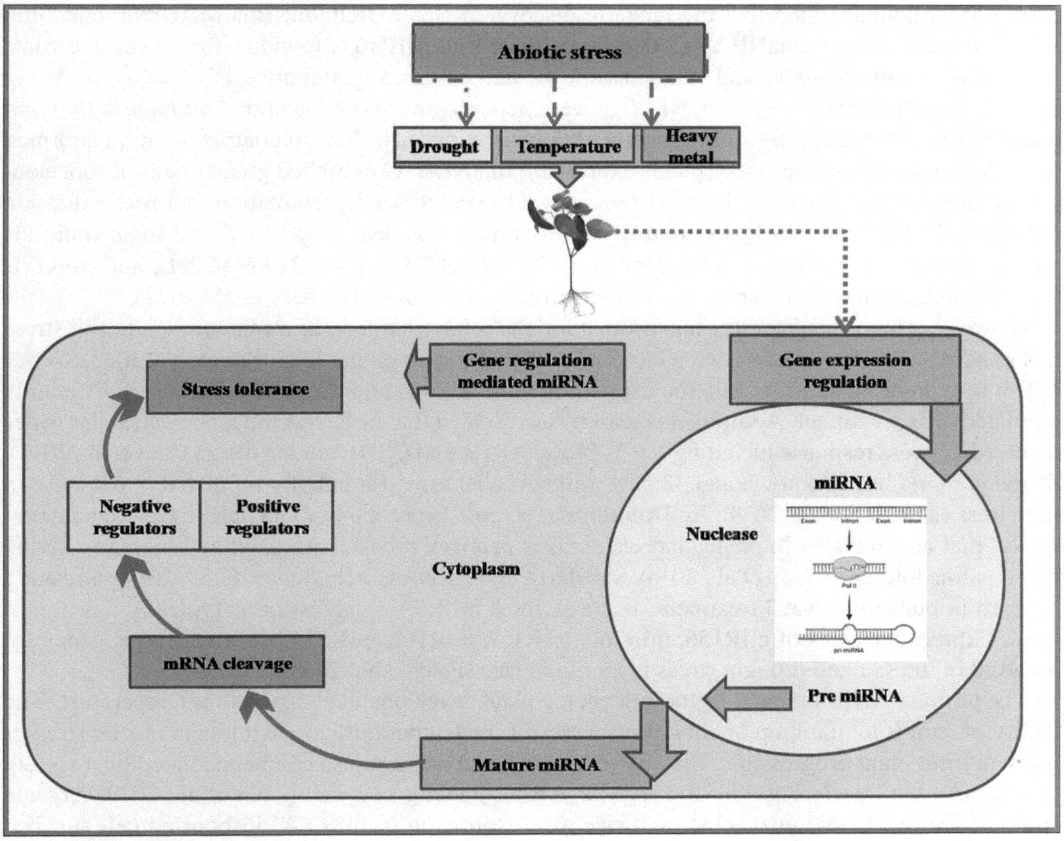

FIGURE 12.1 Molecular mechanisms of miRNAs in abiotic stress response.

to the resistance of plants to drought. NFYA5 has an miR169 target site that directs mRNA breakdown or translation suppression (Ma et al., 2022). Additionally, rice (*Oryza sativa*) that has had microRNA166 knocked down exhibits morphological alterations that increase its tolerance to drought. For instance under drought stress, rice frequently exhibits a leaf-rolling phenotypic trait mainly seen in the miR166 knockdown strain (STTM166). However, overexpressing miR166 can make rice more resistant to drought, as well as cause leaf rolling and a reduction in xylem diameter, which is a typical response of plants to water stress (Zhang et al., 2018). In a study, the apical leaf of plants that had been depriving themselves of water for 14 days was examined for the prediction and characterization of miRNAs and their putative target genes. A total of 81 miRNAs have been recognized, out of which 29 of them are particular to oil palm, including two newly discovered ones (egu-miR28ds and egu-miR29ds) that have just been described. Regarding the expression pattern, 62 miRNAs showed substantially different levels of expression in response to drought stress, with 57 miRNAs downregulated and five upregulated (egu-miR29ds, miR529b, miR396e, egu-miR19sds, and miR159b). The majority of probable miRNA-target genes in oil palm under water scarcity were anticipated to be transcription factors, including NF-Ys, HOXs, and MYBs. A number of putative target genes with comparable behavior under salt and drought conditions were finally identified by the correlation analysis investigation. The lncRNAs and proteins associated with stress tolerance, stress memory, control of ROS signaling, and defense response are encoded by the genes that are increased by these two abiotic stresses (Salgado et al., 2022). Furthermore, in another study by means of 5′ RACE and transient expression in *Nicotiana benthamiana*, Wang et al. (2022) demonstrated that the recently discovered Bna-MIR169m, Bna-MIR169n, and Bna-MIR169t cleave target BnaNF-YAs. The majority of Bna-MIR169s found in the leaves and roots responded to salt, drought, and ABA treatment, according to quantitative PCR analysis (Wang et al., 2022). Bna-MI169n and BnaNF-YA9 were heterologously produced in *Arabidopsis thaliana* plants to test if these controls are transmitted downstream of NF-YA. In contrast to untransformed *Arabidopsis*, under abiotic stress, plants expressing BnaNF-YA9 exhibited greater rates of root elongation and seed germination. Bna-MI169n showed overdue seed germination and root extension (Wang et al., 2023). Similarly, by controlling wax production, leaf senescence, and sugar transport during drought, respectively, stolonifera's miR530/TGA1/CER1, miR319/NAC29L, and miR171/SUC2L modules may be essential for sesame drought resistance (Baghery et al., 2022).

However, a number of studies suggest that miRNAs have a role in how plants adapt to salt stress and enhance their stress tolerance (Gao et al., 2022; Nguyen et al., 2023; Sun et al., 2023). When radish is exposed to saline stress, the expression of 49 known and 22 novel miRNAs is differently regulated after treatment. Additional research has revealed that miRNAs influence particular genes involved in stress response including NF-Y, SPLs, ARFs, and CSD1, among others (Sun et al., 2015). *Oryza glaberri* has 29 known and 32 new miRNAs that are differentially regulated in response to salt stress (Mondal et al., 2018). In *Arabidopsis*, ectopic expression of GhCHR lowers the amount of Na^+ that accumulates in plants and encourages primary root development and biomass, giving plants saline tolerance (Gao et al., 2016). Similarly, 10 miRNAs, including miR169, are significantly reduced in buds after NaCl treatment. In roots, most miRNAs' expression is typically downregulated. Expreesion levels of miR156, miR160, miR167, miR169, miR528, and their target genes are involved in the salt and drought stress response in rice (Peng et al., 2018).

The primary environmental factor influencing plant development and growth is temperature. The ability of fruit is to maintain quality after harvest. Low temperature is a significant abiotic stressor that can limit plant progression. The low-temperature stress response can be mediated by a variety of miRNAs via interfering with the IAA or ABA signaling pathway (Abla et al., 2019; Argenta et al., 2023). In *Arabidopsis*, pine, and rice, overexpression of miRNA156 boosted cell survival and growth rate during cold stress. Targeting OsSPL3, which positively regulates the expression of *OsWRKY71*, a negative regulator of the transcription factors *OsMYB2* and *OsMYB3R-2*; *OsmiR156* enhanced plant cold tolerance (Zhou et al., 2019c). Similarly, in sugarcane six pathways such as calcium signaling pathway, ABC transporter, phosphatidylinositol signaling system, peroxisome, and plant hormone signaling are affected by miRNAs at low temperature response (Yang et al., 2017). Moreover, in the study of poplar, it was discovered that DNA methylation influences the expression of its target genes via controlling the production of miRNA genes, which react to temperature stress and aid plants in surviving in harsh conditions (Ci et al., 2015).

Excess accumulation of heavy metals in plants poses potential risks such as stunting their growth and productivity, enhancing the food chain, harming human health, promoting sickness, and causing cell damage. Both essential and non-essential metal elements are present in this context. Essential metals like zinc, manganese, and copper, as well as non-essential metals like cadmium (Cd), lead (Pb), and mercury (Hg), are necessary for numerous physiological functions in living organisms (Ghazzal et al., 2022; Vega et al., 2022; Alsafran et al., 2023). In a study, it was found that plants overexpressing miR156 showed increased resistance to Cd stress and significantly decreased Cd accumulation in their tissues. This is attributed to the influence of miR156 on Cd uptake/transport gene expression and reactive oxygen species levels, which positively affect Cd stress tolerance (Zhang et al., 2020). In another investigation, MiR408 knockout mutants demonstrated tolerance by regulating the expression of genes involved in the sulfur reduction pathway and affecting the accumulation of sulfate and glutathione. Conversely, plants overexpressing miR408 displayed severe susceptibility to low sulfur (LS), arsenite As(III), and LS+As(III) stresses (Kumar et al., 2023b). Moreover, Nie and colleagues discovered MiRNA167a, novel_miRNA15, and novel_miRNA22, and their targets are implicated in the transport and chelation of chromium (Cr). Additionally, novel_miRNA155, miRNA156a, miRNA164, and miRNA396d were found to play a role in plant detoxification of Cr (Nie et al., 2021).

12.5.1 Key Genes under the Influence of Stress-Responsive MiRNAs

As a result of studies on the physiological and molecular mechanisms of abiotic stress tolerance, a large number of genes that are associated with adaptation to stress have been identified (Ali et al., 2022). It has been demonstrated that methods like microarrays are invaluable for compiling a list of genes connected to stress. Certain stresses are specific to certain genes, while others are shared among different stresses (Ahmad, 2022). Both abiotic and biotic stress responses share many genes (Figure 12.2). This demonstrates the intricacy of plants' stress reactions and adaptations (Tariq et al., 2022). Abiotic stress tolerance is mediated by a cascade of genes that begins with the perception of stress and progresses through transcriptional activation of downstream genes to adaptation and tolerance (Manna et al., 2023). Abiotic stress is largely influenced by microRNAs. Under abiotic stress, it is a key gene expression regulator in plants (Ali et al., 2022). Mature miRNA has a hydroxyl group linked at the 3 end and a phosphate group at the 5 ends. A certain 30 UTR region of mRNA is the target of a miRNA, which causes post-transcriptional regulation to be silenced, leading to mRNA degradation and/or translation suppression (Yu et al., 2019a). In a study it was observed that dehydration, cold, high salinity, and ABA treatments were found to significantly upregulate miR393 in a variety of plant species (Zhang, 2015). Similarly, drought stress decreased the expression of miR169 in *Arabidopsis*, and its target nuclear factor YA5 (NF-YA5) was considerably enhanced by drought stress (Nadarajah and Kumar, 2019). However, in contrast to plants of the wild type, the miRNA169 overexpression were more vulnerable to drought stress by leaf water loss, while the miRNA169a-targeted NF-YA5 overexpression boosted drought stress tolerance (Ni et al., 2013). Similar to this, *Arabidopsis thaliana* dramatically increased resistance to drought stress as a result of the characterized GmNF-YA3 gene, a target of miR169. The researchers also noted that *Arabidopsis's* heightened sensitivity to salt and ABA stressors was caused by the overexpression of GmNF-YA3 (Ni et al., 2013). In addition it was observed that through the downregulation of its four targets, NF-YA1/2/3 and multidrug resistance-associated protein gene 1 (MRP1), tomato miR169 overexpression lines produced drought stress. Comparing plants of the wild kind, tomato plants with constitutive overexpression of the miR169c gene were highly tolerant to drought stress because less water was lost by transpiration, the stomatal opening, and leaves (Rao et al., 2020). However, teosinte branched/cycloidea/proliferating factors (TCP) AsTCP14, AsPCF8, AsPCF6, AsPCF5, and a homolog of a rice NAC (NAM, ATAF, and CUC) domain gene AsNAC60 were at least four putative turf miR319 target genes that were significantly downregulated, according to a gene expression study, which revealed that the improved abiotic stress tolerance was the result of this. Furthermore, in *Agrostis stolonifera*, overexpression of Osa-miR319 increased leaf dimensions, increased stem dimensions, and decreased the number of tillers (Rao et al., 2020; Anwar and Kim, 2020). Furthermore, in *Agrostis stolonifera*, overexpression of Osa-miR319 increased leaf magnitude, increased stem dimensions, and decreased the total number of tillers (Zhou et al., 2013).

In a study it was observed that plants with the Osa-miR319 gene overexpressed have higher cold stress tolerance (4 C) following chilling acclimation (12 C) than plants with the wild-type gene (Sun et al., 2020). However, MiR398 is engaged in plant thermo-tolerance processes, particularly for the defense of the reproductive tissues. According to research, miR398 suppressed the expression of its target CSD (copper/zinc superoxide dismutase) genes (Guan et al., 2013). CCS is a gene which encodes a copper chaperone for CSD1 and CSD2, as well as CSD1 and CSD2. As a result of enhanced buildup of heat stress transcription factors and heat shock proteins as well as a smaller amount of floral damage, the csd1, csd2, and ccs mutant plants showed stronger heat stress tolerance than wild-type plants (Guan et al., 2013). Thus, in the end, we can say that miR398 and the gene that it targets can help crops become more tolerant to heat stress.

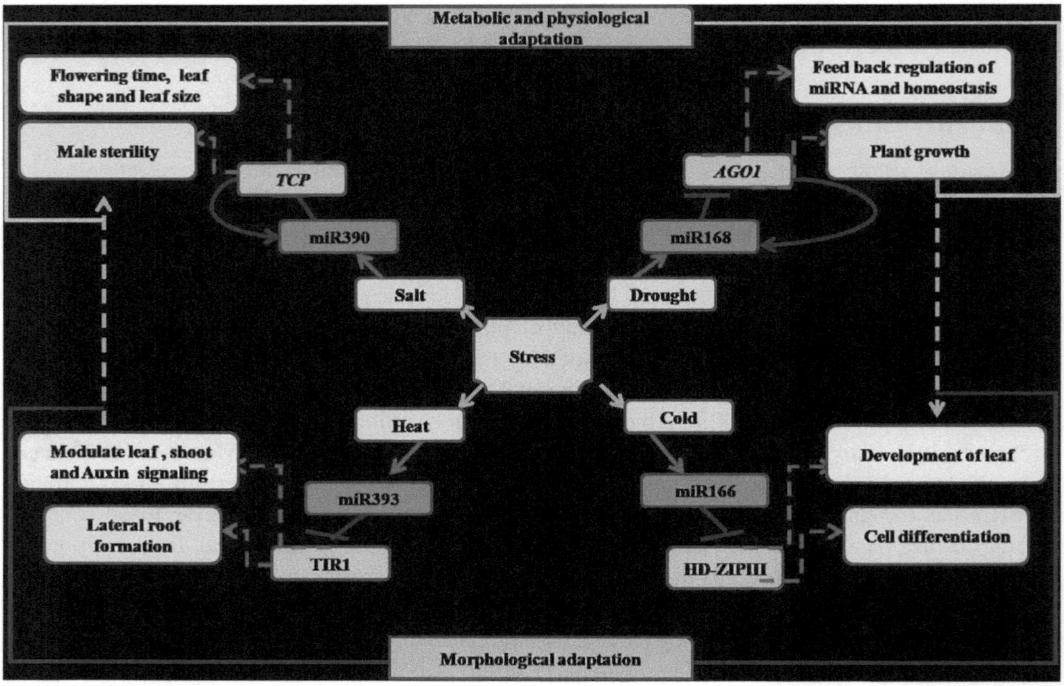

FIGURE 12.2 Key genes under the influence of stress-responsive miRNAs.

12.5.2 Interplay with Other Regulatory Molecular Players

Several miRNAs in cassava have been revealed to mimic lncRNA targets in drought circumstances. Indeed, Ding et al. (2019) proposed that long non-coding RNAs (lncRNAs) in cassava may operate via miRNAs in response to drought stress. For example, they discovered 11 drought-specific differentially expressed lncRNAs that functioned as potential target mimics of 24 miRNAs in cassava under drought stress, with miR156, miR164, miR169, miR172, and miR395 all implicated in abiotic stress response (Wang et al., 2017). MiR156 targets SPL genes, miR164 targets *Myelocytomatosis 2* (*MYC2*) genes, and miR172 targets AP2-like regulatory factor, whose expression is boosted by SPL genes, whereas miR169 and miR398 target MYC2 and chloroplastic superoxide dismutase 2 (CSD2) genes, respectively (Ding et al., 2019). Wu et al. (2019) discovered four lncRNAs with binding sites for two miRNAs (miR156 and miR159) that respond to both ABA and drought stress. Xiao et al. (2019) also discovered 21 possible miRNA precursors in lncRNAs in cassava under drought stress, including miR162, miR166, miR408, miR477, miR530, miR171, miR159, miR535, miR169, and miR167. Differential expression of these miRNAs and their target genes has been linked to drought stress tolerance in plants. Under drought stress circumstances, for example, miR169 modulates stomatal opening and transpiration rate. According to Li et al. (2008), miR169 targeted NF-YA family genes, and overexpression of NF-YA5 and NF-YA3 or downregulation of miR169 improved drought stress tolerance in *Arabidopsis* and soybean, respectively. The miR169 negatively controlled MeNF-YA3 in cultivated cassava cultivars, with lower miR169 expression found in wild cassava progenitor, indicating that the progenitor was more drought tolerant than the cultivars (Chen et al., 2015). Figure 12.3 summarizes the general interplay of miRNAs with other regulatory factors under abiotic stress. Li et al. (2017) found 12 lncRNAs as precursors to 11 known cassava miRNAs, together with miR169d, miR167g, miR166h, miR160d, and miR156g which may be implicated in stress response. They also discovered 16 lncRNAs that might function as miRNA mimics when bound by conserved miRNAs including miR164, miR169, miR2275, and miR1446 (Li et al., 2017). Drought stress substantially increased the expression of miR164, whereas drought stress lowered the expression of miR169 and miR2275. Drought stress elevated the production of lncRNA340, which was followed by a rise in miR169-targeted NF-Y genes following drought treatment. Drought stress also enhanced the abundance of lncRNA119, which was associated with increased mRNA abundance of relevant miR2275 targets including Manes.02G04600 and Manes.14G106000 (Li et al., 2017). TCONS_00068353 also served as a target mimic for miR156k and miR172c, allowing them to modulate many abiotic stress-responsive genes (Li et al., 2017).

Recent studies have found that small interfering RNAs (siRNAs) play a role in conferring resistance to abiotic stress (Sunkar and Zhu, 2004). Li et al. (2009) investigated the effect of RNAi-mediated downregulation of RACK1, a highly conserved scaffold protein with multiple roles in plants, including growth and development. Transgenic rice plants with lower RACK1 expression exhibited increased resistance to drought stress compared to non-transgenic rice plants. The experimental study also found that oxidative stress and malondialdehyde (MDA) generation were much lower in transgenic rice plants than in non-transgenic rice plants, whereas superoxide dismutase activity was significantly higher. This revealed that RACK1 adversely controlled the redox system associated with rice tolerance to drought stress. Yu et al. (2019b) studied the involvement of a dehydrin gene (WZY2) in osmotic pressure resistances in wheat. When exposed to osmotic stress, transgenic wheat with WZY2 downregulation had lower relative water content and oxidative associated enzyme activities and greater MDA content than wild-type bread wheat (Zhengyin No.1). WZY2 overexpression in *Arabidopsis* also boosted drought tolerance, demonstrating that dehydrins are important in reducing osmotic injury. Wang et al. (2009) discovered and identified the *Arabidopsis* hydroxyl pyruvate reductase (AtHPR1) promoter, which is expressed specifically in shoots.

Furthermore, by employing an RNAi construct containing the AtHPR1 promoter to induce conditional and specific downregulation of a farnesyltransferase (FTA) subunit, they successfully safeguarded canola yield against dehydration stress under field conditions. Senthil-Kumar et al. (2007) used virus-induced gene silencing (VIGS) to identify and characterize water deficit-induced peanut genes in *N. benthamiana*. The researchers identified 25 genes in *N. benthamiana* that, when silenced, caused greater membrane injury in plants under water shortage stress, indicating a function in drought resistance. During water shortage stress, homologs of flavonol 3-O-glucosyltransferase (F3OGT), salt inducible protein, alcohol dehydrogenase, and heat shock protein 70 displayed more evident wilting signs than control plants. AtPARP1 or AtPARP2 double-stranded RNA (dsRNA) constructs in contradiction of poly (ADP-ribose) polymerase were converted into *Arabidopsis* and *Brassica napus* paths, which demonstrated enhanced tolerance to a variety of abiotic stressors, including drought (Block et al., 2005). Downregulation of PARP efficiently maintained low ADP ribosylation functioning and enhanced stress resistances due to decreased NAD breakdown, which controlled energy homeostasis. In another part of the study, miRNAs were employed to silence maize PARP1, resulting in increased drought tolerance in transgenic plants (Anami and Inzé, 2010). Tolerance to salt-stress sensitive small RNAs (sRNAs) have been found and used to confer salt stress tolerance to plants utilizing a variety of approaches. Salt-induced transcription factor gene MYB74 was demonstrated to be controlled by a 24-nucleatide extended siRNA via RNA-directed DNA methylation in *Arabidopsis* (Xu et al., 2015). Silencing the floral meristem-specific cytokinin oxidase (OsCKX2) gene in rice using siRNA resulted in the creation of saline-stress-resistant rice transgenic plants with no harvest penalty (Joshi et al., 2018).

Adaptability to low temperatures was improved by RNAi-induced inhibition of the auxin response factor gene (Xia et al., 2014). Heat stress response in plants has been demonstrated to be regulated by modifying the Ca^{2+} signaling pathway or genes encoding different transcription factors connected to floral growth and total yield (Kaur et al., 2019, 2022). Li et al. (2014) discovered that two genes targeted by TAS1-derived siRNA, heat-induced TAS1 target1 (HTT1) and HTT2, have a role in temperature stress in *Arabidopsis* and are upregulated in response to heat shock. TAS1a overexpression reduces thermo-tolerance via targeting HTT genes with tasiRNA. HTT1 and HTT2 upregulation, on the other hand, promotes higher thermo-tolerance by upregulating many

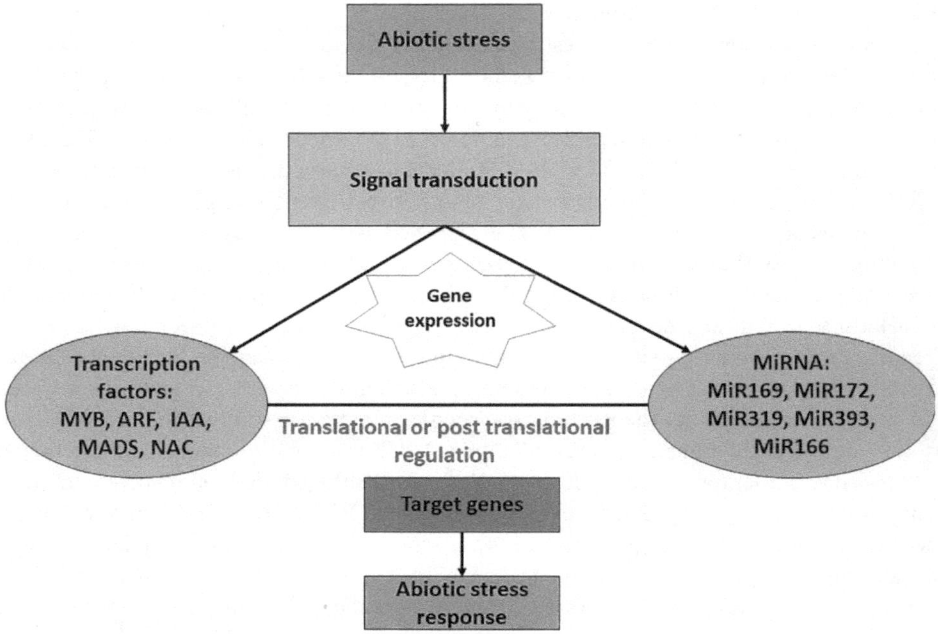

FIGURE 12.3 Interplay of MiRNAs with other regulatory factors. A, agamous, D: deficiencies; ARF, auxin response factors; IAA, indole-3-acetic acid, M, minichromosome maintenance factor 1; MYB, myeloblastosis; S, serum response factor (Gupta et al., 2021).

heat stress transcription factor (Hsf) genes. A group created alternative transgenic tomatoes that altered the expression of the heat stress transcription factor HsfA1, and silencing the HsfA1 gene enhanced plant responsiveness to higher temperatures by decreasing the production of chaperones and HSFs (Mishra et al., 2002). Genetically modified rice containing the phytochelatins synthase (PCS) RNAi construct dramatically decreased cadmium (Cd) buildup (Li et al., 2007). A recent study using RNAi to silence two PCS homolog genes in rice, OsPCS2 and OsPCS1, discovered their functions in arsenic tolerance (Yamazaki et al., 2018). RNAi inhibition of the rice MATE family transporter gene, *OsMATE2*, resulted in lesser arsenic buildup (36.9%–47.8%) in transgenic rice (Das et al., 2018).

12.5.3 Feedback Loops and MiRNA-Mediated Signal Amplification

Drought, salt, severe temperatures, and nutrient deficits are examples of abiotic stressors that can disturb normal cellular processes in plants (Tikoria et al., 2022, 2023a, 2023b). Plants have evolved complicated regulatory systems, such as feedback loops and microRNA mediated signal amplification, to survive and adapt to severe stress circumstances. Figure 12.4 outlines a general feedback loop of miRNAs in response to abiotic stress.

12.5.3.1 Feedback Loops During Abiotic Stress Response

Feedback loops are regulatory systems in which the output of one activity influences another (Ali et al., 2023). Feedback loops can influence multiple signaling pathways and gene expression networks in plants throughout abiotic stress response. Here's how feedback loops function under abiotic stress:

- *ROS signaling loop*: Abiotic stress frequently causes the manufacture of reactive oxygen species in plant cells (Dhiman et al., 2021, 2023). ROS are secondary messengers that can

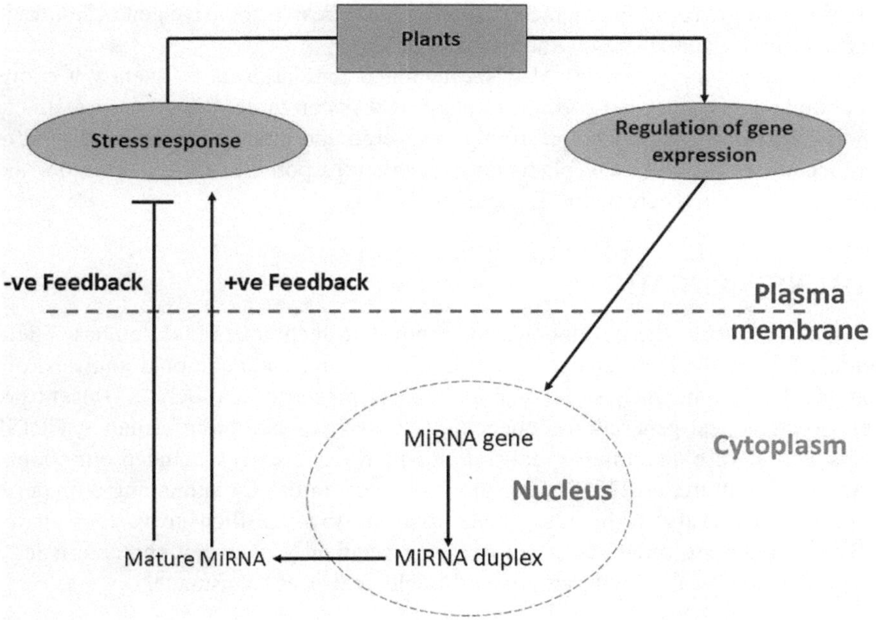

FIGURE 12.4 Feedback loop of miRNA in response to abiotic stress (Wani et al., 2020).

activate stress-responsive gene expression. ROS feedback loops aid in the regulation of ROS levels and the prevention of oxidative damage. ROS, for example, can increase the development of antioxidant enzymes such as superoxide dismutase, which helps neutralize excess ROS (Jain et al., 2018).

- *ABA signaling*: Abscisic acid is a phytohormone that shows an imperative role in drought and other stress responses. Feedback loops govern ABA production and response in ABA signaling (Salvi et al., 2021). For example, ABA stimulates the expression of genes involved in its manufacture, resulting in a positive feedback loop that boosts ABA levels.

12.5.3.2 MiRNA-Mediated Signaling

MicroRNAs (miRNAs) are tiny non-coding RNA molecules that show an imperative function in gene control after transcription (Santosh et al., 2015). MiRNAs can alter the expression of target genes during abiotic stress to either enhance or attenuate the stress signal (Ramakrishnan et al., 2022). The following is an explanation of how miRNA-mediated signal amplification works.

- *MiRNA regulation*: Specific miRNAs are increased or downregulated in plant cells in response to abiotic stress. These stress-responsive miRNAs have the ability to target and suppress the expression of stress-tolerant genes (Shriram et al., 2016).
- *Transcription factors and stress-responsive genes*: Many miRNAs target transcription factors and stress-responsive genes (Budak et al., 2015). MiRNAs can fine-tune the stress response by suppressing the production of certain genes. MiRNAs, for example, can target transcription factors that trigger the expression of stress-inducible genes, thereby lowering their expression (Gao et al., 2022).
- *Stress signal amplification*: MiRNAs can amplify the stress signal by downregulating target genes. For example, if an miRNA targets a negative regulator of stress response,

inhibiting it might result in enhanced expression of stress-responsive genes, improving the plant's ability to handle stress (Zhang et al., 2022a).

- *Complex regulatory networks*: MiRNA-mediated regulation is frequently a component of a complex regulatory network. Multiple miRNAs can target the same or distinct stress response pathway components, resulting in layered and nuanced gene expression regulation. Because of this intricacy, plants may vary their response to various stressors and fine-tune their adaption mechanisms (Yang et al., 2021).

12.6 FUTURE IMPLICATIONS

MicroRNAs have spurred the development of innovative techniques and databases dedicated to their research. Earlier methods, including direct cloning and computational analysis, often faced challenges due to the inherently small size and sparse presence of miRNAs. Recent progress in technologies such as next-generation sequencing (NGS), advanced bioinformatics, CRISPR gene-editing tools, and precise quantitative real-time RT-PCR has greatly expanded our comprehension of miRNAs. In agriculture, miRNAs offer valuable opportunities for enhancing crop performance. They are particularly useful in breeding plants that are more resilient to various environmental stresses. By manipulating miRNAs, scientists can specifically alter key characteristics of crops, leading to improvements in growth rates, overall yield, and stress resistance.

12.6.1 Emerging Techniques for Studying MiRNAs

Due to the key functions that miRNAs play in the development and adaptability of plants, several novel techniques and databases have been created to study miRNAs. Initially, direct cloning and computational methods were the main strategies for learning miRNAs (Mangrauthia et al., 2013). Strategies are based on comparative genomics. However, these procedures are difficult and time-consuming because of miRNAs' lower sizes and prevalence at low amounts (Ražná et al., 2022). The fact that one miRNA may control several genes, which may be controlled by numerous miRNAs, adds to the difficulties. But as a consequence of NGS technology, the development of bioinformatics tools, the accessibility of open-source miRNA databases, CRISPR technologies, and quantitative real-time (q or RT)-PCR, our understanding of MIR genes has rapidly advanced (Zhang and Wang, 2015). Quantitative real-time PCR, which provides excellent levels of sensitivity, precision, and specificity, has been frequently employed to quantify gene expression (Charrrier et al., 2019). As seen in Figure 12.5, current methods for the identification and characterization of miRNAs include biological and bioinformatics techniques. Because short RNA sequencing is focused on tiny RNA transcripts, including miRNAs, it is proving to be more useful. The accuracy in identifying miRNAs and their targets is significantly improved by the use of computational tools and algorithms. These are explained in detail as follows:

A. *MiRNA excessive expression*

The technique is based on upregulation and downregulation of miRNA-based MIR gene transcripts in plants. Overexpressing miRNAs like miR858, miR408, miR399, miR397, miR395, miR319, miR172, miR169, miR164 and miR156 in plants can lead to undesirable traits due to their involvement in complex regulatory networks. Robust overexpression disrupts the fine-tuning of multiple biological pathways, resulting in pleiotropic effects (Djami-Tchatchou et al., 2017). Moreover, excessive buildup of specific miRNAs can negatively impact the expression of crucial target genes, leading to unwanted phenotypes. To address these issues, researchers can use tissue-specific or stress-inducible promoters for more precise control of miRNA overexpression. Another approach is to overexpress target mRNAs resistant to specific miRNAs, mitigating unintended consequences (Ferdous et al., 2017; Trumbo et al., 2015). These strategies aim to minimize the drawbacks associated with manipulating miRNAs in plants.

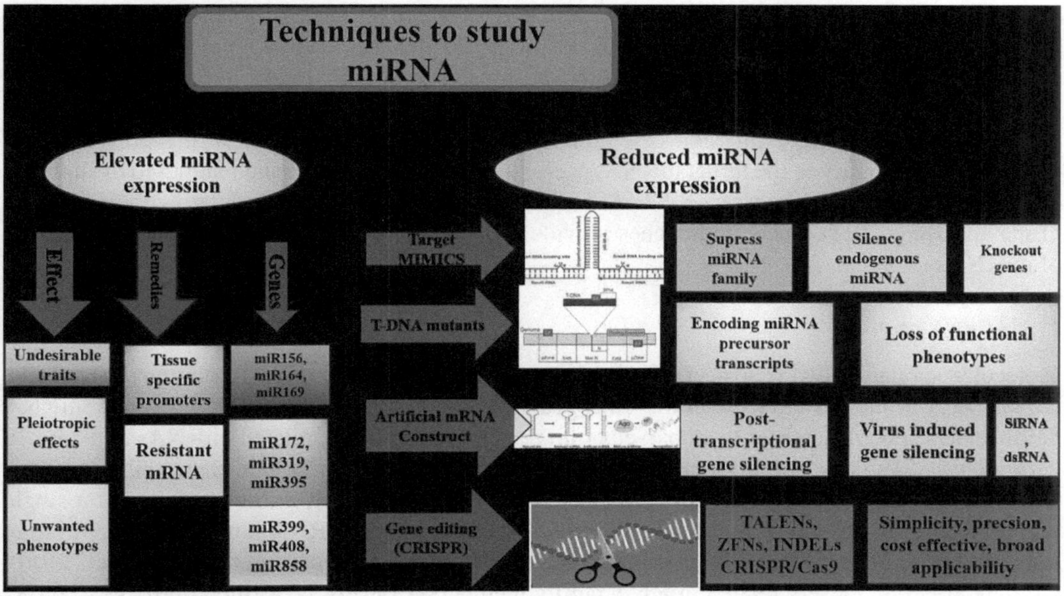

FIGURE 12.5 The figure depicts the emerging techniques in studying, correcting, and modifying the miRNA during various stresses in plants.

B. *T-DNA mutants*

In *Arabidopsis* research, losing a functional T-DNA or transposon insertional mutants have proven invaluable for uncovering the biological roles of genes (O'Malley et al., 2015; Eamens et al., 2011). However, their application faces significant challenges when targeting minor genes or loci within gene-poor sections of the genome, such as those encoding miRNA precursor transcripts. Retrieving such mutants is a highly daunting, time-consuming, and labor-intensive task (Zhou et al., 2017). While a few T-DNA mutants for miRNAs have been generated and utilized in preceding studies, including miR156a, miR156c, miR172a, and miR408 , a substantial limitation persists (Sharma et al., 2023a). These mutations may not necessarily affect the mature miRNA sequence, potentially leading to leaky expression. Moreover, the pool of available miRNA mutants remains quite limited (Yu et al., 2013; Zhang et al., 2013). In summary, the utility of T-DNA mutants in *Arabidopsis* research is undeniable, but the challenge of targeting small genes and miRNA loci within gene-poor regions has persisted. This limitation hinders our ability to comprehensively study miRNA functions, as the mutants generated may not precisely disrupt the mature miRNA sequence, leading to incomplete loss-of-function phenotypes and a scarcity of available mutants.

C. *Artificial miRNA*

A common approach to understanding gene function in plants involves the analysis of loss-of-function alleles or mutant genes. Various strategies for downregulating specific gene functions have been discovered, with a primary focus on the buildup of small interfering RNAs (siRNAs) resulting from double-stranded RNAs (dsRNAs) using a virus-induced gene silencing model (Lee et al., 2012). Plants naturally employ post-transcriptional gene silencing (PTGS) to defend against viruses by producing siRNAs (Yu and Kumar, 2003). However, these methods raise concerns about off-target effects. To address these issues, alternate biotechnological tactics have emerged for fine-tuning miRNA activity. Artificial microRNAs (amiRNAs) are improved versions of endogenous miRNA precursors, where the miRNA:miRNA* duplex is substituted with sequences designed to silence specific genes such as in rice (Warthmann et al., 2008; Parizotto et al., 2004). Additionally, the negative regulation of specific miRNA buildup allows for the achievement of desired traits.

Endogenous target mimicry (eTMs) is an alternative mechanism for modulating miRNA buildup and controlling essential plant processes (Basso et al., 2019). This technology originated from the detection of a target mimic (TM), induced by phosphate starvation1 (IPS1), in *Arabidopsis*, which evolved into short tandem target mimic (STTM) (Othman et al., 2023; Franco-Zorrilla et al., 2007). STTM involves the overexpression of engineered long non-coding RNAs with high nucleotide sequence individuality to the target mRNA, preventing miRNA cleavage and leading to the upregulation of the target gene. In essence, STTM sequesters miRNAs, thereby modulating their function (Tang and Tang, 2013). These approaches provide versatile tools for the functional characterization of miRNAs in plants.

D. MiRNA sponges

Molecular sponges (SPs) represent a technology utilized to impede miRNA function effectively, notably in *Arabidopsis* (Reichel et al., 2015). SPs consist of manifold tandem repeats of miRNA binding sites and are commonly employed to hinder entire families of related miRNAs in plants. In *Arabidopsis*, SPs designed to block miR159 and miR165/166 demonstrated their ability to interrupt miRNA activity (Jia et al., 2015; Reichel et al., 2015). However, a drawback is that a single target mRNA sponge can suppress the expression of a complete miRNA family. Similarly, target mimics (MIMICs) can silence endogenous miRNAs efficiently in *Arabidopsis* (Sharma et al., 2016), but they affect all closely related miRNA family members. Conversely, artificial miRNAs (amiRNAs) have been successfully used to knock out genes in phytopathogens (Tiwari et al., 2014; Kis et al., 2016; Wagaba et al., 2016). Numerous short tandem target mimics targeting miRNAs have found application in model and crop plants, such as tobacco (Zhao et al., 2016), cotton (Liu et al., 2019a), and tomato (Jiang et al., 2020), for functional miRNA analysis. However, when miRNA families share exceedingly similar sequences and overlapping targets, artificial target mimics may not clearly differentiate between different miRNAs (Todesco et al., 2010).

E. Genome editing technologies

Genome-editing techniques have evolved significantly in recent years, with some earlier methods like engineered meganucleases, zinc-finger nucleases (ZFNs), and transcription activator-like effector-based nucleases (TALENs) losing popularity due to their lesser specificity and off-target effects (Wendt et al., 2013). In contrast, the clustered regularly interspaced short palindromic repeats/CRISPR-associated protein 9 (CRISPR/Cas9) technologies has appeared as a revolutionary tool for gene functional characterization (Alamillo et al., 2023).

CRISPR/Cas9 offers several advantages, including simplicity, precision, cost-effectiveness, efficiency, and broad applicability across various organisms (Hong et al., 2021). This technique has been successfully applied in model plants like *Arabidopsis* and tobacco, as well as important crops such as rice, wheat, maize, sorghum, and tomato. Researchers have used CRISPR/Cas9 to introduce insertions/deletions (indels) in pre-miRNA sequences; disorderly miRNA processing or mature miRNA sequences (Sharma et al., 2020a). Examples include the creation of transgene-free soybean plants by targeting miR1514 and miR1509 and the editing of miR858 family members in *Arabidopsis* (Sharma et al., 2023a). Additionally, CRISPR/Cas9-intermediated editing of miRNAs like miR482b and miR482c in tomato conferred resistance to *Phytophthora infestans*, a pathogen causing late blight disease (Hong et al., 2021; Derevnina et al., 2016). Several miRNAs in rice, such as miR408, miR528, miR815, miR820, and miR397b, have been successfully edited using CRISPR/Cas9. While CRISPR/Cas9 presents exciting possibilities for manipulating plant miRNAs, it poses challenges due to potential unintended effects on multiple genes (Zhou et al., 2017, 2019a). Therefore, a comprehensive understanding of miRNA regulation in plant progression, development, and stress responses is essential to design strategies that achieve chosen traits with minimal unwanted side effects in improved plants (Singh et al., 2023). Careful consideration and further research are crucial for harnessing the full potential of miRNA manipulation in plant improvement.

12.6.2 Biotechnological Applications of MiRNA in Agriculture

The exploration of microRNAs in plant biology opens up innovative avenues for crop improvement, especially in developing resilience to abiotic stresses (Xu et al., 2019). This section discusses how miRNA knowledge can be applied in biotechnology for crop enhancement, the creation of stress-resistant transgenic plants, and the challenges faced in practical applications.

By manipulating miRNAs, scientists can target specific genes to modify important crop traits like growth rate, yield, and stress tolerance (Zhou et al., 2013; Chaudhary et al., 2021). Additionally, trait optimization can be done by adjusting miRNA expression that leads to changes in plant characteristics such as size, flowering time, and root development, optimizing them for better agricultural output (Dong et al., 2022a). Furthermore, abiotic stress-resistant transgenic plants developed through building environmental resilience. This is done by modifying the expression of certain miRNAs that can result in crops that are more resistant to environmental stresses like drought, salinity, and extreme temperature fluctuations (Wijerathna-Yapa, 2017). Precise genetic enhancements were also done through miRNA-based techniques that offered a more targeted approach than traditional breeding, allowing for the precise introduction of desired stress resistance traits (Singh et al., 2020). Transferring traits across species was also done through miRNAs from hardy plant species which can be introduced into more sensitive ones, bestowing enhanced tolerance to environmental stresses (Dong et al., 2022a). Leveraging miRNAs for crop improvement and the generation of abiotic stress-resistant plants holds significant promise for the forthcoming future of agriculture (Raza et al., 2023). However, realizing this potential requires navigating a complex landscape of technical challenges, regulatory frameworks, and public acceptance. Continued research and development, along with responsible stewardship and transparent communication, is necessary for harnessing the power of miRNAs in sustainable and widely accepted agricultural practices.

12.7 CONCLUSION AND FUTURE PROSPECTS

This chapter has provided a thorough exploration of the role of microRNAs in plant responses to abiotic stress. We have delved into the intricate ways in which miRNAs assist plants in managing and adapting to environmental challenges such as drought, salt, extreme temperatures, and heavy metal exposure. Through this exploration, we have uncovered the complex networks and regulatory pathways that miRNAs navigate, underscoring their integral role in plant stress physiology. Additionally, the chapter discusses the promising prospects of applying this knowledge in biotechnological advancements, particularly in the realm of creating stress-resistant crops. While the potential benefits are significant, the journey from research to practical application entails overcoming numerous challenges, emphasizing the importance of ongoing research and innovation. As the global climate continues to change, the insights gleaned from miRNA research become increasingly crucial for developing sustainable agricultural practices and ensuring food security. The field remains dynamic, with emerging technologies like CRISPR-Cas9 systems opening up fresh avenues for research and application. Continued investigation into miRNAs is poised to make substantial contributions to our comprehensive understanding of plant biology and the development of solutions for environmental stress management.

12.8 DECLARATION OF COMPETING INTEREST

The authors declare that they have no known competing financial interests or personal relationships that could have appeared to influence the work reported in this book chapter.

REFERENCES

Abla, Merhaba, Huigai Sun, Zhuyun Li, Chunxiang Wei, Fei Gao, Yijun Zhou, and Jinchao Feng. "Identification of miRNAs and their response to cold stress in *Astragalus membranaceus*." *Biomolecules* 9, no. 5 (2019): 182.

Adhikari, Ayan, Doyel Roy, Sinchan Adhikari, Shrabani Saha, Pratyush Kanti Ghosh, Arun Kumar Shaw, and Zahed Hossain. "microRNAomic profiling of maize root reveals multifaceted mechanisms to cope with Cr (VI) stress." *Plant Physiology and Biochemistry* 198 (2023): 107693.

Ahmad, Mushtaq. "Genomics and transcriptomics to protect rice (*Oryza sativa* L.) from abiotic stressors:-pathways to achieving zero hunger." *Frontiers in Plant Science* 13 (2022): 1002596.

Ahmad, Naveed, Muhammad Naeem, Hina Ali, Khulood Fahad Alabbosh, Hamad Hussain, Ismail Khan, Shahrood Ahmed Siddiqui, Asif Ali Khan, and Babar Iqbal. "From challenges to solutions: The impact of melatonin on abiotic stress synergies in horticultural plants via redox regulation and epigenetic signaling." *Scientia Horticulturae* 321 (2023): 112369.

Alamillo, Josefa M., Cristina M. López, Félix J. Martínez Rivas, Fernando Torralbo, Mustafa Bulut, and Saleh Alseekh. "Clustered regularly interspaced short palindromic repeats/CRISPR-associated protein and hairy roots: A perfect match for gene functional analysis and crop improvement." *Current Opinion in Biotechnology* 79 (2023): 102876.

Ali, Baber, Aqsa Hafeez, Muhammad Ammar Javed, Muhammad Siddique Afridi, Hina Ali Abbasi, Ayesha Qayyum, Tayyaba Batool et al. "Role of endophytic bacteria in salinity stress amelioration by physiological and molecular mechanisms of defense: A comprehensive review." *South African Journal of Botany* 151 (2022): 33–46.

Ali, Mohd, Deepak Kumar, Raman Tikoria, Roohi Sharma, Parkirti Parkirti, Vikram Vikram, Kritika Kaushal, and Puja Ohri. "Exploring the potential role of hydrogen sulfide and jasmonic acid in plants during heavy metal stress." *Nitric Oxide* 140–141 (2023): 16–29.

Alsafran, Mohammed, Muhammad Hamzah Saleem, Hareb Al Jabri, Muhammad Rizwan, and Kamal Usman. "Principles and applicability of integrated remediation strategies for heavy metal removal/recovery from contaminated environments." *Journal of Plant Growth Regulation* 42, no. 6 (2023): 3419–3440.

Alzahrani, Saud M., Ibrahim A. Alaraidh, Muhammad A. Khan, Hussein M. Migdadi, Salem S. Alghamdi, and Abdluaziz A. Alsahli. "Identification and characterization of salt-responsive microRNAs in Vicia faba by high-throughput sequencing." *Genes* 10, no. 4 (2019): 303.

Amini, Zohreh, Hassan Salehi, Mehrangiz Chehrazi, Mohammad Etemadi, and Mingying Xiang. "miRNAs and their target genes play a critical role in response to heat stress in *Cynodon dactylon* (L.) Pers." *Molecular Biotechnology* (2023): 1–14.

Anami, Elikana, and Dirk Inzé. "Silencing approach using Poly (ADP-ribose) polymerase gene to improve drought stress tolerance in maize." *Afrika Focus* 23, no. 2 (2010).

Anwar, Ali, and Ju-Kon Kim. "Transgenic breeding approaches for improving abiotic stress tolerance: Recent progress and future perspectives." *International Journal of Molecular Sciences* 21, no. 8 (2020): 2695.

Aravind, Jayaraman, Sharma Rinku, Banduni Pooja, Mittal Shikha, ShirigaKaliyugam, Mallana Gowdra Mallikarjuna, Arun Kumar, Atmakuri Ramakrishna Rao, and Thirunavukkarasu Nepolean. "Identification, characterization, and functional validation of drought-responsive microRNAs in subtropical maize inbreds." *Frontiers in Plant Science* 8 (2017): 941.

Argenta, Luiz Carlos, Rachael Maree Wood, James P. Mattheis, Fabio Rodrigo Thewes, Cristiano Nunes Nesi, and Daniel Alexandre Neuwald. "Factors affecting development of disorders expressed after storage of 'Gala'apple fruit." *Postharvest Biology and Technology* 204 (2023): 112439.

Asgher, Mohd, M. Iqbal R. Khan, Naser A. Anjum, Susheel Verma, Dhiraj Vyas, Tasir S. Per, Asim Masood, and Nafees A. Khan. "Ethylene and polyamines in counteracting heavy metal phytotoxicity: A crosstalk perspective." *Journal of Plant Growth Regulation* 37 (2018): 1050–1065.

Aydinoglu, Fatma. "Elucidating the regulatory roles of microRNAs in maize (*Zea mays* L.) leaf growth response to chilling stress." *Planta* 251, no. 2 (2020): 38.

Azahar, Ikbal, Supriya Ghosh, Ayan Adhikari, Sinchan Adhikari, Doyel Roy, Arun K. Shaw, Kashmir Singh, and Zahed Hossain. "Comparative analysis of maize root sRNA transcriptome unveils the regulatory roles of miRNAs in submergence stress response mechanism." *Environmental and Experimental Botany* 171 (2020): 103924.

Baghery, Mohammad Amin, Seyed Kamal Kazemitabar, Ali Dehestani, Pooyan Mehrabanjoubani, Mohammad Mehdi Naghizadeh, and Ali Masoudi-Nejad. "Insight into gene regulatory networks involved in sesame (*Sesamum indicum* L.) drought response." *Biologia* 77, no. 4 (2022): 1181–1196.

Ballén-Taborda, Carolina, Germán Plata, Sarah Ayling, Fausto Rodríguez-Zapata, Luis Augusto Becerra Lopez-Lavalle, Jorge Duitama, and Joe Tohme. "Identification of cassava microRNAs under abiotic stress." *International Journal of Genomics* 2013 (2013).

Banerjee, Pradipta. "Plant abiotic stress responses and microRNAs." *Advanced Agriculture* (2020): 119–118.

Bartel, David P. "MicroRNAs: Genomics, biogenesis, mechanism, and function." *Cell* 116, no. 2 (2004): 281–297.

Basso, Marcos Fernando, Paulo Cavalcanti Gomes Ferreira, Adilson Kenji Kobayashi, Frank G. Harmon, Alexandre Lima Nepomuceno, Hugo Bruno Correa Molinari, and Maria Fatima Grossi-de-Sa. "Micro RNA s and new biotechnological tools for its modulation and improving stress tolerance in plants." *Plant Biotechnology Journal* 17, no. 8 (2019): 1482–1500.

Batool, Maria, Ali Mahmoud El-Badri, Muhammad Umair Hassan, Yang Haiyun, Wang Chunyun, Yan Zhenkun, Kuai Jie, Bo Wang, and Guangsheng Zhou. "Drought stress in Brassica napus: Effects, tolerance mechanisms, and management strategies." *Journal of Plant Growth Regulation* (2022): 1–25.

Bharti, Ruchi, and Renu Sharma. "Effect of heavy metals: An overview." *Materials Today: Proceedings* 51 (2022): 880–885.

Block, Marc De, Christoph Verduyn, Dirk De Brouwer, and Marc Cornelissen. "Poly (ADP-ribose) polymerase in plants affects energy homeostasis, cell death and stress tolerance." *The Plant Journal* 41, no. 1 (2005): 95–106.

Bonnet, Eric, Jan Wuyts, Pierre Rouzé, and Yves Van de Peer. "Evidence that microRNA precursors, unlike other non-coding RNAs, have lower folding free energies than random sequences." *Bioinformatics* 20, no. 17 (2004): 2911–2917.

Budak, Hikmet, Melda Kantar, Reyyan Bulut, and Bala Ani Akpinar. "Stress responsive miRNAs and isomiRs in cereals." *Plant Science* 235 (2015): 1–13.

Carrió-Seguí, Àngela, Omar Ruiz-Rivero, Laura Villamayor-Belinchón, Sergi Puig, Ana Perea-García, and Lola Peñarrubia. "The altered expression of microRNA408 influences the Arabidopsis response to iron deficiency." *Frontiers in Plant Science* 10 (2019): 324.

Çelik, Özge, and Enes Yağız Akdaş. "Tissue-specific transcriptional regulation of seven heavy metal stress-responsive miRNAs and their putative targets in nickel indicator castor bean (*R. communis* L.) plants." *Ecotoxicology and Environmental Safety* 170 (2019): 682–690.

Chakraborty, Swarnavo, and Aryadeep Roychoudhury. "MicroRNA-mediated regulation of heavy metal stress in plants." *Plant MicroRNAs and Stress Response* (2023): 183–197.

Charrier, Aurélie, Emilie Vergne, Clément Joffrion, Andréa Richer, Nicolas Dousset, and Elisabeth Chevreau. "An artificial miRNA as a new tool to silence and explore gene functions in apple." *Transgenic Research* 28 (2019): 611–626.

Chaudhary, Saurabh, Atul Grover, and Prakash Chand Sharma. "MicroRNAs: Potential targets for developing stress-tolerant crops." *Life* 11, no. 4 (2021): 289.

Chen, Jin, Ao Pan, Shujun He, Pin Su, Xiaoling Yuan, Shengwei Zhu, and Zhi Liu. "Different microRNA families involved in regulating high temperature stress response during cotton (*Gossypium hirsutum* L.) anther development." *International Journal of Molecular Sciences* 21, no. 4 (2020): 1280.

Chen, Miao, Yujuan Zhao, Chunliu Zhuo, Shaoyun Lu, and Zhenfei Guo. "Overexpression of a NF-YC transcription factor from bermudagrass confers tolerance to drought and salinity in transgenic rice." *Plant Biotechnology Journal* 13, no. 4 (2015): 482–491.

Chen, Yong, and Xing Yu. "Endoplasmic reticulum stress-responsive microRNAs are involved in the regulation of abiotic stresses in wheat." *Plant Cell Reports* (2023): 1–20.

Cheng, Chunzhen, Fan Liu, Xueli Sun, Bin Wang, Jiapeng Liu, Xueting Ni, Chunhua Hu et al. "Genome-wide identification of FAD gene family and their contributions to the temperature stresses and mutualistic and parasitic fungi colonization responses in banana." *International Journal of Biological Macromolecules* 204 (2022): 661–676.

Cheng, Xiliu, Qiang He, Sha Tang, Haoran Wang, Xiangxiang Zhang, Mingjie Lv, Huafeng Liu et al. "The miR172/IDS1 signaling module confers salt tolerance through maintaining ROS homeostasis in cereal crops." *New Phytologist* 230, no. 3 (2021): 1017–1033.

Ci, Dong, Yuepeng Song, Min Tian, and Deqiang Zhang. "Methylation of miRNA genes in the response to temperature stress in *Populus simonii*." *Frontiers in Plant Science* 6 (2015): 921.

Das, Natasha, Surajit Bhattacharya, Somnath Bhattacharyya, and Mrinal K. Maiti. "Expression of rice MATE family transporter OsMATE2 modulates arsenic accumulation in tobacco and rice." *Plant Molecular Biology* 98 (2018): 101–120.

Demirkol, Gürkan. "MicroRNA expression under phosphate deficiency stress in red clover (*Trifolium pratense* L.): A three-year field experiment." *New Zealand Journal of Agricultural Research* 65, nos. 4–5 (2022): 290–301.

Denver, Jn Baptiste, and Hemayet Ullah. "miR393s regulate salt stress response pathway in *Arabidopsis thaliana* through scaffold protein RACK1A mediated ABA signaling pathways." *Plant Signaling & Behavior* 14, no. 6 (2019): 1600394.

Derevnina, Lida, Benjamin Petre, Ronny Kellner, Yasin F. Dagdas, Mohammad Nasif Sarowar, Artemis Giannakopoulou, Juan Carlos De la Concepcion et al. "Emerging oomycete threats to plants and animals." *Philosophical Transactions of the Royal Society B: Biological Sciences* 371, no. 1709 (2016): 20150459.

Dhiman, Shalini, Mohd Ibrahim, Kamini Devi, Neerja Sharma, Nitika Kapoor, Ravinderjit Kaur, Nandni Sharma et al. "Biochar assisted remediation of toxic metals and metalloids." In *Handbook of Assisted and Amendment: Enhanced Sustainable Remediation Technology*, pp. 131–162. Wiley, 2021.

Dhiman, Shalini, Arun Dev Singh, Jaspreet Kour, Tamanna Bhardwaj, Kamini Devi, Raman Tikoria, Puja Ohri, Renu Bhardwaj, and Priyanka Sharma. "Redox homeostasis in response to abiotic stresses in plants." In *The Role of Growth Regulators and Phytohormones in Overcoming Environmental Stress*, pp. 15–37. Academic Press, 2023.

Dikilitas, Murat, Simsek, Eray and Karakas, Sema. "Chapter 2 – Stress responsive signaling molecules and genes under stressful environments in plants." In *Plant Signaling Molecules*, edited by M. Iqbal R. Khan, Palakolanu Sudhakar Reddy, Antonio Ferrante, and Nafees A. Khan, pp. 19–42. Woodhead Publishing, 2019.

Ding, Dong, Lifang Zhang, Hang Wang, Zhijie Liu, Zuxin Zhang, and Yonglian Zheng. "Differential expression of miRNAs in response to salt stress in maize roots." *Annals of Botany* 103, no. 1 (2009): 29–38.

Ding, Na, Raul Huertas, Ivone Torres-Jerez, Wei Liu, Bonnie Watson, Wolf-Rüdiger Scheible, and Michael Udvardi. "Transcriptional, metabolic, physiological and developmental responses of switchgrass to phosphorus limitation." *Plant, Cell & Environment* 44, no. 1 (2021): 186–202.

Ding, Yanfei, Shaohua Gong, Yi Wang, Feijuan Wang, Hexigeduleng Bao, Junwei Sun, Chong Cai, Keke Yi, Zhixiang Chen, and Cheng Zhu. "MicroRNA166 modulates cadmium tolerance and accumulation in rice." *Plant Physiology* 177, no. 4 (2018): 1691–1703.

Ding, Yanfei, Yi Wang, Zhihua Jiang, Feijuan Wang, Qiong Jiang, Junwei Sun, Zhixiang Chen, and Cheng Zhu. "MicroRNA268 overexpression affects rice seedling growth under cadmium stress." *Journal of Agricultural and Food Chemistry* 65, no. 29 (2017): 5860–5867.

Ding, Zehong, Weiwei Tie, Lili Fu, Yan Yan, Guanghua Liu, Wei Yan, Yanan Li, Chunlai Wu, Jiaming Zhang, and Wei Hu. "Strand-specific RNA-seq based identification and functional prediction of drought-responsive lncRNAs in cassava." *BMC Genomics* 20, no. 1 (2019): 1–13.

Djami-Tchatchou, Arnaud T., Neeti Sanan-Mishra, Khayalethu Ntushelo, and Ian A. Dubery. "Functional roles of microRNAs in agronomically important plants – potential as targets for crop improvement and protection." *Frontiers in Plant Science* 8 (2017): 378.

Dong, Qingkun, Binbin Hu, and Cui Zhang. "microRNAs and their roles in plant development." *Frontiers in Plant Science* 13 (2022a): 824240.

Dong, Yufei, Mingjia Tang, Zelan Huang, Jianing Song, Jin Xu, Golam Jalal Ahammed, Jingquan Yu, and Yanhong Zhou. "The miR164a-NAM3 module confers cold tolerance by inducing ethylene production in tomato." *The Plant Journal* 111, no. 2 (2022b): 440–456.

Eamens, Andrew L., Claire Agius, Neil A. Smith, Peter M. Waterhouse, and Ming-Bo Wang. "Efficient silencing of endogenous microRNAs using artificial microRNAs in *Arabidopsis thaliana*." *Molecular Plant* 4, no. 1 (2011): 157–170.

Fadiji, Ayomide Emmanuel, Ajar Nath Yadav, Gustavo Santoyo, and Olubukola Oluranti Babalola. "Understanding the plant-microbe interactions in environments exposed to abiotic stresses: An overview." *Microbiological Research* (2023): 127368.

Ferdous, Jannatul, Ryan Whitford, Martin Nguyen, Chris Brien, Peter Langridge, and Penny J. Tricker. "Drought-inducible expression of Hv-miR827 enhances drought tolerance in transgenic barley." *Functional & Integrative Genomics* 17 (2017): 279–292.

Floris, Maïna, Hany Mahgoub, Elodie Lanet, Christophe Robaglia, and Benoît Menand. "Post-transcriptional regulation of gene expression in plants during abiotic stress." *International Journal of Molecular Sciences* 10, no. 7 (2009): 3168–3185.

Fontana, Julia Elise, Guo Wang, Runrun Sun, Huiyun Xue, Qian Li, Jia Liu, Kyle E. Davis et al. "Impact of potassium deficiency on cotton growth, development and potential microRNA-mediated mechanism." *Plant Physiology and Biochemistry* 153 (2020): 72–80.

Francis, Bonia, C. T. Aravindakumar, Philip B. Brewer, and Sibu Simon. "Plant nutrient stress adaptation: A prospect for fertilizer limited agriculture." *Environmental and Experimental Botany* (2023): 105431.

Franco-Zorrilla, José Manuel, Adrián Valli, Marco Todesco, Isabel Mateos, María Isabel Puga, Ignacio Rubio-Somoza, Antonio Leyva, Detlef Weigel, Juan Antonio García, and Javier Paz-Ares. "Target mimicry provides a new mechanism for regulation of microRNA activity." *Nature Genetics* 39, no. 8 (2007): 1033–1037.

Fridovich, Irwin. "Superoxide radical and superoxide dismutases." *Annual Review of Biochemistry* 64, no. 1 (1995): 97–112.

Fu, Ying, Annaliese S. Mason, Yaofeng Zhang, Baogang Lin, Meili Xiao, Donghui Fu, and Huasheng Yu. "MicroRNA-mRNA expression profiles and their potential role in cadmium stress response in *Brassica napus*." *BMC Plant Biology* 19 (2019): 1–20.

Gao, Shuai, Lu Yang, Hou Qing Zeng, Zhao Sheng Zhou, Zhi Min Yang, Hua Li, Di Sun, Fuliang Xie, and Baohong Zhang. "A cotton miRNA is involved in regulation of plant response to salt stress." *Scientific Reports* 6, no. 1 (2016): 19736.

Gao, Zhen, Chao Ma, Chengchao Zheng, Yuxin Yao, and Yuanpeng Du. "Advances in the regulation of plant salt-stress tolerance by miRNA." *Molecular Biology Reports* 49, no. 6 (2022): 5041–5055.

Gentile, Agustina, Lara I. Dias, Raphael S. Mattos, Thaís H. Ferreira, and Marcelo Menossi. "MicroRNAs and drought responses in sugarcane." *Frontiers in Plant Science* 6 (2015): 58.

Ghabooli, Mehdi, Shima Sadat Beheshti Rooy, Ehsan Mohseni Fard, and Rouhollah Karimi. "miR395 is involved in response to cold stress and modulation of sulfate and phosphate deficiency in Grape (*Vitis vinifera*)." *Journal of Plant Molecular Breeding* 7, no. 1 (2019): 56–66.

Ghazzal, Maria, Muhammad Iftikhar Hussain, Zafar Iqbal Khan, M. Habib ur Rahman, Abeer A. El-Habeeb, and Hsi-Hsien Yang. "Chromium poisoning in buffaloes in the vicinity of contaminated pastureland, Punjab, Pakistan." *Sustainability* 14, no. 22 (2022): 15095.

Gielen, Heidi, Tony Remans, Jaco Vangronsveld, and Ann Cuypers. "Toxicity responses of Cu and Cd: The involvement of miRNAs and the transcription factor SPL7." *BMC Plant Biology* 16, no. 1 (2016): 1–16.

Gong, Jian, Dong Li, Hao Li, Huakun Zhou, and Jin Xu. "Identification of manganese-responsive microRNAs in Arabidopsis by small RNA sequencing." *Czech Journal of Genetics and Plant Breeding* 55, no. 2 (2019): 76–82.

Gong, Zhizhong, Liming Xiong, Huazhong Shi, Shuhua Yang, Luis R. Herrera-Estrella, Guohua Xu, Dai-Yin Chao et al. "Plant abiotic stress response and nutrient use efficiency." *Science China Life Sciences* 63 (2020): 635–674.

Gu, Quan, Ziping Chen, Xiuli Yu, Weiti Cui, Jincheng Pan, Gan Zhao, Sheng Xu, Ren Wang, and Wenbiao Shen. "Melatonin confers plant tolerance against cadmium stress via the decrease of cadmium accumulation and reestablishment of microRNA-mediated redox homeostasis." *Plant Science* 261 (2017): 28–37.

Guan, Qingmei, Xiaoyan Lu, Haitao Zeng, Yanyan Zhang, and Jianhua Zhu. "Heat stress induction of mi R 398 triggers a regulatory loop that is critical for thermotolerance in A rabidopsis." *The Plant Journal* 74, no. 5 (2013): 840–851.

Gupta, Kapil, Sujit Kumar Mishra, Shubhra Gupta, Saurabh Pandey, Jogeswar Panigrahi, and Shabir Hussain Wani. "Functional role of miRNAs: key players in soybean improvement." *Phyton* 90, no. 5 (2021): 1339–1362.

Gupta, Rishil, Faryad Khan, Fatmah M. Alqahtani, Mohamed Hashem, and Faheem Ahmad. "Plant growth–promoting rhizobacteria (PGPR) assisted bioremediation of heavy metal toxicity." *Applied Biochemistry and Biotechnology* (2023): 1–29.

Haider, Muhammad Salman, Sudisha Jogaiah, Tariq Pervaiz, Zhao Yanxue, Nadeem Khan, and Jinggui Fang. "Physiological and transcriptional variations inducing complex adaptive mechanisms in grapevine by salt stress." *Environmental and Experimental Botany* 162 (2019): 455–467.

Haiying, Tang, Wang Shubin, Liu Ying, Muhammad Umair Hassan, Song Ying, Guoqin Huang, Mohamed Hashem, Saad Alamri, and Yasser S. Mostafa. "Biochar: A promising soil amendment to mitigate heavy metals toxicity in plants." *Notulae Botanicae Horti Agrobotanici Cluj-Napoca* 50, no. 3 (2022): 12778–12778.

Hassani, Seyedeh Batool, Mojgan Latifi, Sasan Aliniaeifard, Shabnam Sohrabi Bonab, Neda Nasiri Almanghadim, Sara Jafari, Elham Mohebbifar et al. "Response to cadmium toxicity: Orchestration of polyamines and microRNAs in maize plant." *Plants* 12, no. 10 (2023): 1991.

He, Lizhong, Jun Yan, Xiaotao Ding, Haijun Jin, Hongmei Zhang, Jiawei Cui, Qiang Zhou, and Jizhu Yu. "Integrated analysis of transcriptome and microRNAs associated with exogenous calcium-mediated enhancement of hypoxic tolerance in cucumber seedlings (*Cucumis sativus* L.)." *Frontiers in Plant Science* 13 (2023): 994268.

He, Wenqi, Hongyan Liu, Liang Shi, Mei Zhou, Yang Qi, Feng Liu, Xinrong Zhu, Pengwu Zhao, Changlin Xiang, and Yang Shu. "Climate and soil change nutrient element allocation of *Siberian larch* in the *Mongolian semiarid* forest." *Agricultural and Forest Meteorology* 315 (2022): 108825.

He, Xueying, Shirong Guo, Ying Wang, Liwei Wang, Sheng Shu, and Jin Sun. "Systematic identification and analysis of heat-stress-responsive lncRNAs, circRNAs and miRNAs with associated co-expression and ceRNA networks in cucumber (*Cucumis sativus* L.)." *Physiologia Plantarum* 168, no. 3 (2020): 736–754.

He, Zhenghua, Jinfeng Wu, Xiaopeng Sun, and Mingqiu Dai. "The maize clade A PP2C phosphatases play critical roles in multiple abiotic stress responses." *International Journal of Molecular Sciences* 20, no. 14 (2019): 3573.

Hong, Yuhui, Jun Meng, Xiaoli He, Yuanyuan Zhang, Yarong Liu, Chengwei Zhang, Hongyan Qi, and Yushi Luan. "Editing miR482b and miR482c simultaneously by CRISPR/Cas9 enhanced tomato resistance to *Phytophthora infestans*." *Phytopathology®* 111, no. 6 (2021): 1008–1016.

Hou, Yabing, Fangling Jiang, Xiaolan Zheng, and Zhen Wu. "Identification and analysis of oxygen responsive microRNAs in the root of wild tomato (*S. habrochaites*)." *BMC Plant Biology* 19 (2019): 1–13.

Huang, Jing-Hao, Xiong-Jie Lin, Ling-Yuan Zhang, Xian-Da Wang, Guo-Cheng Fan, and Li-Song Chen. "MicroRNA sequencing revealed Citrus adaptation to long-term boron toxicity through modulation of root development by miR319 and miR171." *International Journal of Molecular Sciences* 20, no. 6 (2019): 1422.

Huang, Xiaozhen, Lingyan Hou, Jingjing Meng, Huiwen You, Zhen Li, Zhizhong Gong, Shuhua Yang, and Yiting Shi. "The antagonistic action of abscisic acid and cytokinin signaling mediates drought stress response in Arabidopsis." *Molecular Plant* 11, no. 7 (2018): 970–982.

Iqbal, Shehzad, Xiukang Wang, Iqra Mubeen, Muhammad Kamran, Iqra Kanwal, Gonzalo A. Díaz, Aqleem Abbas et al. "Phytohormones trigger drought tolerance in crop plants: outlook and future perspectives." *Frontiers in Plant Science* 12 (2022): 3378.

Islam, Waqar, Muhammad Adnan, Zhiqun Huang, Guo-dong Lu, and Han Y. H. Chen. "Small RNAs from seed to mature plant." *Critical Reviews in Plant Sciences* 38, no. 2 (2019): 117–139.

Islam, Waqar, Arfa Tauqeer, Abdul Waheed, and Fanjiang Zeng. "MicroRNA mediated plant responses to nutrient stress." *International Journal of Molecular Sciences* 23, no. 5 (2022): 2562.

Jain, Muskan, Preeti Nagar, Parul Goel, Anil Kumar Singh, Sumita Kumari, and Ananda Mustafiz. "Second messengers: Central regulators in plant abiotic stress response." In *Abiotic Stress-Mediated Sensing and Signaling in Plants: An Omics Perspective*, pp. 47–94. Springer, 2018.

Jamla, Monica, Shrushti Joshi, Suraj Patil, Bhumi Nath Tripathi, and Vinay Kumar. "MicroRNAs modulating nutrient homeostasis: A sustainable approach for developing biofortified crops." *Protoplasma* 260, no. 1 (2023): 5–19.

Jia, Xiaoyun, Jie Shen, Hui Liu, Fang Li, Na Ding, Changyong Gao, Sitakanta Pattanaik, Barunava Patra, Runzhi Li, and Ling Yuan. "Small tandem target mimic-mediated blockage of microRNA858 induces anthocyanin accumulation in tomato." *Planta* 242 (2015): 283–293.

Jian, Hongju, B. O. Yang, Aoxiang Zhang, Jinqi Ma, Yiran Ding, Zhiyou Chen, Jiana Li, Xinfu Xu, and Liezhao Liu. "Genome-wide identification of microRNAs in response to cadmium stress in oilseed rape (*Brassica napus* L.) using high-throughput sequencing." *International Journal of Molecular Sciences* 19, no. 5 (2018): 1431.

Jiang, Ning, Jun Cui, Xinxin Hou, Guanglei Yang, Yu Xiao, Lu Han, Jun Meng, and Yushi Luan. "Sl-lncRNA15492 interacts with Sl-miR482a and affects Solanum lycopersicum immunity against *Phytophthora infestans*." *The Plant Journal* 103, no. 4 (2020): 1561–1574.

Jiang, Zheng, Xin Xia, Yiyong Liu, Jie He, and Xu Yang. "Integrated miRNA and mRNA transcriptome analysis reveals eggplant's (*Solanum melongena* L.) responses to waterlogging stress." *Agronomy* 13, no. 9 (2023): 2215.

Joshi, Rohit, Khirod Kumar Sahoo, Amit Kumar Tripathi, Ritesh Kumar, Brijesh Kumar Gupta, Ashwani Pareek, and Sneh Lata Singla-Pareek. "Knockdown of an inflorescence meristem-specific cytokinin oxidase–OsCKX2 in rice reduces yield penalty under salinity stress condition." *Plant, Cell & Environment* 41, no. 5 (2018): 936–946.

Kaur, Ranjeet, Kshitija Sinha, and Rupam Kumar Bhunia. "Can wheat survive in heat? Assembling tools towards successful development of heat stress tolerance in *Triticum aestivum* L." *Molecular Biology Reports* 46 (2019): 2577–2593.

Kaur, Ravinderjit, Nandni Sharma, Raman Tikoria, Mohd Ali, Sandeep Kour, Deepak Kumar, and Puja Ohri. "Insights into biosynthesis and signaling of cytokinins during plant growth, development and stress tolerance." In *Auxins, Cytokinins and Gibberellins Signaling in Plants*, pp. 153–187. Cham: Springer International Publishing, 2022.

Kayihan, Doğa Selin, Ceyhun Kayihan, and Yelda Özden Çiftçi. "Moderate level of toxic boron causes differential regulation of microRNAs related to jasmonate and ethylene metabolisms in *Arabidopsis thaliana*." *Turkish Journal of Botany* 43, no. 2 (2019): 167–172.

Khatabi, Behnam, Siwaret Arikit, Rui Xia, Stephan Winter, Doungous Oumar, Kone Mongomake, Blake C. Meyers, and Vincent N. Fondong. "High-resolution identification and abundance profiling of cassava (*Manihot esculenta* Crantz) microRNAs." *BMC Genomics* 17 (2016): 1–18.

Kis, András, Gergely Tholt, Milán Ivanics, Éva Várallyay, Barnabás Jenes, and Zoltán Havelda. "Polycistronic artificial miRNA-mediated resistance to Wheat dwarf virus in barley is highly efficient at low temperature." *Molecular Plant Pathology* 17, no. 3 (2016): 427–437.

Kumar, Deepak, Rajesh Kumari Manhas, and Puja Ohri. "Deciphering the growth promoting and stress curtailing role of nitric oxide in *Meloidogyne incognita* infested *Solanum lycopersicum* seedlings." *Scientia Horticulturae* 319 (2023a): 112147.

Kumar, Deepak, and Puja Ohri. "Say "NO" to plant stresses: Unravelling the role of nitric oxide under abiotic and biotic stress." *Nitric Oxide* 130 (2022): 36–57.

Kumar, Ravi Shankar, Hiteshwari Sinha, Tapasya Datta, Mehar Hasan Asif, and Prabodh Kumar Trivedi. "microRNA408 and its encoded peptide regulate sulfur assimilation and arsenic stress response in Arabidopsis." *Plant Physiology* 192, no. 2 (2023b): 837–856.

Kushawaha, Akhilesh Kumar, Ambreen Khan, Sudhir Kumar Sopory, and Neeti Sanan-Mishra. "Priming by high temperature stress induces microRNA regulated heat shock modules indicating their involvement in thermopriming response in rice." *Life* 11, no. 4 (2021): 291.

Laubinger, Sascha, Timo Sachsenberg, Georg Zeller, Wolfgang Busch, Jan U. Lohmann, Gunnar Rätsch, and Detlef Weigel. "Dual roles of the nuclear cap-binding complex and SERRATE in pre-mRNA splicing and microRNA processing in *Arabidopsis thaliana*." *Proceedings of the National Academy of Sciences* 105, no. 25 (2008): 8795–8800.

Lee, Rosalind C., Rhonda L. Feinbaum, and Victor Ambros. "The C. elegans heterochronic gene lin-4 encodes small RNAs with antisense complementarity to lin–14." *Cell* 75, no. 5 (1993): 843–854.

Lee, Wing-Sham, Kim E. Hammond-Kosack, and Kostya Kanyuka. "Barley stripe mosaic virus-mediated tools for investigating gene function in cereal plants and their pathogens: Virus-induced gene silencing, host-mediated gene silencing, and virus-mediated overexpression of heterologous protein." *Plant Physiology* 160, no. 2 (2012): 582–590.

Lee, Yoontae, Minju Kim, Jinju Han, Kyu-Hyun Yeom, Sanghyuk Lee, Sung Hee Baek, and V. Narry Kim. "MicroRNA genes are transcribed by RNA polymerase II." *The EMBO Journal* 23, no. 20 (2004): 4051–4060.

Leng, Xiangpeng, Peipei Wang, Pengcheng Zhao, Mengqi Wang, Liwen Cui, Lingfei Shangguan, and Chen Wang. "Conservation of microRNA-mediated regulatory networks in response to copper stress in grapevine." *Plant Growth Regulation* 82 (2017): 293–304.

Li, Da-hong, Hui Liu, Yang Yan-li, Zhen Ping-ping, and Liang Jian-sheng. "Down-regulated expression of RACK1 gene by RNA interference enhances drought tolerance in rice." *Rice Science* 16, no. 1 (2009): 14–20.

Li, Jiang-Chuan, Jiang-Bo Guo, Wen-Zhong Xu, and Mi Ma. "RNA interference-mediated silencing of phytochelatin synthase gene reduce cadmium accumulation in rice seeds." *Journal of Integrative Plant Biology* 49, no. 7 (2007): 1032–1037.

Li, Jing, Qiaoqiao Song, Zhi-Fang Zuo, and Lin Liu. "MicroRNA398: A master regulator of plant development and stress responses." *International Journal of Molecular Sciences* 23, no. 18 (2022): 10803.

Li, Shuxia, Jinxin Liu, Zhongyuan Liu, Xiaorong Li, Feijie Wu, and Yuke He. "Heat-induced tas1 target1 mediates thermotolerance via heat stress transcription factor A1a–directed pathways in Arabidopsis." *The Plant Cell* 26, no. 4 (2014): 1764–1780.

Li, Shuxia, Xiang Yu, Ning Lei, Zhihao Cheng, Pingjuan Zhao, Yuke He, Wenquan Wang, and Ming Peng. "Genome-wide identification and functional prediction of cold and/or drought-responsive lncRNAs in cassava." *Scientific Reports* 7, no. 1 (2017): 45981.

Li, Wen-Xue, Youko Oono, Jianhua Zhu, Xin-Jian He, Jian-Min Wu, Kei Iida, Xiao-Yan Lu, Xinping Cui, Hailing Jin, and Jian-Kang Zhu. "The arabidopsis NFYA5 transcription factor is regulated transcriptionally and posttranscriptionally to promote drought resistance." *The Plant Cell* 20, no. 8 (2008): 2238–2251.

Li, Yajie, Xiaorong Li, Jun Yang, and Yuke He. "Natural antisense transcripts of MIR398 genes suppress microR398 processing and attenuate plant thermotolerance." *Nature Communications* 11, no. 1 (2020a): 5351.

Li, Yanqun, Dexin Kong, Ying Fu, Michael R. Sussman, and Hong Wu. "The effect of developmental and environmental factors on secondary metabolites in medicinal plants." *Plant Physiology and Biochemistry* 148 (2020b): 80–89.

Li, Yong-Fang, Kangning Wei, Menglei Wang, Li Wang, Junxia Cui, Daijing Zhang, Junqiang Guo, Miao Zhao, and Yun Zheng. "Identification and temporal expression analysis of conserved and novel microRNAs in the leaves of winter wheat grown in the field." *Frontiers in Genetics* 10 (2019): 779.

Li, Zhaoxia, Xinrui Zhang, Xiuxia Liu, Yajie Zhao, Baomei Wang, and Juren Zhang. "miRNA alterations are important mechanism in maize adaptations to low-phosphate environments." *Plant Science* 252 (2016): 103–117.

Liang, Wei-Wei, Jing-Hao Huang, Chun-Ping Li, Lin-Tong Yang, Xin Ye, Dan Lin, and Li-Song Chen. "MicroRNA-mediated responses to long-term magnesium-deficiency in Citrus sinensis roots revealed by Illumina sequencing." *BMC Genomics* 18, no. 1 (2017): 1–16.

Lin, Yichun, Luying Zhang, Yuanyuan Zhao, Zewei Wang, Heng Liu, Linrun Zhang, Yuying Zhang et al. "Comparative analysis and functional identification of temperature-sensitive miRNA in *Arabidopsis anthers*." *Biochemical and Biophysical Research Communications* 532, no. 1 (2020): 1–10.

Liu, Guoyuan, Ji Liu, Wenfeng Pei, Xihua Li, Nuohan Wang, Jianjiang Ma, Xinshan Zang et al. "Analysis of the MIR160 gene family and the role of MIR160a_A05 in regulating fiber length in cotton." *Planta* 250 (2019a): 2147–2158.

Liu, Han-Hua, Xin Tian, Yan-Jie Li, Chang-Ai Wu, and Cheng-Chao Zheng. "Microarray-based analysis of stress-regulated microRNAs in *Arabidopsis thaliana*." *RNA* 14, no. 5 (2008): 836–843.

Liu, Wei, Liang Xu, Yan Wang, Hong Shen, Xianwen Zhu, Keyun Zhang, Yinglong Chen, Rugang Yu, Cecilia Limera, and Liwang Liu. "Transcriptome-wide analysis of chromium-stress responsive microRNAs to explore miRNA-mediated regulatory networks in radish (*Raphanus sativus* L.)." *Scientific Reports* 5, no. 1 (2015): 14024.

Liu, Yanrong, Dayong Li, Jianping Yan, Kexin Wang, Hong Luo, and Wanjun Zhang. "MiR319 mediated salt tolerance by ethylene." *Plant Biotechnology Journal* 17, no. 12 (2019b): 2370–2383.

Liu, Zhijie, Sunita Kumari, Lifang Zhang, Yonglian Zheng, and Doreen Ware. "Characterization of miRNAs in response to short-term waterlogging in three inbred lines of *Zea mays*." *PloS One* 7, no. 6 (2012): e39786.

Llave, Cesar, Kristin D. Kasschau, Maggie A. Rector, and James C. Carrington. "Endogenous and silencing-associated small RNAs in plants." *The Plant Cell* 14, no. 7 (2002): 1605–1619.

Loreti, Elena, Federico Betti, Maria Jose Ladera-Carmona, Fabrizia Fontana, Giacomo Novi, Maria Cristina Valeri, and Pierdomenico Perata. "ARGONAUTE1 and ARGONAUTE4 regulate gene expression and hypoxia tolerance." *Plant Physiology* 182, no. 1 (2020): 287–300.

Lu, Shanfa, Ying-Hsuan Sun, Rui Shi, Catherine Clark, Laigeng Li, and Vincent L. Chiang. "Novel and mechanical stress–responsive microRNAs in Populus trichocarpa that are absent from Arabidopsis." *The Plant Cell* 17, no. 8 (2005): 2186–2203.

Ma, Xu, Fei Zhao, and Bo Zhou. "The characters of non-coding RNAs and their biological roles in plant development and abiotic stress response." *International Journal of Molecular Sciences* 23, no. 8 (2022): 4124.

Ma, Yue, Hao Xue, Feng Zhang, Qiu Jiang, Shuang Yang, Pengtao Yue, Feng Wang et al. "The miR156/SPL module regulates apple salt stress tolerance by activating MdWRKY100 expression." *Plant Biotechnology Journal* 19, no. 2 (2021): 311–323.

Ma, Ziming, and Lanjuan Hu. "MicroRNA: A dynamic player from signalling to abiotic tolerance in plants." *International Journal of Molecular Sciences* 24, no. 14 (2023): 11364.

Mangrauthia, Satendra K., Surekha Agarwal, B. Sailaja, M. Sheshu Madhav, and S. R. Voleti. "MicroRNAs and their role in salt stress response in plants." In *Salt Stress in Plants: Signalling, Omics and Adaptations*, pp. 15–46. Springer, Nature, 2013.

Manna, Mrinalini, Balakrishnan Rengasamy, and Alok Krishna Sinha. "Revisiting the role of MAPK signalling pathway in plants and its manipulation for crop improvement." *Plant, Cell & Environment* 46, no. 8 (2023): 2277–2295.

Matthews, Craig, Muhammad Arshad, and Abdelali Hannoufa. "Alfalfa response to heat stress is modulated by microRNA156." *Physiologia Plantarum* 165, no. 4 (2019): 830–842.

Meng, Yijun, Xiaoxia Ma, Dijun Chen, Ping Wu, and Ming Chen. "MicroRNA-mediated signaling involved in plant root development." *Biochemical and Biophysical Research Communications* 393, no. 3 (2010): 345–349.

Min Yang, Zhi, and Jian Chen. "A potential role of microRNAs in plant response to metal toxicity." *Metallomics* 5, no. 9 (2013): 1184–1190.

Mishra, Shravan Kumar, Joanna Tripp, Sybille Winkelhaus, Bettina Tschiersch, Klaus Theres, Lutz Nover, and Klaus-Dieter Scharf. "In the complex family of heat stress transcription factors, HsfA1 has a unique role as master regulator of thermotolerance in tomato." *Genes & Development* 16, no. 12 (2002): 1555–1567.

Mittler, Ron. "Oxidative stress, antioxidants and stress tolerance." *Trends in Plant Science* 7, no. 9 (2002): 405–410.

Molina, Lilia, Jennine Rose Lapis, Nese Sreenivasulu, and Rosa Paula O. Cuevas. "Determination of macronutrient and micronutrient content in rice grains using inductively coupled plasma-optical emission spectrometry (ICP-OES)." In *Rice Grain Quality: Methods and Protocols*, pp. 253–264. Springer, Nature, 2019.

Mondal, Tapan Kumar, Alok Kumar Panda, Hukam C. Rawal, and Tilak Raj Sharma. "Discovery of microRNA-target modules of African rice (*Oryza glaberrima*) under salinity stress." *Scientific Reports* 8, no. 1 (2018): 570.

Nadarajah, Kalaivani, and Ilakiya Sharanee Kumar. "Drought response in rice: The miRNA story." *International Journal of Molecular Sciences* 20, no. 15 (2019): 3766.

Nazir, Faroza, Badar Jahan, Noushina Iqbal, Ashish B. Rajurkar, Manzer H. Siddiqui, and M. Iqbal R. Khan. "Methyl jasmonate influences ethylene formation, defense systems, nutrient homeostasis and carbohydrate metabolism to alleviate arsenic-induced stress in rice (*Oryza sativa*)." *Plant Physiology and Biochemistry* 202 (2023): 107990.

Nazir, Faroza, Laiba Khan, Shahid Umar, and M. Iqbal R. Khan. "Omics-based strategies for improving salt tolerance in rice." In *Current Omics Advancement in Plant Abiotic Stress Biology*, pp. 63–70. Academic Press, 2024.

Nguyen, Duc Quan, Ngoc Lan Nguyen, Van Tung Nguyen, Thi Huong Giang Tran, Thanh Hien Nguyen, Thi Kim Lien Nguyen, and Huy Hoang Nguyen. "Comparative analysis of microRNA expression profiles in shoot and root tissues of contrasting rice cultivars (*Oryza sativa* L.) with different salt stress tolerance." *Plos One* 18, no. 5 (2023): e0286140.

Ni, Zhiyong, Zheng Hu, Qiyan Jiang, and Hui Zhang. "GmNFYA3, a target gene of miR169, is a positive regulator of plant tolerance to drought stress." *Plant Molecular Biology* 82 (2013): 113–129.

Nie, Gang, Zongchao Liao, Minyi Zhong, Jie Zhou, Jiabang Cai, Aiyu Liu, Xia Wang, and Xinquan Zhang. "MicroRNA-mediated responses to chromium stress provide insight into tolerance characteristics of Miscanthus sinensis." *Frontiers in Plant Science* 12 (2021): 666117.

Ning, Li-Hua, Wen-kai Du, Hai-Na Song, Hong-Bo Shao, Wei-Cong Qi, Mohamed Salah Amr Sheteiwy, and De-yue Yu. "Identification of responsive miRNAs involved in combination stresses of phosphate starvation and salt stress in soybean root." *Environmental and Experimental Botany* 167 (2019): 103823.

Njaci, Isaac, Brett Williams, Claudia Castillo-González, Martin B. Dickman, Xiuren Zhang, and Sagadevan Mundree. "Genome-wide investigation of the role of microRNAs in desiccation tolerance in the resurrection grass *Tripogon loliiformis*." *Plants* 7, no. 3 (2018): 68.

Noman, Ali, Shah Fahad, Muhammad Aqeel, Usman Ali, Amanullah, Sumera Anwar, Shahbaz Khan Baloch, and Madiha Zainab. "miRNAs: Major modulators for crop growth and development under abiotic stresses." *Biotechnology Letters* 39 (2017): 685–700.

Noor, Iqra, Hamza Sohail, Jingxian Sun, Muhammad Azher Nawaz, Guohuai Li, Mirza Hasanuzzaman, and Junwei Liu. "Heavy metal and metalloid toxicity in horticultural plants: Tolerance mechanism and remediation strategies." *Chemosphere* 303 (2022): 135196.

O'Brien, Jacob, Heyam Hayder, Yara Zayed, and Chun Peng. "Overview of microRNA biogenesis, mechanisms of actions, and circulation." *Frontiers in Endocrinology* 9 (2018): 402.

O'Malley, Ronan C., Cesar C. Barragan, and Joseph R. Ecker. "A user's guide to the Arabidopsis T-DNA insertion mutant collections." In *Plant Functional Genomics: Methods and Protocols*, pp. 323–342. Springer, 2015.

Orek, Charles, Ken Mulupi, and Irene Koima. "A review of the roles of non-coding RNAs associated with drought stress response in Cassava (*Manihot esculenta* Crantz)." *Journal of Advances in Biology & Biotechnology* 26, no. 7 (2023): 1–10.

Othman, Syed Muhammad Iqbal Syed, Arif Faisal Mustaffa, M. Hafiz Che-Othman, Abdul Fatah A. Samad, Hoe-Han Goh, Zamri Zainal, and Ismanizan Ismail. "Overview of repressive miRNA regulation by short tandem target mimic (STTM): Applications and impact on plant biology." *Plants* 12, no. 3 (2023): 669.

Pagano, Luca, Riccardo Rossi, Laura Paesano, Nelson Marmiroli, and Marta Marmiroli. "miRNA regulation and stress adaptation in plants." *Environmental and Experimental Botany* 184 (2021): 104369.

Pan, LN. "Epigenetic regulation of abiotic stress response in plants to improve the stress tolerance." *Yi Chuan= Hereditas* 35, no. 6 (2013): 745–751.

Pant, Poonam, Sudip Pandey, and Stefano Dall'Acqua. "The influence of environmental conditions on secondary metabolites in medicinal plants: A literature review." *Chemistry & Biodiversity* 18, no. 11 (2021): e2100345.

Parizotto, Eneida Abreu, Patrice Dunoyer, Nadia Rahm, Christophe Himber, and Olivier Voinnet. "In vivo investigation of the transcription, processing, endonucleolytic activity, and functional relevance of the spatial distribution of a plant miRNA." *Genes & Development* 18, no. 18 (2004): 2237–2242.

Park, Wonkeun, Junjie Li, Rentao Song, Joachim Messing, and Xuemei Chen. "CARPEL FACTORY, a Dicer homolog, and HEN1, a novel protein, act in microRNA metabolism in *Arabidopsis thaliana*." *Current Biology* 12, no. 17 (2002): 1484–1495.

Parmar, Shaifaly, Sachin Ashruba Gharat, Ravichandra Tagirasa, Tilak Chandra, Lambodar Behera, Sushant Kumar Dash, and Birendra Prasad Shaw. "Identification and expression analysis of miRNAs and elucidation of their role in salt tolerance in rice varieties susceptible and tolerant to salinity." *PLoS One* 15, no. 4 (2020): e0230958.

Patanun, Onsaya, Manassawe Lertpanyasampatha, Punchapat Sojikul, Unchera Viboonjun, and Jarunya Narangajavana. "Computational identification of microRNAs and their targets in cassava (*Manihot esculenta* Crantz.)." *Molecular Biotechnology* 53 (2013): 257–269.

Pegler, Joseph L., Jackson M. J. Oultram, Duc Quan Nguyen, Christopher P. L. Grof, and Andrew L. Eamens. "MicroRNA-mediated responses to cadmium stress in *Arabidopsis thaliana*." *Plants* 10, no. 1 (2021): 130.

Peng, T., H. Wen, Y. Zhao, B. Wang, Y. Jin, H. Sun, and Q. Zhao. "Identification and expressions analysis of rice miRNA related to salt and drought stresses." *Acta Agric Boreali-Sinica* 8 (2018): 2114.

Perković, Stanislava, Carsten Paul, Filip Vasić, and Katharina Helming. "Human health and soil health risks from heavy metals, micro (nano) plastics, and antibiotic resistant bacteria in agricultural soils." *Agronomy* 12, no. 12 (2022): 2945.

Pervaiz, Tariq, Muhammad Waqas Amjid, Ashraf El-kereamy, Shi-Hui Niu, and Harry X. Wu. "MicroRNA and cDNA-microarray as potential targets against abiotic stress response in plants: Advances and prospects." *Agronomy* 12, no. 1 (2021): 11.

Phookaew, P., S. Netrphan, P. Sojikul, and J. Narangajavana. "Involvement of miR164-and miR167-mediated target gene expressions in responses to water deficit in cassava." *Biologia Plantarum* 58 (2014): 469–478.

Ramakrishnan, Muthusamy, Pradeep K. Papolu, Lakkakula Satish, Kunnummal Kurungara Vinod, Qiang Wei, Anket Sharma, Abolghassem Emamverdian, Long-Hai Zou, and Mingbing Zhou. "Redox status of the plant cell determines epigenetic modifications under abiotic stress conditions and during developmental processes." *Journal of Advanced Research* 42 (2022): 99–116.

Rao, Sombir, Sonia Balyan, Sarita Jha, and Saloni Mathur. "Novel insights into expansion and functional diversification of MIR169 family in tomato." *Planta* 251 (2020): 1–17.

Raza, Ali. "Eco-physiological and biochemical responses of rapeseed (*Brassica napus* L.) to abiotic stresses: Consequences and mitigation strategies." *Journal of Plant Growth Regulation* 40, no. 4 (2021): 1368–1388.

Raza, Ali, Sidra Charagh, Benjamin Karikari, Rahat Sharif, Vivek Yadav, Muhammad Salman Mubarik, Madiha Habib et al. "miRNAs for crop improvement." *Plant Physiology and Biochemistry* (2023): 107857.

Ražná, Katarína, Ľubomír Harenčár, and Matúš Kučka. "The involvement of microRNAs in plant lignan biosynthesis – current view." *Cells* 11, no. 14 (2022): 2151.

Reichel, Marlene, Yanjiao Li, Junyan Li, and Anthony A. Millar. "Inhibiting plant micro RNA activity: Molecular SPONGE s, target MIMIC s and STTM s all display variable efficacies against target micro RNAs." *Plant Biotechnology Journal* 13, no. 7 (2015): 915–926.

Reinhart, Brenda J., Earl G. Weinstein, Matthew W. Rhoades, Bonnie Bartel, and David P. Bartel. "MicroRNAs in plants." *Genes & Development* 16, no. 13 (2002): 1616–1626.

Rogans, Sarah Jane, and Chrissie Rey. "Unveiling the micronome of Cassava (*Manihot esculenta* Crantz)." *PLoS One* 11, no. 1 (2016): e0147251.

Salgado, Fernanda Ferreira, Thalliton Luiz Carvalho Da Silva, Letícia Rios Vieira, Vivianny Nayse Belo Silva, André Pereira Leão, Marcos Mota do Carmo Costa, Roberto CoitiTogawa, Carlos Antonio Ferreira De Sousa, Priscila Grynberg, and Manoel Teixeira Souza Jr. "The early response of oil palm (*Elaeis guineensis* Jacq.) plants to water deprivation: Expression analysis of miRNAs and their putative target genes, and similarities with the response to salinity stress." *Frontiers in Plant Science* 13 (2022): 970113.

Salvi, Prafull, Mrinalini Manna, Harmeet Kaur, Tanika Thakur, Nishu Gandass, Deepesh Bhatt, and Mehanathan Muthamilarasan. "Phytohormone signaling and crosstalk in regulating drought stress response in plants." *Plant Cell Reports* 40 (2021): 1305–1329.

Santosh, Baby, Akhil Varshney, and Pramod Kumar Yadava. "Non-coding RNAs: Biological functions and applications." *Cell Biochemistry and Function* 33, no. 1 (2015): 14–22.

Senthil-Kumar, M., Geetha Govind, Li Kang, Kirankumar S. Mysore, and M. Udayakumar. "Functional characterization of *Nicotiana benthamiana* homologs of peanut water deficit-induced genes by virus-induced gene silencing." *Planta* 225 (2007): 523–539.

Sepúlveda-García, E. B., J. F. Pulido-Barajas, and A. A. Huerta-Heredia. "Differential expression of maize and teosinte micrornas under submergence, drought, and alternated stress." *Plants* 9 (2020): 1–20.

Shah, Iftikhar Hussain, Muhammad Aamir Manzoor, Irfan Ali Sabir, Muhammad Ashraf, Fazal Haq, Samiah Arif, Muhammad Abdullah, Qingliang Niu, and Yidong Zhang. "Genome-wide identification and comparative analysis of MATE gene family in Cucurbitaceae species and their regulatory role in melon (*Cucumis melo*) under salt stress." *Horticulture, Environment, and Biotechnology* 63, no. 4 (2022): 595–612.

Shaik, N. A., K. Nasser, A. Mohammed, A. Mujalli, A. A. Obaid, A. A. El-Harouni, R. Elango, and B. Banaganapalli. "Identification of miRNA–mRNA–TFs regulatory network and crucial pathways involved in asthma through advanced systems biology approaches." *Plos One* 17, no. 10 (2022): e0271262.

Shan, T., R. Fu, Y. Xie, Q. Chen, Y. Wang, Zh Li, X. Song, P. Li, and B. Wang. "Regulatory mechanism of maize (*Zea mays* L.) miR164 in salt stress response." *Russian Journal of Genetics* 56 (2020): 835–842.

Shang, Renfu, Seungjae Lee, Gayan Senavirathne, and Eric C. Lai. "microRNAs in action: biogenesis, function and regulation." *Nature Reviews Genetics* (2023): 1–18.

Sharma, Ashish, Poorwa Kamal Badola, Chitra Bhatia, Deepika Sharma, and Prabodh Kumar Trivedi. "Primary transcript of miR858 encodes regulatory peptide and controls flavonoid biosynthesis and development in Arabidopsis." *Nature Plants* 6, no. 10 (2020a): 1262–1274.

Sharma, Ashish, Himanshi Gautam, and Prabodh Kumar Trivedi. "Genetic manipulation of microRNAs: Approaches and limitations." *Journal of Plant Biochemistry and Biotechnology* (2023a): 1–13.

Sharma, Ashutosh, Luis M. Ruiz-Manriquez, Francisco I. Serrano-Cano, Paula Roxana Reyes-Pérez, Cynthia Karina Tovar Alfaro, Yulissa Esmeralda Barrón Andrade, Ana Karen Hernández Aros, Aashish Srivastava, and Sujay Paul. "Identification of microRNAs and their expression in leaf tissues of guava (*Psidium guajava* L.) under salinity stress." *Agronomy* 10, no. 12 (2020b): 1920.

Sharma, Deepika, Manish Tiwari, Ashutosh Pandey, Chitra Bhatia, Ashish Sharma, and Prabodh Kumar Trivedi. "MicroRNA858 is a potential regulator of phenylpropanoid pathway and plant development." *Plant Physiology* 171, no. 2 (2016): 944–959.

Sharma, Nandni, Kanika Khanna, Shivam Jasrotia, Deepak Kumar, Renu Bhardwaj, and Puja Ohri. "Metabolites and chemical agents in the plant roots: An overview of their use in plant-parasitic nematode management." *Nematology* 25, no. 3 (2023b): 243–257.

Sharma, Nandni, Sandeep Kour, Deepak Kumar, Ravinderjit Kaur, Anjali Khajuria, and Puja Ohri. "Role of brassinosteroids (BRs) in modulating antioxidative defense mechanism in plants growing under abiotic and biotic stress conditions." In *Antioxidant Defense in Plants: Molecular Basis of Regulation*, pp. 325–367. Singapore: Springer Nature Singapore, 2022.

Shen, Chuang, Ying-Ying Huang, Jun-Liang Xin, Chun-Tao He, and Zhong-Yi Yang. "A novel microRNA IamiR-4–3p from water spinach (*Ipomoea aquatica* Forsk.) increased Cd uptake and translocation in *Arabidopsis thaliana*." *Environmental Science and Pollution Research* 29, no. 27 (2022): 41375–41385.

Shi, Xiaopu, Fangling Jiang, Junqin Wen, and Zhen Wu. "Overexpression of *Solanum habrochaites* microRNA319d (sha-miR319d) confers chilling and heat stress tolerance in tomato (*S. lycopersicum*)." *BMC Plant Biology* 19 (2019): 1–17.

Shin, Sang-Yoon, Jin Seo Jeong, Jae Yun Lim, Taewook Kim, June Hyun Park, Ju-Kon Kim, and Chanseok Shin. "Transcriptomic analyses of rice (*Oryza sativa*) genes and non-coding RNAs under nitrogen starvation using multiple omics technologies." *BMC Genomics* 19 (2018): 1–20.

Shinde, Harshraj, Ambika Dudhate, Lakshay Anand, Daisuke Tsugama, Shashi K. Gupta, Shenkui Liu, and Tetsuo Takano. "Small RNA sequencing reveals the role of pearl millet miRNAs and their targets in salinity stress responses." *South African Journal of Botany* 132 (2020): 395–402.

Shinozaki, Kazuo, and Kazuko Yamaguchi-Shinozaki. "Molecular responses to dehydration and low temperature: Differences and cross-talk between two stress signaling pathways." *Current Opinion in Plant Biology* 3, no. 3 (2000): 217–223.

Shriram, Varsha, Vinay Kumar, Rachayya M. Devarumath, Tushar S. Khare, and Shabir H. Wani. "MicroRNAs as potential targets for abiotic stress tolerance in plants." *Frontiers in Plant Science* 7 (2016): 817.

Shukla, Pushp Sheel, Tudor Borza, Alan T. Critchley, David Hiltz, Jeff Norrie, and Balakrishnan Prithiviraj. "Ascophyllum nodosum extract mitigates salinity stress in *Arabidopsis thaliana* by modulating the expression of miRNA involved in stress tolerance and nutrient acquisition." *PLoS One* 13, no. 10 (2018): e0206221.

Sihag, Pooja, Vijeta Sagwal, Anuj Kumar, Priyanka Balyan, Reyazul Rouf Mir, Om Parkash Dhankher, and Upendra Kumar. "Discovery of miRNAs and development of heat-responsive miRNA-SSR markers for characterization of wheat germplasm for terminal heat tolerance breeding." *Frontiers in Genetics* 12 (2021): 699420.

Singh, Archana, Deepti Jain, Jyotsna Pandey, Manisha Yadav, Kailash C. Bansal, and Indrakant K. Singh. "Deciphering the role of miRNA in reprogramming plant responses to drought stress." *Critical Reviews in Biotechnology* 43, no. 4 (2023): 613–627.

Singh, Puja, Prasanna Dutta, and Debasis Chakrabarty. "miRNAs play critical roles in response to abiotic stress by modulating cross-talk of phytohormone signaling." *Plant Cell Reports* 40, no. 9 (2021): 1617–1630.

Singh, Roshan Kumar, Ashish Prasad, Mehanathan Muthamilarasan, Swarup K. Parida, and Manoj Prasad. "Breeding and biotechnological interventions for trait improvement: status and prospects." *Planta* 252 (2020): 1–18.

Singroha, Garima, Pradeep Sharma, and Ramanjulu Sunkur. "Current status of microRNA-mediated regulation of drought stress responses in cereals." *Physiologia Plantarum* 172, no. 3 (2021): 1808–1821.

Srivastava, Vikas, Aksar Ali Chowdhary, Praveen Kumar Verma, Shakti Mehrotra, and Sonal Mishra. "Hydrogen sulfide-mediated mitigation and its integrated signaling crosstalk during salinity stress." *Physiologia Plantarum* 174, no. 1 (2022): e13633.

Sun, Fenglong, Guanghui Guo, Jinkun Du, Weiwei Guo, Huiru Peng, Zhongfu Ni, Qixin Sun, and Yingyin Yao. "Whole-genome discovery of miRNAs and their targets in wheat (*Triticum aestivum* L.)." *BMC Plant Biology* 14, no. 1 (2014): 1–17.

Sun, Mingzhe, Yang Shen, Junkai Yang, Xiaoxi Cai, Hongyu Li, Yanming Zhu, Bowei Jia, and Xiaoli Sun. "miR535 negatively regulates cold tolerance in rice." *Molecular Breeding* 40 (2020): 1–12.

Sun, Xi, Hong-Xiang Zheng, Simin Li, Yinping Gao, Yingying Dang, Zengting Chen, Fenghui Wu, Xuemei Wang, Qi Xie, and Na Sui. "MicroRNAs balance growth and salt stress responses in sweet sorghum." *The Plant Journal* 113, no. 4 (2023): 677–697.

Sun, Xiaochuan, Liang Xu, Yan Wang, Rugang Yu, Xianwen Zhu, Xiaobo Luo, Yiqin Gong et al. "Identification of novel and salt-responsive miRNAs to explore miRNA-mediated regulatory network of salt stress response in radish (*Raphanus sativus* L.)." *BMC Genomics* 16, no. 1 (2015): 1–16.

Sunkar, Ramanjulu, Yong-Fang Li, and Guru Jagadeeswaran. "Functions of microRNAs in plant stress responses." *Trends in Plant Science* 17, no. 4 (2012): 196–203.

Sunkar, Ramanjulu, and Jian-Kang Zhu. "Novel and stress-regulated microRNAs and other small RNAs from Arabidopsis." *The Plant Cell* 16, no. 8 (2004): 2001–2019.

Taghvaei, Mohammad Mahdi, Habibollah Samizadeh Lahiji, and Mohammad Mohsenzadeh Golfazani. "Evaluation of expression changes, proteins interaction network, and microRNAs targeting catalase and superoxide dismutase genes under cold stress in rapeseed (*Brassica napus* L.)." *OCL* 29 (2022): 3.

Tang, Guiliang, and Xiaoqing Tang. "Short tandem target mimic: A long journey to the engineered molecular landmine for selective destruction/blockage of microRNAs in plants and animals." *Journal of Genetics and Genomics* 40, no. 6 (2013): 291–296.

Tariq, Rezwan, Ammara Hussain, Arslan Tariq, Muhammad Hayder Bin Khalid, Imran Khan, Huseyin Basim, and Pär K. Ingvarsson. "Genome-wide analyses of the mung bean NAC gene family reveals orthologs, co-expression networking and expression profiling under abiotic and biotic stresses." *BMC Plant Biology* 22, no. 1 (2022): 343.

Teng, Lidong, Xueqing Zhang, Runfeng Wang, Kaina Lin, Meng Zeng, Hao Chen, and Fangbin Cao. "miRNA transcriptome reveals key miRNAs and their targets contributing to the difference in Cd tolerance of two contrasting maize genotypes." *Ecotoxicology and Environmental Safety* 256 (2023): 114881.

Thornburg, Thomas Elliott, Jia Liu, Qian Li, Huiyun Xue, Guo Wang, Lijie Li, Julia Elise Fontana et al. "Potassium deficiency significantly affected plant growth and development as well as microRNA-mediated mechanism in wheat (*Triticum aestivum* L.)." *Frontiers in Plant Science* 11 (2020): 1219.

Tikoria, Raman, Arvinder Kaur, and Puja Ohri. "Physiological, biochemical and structural changes in tomato plants by vermicompost application in different exposure periods under glass house conditions." *Plant Physiology and Biochemistry* 197 (2023a): 107656.

Tikoria, Raman, Arvinder Kaur, and Puja Ohri. "Amelioration of oxidative stress and growth enhancement by application of vermicompost via modulating phyto-constituents in tomato plants during nematode stress." *Journal of Soil Science and Plant Nutrition* (2023b): 1–17.

Tikoria, Raman, Arvinder Kaur, and Puja Ohri. "Modulation of various phytoconstituents in tomato seedling growth and *Meloidogyne incognita*–induced stress alleviation by vermicompost application." *Frontiers in Environmental Science* 10 (2022): 891195.

Tiwari, Manish, Deepika Sharma, and Prabodh Kumar Trivedi. "Artificial microRNA mediated gene silencing in plants: Progress and perspectives." *Plant Molecular Biology* 86 (2014): 1–18.

Tiwari, Sanjesh, Anuradha Patel, Neeraj Pandey, Amandeep Raju, Madhulika Singh, and Sheo Mohan Prasad. "Deficiency of essential elements in crop plants." In K. Mishra, P. K. Tandon, and S. Srivastava (Eds.), *Sustainable Solutions for Elemental Deficiency and Excess in Crop Plants*, pp. 19–52. Springer, 2020.

Todesco, Marco, Ignacio Rubio-Somoza, Javier Paz-Ares, and Detlef Weigel. "A collection of target mimics for comprehensive analysis of microRNA function in *Arabidopsis thaliana*." *PLoS Genetics* 6, no. 7 (2010): e1001031.

Treiber, Thomas, Nora Treiber, and Gunter Meister. "Regulation of microRNA biogenesis and its crosstalk with other cellular pathways." *Nature Reviews Molecular Cell Biology* 20, no. 1 (2019): 5–20.

Trumbo, Jennifer Lynn, Baohong Zhang, and Charles Neal Stewart Jr. "Manipulating micro RNA s for improved biomass and biofuels from plant feedstocks." *Plant Biotechnology Journal* 13, no. 3 (2015): 337–354.

Ulhassan, Zaid, Javaid Akhter Bhat, Weijun Zhou, Ahmed M. Senan, Pravej Alam, and Parvaiz Ahmad. "Attenuation mechanisms of arsenic induced toxicity and its accumulation in plants by engineered nanoparticles: A review." *Environmental Pollution* 302 (2022): 119038.

Ullah, Izhar, Ehab AR Kamel, Syed Tanveer Shah, Abdul Basit, Heba I. Mohamed, and Muhammad Sajid. "Application of RNAi technology: A novel approach to navigate abiotic stresses." *Molecular Biology Reports* 49, no. 11 (2022): 10975–10993.

Valdés-López, Oswaldo, S. Samuel Yang, Rosaura Aparicio-Fabre, Peter H. Graham, José Luis Reyes, Carroll P. Vance, and Georgina Hernández. "MicroRNA expression profile in common bean (*Phaseolus vulgaris*) under nutrient deficiency stresses and manganese toxicity." *New Phytologist* 187, no. 3 (2010): 805–818.

Vega, Andrea, Ninoska Delgado, and Michael Handford. "Increasing heavy metal tolerance by the exogenous application of organic acids." *International Journal of Molecular Sciences* 23, no. 10 (2022): 5438.

Vega, Andrea, José Antonio O'Brien, and Rodrigo A. Gutiérrez. "Nitrate and hormonal signaling crosstalk for plant growth and development." *Current Opinion in Plant Biology* 52 (2019): 155–163.

Visentin, Ivan, Chiara Pagliarani, Eleonora Deva, Alessio Caracci, Veronika Turečková, Ondrej Novák, Claudio Lovisolo, Andrea Schubert, and Francesca Cardinale. "A novel strigolactone-miR156 module controls stomatal behaviour during drought recovery." *Plant, Cell & Environment* 43, no. 7 (2020): 1613–1624.

Wagaba, Henry, Basavaprabhu L. Patil, Settumba Mukasa, Titus Alicai, Claude M. Fauquet, and Nigel J. Taylor. "Artificial microRNA-derived resistance to Cassava brown streak disease." *Journal of Virological Methods* 231 (2016): 38–43.

Wan, Jiong, Shujun Meng, Qiyue Wang, Jiawen Zhao, Xiaoqian Qiu, Liangfa Wang, Juan Li et al. "Suppression of microRNA168 enhances salt tolerance in rice (*Oryza sativa* L.)." *BMC Plant Biology* 22, no. 1 (2022): 1–11.

Wang, B., Y. F. Sun, N. Song, X. J. Wang, H. Feng, L. L. Huang, and Z. S. Kang. "Identification of UV-B-induced microRNAs in wheat." *Genetics and Molecular Research* 12, no. 4 (2013): 4213–4221.

Wang, Ji, Zhaosheng Zhou, Qing Tao, Xiuping Chen, Chen Shui, Xuyang Ren, Ling Yu, and Mingxiang Liang. "Brassica napus miR169 regulates BnaNF-YA in salinity, drought and ABA responses." *Environmental and Experimental Botany* 199 (2022): 104882.

Wang, Jianwen, Youju Ye, Meng Xu, Liguo Feng, and Li-an Xu. "Roles of the SPL gene family and miR156 in the salt stress responses of tamarisk (*Tamarix chinensis*)." *BMC Plant Biology* 19 (2019a): 1–11.

Wang, Jingjing, Xianwen Meng, Oxana B. Dobrovolskaya, Yuriy L. Orlov, and Ming Chen. "Non-coding RNAs and their roles in stress response in plants." *Genomics, Proteomics & Bioinformatics* 15, no. 5 (2017): 301–312.

Wang, Ling, Xinyang Bai, Yuanjinzi Qiao, Lili Si, Zidi Yu, Chenyang Ni, Tianjiao Li, Chengjin Guo, and Kai Xiao. "tae-miR9674a, a microRNA member of wheat, confers plant drought and salt tolerance through modulating the stomata movement and ROS homeostasis." *Plant Biotechnology Reports* 17, no. 4 (2023): 471–488.

Wang, Wei, Dan Liu, Dongdong Chen, Yingying Cheng, Xiaopei Zhang, Lirong Song, Mengjiao Hu, Jie Dong, and Fafu Shen. "MicroRNA414c affects salt tolerance of cotton by regulating reactive oxygen species metabolism under salinity stress." *RNA Biology* 16, no. 3 (2019b): 362–375.

Wang, Yang, Michelle Beaith, Maryse Chalifoux, Jifeng Ying, Tina Uchacz, Carlene Sarvas, Rebecca Griffiths, Monika Kuzma, Jiangxin Wan, and Yafan Huang. "Shoot-specific down-regulation of protein farnesyl-transferase (α-subunit) for yield protection against drought in canola." *Molecular Plant* 2, no. 1 (2009): 191–200.

Wani, Shabir Hussain, Vinay Kumar, Tushar Khare, Prateek Tripathi, Tariq Shah, Chopperla Ramakrishna, Supriya Aglawe, and Satendra Kumar Mangrauthia. "miRNA applications for engineering abiotic stress tolerance in plants." *Biologia* 75 (2020): 1063–1081.

Warthmann, Norman, Hao Chen, Stephan Ossowski, Detlef Weigel, and Philippe Hervé. "Highly specific gene silencing by artificial miRNAs in rice." *PLoS One* 3, no. 3 (2008): e1829.

Wen, Fen Liang, Yang Yue, Teng Fei He, Xiu Mei Gao, Zhao Sheng Zhou, and Xiao Hua Long. "Identification of miR390-TAS3-ARF pathway in response to salt stress in *Helianthus tuberosus* L." *Gene* 738 (2020): 144460.

Wendt, Toni, Preben Bach Holm, Colby G. Starker, Michelle Christian, Daniel F. Voytas, Henrik Brinch-Pedersen, and Inger Bæksted Holme. "TAL effector nucleases induce mutations at a pre-selected location in the genome of primary barley transformants." *Plant Molecular Biology* 83 (2013): 279–285.

Wijerathna-Yapa, Akila. "Transgenic plants: Resistance to abiotic and biotic stresses." *Journal of Agriculture and Environment for International Development (JAEID)* 111, no. 1 (2017): 245–275.

Wilczynska, Ania, and Martin Bushell. "The complexity of miRNA-mediated repression." *Cell Death & Differentiation* 22, no. 1 (2015): 22–33.

Wu, Chunlai, Zehong Ding, Mingjie Chen, Guangxiao Yang, Weiwei Tie, Yan Yan, Jian Zeng, Guangyuan He, and Wei Hu. "Identification and functional prediction of lncRNAs in response to PEG and ABA treatment in cassava." *Environmental and Experimental Botany* 166 (2019): 103809.

Wulandhari, Lili Ayu, Alexander Agung Santoso Gunawan, Arie Qurania, PrihastutiHarsani, Triastinurmiatiningsih Ferdy Tarawan, and Riska Fauzia Hermawan. "Plant nutrient deficiency detection using deep convolutional neural network." *ICIC Express Letters* 13, no. 10 (2019): 971–977.

Xia, Jing, Changying Zeng, Zheng Chen, Kevin Zhang, Xin Chen, Yufei Zhou, Shun Song et al. "Endogenous small-noncoding RNAs and their roles in chilling response and stress acclimation in Cassava." *BMC Genomics* 15, no. 1 (2014): 1–19.

Xiao, Liang, Xiao-Hong Shang, Sheng Cao, Xiang-Yu Xie, Wen-Dan Zeng, Liu-Ying Lu, Song-Bi Chen, and Hua-Bing Yan. "Comparative physiology and transcriptome analysis allows for identification of lncRNAs imparting tolerance to drought stress in autotetraploid cassava." *BMC Genomics* 20, no. 1 (2019): 1–15.

Xie, Zhixia, Ping Zhang, Jingjing Zhao, Ruling Wang, Jianping Gao, and Jin Xu. "Identification of microRNAs from Zn-treated Solanum nigrum roots by small RNA sequencing." *Acta Physiologiae Plantarum* 39 (2017): 1–5.

Xing, Lijuan, Ming Zhu, Mingda Luan, Min Zhang, Lian Jin, Yueping Liu, Junjie Zou, Lei Wang, and Miaoyun Xu. "miR169q and NUCLEAR FACTOR YA8 enhance salt tolerance by activating PEROXIDASE1 expression in response to ROS." *Plant Physiology* 188, no. 1 (2022): 608–623.

Xu, Jin, Qin-Min Hou, Tushar Khare, Sandeep Kumar Verma, and Vinay Kumar. "Exploring miRNAs for developing climate-resilient crops: A perspective review." *Science of the Total Environment* 653 (2019): 91–104.

Xu, Rui, Yuhan Wang, Hao Zheng, Wei Lu, Changai Wu, Jinguang Huang, Kang Yan, Guodong Yang, and Chengchao Zheng. "Salt-induced transcription factor MYB74 is regulated by the RNA-directed DNA methylation pathway in Arabidopsis." *Journal of Experimental Botany* 66, no. 19 (2015): 5997–6008.

Yaashikaa, P. R., P. Senthil Kumar, S. Jeevanantham, and R. Saravanan. "A review on bioremediation approach for heavy metal detoxification and accumulation in plants." *Environmental Pollution* 301 (2022): 119035.

Yamazaki, Shinichi, Yosuke Ueda, Aya Mukai, Kumiko Ochiai, and Toru Matoh. "Rice phytochelatin synthases Os PCS 1 and Os PCS 2 make different contributions to cadmium and arsenic tolerance." *Plant Direct* 2, no. 1 (2018): e00034.

Yang, Hua, Rui Fang, Ling Luo, Wei Yang, Qiong Huang, Chunlin Yang, Wenkai Hui, Wei Gong, and Jingyan Wang. "Uncovering the mechanisms of salicylic acid-mediated abiotic stress tolerance in horticultural crops." *Frontiers in Plant Science* 14 (2023).

Yang, Qiuying, Liande Li, Zhihong Xue, Qiaohong Ye, Lin Zhang, Shaojie Li, and Yi Liu. "Transcription of the major Neurospora crassa microRNA–like small RNAs relies on RNA polymerase III." *PLoS Genetics* 9, no. 1 (2013): e1003227.

Yang, Tongwen, Lingui Xue, and Lizhe An. "Functional diversity of miRNA in plants." *Plant Science* 172, no. 3 (2007): 423–432.

Yang, Xiaoqian, Lichun Zhang, Yuzhang Yang, Markus Schmid, and Yanwei Wang. "miRNA mediated regulation and interaction between plants and pathogens." *International Journal of Molecular Sciences* 22, no. 6 (2021): 2913.

Yang, Yuting, Xu Zhang, Yachun Su, Jiake Zou, Zhoutao Wang, Liping Xu, and Youxiong Que. "miRNA alteration is an important mechanism in sugarcane response to low-temperature environment." *BMC Genomics* 18, no. 1 (2017): 1–18.

Yousuf, Peerzada Yasir, Peerzada Arshid Shabir, and Khalid Rehman Hakeem. "miRNAomic approach to plant nitrogen starvation." *International Journal of Genomics* 2021 (2021).

Yu, Chao, Yutong Chen, Yaqian Cao, Huamin Chen, Jichun Wang, Yong-Mei Bi, Fang Tian et al. "Overexpression of miR169o, an overlapping microRNA in response to both nitrogen limitation and bacterial infection, promotes nitrogen use efficiency and susceptibility to bacterial blight in rice." *Plant and Cell Physiology* 59, no. 6 (2018): 1234–1247.

Yu, H., and P. P. Kumar. "Post-transcriptional gene silencing in plants by RNA." *Plant Cell Reports* 22 (2003): 167–174.

Yu, Jingjing, Dan Su, Dongjing Yang, Tingting Dong, Zhonghou Tang, Hongmin Li, Yonghua Han, Zongyun Li, and Baohong Zhang. "Chilling and heat stress-induced physiological changes and microRNA-related mechanism in sweetpotato (*Ipomoea batatas* L.)." *Frontiers in Plant Science* 11 (2020): 687.

Yu, Sha, Li Cao, Chuan-Miao Zhou, Tian-Qi Zhang, Heng Lian, Yue Sun, Jianqiang Wu, Jirong Huang, Guodong Wang, and Jia-Wei Wang. "Sugar is an endogenous cue for juvenile-to-adult phase transition in plants." *elife* 2 (2013): e00269.

Yu, Yuehua, Zhiyong Ni, Yi Wang, Huina Wan, Zheng Hu, Qiyan Jiang, Xianjun Sun, and Hui Zhang. "Overexpression of soybean miR169c confers increased drought stress sensitivity in transgenic *Arabidopsis thaliana*." *Plant Science* 285 (2019a): 68–78.

Yu, Zhengyang, Xin Wang, Xiaoqian Mu, and Linsheng Zhang. "RNAi mediated silencing of dehydrin gene WZY2 confers osmotic stress intolerance in transgenic wheat." *Functional Plant Biology* 46, no. 10 (2019b): 877–884.

Yuan, Shuangrong, Junming Zhao, Zhigang Li, Qian Hu, Ning Yuan, Man Zhou, Xiaoxia Xia et al. "MicroRNA396-mediated alteration in plant development and salinity stress response in creeping bentgrass." *Horticulture Research* 6 (2019).

Zeng, Changying, Wenquan Wang, Yun Zheng, Xin Chen, Weiping Bo, Shun Song, Weixiong Zhang, and Ming Peng. "Conservation and divergence of microRNAs and their functions in Euphorbiaceous plants." *Nucleic Acids Research* 38, no. 3 (2010): 981–995.

Zeng, Houqing, Xin Zhang, Ming Ding, and Yiyong Zhu. "Integrated analyses of miRNAome and transcriptome reveal zinc deficiency responses in rice seedlings." *BMC Plant Biology* 19, no. 1 (2019): 1–18.

Zhan, Jingjing, Yangyang Diao, Guo Yin, Muhammad Sajjad, Xi Wei, Zhengying Lu, and Ye Wang. "Integration of mRNA and miRNA analysis reveals the molecular mechanism of cotton response to salt stress." *Frontiers in Plant Science* 12 (2021): 767984.

Zhang, Baohong. "MicroRNA: A new target for improving plant tolerance to abiotic stress." *Journal of Experimental Botany* 66, no. 7 (2015): 1749–1761.

Zhang, Baohong, Xiaoping Pan, George P. Cobb, and Todd A. Anderson. "Plant microRNA: a small regulatory molecule with big impact." *Developmental Biology* 289, no. 1 (2006): 3–16.

Zhang, Baohong, and Qinglian Wang. "MicroRNA-based biotechnology for plant improvement." *Journal of Cellular Physiology* 230, no. 1 (2015): 1–15.

Zhang, Feiyan, Jiangwei Yang, Ning Zhang, Jiahe Wu, and Huaijun Si. "Roles of microRNAs in abiotic stress response and characteristics regulation of plant." *Frontiers in Plant Science* 13 (2022a): 919243.

Zhang, Yingying, Ye Zhou, Weimin Zhu, Junzhong Liu, and Fang Cheng. "Non-coding RNAs fine-tune the balance between plant growth and abiotic stress tolerance." *Frontiers in Plant Science* 13 (2022b): 965745.

Zhang, Jinshan, Hui Zhang, Ashish Kumar Srivastava, Yujie Pan, Jinjuan Bai, Jingjing Fang, Huazhong Shi, and Jian-Kang Zhu. "Knockdown of rice microRNA166 confers drought resistance by causing leaf rolling and altering stem xylem development." *Plant Physiology* 176, no. 3 (2018): 2082–2094.

Zhang, Li, Yun Zheng, Guru Jagadeeswaran, Yongfang Li, Kanchana Gowdu, and Ramanjulu Sunkar. "Identification and temporal expression analysis of conserved and novel microRNAs in Sorghum." *Genomics* 98, no. 6 (2011): 460–468.

Zhang, Lipeng, Dongying Fan, Hui Li, Qiuju Chen, Zhen Zhang, Minying Liu, Jingjing Liu et al. "Characterization and identification of grapevine heat stress-responsive microRNAs revealed the positive regulated function of vvi-miR167 in thermostability." *Plant Science* 329 (2023): 111623.

Zhang, Lu, Han Ding, Hailing Jiang, Huasen Wang, Kexin Chen, Jinju Duan, Shengjun Feng, and Gang Wu. "Regulation of cadmium tolerance and accumulation by miR156 in Arabidopsis." *Chemosphere* 242 (2020): 125168.

Zhang, Yingting, Junjie Yang, Lijuan Zhu, Jinyu Xue, Hailiang Hu, Jiebing Cui, and Jin Xu. "Identification of microRNAs and their target genes related to needle discoloration of evergreen tree Chinese cedar (*Cryptomeria fortunei*) in cold winters." *Planta* 254 (2021): 1–21.

Zhang, Yu-Chan, Yang Yu, Cong-Ying Wang, Ze-Yuan Li, Qing Liu, Jie Xu, Jian-You Liao et al. "Overexpression of microRNA OsmiR397 improves rice yield by increasing grain size and promoting panicle branching." *Nature Biotechnology* 31, no. 9 (2013): 848.

Zhao, Jinping, Qingtao Liu, Pu Hu, Qi Jia, Na Liu, Kangquan Yin, Ye Cheng, Fei Yan, Jianping Chen, and Yule Liu. "An efficient Potato virus X-based microRNA silencing in *Nicotiana benthamiana*." *Scientific Reports* 6, no. 1 (2016): 20573.

Zheng, Yun, Vandana Hivrale, Xiaotuo Zhang, Babu Valliyodan, Christine Lelandais-Brière, Andrew D. Farmer, Gregory D. May, Martin Crespi, Henry T. Nguyen, and Ramanjulu Sunkar. "Small RNA profiles in soybean primary root tips under water deficit." *BMC Systems Biology* 10 (2016): 1–10.

Zhou, Bo, Xiang Gao, and Fei Zhao. "Integration of mRNA and miRNA analysis reveals the post-transcriptional regulation of salt stress response in *Hemerocallis fulva*." *International Journal of Molecular Sciences* 24, no. 8 (2023): 7290.

Zhou, Jianping, Kejun Deng, Yan Cheng, Zhaohui Zhong, Li Tian, Xu Tang, Aiting Tang et al. "CRISPR-Cas9 based genome editing reveals new insights into microRNA function and regulation in rice." *Frontiers in Plant Science* 8 (2017): 1598.

Zhou, Jianping, Zhaohui Zhong, Hongqiao Chen, Qian Li, Xuelian Zheng, Yiping Qi, and Yong Zhang. "Knocking out MicroRNA genes in rice with CRISPR-Cas9." In *Plant Genome Editing with CRISPR Systems: Methods and Protocols*, pp. 109–119. Springer, Nature, 2019a.

Zhou, Min, Shigang Zheng, Rong Liu, Lu Lu, Chihong Zhang, Lei Zhang, Levi Yant, and Yu Wu. "The genome-wide impact of cadmium on microRNA and mRNA expression in contrasting Cd responsive wheat genotypes." *BMC Genomics* 20 (2019b): 1–19.

Zhou, Rong, Lingpeng Kong, Xiaqing Yu, Carl-Otto Ottosen, Tongmin Zhao, Fangling Jiang, and Zhen Wu. "Oxidative damage and antioxidant mechanism in tomatoes responding to drought and heat stress." *Acta Physiologiae Plantarum* 41 (2019c): 1–11.

Zhou, Man, Dayong Li, Zhigang Li, Qian Hu, Chunhua Yang, Lihuang Zhu, and Hong Luo. "Constitutive expression of a miR319 gene alters plant development and enhances salt and drought tolerance in transgenic creeping bentgrass." *Plant Physiology* 161, no. 3 (2013): 1375–1391.

Zhou, Man, and Hong Luo."MicroRNA-mediated gene regulation: Potential applications for plant genetic engineering."*Plant Molecular Biology* 83 (2013): 59–75.

Zhou, Rong, Xiaqing Yu, Carl-Otto Ottosen, Tingling Zhang, Zhen Wu, and Tongmin Zhao. "Unique miRNAs and their targets in tomato leaf responding to combined drought and heat stress." *BMC Plant Biology* 20, no. 1 (2020): 1–10.

Zhu, Hong, Chengjie Chen, Jun Zeng, Ze Yun, Yuanlong Liu, Hongxia Qu, Yueming Jiang, Xuewu Duan, and Rui Xia. "Micro RNA 528, a hub regulator modulating ROS homeostasis via targeting of a diverse set of genes encoding copper-containing proteins in monocots." *New Phytologist* 225, no. 1 (2020): 385–399.

Zhu, Hong, Yu Zhang, Ruifang Tang, Hongxia Qu, Xuewu Duan, and Yueming Jiang. "Banana sRNAome and degradome identify microRNAs functioning in differential responses to temperature stress." *BMC Genomics* 20, no. 1 (2019): 1–16.

13 Manipulation of Metabolic Pathways in Regulation and Development of Nutritionally Enriched Crops under Abiotic Stress Factors

Ghazala Ambrin and Rashke Eram

13.1 INTRODUCTION

Crop plants face long-term abiotic stresses such as severe temperatures, hyper salinity, drought, flooding, and metal toxicity throughout their life cycle, primarily driven by environmental changes, pollution, and global warming that impact plant growth, nutrition, and production (Khan et al., 2021a, 2023a, 2023b; Nazir et al., 2024). These stresses result in economic setbacks in the agricultural sector by causing substantial financial and production losses (Ghosh and Majee, 2023). The stresses mainly affect the seed-filling stage by altering source-sink activities and eliciting undesired biochemical changes, which impact the nutritional characteristics of the crops (Hütsch et al., 2019; Nazir et al., 2023a). Drought affects plant physiology by reducing photosynthesis and leaf water potential and impeding nutrient absorption, membrane integrity, and antioxidant metabolism (Ashraf and Harris, 2013; Abbas et al., 2018; Khan et al., 2021a). Meanwhile, salt stress reduces water and nutrient uptake, producing osmotic and ionic stress and restricting cell expansion, tissue growth, and overall plant development. This results in stunted growth, decreased leaf area, impaired plant size, nutritional imbalances, and reduced biomass production (Derakhshani et al., 2020). Also, non-perennial crops with a shorter life cycle are deeply affected by heat stress and undergo impaired crop development and oxidative damage (Ullah et al., 2022).

Moreover, population expansion and climate change exacerbate the issue, causing essential grains such as rice, barley, maize, wheat, and sorghum to suffer disproportionately as the crops lose the ability to absorb essential nutrients and experience low yield (Myers et al., 2017). Less availability of nutrients such as nitrogen, phosphorus, potassium, and micronutrients negatively affect the growth and development of crops. At the same time, long-term nutrition stress leads to the vicious cycle of diminishing agricultural productivity, exacerbating hunger and malnutrition. Additionally, a large portion of the population currently suffers from malnourishment, and a diet dependent on primary crops causes serious health concerns such as childhood mortality and stunted growth, underscoring the critical need for access to nutrient-rich foods (Rao and Annadana, 2017). Given these challenges, the strategic development of plants fortified with nutritional components under abiotic stress is emerging as a critical strategy for achieving food and nutritional security (Singhal et al., 2023; Khan et al., 2021b).

Biofortification of crops requires a comprehensive understanding of fundamental biological processes, encompassing intricate molecular and biochemical mechanisms (Oguz et al., 2022). Additionally, studying the regulatory mechanisms governing commercially and nutritionally essential cereals is crucial. Despite undeniable challenges inherent in pathway manipulation, recent

DOI: 10.1201/9781032630380-13

advancements in molecular biotechnology, genetic engineering, single-cell omics, and genome editing present promising avenues. These technologies enable targeted genetic modifications, optimizing the synthesis of vital compounds (Shelake et al., 2023; Toseef et al., 2023). Furthermore, the advent of synthetic biology (synbio)-based techniques, notably CRISPR-Cas9, holds immense potential. These techniques not only offer the capability to generate stress-resistant crops but also to enhance nutrient density in cereals. This aligns with a global shift in GMO regulations, reflecting a concerted effort to address both food security and malnutrition (Kumar et al., 2023). In essence, manipulating metabolic pathways and understanding their regulation emerges as a vital approach to enhance nutritional content under challenging environmental conditions. To delve deeper into this critical aspect, the chapter explores the manipulation and regulation of metabolic pathways aimed at enhancing the nutritional quality of crops amidst abiotic stress conditions, providing insights into the studies conducted in pursuit of this goal.

13.2 ROLE OF NUTRIENTS IN STRESS RESPONSE

The basic mineral elements, such as nitrogen, carbon, and phosphorus, control plant growth and development and synthesize the fundamental biomolecules necessary for high grain yields (Ahmed et al., 2020; Nazir et al., 2023b). Potassium and magnesium are crucial to photosynthesis and photoassimilate translocation (Koch et al., 2019). Potassium is also a critical osmolyte that modulates stomatic conductivity (Kumari et al., 2021), while magnesium helps absorb light, produce chlorophyll, and activate enzymes. Furthermore, these mineral elements are essential to the cell structure and impact plant reproduction and energy transfer activities (Hasanuzzaman et al., 2018; Johnson et al., 2022). Iron, zinc, and boron play various roles in physiological processes, ranging from photosynthesis to the production of growth hormones, and maintain the optimal level for the maximum growth of plants (Shireen et al., 2018). In stress-prone soils, absorbing these minerals becomes an important means of stable development and survival in difficult conditions (Rane et al., 2021). The regulation of the stress response by essential minerals and nutrients is a nuanced and integrated process. Nitrogen, being a basic component of proteins and amino acids, promotes the synthesis of stress-responsive proteins, thereby protecting cells (Zhou et al., 2022). Phosphorus enhances energy metabolism and strengthens cell membranes, crucial to maintaining plant functions under stress. It is also an essential element for the stability of ATP and membranes and regulates osmotic potential and stomatal conductivity, optimizes water use efficiency, and reduces drought stress (Khan et al., 2023). Calcium, the secondary messenger of stress signaling, influences gene expression, stomatal regulation, and cell wall strengthening, while magnesium, a cofactor of photosynthesis and nucleic acid metabolism enzymes, maintains crucial physiological processes under stress (Kleczkowski and Igamberdiev, 2021; Kour et al., 2023). These nutrients jointly assemble a complex reaction of changes in gene expression, enzyme activity, and cellular structure, ultimately improving plant resistance to various environmental constraints (Farooq et al., 2009). Table 13.1 provides an overview of the function of these mineral elements. Additionally, plants have complex metabolic networks to ensure the acquisition and efficient use of important minerals, which are the key processes for their growth, development, and ability to deal with environmental challenges (Pilon-Smits et al., 2009). Specialized transporters and channels have evolved to facilitate the mineral absorption from the soil, and these pathways have been finely regulated under stress conditions such as nutrient shortages and salinity to optimize resource use. Ion transporters embedded in cell membranes play a key role in maintaining mineral homeostasis in plant cells, and their expression is modulated under stress to handle environmental imbalances (Jeon, 2003; Pinto and Ferreira, 2015; Gupta et al., 2016). Plant metabolism systems adapt to stress, including changes in the production of secondary metabolites that act as signal molecules and contribute to stress tolerance (Isah, 2019). In the soil region influenced by plant roots, metabolic pathways are crucial to the acquisition of nutrients. Root exudation shapes the microbial community and influences the availability of nutrients (Canarini et al., 2019). Stress signaling and hormonal regulation, especially hormones such as abscisic acid and

TABLE 13.1
The Role of Mineral Elements in Stress Responses

Mineral Element	Role in Plant Stress Response
Nitrogen	Synthesis of stress-responsive proteins
Carbon	Synthesis of fundamental molecules
Phosphorus	Strengthens cell membranes, enhances energy metabolism, regulates osmotic potential
Potassium	Crucial for photosynthesis, osmolyte for stomatic conductivity
Magnesium	Absorbs light, produces chlorophyll, activates enzymes
Iron, Zinc, Boron	Production of growth hormones, essential for maximum growth
Calcium	Influences gene expression, stomatal regulation, cell wall strengthening

ethylene, are closely related to nutrient absorption pathways in order to coordinate responses for the optimal use of resources (Ali and Baloch, 2020). In addition, plants activate adaptive biochemical pathways under stress; for example, they improve the production of siderophores in iron-deficiency environments (Pijuan et al., 2023). Understanding these metabolic pathways associated with mineral absorption under stress conditions provides valuable insights into plant adaptation strategies, improving the efficiency of nutrient absorption in stress-tolerant crops and contributing to sustainable agriculture under challenging conditions.

13.3 REGULATION OF METABOLIC PATHWAYS TO IMPROVE NUTRIENT UPTAKE UNDER ABIOTIC STRESS

Stress-responsive genes, transcription factor regulation, hormonal signaling, and epigenetic modifications are integral components of the regulatory mechanisms involved in improving nutrient uptake in crops under stress conditions. The integration of these regulatory elements contributes to the development of stress-resistant crop varieties with improved nutrient utilization efficiency.

13.3.1 Transcription Factors

The transcription factor (TF) family NAC (NAM, ATAF1/2, CUC2), basic leucine zipper (bZIP), myeloblastosis (MYB), WRKY, MADS-box, APETALA2/ethylene receptor (AP2/ERF), phosphate starvation response 1 (PHR1), basic helix-loop-helix (bHLH), auxin receptor factor (ARF), and teosintebranched1/cycloidea/PCF (TCP) regulate the absorption, transport, and assimilation activities, which is crucial to developing nutrient-rich crops under stress. The transcription factors of WRKY (Chen et al., 2022) mainly regulate the expression of nutrients-linked genes in response to the mitogen-activated protein kinase (MAPK) cascade, calcium-dependent protein kinase, and hormonal signals such as abscisic acid (ABA). On the other hand, NAC transcription factors control ethylene, ABA, and reactive oxygen species and direct the response to nutritional stress (Kou et al., 2021). MYB transcription factors affect nutritional quality by regulating secondary metabolism and phytochemical production (Cao et al., 2020) and posttranslational changes. Meanwhile, MPK-mediated phosphorylation improves MYB stability and regulates stress reactions and anthocyanin production. Ring-type E3 ligases such as COP1 and MIEL1 degrade MYB, but SIZ1 SUMOylation protects MYB30 from degradation by ABA signaling (Wang et al., 2021). Further, bHLH transcription factors regulate the intake and balance of iron and respond to the signals of iron availability and deficiency (Gao et al., 2019). Also, MADS-box transcription factors improve the transportation and assimilation of nutrients under hormone and environmental stimulation regulation (Castelán-Muñoz et al., 2019), and AP2/ERF gets triggered by ethylene signaling (Xie et al., 2022). It is important to note that these TF regulatory systems include hormonal and environmental signals, cross-talk between pathways, and epigenetic regulatory systems, providing a comprehensive understanding of crop nutritional quality improvement solutions under abiotic stress.

13.3.2 Stress-Responsive Genes

Stress-sensitive genes emerge as key players in developing nutrient-rich crop plants under abiotic stress, as they modulate physiological and biochemical changes in response (Dikilitas et al., 2019). For example, DREB/CBF genes regulate adaptation to nutritional stress in drought conditions (Yang et al., 2021). Similarly, *RD29A*, a well-known stress-responsive gene activated by ABA-responsive elements, responds to various stress factors, including drought and salinity, and affects nutrient pathways (Lee et al., 2016). Additionally, ion homeostasis and salt tolerance genes (*SOS1*, *NHX*, and *HKT1*) influence nutrient absorption and transport in salty environments (Rao et al., 2021). Further, *P5CS*, involved in proline biosynthesis, affects nutrient utilization and stress adaptation and is influenced by environmental problems (Meena et al., 2019). Moreover, SUMO E3 ligase *SIZ1* affects transcription factors, protein stability, and stress responses, including nutrient-related pathways (Zhou et al., 2019). Furthermore, metabolic genes also affect the response of nutrition to stress, especially in the case of heavy metal pollution, which is controlled by transcription factors related to metals (Ghori et al., 2019).

It is important to note that various transcription factors, including but not limited to members of the *NAC*, *bZIP*, *MYB*, *WRKY*, and *AP2/ERF* families, participate in regulating stress-responsive genes. These transcription factors function as master regulators, directing the expression of downstream stress-responsive genes. Moreover, these genes are governed by epigenetic regulation, post-translational modifications, and hormone signaling pathways (Han et al., 2022; Sahu et al., 2013). For example, under stress, abscisic acid signaling pathways activate and enable the expression of ABA-responsive element (*ABRE*) – binding transcription factors, such as those in the *bZIP* family (Hussain et al., 2021). Understanding and controlling the expression of these stress-responsive genes is a promising frontier for nutrient enrichment because it directly affects a plant's ability to maintain appropriate metabolic pathways for nutrient absorption, transport, and assimilation (Dikilitas et al., 2019). Knowledge of these genes and their regulation can increase crop resistance to abiotic stress and improve nutrition profiles.

13.3.3 Hormonal Regulation

Hormonal regulation is a key factor that helps crops to enrich nutrient content under abiotic stress. Well-known plant hormones, such as abscisic acid, jasmonic acid (JA), auxins, ethylene, cytokinins, gibberellins, brassinosteroids, and salicylic acid play essential roles in these processes. Auxins influence root growth and nutrient uptake, while ethylene mediates stress responses. Gibberellins affect starch and sugar metabolism, and cytokinins influence nitrogen absorption and nutrient translocation. Brassinosteroids impact nutrient transport and allocation, while salicylic acid, known for its role in pathogen defense, also affects responses to abiotic stress (Fahad et al., 2015). Among these hormones, ABA and JA are particularly important in guiding plant responses to various environmental stimuli. ABA functions as a signaling molecule, intricately adjusting metabolic pathways involved in nutrient absorption and utilization under drought and salinity conditions. Meanwhile, JA regulates the metabolic pathways involved in the biosynthesis of secondary metabolites. Additionally, a crosstalk between the ABA and the JA signaling pathways occurs throughout the integrated stress response (Liu et al., 2014). Understanding and regulating these hormones can open new opportunities for improving crop tolerance, nutritional quality, and overall productivity.

13.3.4 Epigenetic Regulation

Epigenetic regulation controls the complex metabolic regulation of plant crops exposed to abiotic stress. In stress conditions, DNA methylation dynamically affects phosphate homeostasis by changing the expression of phosphate transporters (S. Kumar et al., 2022). Meanwhile, histone changes, such as acetylation, influence the nutritional regulation of *Arabidopsis* by modulating genes involved in iron absorption, such as *ferric reduction oxidase 2* (Secco et al., 2017). Furthermore, miRNAs such as miR399 in soybeans participate in post-transcriptional gene silencing and affect phosphate homeostasis by targeting *phosphate2* transcripts (Fan et al., 2021). Under drought, *WRKY* genes in maize undergo histone changes (especially H3K4me3), which affects downstream metabolic pathways (Forestan et al., 2020). Furthermore, changes in DNA methylation patterns suggest a transgenerational response to stress in crops such as wheat and may impact gene relations with nutrients (Kong et al., 2020). Identifying and regulating these epigenetic pathways can help precision breeding and provide strategies to improve crop resistance, nutritional content, and overall production in adverse environmental conditions.

13.4 STRATEGIES FOR MANIPULATING METABOLIC PATHWAYS

The metabolic pathways are manipulated through a lot of approaches, among which genetic engineering, marker-assisted breeding, CRISPR/Cas9 technology, and omics approaches for pathway analysis play crucial roles.

13.4.1 Overexpression of Key Enzymes

The targeted regulation of metabolic pathways through genetic engineering, especially the overexpression of essential enzymes, has become a strategic and precise technique for increasing the nutritional content of crops under abiotic stress conditions. This method focuses on increasing the production of essential compounds needed to resist stress and improve nutritional quality by identifying key enzymes in the nutrient-synthesizing pathway (Ambrin et al., 2020). The overexpression of stress-tolerant enzymes with dual functions that contribute to both plant nutritional production and abiotic stress tolerance may be a promising strategy (Zhang et al., 2020a). For example, the overexpression of trehalose phosphate phosphatase in maize phloem-associated cells increases *SnRK1* activity, the expression of the SWEET sugar transporter gene, and the flow of sucrose into the grain (Oszvald et al., 2018). Similarly, stress-responsive metabolites such as proline and betaine also show promising results (Omari Alzahrani, 2021; Ghosh et al., 2022). Furthermore, the overexpression of dual-function enzymes identifies enzymes involved in stress-reaction pathways and critical nutrient production, providing a compressive approach (Kavi Kishor et al., 2020). Additionally, the overexpression of TFs, particularly those involved in signal pathways such as ABA, can lead to coordinated responses that influence stress tolerance and nutrient metabolism (Zhang et al., 2023).

13.4.2 Gene Knockout Strategies

Manipulation of metabolic pathways through genetic engineering extends beyond overexpression and includes gene knockout studies that offer a targeted approach to increase nutritional content in crops under abiotic stress. The specific genes in the metabolic pathway associated with the synthesis of nutrients can be targeted to produce essential compounds. Also, stress inhibitor knockouts can allocate resources to plants more efficiently under stress conditions to increase the accumulation of nutrients (Sathee et al., 2022). In addition, the knockout of genes associated with stress-induced enzymes can redirect metabolic flows to produce nutritionally beneficial compounds. Furthermore, it can also target pathways involved in stress-induced nutrient degradation, reducing critical nutrient loss under adverse conditions (Zhang et al., 2020a).

13.4.3 Marker-Assisted Breeding for Stress Tolerance

Marker-assisted breeding is a viable method for genetic engineering that incorporates stress-tolerant molecular markers into breeding programs. This approach generates genetically selected plants with desirable characteristics. By discovering and integrating markers related to necessary enzymes and genes involved in stress-responsive metabolic pathways, the production of nutrient rich crop varieties can be accelerated. For example, markers associated with stress-tolerant enzymes or transcription factors that regulate stress and nutrient synthesis pathways can guide the selection of plants that cope with environmental challenges while preserving the nutritive quality. This precise and effective technology supports the development of stress resistance and nutritional quality of crops and aligns with the broader goal of sustainable agriculture amidst climate change (Razzaq et al., 2019).

13.4.4 CRISPR/Cas9 Technology

CRISPR/Cas9 technology has transformed genetic engineering by allowing for unprecedented accuracy in manipulating metabolic pathways to boost agricultural resilience and nutritional value in abiotic stress, where environmental challenges can affect crop productivity and nutritional quality. This groundbreaking technique enables researchers to precisely target and change specific genes involved in nutrition synthesis, stress response, and other metabolic processes (Liu et al., 2021).

13.4.5 Omics Approaches for Pathway Analysis

Omics methods, including genomics, transcriptomics, proteomics, and metabolomics, play a critical role in manipulating metabolic pathways. Genomics facilitates the identification of essential genes involved in nutrient production and stress response more efficiently through methods such as genome-wide association studies (GWAS) and quantitative trait locus (QTL) mapping (Sinha et al., 2023). Transcriptomics reveals dynamic gene expression patterns under stress, influencing the selection of candidates for genetic modification (Afzal et al., 2023). Meanwhile, proteomics provides information about functional proteins engaged in metabolic pathways, bridging the gap between gene expression and protein function (Liu et al., 2019). Additionally, metabolomics gives a snapshot of the plant's metabolic status, assisting in identifying essential compounds associated with nutrient content and stress responses (Sun et al., 2020). Combining multiple omics methods allows for a compressive understanding of metabolic networks, guiding targeted strategies such as genetic modifications. This knowledge-driven approach has enormous potential for producing crops with higher nutritional quality and abiotic stress resilience, addressing critical areas of global food security (Joshi et al., 2024).

13.5 STUDIES FOR IMPROVED NUTRITIVE VALUES OF CROPS UNDER ABIOTIC STRESS

Many studies successfully manipulate plant metabolic pathways, using advanced biotechnological and molecular methods to increase nutrient uptake and crop nutritional quality. Marker-assisted breeding (MAB) has successfully transferred essential nutritional features such as quality protein maize (QPM) and pro-vitamin A content (regulated by significant genes in maize) to elite hybrids for generating commercial cultivars like *Pusa Vivek QPM-9 Improved*, *Pusa Vivek Hybrid-27 Improved*, *Pusa HQPM5 Improved*, and *Pusa HQPM7 Improved* (Hossain et al., 2018; Prasanna et al., 2020). Similarly, a soybean variety, NRC 142 , free from KTI and lipoxygenase 2, and a KTI-free variety, UEL 175 , have been developed through marker-assisted breeding (Gaikwad et al., 2020). However, limited success in bread wheat has been achieved with the introgression of the critical gene *Gpc-B1* (related to high protein content) utilizing MAB (Vishwakarma et al., 2016). Also, extensive use of MAB for producing nutritionally rich commercial cultivars in many food crops confronts obstacles owing to traits controlled by several genes with little individual impact in expression. In bread wheat, for example, large QTLs for zinc concentration demonstrated minimal genetic diversity, making introgression of these QTLs into elite backgrounds difficult (Velu et al., 2018).

Overexpression and knockout studies provide precise control over target genes, allowing for targeted manipulation of metabolic pathways for developing nutritive crops under stress, surpassing marker-assisted breeding, which is less precise and effective for traits controlled by multiple genes (Gaikwad et al., 2020). Notably, the overexpression of *PmPT1*, a phosphate transporter, in tobacco demonstrated increased phosphorus accumulation in roots and shoots, elevated chlorophyll levels and antioxidant enzyme activity, higher dry weight, enhanced soluble sugar and protein contents, and altered malondialdehyde (MDA) levels (Zhang et al., 2020b). An RNA interference study silenced critical enzymes of phytate production in rice, namely inositol-*1,3,4,5,6-pentakisphosphate 2-kinase (IPK1)*, *inositol triphosphate kinases (ITPK) homolog (OsITP/6K-1)*, and *myo-inositol-3-phosphate synthase*, reducing phytate levels and improving the bioavailability of essential minerals like iron in rice grains (Karmakar et al., 2020). In another study, overexpression of *Arabidopsis AtIRT1*, *AtNAS1*, and bean ferritin in rice increased iron and zinc accumulation (Boonyaves et al., 2017). The introduction of the *appA* (phytase) gene from *Escherichia coli* led to a significant two- to three-fold increase in iron and zinc (Bhattacharya et al., 2019). Singh et al. (2017) reported simultaneous expression of FERRITIN, *nicotianamine synthase* (NAS), *carotene desaturase* (CRTI), and *phytoene synthase* (PSY) genes in rice, resulting in multinutrient-enriched biofortified transgenic rice with increased zinc, iron, and carotene contents. Genetic engineering of the *TaIPK1* gene in wheat increased iron and zinc content, highlighting the importance of genetic manipulations in wheat biofortification under stress (Ibrahim et al., 2021). Also, manipulating the *teosinte branched1* (*OsTB1*) gene in rice improved grain production under low phosphorus conditions (Ishizaki et al., 2023). Genetic engineering of nitrogen transporters, *NRT1.1b*, *NAR2.1*, and *NRT2.3a*, in rice improved the efficiency of nitrogen usage in rice, highlighting the critical function of nitrogen uptake pathways in crop nutrition (Chen et al., 2020). Furthermore, Jeong et al. (2022) overexpressed the *OsMYBR22/OsRVE1* transcription factor in rice, resulting in increased levels of chloroplast-biosynthesized metabolites such as chlorophylls, amino acid derivatives (e.g., GABA), amino acids (lysine and threonine), and carotenoids in rice grains.

Studies focusing on CRISPR-Cas9 have gained significant attention in recent years due to the genome editing tool's precision in DNA-level modifications, targeted editing capabilities, flexibility, time efficiency, and potential for creating heritable genetic changes. For instance, Santosh Kumar et al. (2020) used CRISPR/Cas9-mediated genome editing to manipulate the drought and salt tolerance gene (DST) in rice, resulting in increased grain yield, broader leaves, lower stomatal density, and improved mineral nutrient profiling under salt and drought stress. CRISPR/Cas9 selectively targeted the rice *abnormal cytokinin response1 repressor1* (ARE1) gene and altered the *ARE1* homolog in wheat, resulting in improved nitrogen utilization efficiency (NUE) under nitrogen-limiting conditions (Zhang et al., 2021). *OsZIP9*, a zinc-iron-regulated transporter-like protein, was knocked down in rice using this tool, revealing decreased zinc levels in roots and shoots, demonstrating the gene's critical involvement of the gene in zinc transport in rice. Additionally, targeting the rice *GW2* gene resulted in enhanced grain architecture and higher quantities of iron, zinc, potassium, phosphorus, and calcium (Achary and Reddy, 2021). Phytic acid levels were reduced using CRISPR/Cas9 in soybean seeds, enhancing their nutritional value (Song et al., 2022). Moreover, Takahashi et al. (2014) and Tang et al. (2017) altered the *OsNramp5* gene in rice using CRISPR/Cas9, reporting increased iron availability and decreased cadmium accumulation. These studies demonstrate the potential of CRISPR/Cas9 in addressing multiple stress factors simultaneously and improving nutritional content in crop plants.

Furthermore, various attempts to improve crop nutritional quality have also been made through omics approaches. Detailed insights into the key QTLs, genes, and metabolic pathways involved in nutrient uptake in grains and legumes have been studied. QTLs influencing fatty acids, protein content, and oil content were discovered in groundnuts (Roorkiwal et al., 2021). Park et al. (2019) discovered 14 QTLs for grain quality in rice, and Ruan et al. (2020) discovered QTLs for gluten strength in wheat. QTLs for seed iron, zinc, oil, and protein content were found in soybeans (Huang et al., 2020). Genotype by sequencing (GBS) and GWAS mapping was used by Puranik et al. (2020)

and T. Singhal et al. (2021) to uncover marker-trait relationships for grain nutritional content in finger and pearl millet, respectively. Moreover, Wu et al. (2021) reported a study of cis-regulatory elements and Qutub et al. (2021) reported the application of metabolomics for enhancing carotene content in maize. Furthermore, Roorkiwal et al. (2021) also emphasized the integration of multi-omics technologies, notably in legumes, to target genes and metabolic pathways involved in nutrient uptake and assimilation. Collectively, the studies utilizing different modern approaches underscore the importance of manipulation of metabolic pathways to improve crop nutrition content under stress conditions, addressing vital elements such as nitrogen, iron, and zinc.

13.6 CONCLUSION AND FUTURE PROSPECTS

Traditional breeding, genomics-based breeding, and genetic engineering have provided valuable insights into living processes. However, persistent yield gaps highlight the need for more innovative approaches. Although the success of single-gene targeting in stress tolerance is acknowledged, its limitations arise from the compartmentalization and complex network of genes and pathways. Therefore, a holistic approach to manipulating metabolic pathways that considers the integration of metabolic flux, transcription factors, and co-expressed genes is needed. This approach can offer valuable insights into genome-wide stress-regulation mechanisms. A comprehensive understanding of these mechanisms is pivotal for developing robust crops with enhanced nutritional profiles, which will contribute to global food security under challenging environmental conditions.

13.7 AUTHORSHIP CONTRIBUTION

Conceptualization, G.A.; Writing – Original draft preparation, G.A. and R.E.; Writing – Review and editing, G.A. and R.E. All authors have read and agreed to the published version of the book chapter.

13.8 DECLARATION OF COMPETING INTEREST

The authors declare that they have no known competing financial interests or personal relationships that could have appeared to influence the work reported in this book chapter.

REFERENCES

Abbas, Tahir, Rizwan, Muhammad, Ali, Shafaqat, et al. "Biochar Application Increased the Growth and Yield and Reduced Cadmium in Drought Stressed Wheat Grown in an Aged Contaminated Soil." *Ecotoxicology and Environmental Safety* 148 (2018): 825–33.

Achary, V. Mohan Murali, and Reddy, Malireddy K. "CRISPR-Cas9 Mediated Mutation in GRAIN WIDTH and WEIGHT2 (GW2) Locus Improves Aleurone Layer and Grain Nutritional Quality in Rice." *Scientific Reports* 11, no. 1 (2021): 21941.

Afzal, Muhammad, Alghamdi, Salem S., Khan, Muhammad Altaf, et al. "Transcriptomic Analysis Reveals Candidate Genes Associated with Salinity Stress Tolerance During the Early Vegetative Stage in Fababean Genotype, Hassawi–2." *Scientific Reports* 13, no. 1 (2023): 21223.

Ahmed, Mukhtar, Hasanuzzaman, Mirza, Raza, Muhammad Ali, et al. "Plant Nutrients for Crop Growth, Development and Stress Tolerance." In *Sustainable Agriculture in the Era of Climate Change*, edited by Rajib Roychowdhury, Shuvasish Choudhury, Mirza Hasanuzzaman, and Sangeeta Srivastava, 43–92. Cham: Springer International Publishing, 2020.

Ali, Shahid, and Baloch, Abdul Majeed. "Overview of Sustainable Plant Growth and Differentiation and the Role of Hormones in Controlling Growth and Development of Plants Under Various Stresses." *Recent Patents on Food, Nutrition and Agriculture* 11, no. 2 (2020): 105–14.

Ambrin, Ghazala, Ali, Hayssam M., and Ahmad, Altaf. "Metabolic Regulation Analysis of Ajmalicine Biosynthesis Pathway in *Catharanthus roseus* (L.) G. Don Suspension Culture Using Nanosensor." *Processes* 8, no. 5 (2020): 589.

Ashraf, M., and Harris, P. J. C. "Photosynthesis Under Stressful Environments: An Overview." *Photosynthetica* 51, no. 2 (2013): 163–90.

Bhattacharya, Sananda, Sengupta, Shinjini, Karmakar, Aritra, et al. "Genetically Engineered Rice with appA Gene Enhanced Phosphorus and Minerals." *Journal of Plant Biochemistry and Biotechnology* 28, no. 4 (2019): 470–482.

Boonyaves, Kulaporn, Wu, Ting-Ying, Gruissem, Wilhelm, et al. "Enhanced Grain Iron Levels in Rice Expressing an Iron-Regulated Metal Transporter, Nicotianamine Synthase, and Ferritin Gene Cassette." *Frontiers in Plant Science* 8 (2017).

Canarini, Alberto, Kaiser, Christina, Merchant, Andrew, et al. "Root Exudation of Primary Metabolites: Mechanisms and Their Roles in Plant Responses to Environmental Stimuli." *Frontiers in Plant Science* 10 (2019).

Cao, Yunpeng, Li, Kui, Li, Yanli, et al. "MYB Transcription Factors as Regulators of Secondary Metabolism in Plants." *Biology* 9, no. 3 (2020): 61.

Castelán-Muñoz, Natalia, Herrera, Joel, Cajero-Sánchez, Wendy, et al. "MADS-Box Genes Are Key Components of Genetic Regulatory Networks Involved in Abiotic Stress and Plastic Developmental Responses in Plants." *Frontiers in Plant Science* 10 (2019).

Chen, Jingguang, Liu, Xiaoqin, Liu, Shuhua, et al. "Co-Overexpression of OsNAR2.1 and OsNRT2.3a Increased Agronomic Nitrogen Use Efficiency in Transgenic Rice Plants." *Frontiers in Plant Science* 11 (2020).

Chen, Yao, Kong, Xiangge, Yang, Le, et al. "Genome-Wide Identification of WRKY Family Genes and the Expression Profiles in Response to Nitrogen Deficiency in Poplar." *Genes* 13, no. 12 (2022): 2324.

Derakhshani, Zaynab, Bhave, Mrinal, and Shah, Rohan M. "Metabolic Contribution to Salinity Stress Response in Grains of Two Barley Cultivars with Contrasting Salt Tolerance." *Environmental and Experimental Botany* 179 (2020): 104229.

Dikilitas, Murat, Simsek, Eray and Karakas, Sema. "Chapter 2 – Stress Responsive Signaling Molecules and Genes Under Stressful Environments in Plants." In *Plant Signaling Molecules*, edited by M. Iqbal R. Khan, Palakolanu Sudhakar Reddy, Antonio Ferrante, and Nafees A. Khan, 19–42. Woodhead Publishing, 2019.

Fahad, Shah, Nie, Lixiao, Chen, Yutiao, et al. "Crop Plant Hormones and Environmental Stress." In *Sustainable Agriculture Reviews: Volume 15*, edited by Eric Lichtfouse, 371–400. Cham: Springer International Publishing, 2015.

Fan, Kejing, Wong-Bajracharya, Johanna, Lin, Xiao, et al. "Differentially Expressed microRNAs That Target Functional Genes in Mature Soybean Nodules." *The Plant Genome* 14, no. 2 (2021): e20103.

Farooq, M., Wahid, A., Kobayashi, N., et al. "Plant Drought Stress: Effects, Mechanisms and Management." In *Sustainable Agriculture*, edited by Eric Lichtfouse, Mireille Navarrete, Philippe Debaeke, Souchere Véronique, and Caroline Alberola, 153–88. Dordrecht, The Netherlands: Springer Netherlands, 2009.

Forestan, Cristian, Farinati, Silvia, Zambelli, Federico, et al. "Epigenetic Signatures of Stress Adaptation and Flowering Regulation in Response to Extended Drought and Recovery in Zea Mays." *Plant, Cell and Environment* 43, no. 1 (2020): 55–75.

Gaikwad, Kiran B., Rani, Sushma, Kumar, Manjeet, et al. "Enhancing the Nutritional Quality of Major Food Crops Through Conventional and Genomics-Assisted Breeding." *Frontiers in Nutrition* 7 (2020).

Gao, Fei, Robe, Kevin, Gaymard, Frederic, et al. "The Transcriptional Control of Iron Homeostasis in Plants: A Tale of bHLH Transcription Factors?" *Frontiers in Plant Science* 10 (2019).

Ghori, N.-H., Ghori, T., Hayat, M. Q., et al. "Heavy Metal Stress and Responses in Plants." *International Journal of Environmental Science and Technology* 16, no. 3 (March 1, 2019): 1807–28.

Ghosh, Shraboni, and Majee, Manoj. "Chapter Fifteen – Protein l-isoAspartyl Methyltransferase (PIMT) and Antioxidants in Plants." In *Vitamins and Hormones*, edited by Gerald Litwack, Vol. 121, pp. 413–32. Antioxidants. Academic Press, 2023.

Ghosh, U. K., Islam, M. N., Siddiqui, M. N., et al. "Proline, a Multifaceted Signalling Molecule in Plant Responses to Abiotic Stress: Understanding the Physiological Mechanisms." *Plant Biology* 24, no. 2 (2022): 227–39.

Gupta, Neha, Ram, Hari, and Kumar Balwinder. "Mechanism of Zinc Absorption in Plants: Uptake, Transport, Translocation and Accumulation." *Reviews in Environmental Science and Bio/Technology* 15, no. 1 (2016): 89–109.

Han, Danlu, Yu, Zhibo, Lai, Jianbin, et al. "Post-Translational Modification: A Strategic Response to High Temperature in Plants." *aBIOTECH* 3, no. 1 (2022): 49–64.

Hasanuzzaman, Mirza, Borhannuddin, M. H. M., Kamrun Nahar, Bhuyan, et al. "Potassium: A Vital Regulator of Plant Responses and Tolerance to Abiotic Stresses." *Agronomy* 8, no. 3 (2018): 31.

Hossain, Firoz, Muthusamy, Vignesh, Pandey, Neha, et al. “Marker-Assisted Introgression of Opaque2 Allele for Rapid Conversion of Elite Hybrids into Quality Protein Maize.” *Journal of Genetics* 97, no. 1 (2018): 287–98.

Huang, Jinghua, Ma, Qibin, Cai, Zhandong, et al. “Identification and Mapping of Stable QTLs for Seed Oil and Protein Content in Soybean [*Glycine max* (L.) Merr.].” *Journal of Agricultural and Food Chemistry* 68, no. 23 (2020): 6448–60.

Hussain, Quaid, Asim, Muhammad, Zhang, Rui, et al. “Transcription Factors Interact with ABA Through Gene Expression and Signaling Pathways to Mitigate Drought and Salinity Stress.” *Biomolecules* 11, no. 8 (2021): 1159.

Hütsch, Birgit W., Jahn, Dana, and Schubert, Sven. “Grain Yield of Wheat (*Triticum Aestivum* L.) under Long-Term Heat Stress Is Sink-Limited with Stronger Inhibition of Kernel Setting than Grain Filling.” *Journal of Agronomy and Crop Science* 205, no. 1 (2019): 22–32.

Ibrahim, Saira, Saleem, Bilal, Rehman, Nazia, et al. “CRISPR/Cas9 Mediated Disruption of Inositol Pentakisphosphate 2-Kinase 1 (TaIPK1) Reduces Phytic Acid and Improves Iron and Zinc Accumulation in Wheat Grains.” *Journal of Advanced Research* 37 (July 14, 2021): 33–41.

Isah, Tasiu. “Stress and Defense Responses in Plant Secondary Metabolites Production.” *Biological Research* 52 (2019).

Ishizaki, Takuma, Ueda, Yoshiaki, Takai, Toshiyuki, et al. “In-Frame Mutation in Rice TEOSINTE BRANCHED1 (OsTB1) Improves Productivity Under Phosphorus Deficiency.” *Plant Science* 330 (2023): 111627.

Jeon, Kwang W. *International Review of Cytology: A Survey of Cell Biology*. Elsevier, 2003.

Jeong, Ye Sol, Choi, Heebak, Kim, Jae Kwang, et al. “Overexpression of OsMYBR22/OsRVE1 Transcription Factor Simultaneously Enhances Chloroplast-Dependent Metabolites in Rice Grains.” *Metabolic Engineering* 70 (2022): 89–101.

Johnson, Riya, Vishwakarma, Kanchan, Hossen, Md. Shahadat, et al. “Potassium in Plants: Growth Regulation, Signaling, and Environmental Stress Tolerance.” *Plant Physiology and Biochemistry* 172 (2022): 56–69.

Joshi, Shrushti, Patil, Suraj, Shaikh, Alisha, et al. “Modern Omics Toolbox for Producing Combined and Multifactorial Abiotic Stress Tolerant Plants.” *Plant Stress* 11 (March 1, 2024): 100301.

Karmakar, Aritra, Bhattacharya, Sananda, Sengupta, Shinjini, et al. “RNAi-Mediated Silencing of ITPK Gene Reduces Phytic Acid Content, Alters Transcripts of Phytic Acid Biosynthetic Genes, and Modulates Mineral Distribution in Rice Seeds.” *Rice Science* 27, no. 4 (July 1, 2020): 315–28.

Kavi Kishor, Polavarapu, Suravajhala, Renuka, Guddimalli, Rajasheker, et al. “Lysine, Lysine-Rich, Serine, and Serine-Rich Proteins: Link Between Metabolism, Development, and Abiotic Stress Tolerance and the Role of ncRNAs in Their Regulation.” *Frontiers in Plant Science* 11 (2020).

Khan, Fahad, Siddique, Bakar, Abu, Shabala, Sergey, et al. “Phosphorus Plays Key Roles in Regulating Plants’ Physiological Responses to Abiotic Stresses.” *Plants* 12, no. 15 (2023): 2861.

Khan, M. Iqbal R., Ashfaque, Farha, Chhillar, Himanshu, Irfan, Mohammad, and Khan, Nafees A. “The Intricacy of Silicon, Plant Growth Regulators and Other Signaling Molecules for Abiotic Stress Tolerance: An Entrancing Crosstalk Between Stress Alleviators.” *Plant Physiology and Biochemistry* 162 (2021a): 36–47.

Khan, M. Iqbal R., Palakolanu, Sudhakar R., Chopra, Priyanka, Rajurkar, Ashish B., Gupta, Ravi, Iqbal, Noushina, and Maheshwari, Chirag. “Improving Drought Tolerance in Rice: Ensuring Food Security Through Multi-Dimensional Approaches.” *Physiologia Plantarum* 172, no. 2 (2021b): 645–668.

Khan, M. Iqbal R., Nazir, Faroza, Maheshwari, Chirag, Chopra, Priyanka, Chhillar, Himanshu, and Sreenivasulu, Nese. “Mineral Nutrients in Plants Under Changing Environments: A Road to Future Food and Nutrition Security.” *The Plant Genome* 16, no. 4 (2023a): e20362.

Khan, M. Iqbal R., Kumari, Sarika, Nazir, Faroza, Khanna, Risheek Rahul, Gupta, Ravi, and Chhillar, Himanshu. “Defensive Role of Plant Hormones in Advancing Abiotic Stress-Resistant Rice Plants.” *Rice Science* 30, no. 1 (2023b): 15–35.

Kleczkowski, Leszek A., and Igamberdiev, Abir U. “Magnesium Signaling in Plants.” *International Journal of Molecular Sciences* 22, no. 3 (2021): 1159.

Koch, Mirjam, Busse, Matthies, Naumann, Marcel, et al. “Differential Effects of Varied Potassium and Magnesium Nutrition on Production and Partitioning of Photoassimilates in Potato Plants.” *Physiologia Plantarum* 166, no. 4 (2019): 921–35.

Kong, Lingyao, Liu, Yanna, Wang, Xiaoyu, et al. “Insight into the Role of Epigenetic Processes in Abiotic and Biotic Stress Response in Wheat and Barley.” *International Journal of Molecular Sciences* 21, no. 4 (2020): 1480.

Kou, Xiao Hong, Zhou, Jia Qian, Wu, Cai E., et al. “The Interplay Between ABA/Ethylene and NAC TFs in Tomato Fruit Ripening: A Review.” *Plant Molecular Biology* 106, no. 3 (2021): 223–38.

Kour, Jaspreet, Khanna, Kanika, Singh, Arun Dev, et al. "Calcium's Multifaceted Functions: From Nutrient to Secondary Messenger During Stress." *South African Journal of Botany* 152 (2023): 247–63.

Kumar, Manoj, Prusty, Manas Ranjan, Pandey, Manish K., et al. "Application of CRISPR/Cas9-Mediated Gene Editing for Abiotic Stress Management in Crop Plants." *Frontiers in Plant Science* 14 (2023).

Kumar, Suresh, Seem, Karishma, Kumar, Santosh, et al. "Pup1 QTL Regulates Gene Expression Through Epigenetic Modification of DNA Under Phosphate Starvation Stress in Rice." *Frontiers in Plant Science* 13 (2022).

Kumari, Sarika, Himanshu Chhillar, Priyanka Chopra, Risheek Rahul Khanna, and M. Iqbal R. Khan. "Potassium: A Track to Develop Salinity Tolerant Plants." *Plant Physiology and Biochemistry* 167 (2021): 1011–1023.

Lee, Sang Y., Boon, Neville J., Webb, Alex A. R., et al. "Synergistic Activation of RD29A Via Integration of Salinity Stress and Abscisic Acid in *Arabidopsis thaliana*." *Plant and Cell Physiology* 57, no. 10 (2016): 2147–60.

Liu, Ning, Ding, Yong, Fromm, Michael, et al. "Different Gene-Specific Mechanisms Determine the 'Revised-Response' Memory Transcription Patterns of a Subset of *A. Thaliana* Dehydration Stress Responding Genes." *Nucleic Acids Research* 42, no. 9 (2014): 5556–66.

Liu, Qier, Yang, Fan, Zhang, Jingjuan, et al. "Application of CRISPR/Cas9 in Crop Quality Improvement." *International Journal of Molecular Sciences* 22, no. 8 (2021): 4206.

Liu, Yahui, Lu, Song, Liu, Kefu, et al. "Proteomics: A Powerful Tool to Study Plant Responses to Biotic Stress." *Plant Methods* 15, no. 1 (2019): 135.

Meena, Mukesh, Divyanshu, Kumari, Kumar, Sunil, et al. "Regulation of L-Proline Biosynthesis, Signal Transduction, Transport, Accumulation and its Vital Role in Plants During Variable Environmental Conditions." *Heliyon* 5, no. 12 (2019): e02952.

Myers, Samuel S., Smith, Matthew R., Guth, Sarah, et al. "Climate Change and Global Food Systems: Potential Impacts on Food Security and Undernutrition." *Annual Review of Public Health* 38 (2017): 259–77.

Nazir, Faroza, Badar Jahan, Noushina Iqbal, Ashish B. Rajurkar, Manzer H. Siddiqui, and M. Iqbal R. Khan. "Methyl Jasmonate Influences Ethylene Formation, Defense Systems, Nutrient Homeostasis and Carbohydrate Metabolism to Alleviate Arsenic-Induced Stress in Rice (*Oryza sativa*)." *Plant Physiology and Biochemistry* 202 (2023a): 107990.

Nazir, Faroza, Moksh Mahajan, Sayeda Khatoon, Mohammed Albaqami, Farha Ashfaque, Himanshu Chhillar, Priyanka Chopra, and M. Iqbal R. Khan. "Sustaining Nitrogen Dynamics: A Critical Aspect for Improving Salt Tolerance in Plants." *Frontiers in Plant Science* 14 (2023b): 1087946.

Nazir, Faroza, Poor Peter, Ravi Gupta, Sarika Kumari, Kashif Nawaz, and M. Iqbal R. Khan. "Plant Hormone Ethylene: A Leading Edge in Conferring Drought Stress Tolerance." *Physiologia Plantarum* 176, no. 1 (2024): e14151.

Oguz, Muhammet Murat Aycan, Cagri, Oguz, Ezgi, et al. "Drought Stress Tolerance in Plants: Interplay of Molecular, Biochemical and Physiological Responses in Important Development Stages." *Physiologia* 2, no. 4 (2022): 180–97.

Omari Alzahrani, Fatima. "Metabolic Engineering of Osmoprotectants to Elucidate the Mechanism(s) of Salt Stress Tolerance in Crop Plants." *Planta* 253, no. 1 (2021): 24.

Oszvald, Maria, Primavesi, Lucia F., Griffiths, Cara A., et al. "Trehalose 6-Phosphate Regulates Photosynthesis and Assimilate Partitioning in Reproductive Tissue." *Plant Physiology* 176, no. 4 (2018): 2623–38.

Park, Seul-Gi, Park, Hyun-Su, Baek, Man-Kee, et al. "Improving the Glossiness of Cooked Rice, an Important Component of Visual Rice Grain Quality." *Rice (New York, N.Y.)* 12, no. 1 (2019): 87.

Pijuan, Jordi, Moreno, David F., Yahya, Galal, et al. "Regulatory and Pathogenic Mechanisms in Response to Iron Deficiency and Excess in Fungi." *Microbial Biotechnology* 16, no. 11 (2023): 2053–71.

Pilon-Smits, A. H., Elizabeth, Quinn, Colin F., et al. "Physiological Functions of Beneficial Elements." *Current Opinion in Plant Biology*, Physiology and Metabolism, 12, no. 3 (2009): 267–74.

Pinto, Edgar, and Ferreira, Isabel M. P. L. V. O. "Cation Transporters/Channels in Plants: Tools for Nutrient Biofortification." *Journal of Plant Physiology* 179 (2015): 64–82.

Prasanna, Boddupalli M., Palacios-Rojas, Natalia, Hossain, Firoz, et al. "Molecular Breeding for Nutritionally Enriched Maize: Status and Prospects." *Frontiers in Genetics* 10 (2020).

Puranik, Swati, Sahu, Pranav P., Beynon, Sarah, et al. "Genome-Wide Association Mapping and Comparative Genomics Identifies Genomic Regions Governing Grain Nutritional Traits in Finger Millet (*Eleusine coracana* L. Gaertn.)." *Plants, People, Planet* 2, no. 6 (2020): 649–62.

Qutub, Maqbool, Chandran, Sarankumar, Rathinavel, Krishnakumar, et al. "Improvement of a Yairipok Chujak Maize Landrace from North Eastern Himalayan Region for β-Carotene Content through Molecular Marker-Assisted Backcross Breeding." *Genes* 12, no. 5 (2021): 762.

Rane, Jagadish, Singh, Ajay Kumar, Kumar, Mahesh, et al. "The Adaptation and Tolerance of Major Cereals and Legumes to Important Abiotic Stresses." *International Journal of Molecular Sciences* 22, no. 23 (2021): 12970.

Rao, Chavali Kameswara, and Annadana, Seetharam. "Nutrient Biofortification of Staple Food Crops: Technologies, Products and Prospects." In *Phytonutritional Improvement of Crops*, 113–83. John Wiley and Sons, Ltd, 2017.

Rao, Yalaga Rama, Ansari, Mohammad Wahid, Sahoo, Ranjan Kumar, et al. "Salicylic Acid Modulates ACS, NHX1, Sos1 and HKT1;2 Expression to Regulate Ethylene Overproduction and Na+ Ions Toxicity That Leads to Improved Physiological Status and Enhanced Salinity Stress Tolerance in Tomato Plants Cv. Pusa Ruby." *Plant Signaling and Behavior* 16, no. 11 (2021): 1950888.

Razzaq, Ali, Sadia, Bushra, Raza, Ali et al. "Metabolomics: A Way Forward for Crop Improvement." *Metabolites* 9, no. 12 (2019): 303.

Roorkiwal, Manish, Pandey, Sarita, Thavarajah, Dil, et al. "Molecular Mechanisms and Biochemical Pathways for Micronutrient Acquisition and Storage in Legumes to Support Biofortification for Nutritional Security." *Frontiers in Plant Science* 12 (2021). https://www.frontiersin.org/articles/10.3389/fpls.2021.682842.

Ruan, Yuefeng, Yu, Bianyun, Knox, Ron E., Singh, Asheesh K., DePauw, Ron, Cuthbert, Richard, Zhang, Wentao, et al. "High Density Mapping of Quantitative Trait Loci Conferring Gluten Strength in Canadian Durum Wheat." *Frontiers in Plant Science* 11 (2020). https://www.frontiersin.org/articles/10.3389/fpls.2020.00170.

Sahu, Pranav Pankaj, Pandey, Garima, Sharma, Namisha, et al. "Epigenetic Mechanisms of Plant Stress Responses and Adaptation." *Plant Cell Reports* 32, no. 8 (2013): 1151–59.

Santosh Kumar, V. V., Verma, Rakesh Kumar, Yadav, Shashank Kumar, et al. "CRISPR-Cas9 Mediated Genome Editing of Drought and Salt Tolerance (OsDST) Gene in Indica Mega Rice Cultivar MTU1010." *Physiology and Molecular Biology of Plants* 26, no. 6 (2020): 1099–110.

Sathee, Lekshmy, Jagadhesan, B., Pandesha, Pratheek H., et al. "Genome Editing Targets for Improving Nutrient Use Efficiency and Nutrient Stress Adaptation." *Frontiers in Genetics* 13 (2022).

Secco, David, Whelan, James, Rouached, Hatem, et al. "Nutrient Stress-Induced Chromatin Changes in Plants." *Current Opinion in Plant Biology, Cell Signalling and Gene Regulation* 39 (2017): 1–7.

Shelake, Rahul Mahadev, Jadhav, Amol Maruti, Bhosale, Pritam Bhagwan, et al. "Unlocking Secrets of Nature's Chemists: Potential of CRISPR/Cas-Based Tools in Plant Metabolic Engineering for Customized Nutraceutical and Medicinal Profiles." *Plant Physiology and Biochemistry* 203 (2023): 108070.

Shireen, Fareeha, Nawaz, Muhammad Azher, Chen, Chen, et al. "Boron: Functions and Approaches to Enhance its Availability in Plants for Sustainable Agriculture." *International Journal of Molecular Sciences* 19, no. 7 (2018): 1856.

Singh, Simrat Pal, Gruissem, Wilhelm, and Bhullar, Navreet K. "Single Genetic Locus Improvement of Iron, Zinc and β-Carotene Content in Rice Grains." *Scientific Reports* 7 (2017): 6883.

Singhal, Rajesh Kumar, Fahad, Shah, Kumar, Pawan, et al. "Beneficial Elements: New Players in Improving Nutrient Use Efficiency and Abiotic Stress Tolerance." *Plant Growth Regulation* 100, no. 2 (2023): 237–65.

Singhal, Tripti, Satyavathi, C. Tara, Singh, S. P., et al. "Multi-Environment Quantitative Trait Loci Mapping for Grain Iron and Zinc Content Using Bi-Parental Recombinant Inbred Line Mapping Population in Pearl Millet." *Frontiers in Plant Science* 12 (2021).

Sinha, Mayank Kumar, Aski, Muraleedhar S., Mishra, Gyan Prakash, et al. "Genome Wide Association Analysis for Grain Micronutrients and Anti-Nutritional Traits in Mungbean [*Vigna radiata* (L.) R. Wilczek] Using SNP Markers." *Frontiers in Nutrition* 10 (2023).

Song, Ji Hyeon, Shin, Gilok, Kim, Hye Jeong, et al. "Mutation of GmIPK1 Gene Using CRISPR/Cas9 Reduced Phytic Acid Content in Soybean Seeds." *International Journal of Molecular Sciences* 23, no. 18 (2022): 10583.

Sun, Wenli, Chen, Zican, Hong, Jun, et al. "Promoting Human Nutrition and Health through Plant Metabolomics: Current Status and Challenges." *Biology* 10, no. 1 (2020): 20.

Takahashi, Ryuichi, Ishimaru, Yasuhiro, Shimo, Hugo, et al. "From Laboratory to Field: OsNRAMP5-Knockdown Rice Is a Promising Candidate for Cd Phytoremediation in Paddy Fields." *PLOS One* 9, no. 6 (2014): e98816.

Tang, Xu, Lowder, Levi G., Zhang, Tao, et al. "A CRISPR–Cpf1 System for Efficient Genome Editing and Transcriptional Repression in Plants." *Nature Plants* 3, no. 3 (2017): 1–5.

Toseef, Muhammad, Petinrin, Olutomilayo Olayemi, Wang, Fuzhou, et al. "Deep Transfer Learning for Clinical Decision-Making Based on High-Throughput Data: Comprehensive Survey with Benchmark Results." *Briefings in Bioinformatics* 24, no. 4 (2023): bbad254.

Ullah, Aman, Nadeem, Faisal, Nawaz, Ahmad, et al. "Heat Stress Effects on the Reproductive Physiology and Yield of Wheat." *Journal of Agronomy and Crop Science* 208, no. 1 (2022): 1–17.

Velu, Govindan, Singh, Ravi Prakash, Crespo-Herrera, Leonardo, et al. "Genetic Dissection of Grain Zinc Concentration in Spring Wheat for Mainstreaming Biofortification in CIMMYT Wheat Breeding." *Scientific Reports* 8, no. 1 (2018): 13526.

Vishwakarma, Manish K., Arun, B., Mishra, V. K., et al. "Marker-Assisted Improvement of Grain Protein Content and Grain Weight in Indian Bread Wheat." *Euphytica* 208, no. 2 (2016): 313–21.

Wang, Xiaopei, Niu, Yanli, and Zheng, Yuan. "Multiple Functions of MYB Transcription Factors in Abiotic Stress Responses." *International Journal of Molecular Sciences* 22, no. 11 (2021): 6125.

Wu, Xiaowei, Liang, Yan, Gao, Hengbin, et al. "Enhancing Rice Grain Production by Manipulating the Naturally Evolved Cis-Regulatory Element-Containing Inverted Repeat Sequence of OsREM20." *Molecular Plant* 14, no. 6 (2021): 997–1011.

Xie, Wei, Ding, Chaoqing, Hu, Haitao, et al. "Molecular Events of Rice AP2/ERF Transcription Factors." *International Journal of Molecular Sciences* 23, no. 19 (2022): 12013.

Yang, Xinyi, Lu, Meiqi, Wang, Yufei, et al. "Response Mechanism of Plants to Drought Stress." *Horticulturae* 7, no. 3 (2021): 50.

Zhang, Han, Liu, Shujing, Ren, Tianmeng, et al. "Crucial Abiotic Stress Regulatory Network of NF-Y Transcription Factor in Plants." *International Journal of Molecular Sciences* 24, no. 5 (2023): 4426.

Zhang, Heng, Zhao, Yang, and Zhu, Jian-KangJian-Kang. "Thriving under Stress: How Plants Balance Growth and the Stress Response." *Developmental Cell* 55, no. 5 (2020a): 529–43.

Zhang, Ting, Hong, Yi, and Wen, Xiao Peng. "Over-Expression of Masson Pine PmPT1 Gene in Transgenic Tobacco Confers Tolerance Enhancement to Pi Deficiency by Ameliorating P Level and the Antioxidants." *Plant Molecular Biology Reporter* 38, no. 2 (2020b): 238–49.

Zhang, Jiahui, Zhang, Huating, Li, Shaoya, et al. "Increasing Yield Potential Through Manipulating of an ARE1 Ortholog Related to Nitrogen Use Efficiency in Wheat by CRISPR/Cas9." *Journal of Integrative Plant Biology* 63, no. 9 (2021): 1649–63.

Zhou, Li-Jie, Zhang, Chun-Ling, Zhang, Rui-Fen, et al. "The SUMO E3 Ligase MdSIZ1 Targets MdbHLH104 to Regulate Plasma Membrane H+-ATPase Activity and Iron Homeostasis." *Plant Physiology* 179, no. 1 (2019): 88–106.

Zhou, Xianrong, Joshi, Shrushti, Patil, Suraj, et al. "Reactive Oxygen, Nitrogen, Carbonyl and Sulfur Species and Their Roles in Plant Abiotic Stress Responses and Tolerance." *Journal of Plant Growth Regulation* 41, no. 1 (2022): 119–42.

14 Mitigating High-Temperature Stress in Plants

Strategies for Sustaining Crop Growth and Yield

Shreya Chhabra, Chirag Maheshwari, Shahid Umar, and M. Iqbal R. Khan

14.1 INTRODUCTION

In the present scenario, global warming is the most apparent event that adversely affects eco-biodiversity, a consequence of climate change. Climate change along with many abiotic stresses like high temperature (HT) critically affect the agricultural production of the entire planet. According to the forecasts, by the end of the 21st century, a global rise of about 2.23°C–6.63°C in temperature may happen (EPA, 2011). The excessive usage of traditional energy sources is the key reason for the emission of greenhouse gases like carbon dioxide (CO_2), methane (CH_4), and nitrous oxide (N_2O), and this is made worse by the industrial emission of chlorofluorocarbons (CFCs) and other various anthropogenic activities. Greenhouse gases, mainly, trap heat radiated by the earth's surface and therefore enhance global mean temperature (Kweku et al., 2018). This rise in temperature results in a substantial change in the growing season of crops by fluctuating the threshold temperature for the beginning (i.e. sowing) and end (i.e. crop maturity) of the season (Farhad et al., 2023). Environmental stress caused by high-temperature is currently acknowledged as the leading hazard for agricultural productivity worldwide (Zandalinas et al., 2018).

Temperature is a key factor in determining yield, as it regulates growth. Even a slight increase of 1°C above the threshold in ambient air temperature can induce heat stress in plants, leading to irreversible damage to their growth and functions (Teixeira et al., 2013). Temperature stress has significantly impacted many crop plants, resulting in substantial yield losses (Bita and Gerats, 2013). Rise (1°C) is predicted to lessen the yield of rice, wheat, and maize by 10%–20%, 6%, and 21%–31% respectively (Wang et al., 2019; Asseng et al., 2015; Yang et al., 2017). This hasty increase in temperature followed by unexpected weather conditions is a major concern to agriculturists and researchers worldwide (Hasanuzzaman et al., 2013a). In India, a loss of 10% to 40% in the yield of crops is predicted by 2100 as a result of heat stress (Aggarwal, 2008). Therefore, HT stress has emerged as an immense menace to successful crop production globally (Lobell and Asner, 2003).

Rising temperatures and their effects on many crop plants are a major concern for global food security. As non-motile organisms, plants cannot move to conditions that are more favourable and must face the potentially lethal consequences of high temperatures (Lobell and Field, 2007). Heat stress negatively affects the growth and development of plants in various ways and ultimately impacts yield (Hasanuzzaman et al., 2012). The severity of damage caused by high temperatures depends on three factors: the degree of temperature, the duration of exposure, and the nature of the crop plant. Elevated temperatures generally affect every aspect of plant processes, such as seed germination, reproduction, growth, development, yield, and the stability of many proteins, RNA species, membranes, and cytoskeletal structures (Hasanuzzaman et al., 2013b; Lobell et al., 2011). Therefore, it brings about substantial changes in the morpho-anatomical, physiological, biochemical, and molecular levels of plants.

DOI: 10.1201/9781032630380-14

Plants, to some extent, are able to mitigate high-temperature stress conditions by creating some physical modifications within the plant as well as frequently by generating signals for altering metabolism. Various tolerance strategies are developed in plants endogenously and many compounds can be provided exogenously to plants for the development of tolerance against high temperature. With the aid of phytohormones, osmotic adjustments, and manipulation of nutrients, plants can develop tolerance against the elevated temperature stress. Phytohormones have been acknowledged as a strong tool for reducing the deleterious effects of abiotic stresses including HT stress in crop species (Khan et al., 2015), while osmotic adjustments help plants to maintain cell turgor. Furthermore, the development of tolerant crop lines by plant scientists and agriculturalists through the gene editing process is a boon for humans, especially farmers.

In this chapter, we summarize recent findings on the global impact of high temperatures on agricultural crops, the changes in plant processes in response to high temperatures, and the tolerance strategies developed by plants to mitigate these harmful effects (Figure 14.1). The primary

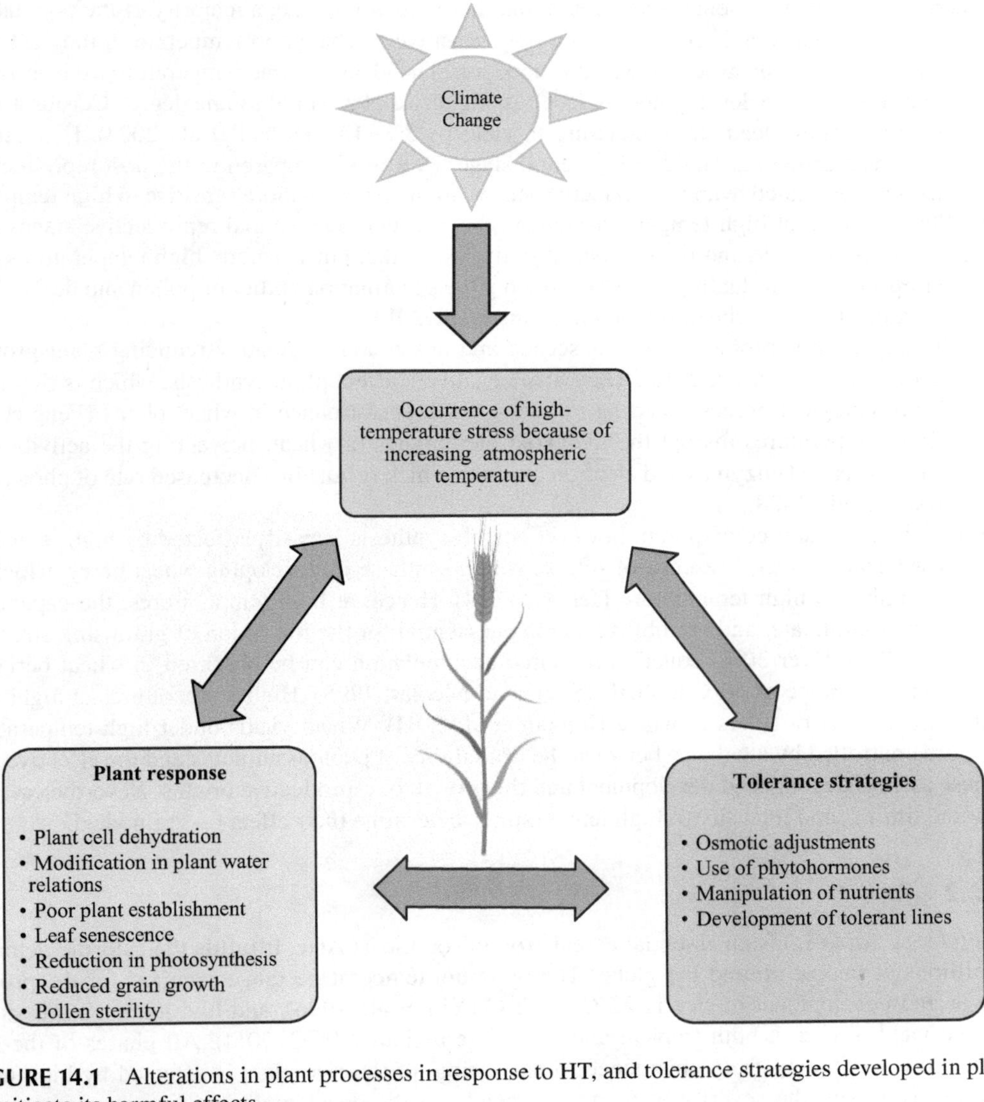

FIGURE 14.1 Alterations in plant processes in response to HT, and tolerance strategies developed in plants to mitigate its harmful effects.

focus of this study is to enhance our understanding of the detrimental effects of high-temperature stress on various crops and to elucidate potential strategies for modulating plant responses to high temperatures.

14.2 IMPACT OF HIGH-TEMPERATURE STRESS ON VARIOUS CROP PLANTS

A major concern for researchers is that the yield and productivity of almost every crop is affected by rising temperature, especially staple crops on which a majority of the world's population depends. Various effects of HT stress have been reported on numerous crops.

14.2.1 Wheat

Wheat (*Triticum aestivum* L.) is an important staple food grain which is cultivated globally for its starchy grain. It is the leading contributor to the world's grain production and trade, which are around 30% and 50% respectively (Sendhil et al., 2023). It is the main cereal crop for satisfying the energy demands of the emerging population, therefore worldwide, a majority of the population depends on it. Being a rabi crop, it is extremely sensitive to change in temperature, thus HT has become a crucial limiting aspect in wheat production. The ideal daytime temperature for its growth throughout reproductive development is 15°C. A slight rise of as small as one degree Celsius above this ideal temperature leads to a lessening in yield by 3%–4% (Farhad et al., 2023). Heat stress can affect wheat growth at any developmental stage. However, compared to the post-reproductive stages, the pre-reproductive and reproductive stages are moderately more sensitive to high temperatures. Short periods of high temperature during the pre-reproductive and reproductive stages can significantly reduce yield and the number of grains per spike. Furthermore, high temperatures can directly impact the reproductive stage by reducing the germination ability of pollen and decreasing the percentage of pollen tube development (Yang et al., 2013).

High temperatures promote leaf senescence and abscission, ultimately reducing plant growth (Djanaguiraman and Prasad, 2010). They also negatively affect photosynthesis, which is the most sensitive physiological process, leading to poor growth performance in wheat plants (Feng et al., 2014). High temperatures disrupt the thylakoid membranes in wheat, preventing the activities of membrane-associated enzymes and electron carriers, which results in a decreased rate of photosynthesis (Ristic et al., 2008).

Wheat is a rich source of starch; however, starch synthesis is easily affected by high-temperature stress because of the presence of soluble starch synthase in developing wheat berry, which is very susceptible to high temperature (Jenner, 1994). Hence, at high temperatures, the capability to produce, accumulate, and remobilize starch is essential for the resolution of grain sink strength (Ni et al., 2017). Over 30% reduction in starch accumulation can be observed in wheat berry at the temperature range of 30°C to 40°C (Stone and Nicolas, 1995). High temperatures at night can reduce the number of tillers in wheat (Johkan et al., 2011). Wheat yields under high-temperature stress are controlled by a balance between the availability of photoassimilates and the effective use of these assimilates for seed development and the growth of reproductive organs. Nevertheless, the duration, timing, and intensity of high temperatures determine their effect on grain yield.

14.2.2 Rice

Rice (*Oryza sativa* L.) is an essential cereal crop in south-east Asia. It fulfils the calorific demand of millions of people around the globe. The optimum temperature range required for the proper growth and development of rice is 27°C to 32°C (Yin et al., 1996), and high-temperature stress cuts the yield of rice in both tropical and temperate regions (IPCC, 2014). All phases of the life cycle, growth initiating from emergence to harvesting, of the rice plant is affected by high temperature. However, the severity of damage depends on the developmental stage at which exposure to high temperature occurs (Wahid et al., 2007). A rise in temperature undesirably affects the quantitative traits like germination rate and percentage, root and shoot length, seed vigour, and viability (Cao and Zhao, 2008).

The reproductive stages including anthesis, fertilization, and to a slighter extent microsporogenesis are most prone to elevated temperature in rice crop (Farrell et al., 2006). High temperature alters the biochemical characteristics of pollen such as enhancement in sugar content, reduction in protein concentration, and decrease in membrane stability index (MSI) (Das et al., 2014). High temperature at anthesis stage, or succeeding stage, negatively impacts the pollen germination with stunted pollen tube growth and lessened anther dehiscence.

Temperature exposure greater than the optimum range reduces spikelet fertility, as high temperatures diminish the ability of pollen grains to swell, leading to poor thecae dehiscence (Matsui et al., 2000). In rice, temperatures above 35°C increase spikelet sterility (Matsui et al., 1997). A small increase in temperature by 1°C causes nearly a 10% reduction in the grain yield of rice, primarily due to the decrease in the number of spikelets per plant (Peng et al., 2004).

High-temperature stress during the grain-filling period limits the translocation of photoassimilates (Farooq et al., 2011) and reduces the accumulation of photosynthates in grains (Morita et al., 2005). Moreover, it also decreases the extent of grain filling in rice, ultimately resulting in lower grain weight and poor grain yield.

14.2.3 Maize

Maize (*Zea mays* L.) is the third most important staple food crop and another member of family Poaceae. HT stress adversely impacts on the growth, development, and yield of the maize plant. Elevated temperature affects almost every growth stage of maize throughout its life cycle. It influences the metabolic functions, enzyme reactions, and structural integrity, especially the membrane structure and properties.

HT impairs the thermostability of thylakoid membrane where the photosystems (both PSI and PSII) and other electron transfer factors are located, ultimately affecting the photosynthesis process. Therefore, it is important to maintain the stability of the thylakoid membrane under stress conditions for photosynthesis (Blum, 1988).

Zhang et al. (2013) reported that in maize leaves, a severe decline in electron transfer activity and a faster decrease in the photochemical activities of PSI and PSII are the two main reasons for the reduction in photosynthetic capacity. A reduction in chlorophyll a and chlorophyll b contents also limits the net photosynthetic rate in maize plants during heat stress (Karim et al., 1999; Zhu et al., 2011). Inhibition of photosynthesis results in a lower supply of photosynthates to the kernel, which in turn reduces grain weight and yield.

High-temperature stress results in water deficit conditions in maize plants, as it adversely impacts root water-absorbing capacity, leaf relative water content, leaf water potential, and stomatal conductance (Tao et al., 2016). The latter affects the grain-filling stage, as the weight of the maize kernel is correlated with the water supply at the time of rapid grain filling (Borrás and Westgate, 2006).

14.2.4 Barley

Barley (*Hordeum vulgare* L.), a widely cultivated cereal grain worldwide, is used for malting, human consumption, and stock feed, and it also serves as a raw material for starch as well as ethanol production. Unfavourable temperature affects the yield and numerous quality parameters of this crop (Clausen et al., 2011).

High temperature has inimical effects on the grain filling periods, by way of reduction in filling duration, perhaps because of impairment in the enzyme activity involved in starch biosynthesis (Jenner, 1994). Elevated temperature decreases starch accumulation in barley (Tester et al., 1991). Super-optimal growth temperatures decrease the activity of soluble starch synthase and thereupon lessen the starch granule number and size in barley (Myllärinen et al., 1998).

14.2.5 Legumes

Legumes are the members of family Leguminosae/Fabaceae. They are well known for their protein content, providing health and nutritional benefits, and play an important role in accomplishing agricultural sustainability. In worldwide crop production, legumes rank third after cereals and oilseeds (Popelka et al., 2004). Legume crops can be classified into two groups on the basis of their growing season: warm-season legumes and cool-season legumes (Miller et al., 2002). Warm-season legumes are those which grow in hot and moist climatic conditions, like mungbean (*Vigna radiata* var. *radiata*), cowpea (*Vigna unguiculata*), common bean (*Phaseolus* spp.), pigeon pea (*Cajanus cajan*), and urd bean (*Vigna mungo*) (Singh and Singh, 2011). Meanwhile, cool-season legumes include dry pea (*Pisum sativum*), chickpea (*Cicer arietinum*), lupin (*Lupinus* spp.), grass pea (*Lathyrus sativus*), lentil (*Lens culinaris*), and broad bean (*Vicia faba*) (Andrews and Hodge, 2010). In contrast to warm-season legumes, cool-season legumes are more vulnerable to elevated temperatures.

High temperatures have harmful effects on the morpho-physiological and reproductive development of legumes. However, among the crop growth sequences, the propagative stage is more sensitive to high-temperature stress than the vegetative stage is. Physiological processes like seed germination show a decrease in many legumes, such as peas, soybeans, and chickpeas, under high-temperature stress (Ren et al., 2009; Nemeskeri, 2004; Kaushal et al., 2011; Piramila et al., 2012). Carbon assimilation processes such as photorespiration and respiration are disrupted significantly, which might result in smaller and distorted organelles (Barnabás et al., 2008). HT negatively impacts photosynthesis, by inducing the reduction of chlorophyll content in chickpea (Kumar et al., 2013) and beans (Petkova et al., 2007). Plant water relations and membrane permeability are also altered under elevated temperature.

Leguminous plants are well recognized for their symbiotic association with rhizobia to fix nitrogen. Nearly 40% of biologically fixed nitrogen is contributed by legumes, which not only enhances soil fertility but also reduces reliance on nitrogen fertilizers (Yadav, 2008). High temperatures affect all phases of the legume-rhizobium symbiosis (Hungria and Vargas, 2000; Yadav and Nehra, 2013). Symbiotic bacteria need an optimum temperature for the process of nitrogen fixation; unfavourable temperatures hinder nodulation, impede nodule development, and prolong root infection (Drouin et al., 2000).

There are several reports that confirm that high temperature impacts the reproductive development in legumes, including pea (Guilioni et al., 1997), chickpea (Kaushal et al., 2013), mungbean (Tzudir et al., 2014), cowpea (Ahmed et al., 1992), and common bean (Vara Prasad et al., 2002). Negative effects of HT on reproductive growth like depletion of flowers, buds, fruits, seeds, and pods are observed which markedly reduced the yield potential (Thakur et al., 2010; Kaushal et al., 2016). Moreover, HT lessens the number of flowering branches and ultimately the number of flowers per plant (Young et al., 2004). Reduced seed filling was detected due to inadequate accumulation of photosynthates leading to a decrease in seed size (Kumar et al., 2016).

14.3 IMPACT OF HIGH TEMPERATURE ON PLANT GROWTH AND DEVELOPMENT

The consequence of high temperature is determined by duration, amount, and rate of temperature simultaneously (Wahid et al., 2007; Zinn et al., 2010). Elevated temperature induces various alterations in plants and affects entire aspects of plant processes.

14.3.1 Morpho-Anatomical Changes

High temperature severely affects plant morphology and anatomy. It diminishes shoot growth resulting to short plant height. In wheat, plant height was shifted from 66.4–73.9 cm under regular environmental conditions to 55.7–82.3 cm under heat-treated conditions (Rahman, 2004). As cell elongation rate altered with HT, it led to a reduction in the leaf weight and size (Prasad et al., 2006).

The morphological indicators of high-temperature stress include leaf senescence and abscission, sunburns and scorching of branches and stems, leaves, and twigs, inhibition of shoot and root growth, fruit discoloration, and damage (Guilioni et al., 1997; Vollenweider and Gunthardt-Goerg, 2005) (Figure 14.2). Necrosis, leaf rolling and drying, and damage to leaf tips and margins have been observed in sugarcane under high-temperature stress (Srivastava et al., 2012).

Stomatal closure, diminution of plant water, enhanced stomatal and trichomatous densities, larger xylem vessels, and reduced cell size are common heat-induced characteristics in plants and help them grapple with high temperatures (Waraich et al., 2012; Khalil et al., 2018). Several morphological adaptations occur in plants to survive under high-temperature conditions. Mechanisms like modifications in leaf orientation, changes in membrane lipid compositions, and transpirational cooling help to avoid high-temperature stress. Some plants growing in hot weather can avoid stress by absorbing less solar radiation due to the presence of small hairy structures (tomentose) forming a thick coat on the outer surface of a leaf, in addition to the cuticle, which is a protective waxy film (Hasanuzzuman et al., 2013a).

14.3.2 Physiological Changes

Physiological processes in plants are adversely affected by alterations in environmental temperature. Respiration, photosynthesis, transpiration, seed germination, plant development, and almost all processes are influenced by high-temperature stress (Takeoka et al., 1991; Stone, 2001). Elevated temperature also affects the water status of a leaf, stomatal conductance, and intercellular CO_2 concentration. Moreover, major changes have been seen in chloroplasts, such as different structural organizations of thylakoids, swelling of grana, and loss of grana stacking under high temperature (Rodríguez et al., 2005).

In plants, respiration first increases with the rise in temperature but only up to a certain level, after which it starts declining (Wahid et al., 2007). Similarly, though photorespiration is less sensitive to temperature, it also starts declining when temperature rises above a critical level.

Photosynthesis is among the finest temperature-sensitive process in plants (Yamamoto et al., 2008). Photosynthetic capacity of C3 plants is much more influenced by high temperatures than that of C4 plants (Crafts-Brandner and Salvucci, 2002). A decline in total photosynthetic pigments and reduction in the activity of photosystem II (PSII) have also been observed during HT (Camejo et al., 2005). Stomatal closure is one of the reasons for lessened photosynthesis affecting the intercellular concentration of CO_2. Rubisco, being the most abundant photosynthetic enzyme, is very much sensitive to temperature alterations in the environment. Its activity is inhibited under high temperature in various plants, including rice (*Oryza sativa*), wheat (*Triticum aestivum*), soybean (*Glycine max*), and spinach (*Spinacea oleracea*) (Salvucci et al., 2001; Gesch et al., 2003; Kim and Portis, 2005).

HT greatly affects the activity of many essential enzymes belonging to sucrose and starch biosynthesis like sucrose phosphate synthase and ADP-glucose pyrophosphorylase, and invertase is reduced (Rodríguez et al., 2005). When high-temperature conditions become prolonged, they lead to a diminution of carbohydrate reserves as well as plant starvation (Djanaguiraman et al., 2009).

Plant water status is a key variable under elevated temperature (Mazorra et al., 2002). A significant drop in leaf water potential occurs when temperature rises. HT stress is also linked with reduced water availability under field conditions as soil moisture decreases with rising temperature (Radwan, 2007). Water potential is altered in sugarcane leaf upon exposure to heat even in the presence of optimal conditions of relative humidity and adequate soil water supply, indicating the heat effect on root hydraulic conductance (Wahid and Close, 2007).

Temperature is a vital factor in the seed germination process (Jagdish et al., 2014). Negative impacts of high temperatures have been observed in the seed germination of various crops (Kumar et al., 2011). The optimum temperature required for seed germination varies from crop to crop, but it usually ranges from 15°C to 30°C (Saitoh, 2008; Johkan et al., 2011). Reduced germination percentage, abnormal seedlings, decreased seedling vigour index, and reduced growth of the radicle and plumule in germinated seedlings are high-temperature effects observed in many cultivated plant

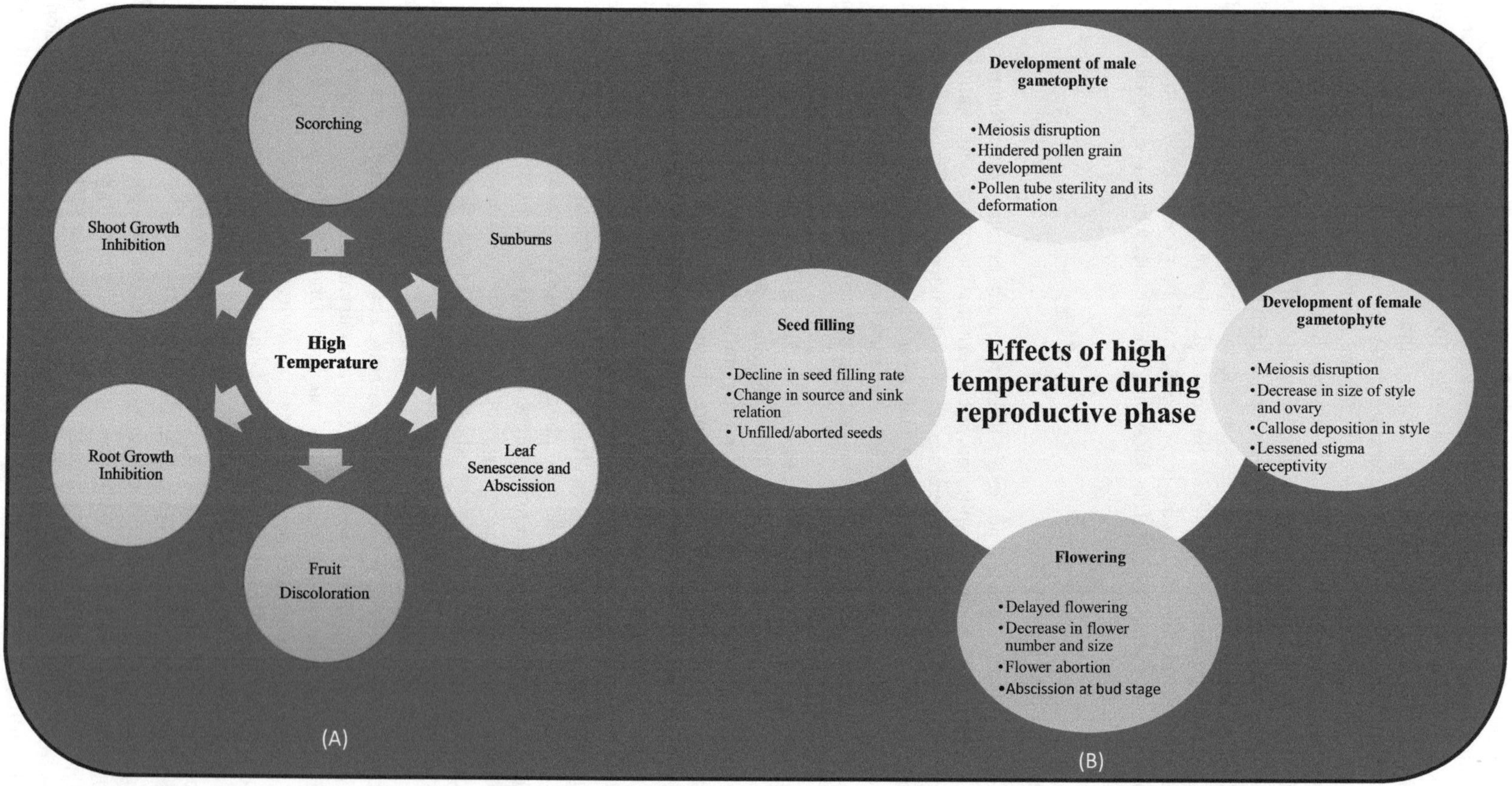

FIGURE 14.2 (A) High-temperature effects on plant morphology. (B) Effects of high-temperature stress on various reproductive development stages.

varieties (Piramila et al., 2012; Toh et al., 2008). High temperatures also adversely affect the grain-filling stage by limiting the dry matter transportation rate from vegetative organs to the kernel.

14.3.3 Biochemical Changes

Many biochemical alterations ensue in plants during high-temperature stress. Plasma membranes are the prime sites of injury under HT because they are in direct contact with high-intensity heat waves, therefore considered as the most heat sensitive (Wise et al., 2004). Elevated temperature severely affects the structure and function of membrane. Thereupon, it increases the membrane's fluidity due to the denaturation of protein and increases the amount of unsaturated fatty acids, resulting into a phase transition from solid gel structure to a flexible liquid crystalline structure (Wahid et al., 2007). Consequently, heat-stress-inflicted membrane impairment increases electrolyte leakage (Bhandari et al., 2017).

Various metabolic pathways produce reactive oxygen species (ROS) as a by-product in different plant organelles under HT. ROS are extremely lethal and can lead to yield loss and can damage the biomolecules including carbohydrates, proteins, DNA, and lipids, which results in plant cell death (Navrot et al., 2007; Khan and Singh, 2008). The major sites of ROS generation are PSI and PSII reaction centres in chloroplast, mitochondria, and peroxisomes (Rio et al., 2006). In wheat, oxidative stress is induced by high-temperature stress (33°C) which damages the membrane properties and causes a degradation of proteins and a deactivation of enzymes (Savicka and Škute, 2010). However, plants have evolved precise strategies to overcome the detrimental effects of ROS. Although ROS are produced under both optimal and stress conditions in cells, their levels can become exceedingly high under stressed conditions (Alscher et al., 2002). To minimize the effects of ROS, the production of antioxidant enzymes increases. These antioxidant enzymes provide protection against the damaging effects of ROS by detoxifying them. Antioxidant defences are of two types: enzymatic and non-enzymatic. The antioxidant system consists of enzymes such as catalase (CAT), glutathione reductase (GR), superoxide dismutase (SOD), and ascorbate peroxidase (APX), along with non-enzymatic antioxidants like glutathione, ascorbate, carotenoids, and α-tocopherol (Foyer et al., 1998). The activity of these enzymes increases with rising temperatures, helping to maintain low ROS levels.

14.3.4 Molecular Changes

High temperature can trigger several mechanisms of defensive gene expression in plants that were not expressed under regular environmental conditions (Feder, 2006). Plants express several stress-responsive genes in response to severe environmental stresses. These genes are regulated via various transcription factors via binding to the cis-acting promoters of the heat shock genes (Nakashima et al., 2009; Dubos et al., 2010; Mizoi et al., 2012). Sudden alterations in genotypic expressions cause an enhancement in the synthesis of protein groups, named heat-shock proteins (HSPs) and sometimes also recognized as stress-induced proteins (Gupta et al., 2010). High-temperature stress also causes denaturation of proteins. HSPs serve as molecular chaperones to repair and maintain the companion protein structure. Meanwhile, they also target non-native proteins and incorrectly aggregated proteins for degradation and elimination from cells (Reis et al., 2012). The primary function of HSPs is to regulate the accurate folding and confirmation of structural (cell membrane) as well as functional (enzyme) proteins under HT conditions (Wang et al., 2004).

14.4 TOLERANCE STRATEGIES

Thermotolerance is generally defined by the plant's ability to grow, develop, and produce economic yield in HT stress. Plant tolerance for high temperature varies from species to species despite their presence in the same geographical region. Plants have various tolerance mechanisms to diminish heat-induced damages. The use of numerous exogenous protectants potentially helps plants to mitigate the adverse effects of HT and promote growth. Furthermore, heat-tolerance capacity can also be enhanced by developing thermotolerant lines.

14.4.1 Role of Phytohormones

Phytohormones play a vital role in helping plants cope with varying environmental conditions by facilitating proper growth and development, nutrient allocation and partitioning, and source-sink transitions (Khan et al., 2013a). Coordination of both endogenous and exogenous signals regulates plant growth and development. Phytohormones, as endogenous messenger molecules, play a key role in plant growth and development and provide physiological protection against high-temperature stress. High temperatures alter hormone stability, homeostasis, biosynthesis, compartmentalization, and content in plants (Maestri et al., 2002). Various phytohormones which were exogenously applied were found effective for the mitigation of high-temperature stress. To date, all major phytohormones such as salicylic acid, abscisic acid, gibberellins, auxin, cytokinins, ethylene, jasmonic acid, and brassinosteroids play a crucial role in plant cell response and plant stress signalling (Mittler et al., 2012; Xia et al., 2015). Targeted manipulation of the phytohormone pool provides better efficacy for modifying the plants' response towards heat stress (Mackova et al., 2013).

14.4.1.1 Salicylic Acid

Salicylic acid (SA) is a phenolic compound formed by phenylpropanoid metabolism and is very effective in imparting thermotolerance (Z. C. Yuan et al., 2008). SA alters activity of various enzymes and upsurges the level of chlorophyll and carotenoids as well as photosynthetic rates (Sita et al., 2017). An SA-mediated signalling pathway develops tolerance against high-temperature stress in a broad range of plant species such as *Arabidopsis thaliana*, grapevines, potato, tobacco, and some others (Li et al., 2010; Horváth et al., 2007). Exogenously applied 0.1 mM SA has persuaded intrinsic thermotolerance in grapevines by reducing electrolyte leakage and thiobarbituric acid reactive substances (TBARS) (Wang and Li, 2006). SA pre-treatment in grapevine leaves alleviated the high-temperature stress-induced lessening in photosynthesis via maintaining a higher Rubisco activation state and boosting photosynthesis II efficiency (Wang et al., 2010). These SA effects are coupled with accelerating levels of HSP21 mainly at the time of recovery period. In wheat, pre-treatment with SA increases protein kinase activity and also inhibits the D1 protein degradation under high-temperature and light stress (Zhao et al., 2011). In rice, seedlings' pre-treatment with SA (0.5 mM) prevented the high-temperature-induced electrolyte osmosis, lessening MDA content and the rate of superoxide anion radicle production. However, the soluble sugar, soluble protein, proline, hydrogen peroxide, ASA, and GSH contents rise with SA pre-treatment under high-temperature stress in rice seedlings (Hasanuzzaman et al., 2013a). Exogenous application of SA improves thermotolerance in *Arabidopsis* by inducing gene expression of several heat shock proteins (Clarke et al., 2009).

14.4.1.2 Abscisic Acid

Abscisic acid (ABA) is commonly acknowledged as a stress hormone, due to its accumulation in plants during stress conditions. Some earlier findings have shown that exogenously applied ABA confers thermotolerance in crops (Ding et al., 2010). Exogenous ABA aids the upsurge in growth and also increases the endogenous concentration of ABA and osmolytes. ABA treatment diminishes oxidative damage in plants under high-temperature stress which was specified by reduced hydrogen peroxide and MDA content (Hasanuzzaman et al., 2013a). It has been reported that watermelon hypocotyls treated at 60°C temperature resulted in a higher concentration of ABA as compared to control hypocotyls (Demir and Van de Venter, 2000). In *Pisum sativum*, plant exposure to high temperature was followed by a rapid increase in ABA levels within 10 minutes which shows its involvement in high-temperature sensing and acclimation (Liu et al., 2006). In the high-temperature stress response, stomatal aperture regulation is a significant adaptive strategy. During HT stress in tobacco, plants enhance the catabolism of ABA which increases stomatal conductance. Therefore, a rise in temperature beyond physiological optimum first stimulates transpiration in order to help the plant in lowering its leaf temperature; nevertheless, higher transpiration creates water deficit conditions that persuade ABA level for the closure of stomata (Mackova et al., 2013).

14.4.1.3 Auxin

Auxin is well known for its role in cell division and elongation. Since the last decade, its role in adaptive growth in response to high temperature has been established (Ahammed et al., 2016). Under HT stress, exogenous application of auxin persuades proliferation of anther cells and can also reverse the male sterility induced from elevated temperature. It shows that high-temperature-induced reduction of auxin in tissues is the chief cause of damage during reproductive development (Sakata et al., 2010; Oshino et al., 2011).

14.4.1.4 Gibberellins

Gibberellins (GAs) are natural diterpenoids generally involved in important plant processes like seed germination, flowering induction, vegetative growth, and fruit development (Sun and Gubler, 2004). They are considered as inhibitors of leaf senescence and are able to delay or avoid leaf yellowing. In *Arabidopsis*, exogenously applied GA_3 (50 μM) diminishes the high-temperature-induced inhibition effects on germination and seedling development (Alonso-Ramirez et al., 2009; Ahammed et al., 2016). Moreover, under high temperature, the GA pathway also enhances phytochrome-interacting factor 4 (PIF4) activity at the post-translational levels (Stavang et al., 2006).

14.4.1.5 Cytokinins

Cytokinins are primarily involved in numerous aspects of plant growth and development. Transpiration is a pivotal process which is responsible for leaf cooling under high-temperature stress. Cytokinins enhance the transpiration rate by stimulating stomatal opening. Under high-temperature stress, numerous heat-responsive proteins are induced by cytokinins, which include various small HSPs and glycine-rich protein (Ahammed et al., 2016). In grass, endogenous level of cytokinins rises from overexpression of isopentenyltransferase (IPT)-cytokinin biosynthetic gene to improve thermotolerance (Xing et al., 2009). Furthermore, transgenic *Arabidopsis* with low cytokinin levels is susceptible to high-temperature stress as compared to transgenic *Arabidopsis* with high cytokinin levels, which shows increase thermotolerance during flowering (Ahammed et al., 2016).

14.4.1.6 Brassinosteroids

Brassinosteroids (BRs) are steroidal phytohormones which help plants to respond against several environmental stresses including high-temperature stress. BRs stimulate improvement in the basal thermotolerance in *Brassica napus* which is also linked with an increase in HSP synthesis and accumulation (Dhaubhadel et al., 2002). In tomato leaves, exogenous application of BRs lessens the high-temperature effects by enhancing the stomatal conductance and net photosynthetic rate and maximizing the carboxylation rate of Rubisco (Ogweno et al., 2002). The optimistic effects of BRs were also witnessed in bean plants subjected to high-temperature stress. Bean plants which were sprayed with BRs (25 and 50 mgL^{-1}) have improved vegetative growth, enhanced total production, and a higher quality of pods under high temperature (El-Bassiony et al., 2012).

14.4.1.7 Ethylene

Ethylene is an essential phytohormone which takes part in several plant developmental processes. Karagezov (2002) reported that ethylene aids in reducing the adverse effects of HT stress. The crucial element of ethylene signalling and response pathway are ethylene response factors (ERFs), which are part of the transcription factor superfamily APETALA2(AP2)/ERF. ERFs act as the main regulatory hub and can bind to dehydration-responsive elements to trigger the stress-responsive genes transcription in order to enhance thermotolerance in plants (Muller and Munne-Bosch 2015). Even though ethylene concentration rises during high-temperature stress, exogenous application of ethylene cannot confer thermotolerance in *Arabidopsis* (Clarke et al., 2009).

14.4.1.8 Jasmonic Acid

Jasmonic acid (JA) regulates plant responses to various types of biotic and abiotic stresses. Grape seedlings' treatment with 50 μM JA solution can diminish the stress effects under high-temperature stress (42°C) by enhancing the antioxidant enzymes' (CAT, SOD, and POD) upregulation (Chen et al., 2006). In wild-type *Arabidopsis*, thermotolerance can be improved by exogenously applied JA. Furthermore, in *Capsicum annuum*, exogenous application of JA, ethylene, and ABA induced expression of CaWRKY6, which plays a crucial role in high-temperature response (Cai et al., 2015). Phytohormone-induced stress tolerance might be mediated by CaWRKY6 being a positive regulator of high-temperature stress tolerance.

14.4.2 Manipulation of Nutrients

Plant mineral nutrients play a vital role in augmenting plant resistance to high-temperature stress and other environmental stresses (Marschner, 1995).

14.4.2.1 Macronutrients

Nitrogen is an important macronutrient and has a dynamic role in providing thermotolerance. As light intensity increases with rising temperature, it affects the nutrient uptake in plants and adversely impacts plant growth. Among all nutrients, nitrogen plays a chief role in photosynthetic carbon assimilation and consumption of absorbed light energy (Kato et al., 2003; Huang et al., 2004). Nitrogen-rich leaves have a greater tolerance for photo-oxidative damage under high light intensity than do nitrogen-deficient leaves (Kato et al., 2003). The thylakoid membrane of chloroplast helps to avoid photo-oxidative damage by its defence mechanism, which includes the dissipation of excess light energy as heat (Waraich et al., 2012). Additionally, in leaves which are N-deficient, non-utilized light energy is the major cause of photo-oxidative damage. It has been reported by Zhu et al. (2000) that bean plants grown in the presence of nitrate have greater tolerance to photo damage than do those grown in the presence of ammonium. Subsequently, higher levels of lipid peroxidation and antioxidant enzyme have been seen in plants growing with ammonium supply under high light intensity. Nitric oxide (NO), a form of nitrogen, is a highly reactive molecule and has a significant role in providing tolerance to plants from abiotic stresses including high temperature (Hasanuzzaman et al., 2012). NO works as a signalling molecule and participates in many plant physiological processes like closure of stomata, cell senescence, program cell death, and seed germination (Hasanuzzaman et al., 2013a; Waraich et al., 2012). According to a report, northern blot analysis confirmed the protective role of NO on chloroplast against oxidative damage under high-temperature stress by persuading the gene expression encoding HSP26 (Uchida et al., 2002). NO may also act as antioxidants directly protect the plant against ROS generation in stressful conditions (Wendehenne et al., 2001).

Potassium (K) is another essential macronutrient which has a key role in many plant physiological processes like photosynthesis, photosynthate translocation from source to sink organs, enzyme activation, and maintain turgidity (Mengel and Kirkby, 2001). Its deficiency causes a severe decrease in all these processes and shows symptoms like chlorosis and necrosis in plants. Reduction in photosynthetic CO_2 fixation leads to excess production of photosynthetic electrons which increases the NADPH oxidase activity and results into

ROS production (Waraich et al., 2011). Plants with a higher supply of K are more protected against oxidative stress harms than those having a lower supply of K (Waraich et al., 2012).

Magnesium (*Mg*), a macronutrient, takes part in various biochemical and physiological processes to facilitate growth and development (Waraich et al., 2011). A significant role of Mg has been seen in plant photosynthesis (Waraich et al., 2012). Mg deficiency or excess can both affect photosynthesis, and even a small deviation in Mg level can directly devastate chlorophyll enzymes (Shaul, 2002). Under temperature stress, ROS production occurs in many organelles as a by-product of numerous metabolic pathways. Lack of minerals in a plant can also cause oxidative damage. It has been reported in maize, pepper, beans, and mulberry that Mg upsurges the concentration and activities of antioxidant enzymes to protect the plant from oxidative damage (Tewari et al., 2004; Anza et al., 2005; Cakmak, 1994; Tewari et al., 2006). Magnesium enhances the root surface area and growth which helps in better uptake by root of water and nutrients. Mg is also a central element in the structure of chlorophyll. Being the chlorophyll constituent, magnesium increases the sucrose quantity and also its transportation from leaves to roots (Waraich et al., 2011). Thus, improvement of Mg status in plants leads to an incrementally higher photosynthetic rate by conserving the structure of chloroplast which ultimately enhances productivity.

Calcium (*Ca*) is a vital nutrient element that is well known for several structural roles in plants under both stressful and optimal conditions (White and Broadley, 2003). In environmental stress, calcium regulates numerous physiological processes at the cellular, tissue, and molecular levels that impact plant growth and response (Waraich et al., 2011). Under HT stress, antioxidant activity and thermotolerance are stimulated by calcium. Application of Ca increases the activity of SOD, CAT, and POD and alleviates the ROS in plant cells. Calcium supply can also increase the photosynthetic rate under high-temperature stress. Exogenous application of Ca on wheat plants growing under high-temperature stress reduces the oxidative damage and enhances the amino acid contents (Goswami et al., 2015).

14.4.2.2 Micronutrients

Manganese (*Mn*) is a necessary element required in nitrogen metabolism, photosynthesis, and the formation of other compounds needed for plant metabolism (Waraich et al., 2012). Mn by increasing nitrogen metabolism and photosynthesis helps the plant to reduce the detrimental effects of high temperature. Mn application causes a reduction in necrotic spots, interveinal chlorosis, and premature leaf abscission. In plants grown under high temperature, various enzymes of oxidation-reduction, hydrolytic, and decarboxylation reactions are activated with Mn involvement (Marschner, 1995; Turhan et al., 2006).

Boron (*B*), in various biochemical and physiological processes like nitrogen metabolism, photosynthesis, membrane function, cell wall biosynthesis, cell elongation, cell division, and uracil synthesis, is involved either directly or indirectly during plant growth (Marschner, 1995; Zhao and Oosterhuis, 2002). Boron application can decrease ROS damage by enhancing the antioxidant activities in plants under temperature stress (Waraich et al., 2012). Furthermore, in-plant body boron also increases the sugar transport which in turn improves the seed formation and its germination. Reduction in phenolic compounds could occur in leaves with boron applications. Consequently, boron alleviates ROS production, heightens the rate of photosynthesis, and minimizes cell damage (Waraich et al., 2011).

Selenium (*Se*), being a trace element, is a stress protectant under several environmental stresses including high temperature. During an experiment, it was observed in cucumber that foliar spray of 8 μM Se maintains stomatal conductance and improves photosynthetic and transpiration rate under high temperature (Balal et al., 2016). In sorghum, foliar spray of Se (75 mg L^{-1}) alters the adverse effects of high temperature by increasing the antioxidant enzyme activities that reduce membrane damage and eventually improve grain yield (Djanaguiraman et al., 2010). Likewise, exogenous application of selenium enhances and improves the chlorophyll content, osmotic adjustment capacity, and antioxidant activities in wheat and also reduces the electrolyte leakage, hydrogen peroxide contents, and MDA under HT stress (Iqbal et al., 2015). Under UV-induced oxidative stress, selenium enhances plant tolerance. Se also helps to delay senescence and stimulate aging seedlings' growth (Hartikainen et al., 2000; Xue et al., 2001).

14.4.3 Osmotic Adjustment

Plants have a well-developed defensive mechanism known as osmoregulation which is mediated by various osmolytes. Osmotic adjustment or osmoregulation is the alleviation of osmotic potential due to net solute accumulation in response to water scarcity conditions (Zhang et al., 1999). It plays a vital role in maintaining water relations and growth in plants. High-temperature stress and drought both influence plant water relations (Mazorra et al., 2002). Plants under high-temperature stress experience an enhanced evaporative demand, which is linked with a rising transpiration rate and increased water transportation (Khalil et al., 2018). Numerous osmotically active constituents, particularly molecules and ions like organic acids, soluble sugars, glycine betaine, proline, amines, calcium, potassium, and chloride ions, are overproduced and accumulated in plants under high-temperature-induced osmotic stress. These compounds help plants maintain turgor and protect them from adverse environmental conditions by protecting biological membranes, detoxifying reactive oxygen species, adjusting cellular osmotic balance, and stabilizing enzymes and proteins (Mishra and Saxena, 2009; Hare et al., 1998; Bohnert and Jensen, 1996). As heat-sensitive plants are not able to accumulate these substances, thermotolerance can be enhanced by exogenous application of these osmolytes (Rasheed et al., 2011; Sakamoto and Murata, 2002). There is a direct correlation between osmotic adjustment and osmolyte accumulations in various concentrations depending on the growth stage at which stress occurred, plant type, and stress severity (Shao et al., 2006). It has been reported by Morgan (1990) that in water deficit conditions osmotic adjustment is essential to postpone dehydration stress.

Soluble sugars play an essential role in both plant structure and metabolism (Khan et al., 2013b). Sepú lveda and Kliewer (1986) reported that when grapevines are exposed to high temperature, sucrose concentration has been enhanced in most of the plants that are protected from stress. Proline generally accumulates in new leaves and is considered as the most crucial solute against abiotic stresses (Perez-Perez et al., 2009). Proline adjusts osmotic balance in cells, stabilizes enzymes, protects biological membranes, and detoxifies excess reactive oxygen species (Iqbal et al., 2014). Glycine betaine (GB) is also a major osmolyte that provides protection against various stresses, including osmotic stress (Munns, 2005) and high-temperature stress (Wang et al., 2010). GB also adjusts osmotic balance in cells, stabilizes membrane integrity, protects Rubisco activity, prevents polypeptide dissociation from the PSII complex, and detoxifies lethal ions (Ashraf and Foolad, 2007). Exogenously applied GB, proline, and trehalose (10 μM) on chickpea plants under heat-stressed conditions promoted plant growth (Kumar et al., 2012). Nonetheless, further studies are still needed to understand the changes in plant osmoregulation under high-temperature stress.

14.4.4 Development of Tolerant Lines by Using Several Molecular Approaches

Developments in genome editing tools provide different opportunities for improving crop varieties to mitigate the adverse effects of high-temperature stress. In order to cope with HT stress, breeding climate-smart crops that are thermotolerant would be a sustainable approach using numerous genetic engineering as well as transgenic methods. Thermotolerance can also be enhanced by constitutive expressions of precise proteins.

Lee et al. (1995) in *Arabidopsis* plant made a modification in transcription factor (AtHSF1) liable for heat shock proteins and successfully changed the HSPs' expression level. In this way they produced an HT tolerant transgenic *Arabidopsis.* In rice, several successful transgenic cases were also reported where thermotolerance had been enhanced after the integration of HSP genes. *Arabidopsis* Hsp 101 gene's successful overexpression in transgenic rice improved heat tolerance (Katiyar-Agarwal et al., 2003). Furthermore, sHSP 17.7 overexpression confers thermotolerance to rice plants (Murakami et al., 2004). Yang et al. (2005) proposed that the transformation of the BADH gene responsible for glycinebetaine overproduction will improve HT tolerance in plants. Transformation of tobacco with Rubisco activase gene provides thermotolerance and helps plants protect their photosynthetic apparatus (Sharkey et al., 2001). Additionally, many successful results were found in developing transgenic plants with thermotolerance by genetically improving the proteins participating in photosynthetic reactions, osmoregulation, ROS detoxification, and the protein biosynthesis process (Hasanuzzaman et al., 2013a). Further studies are needed to identify genes linked to heat tolerance in crops in order to secure the development of thermotolerant transgenics.

Genome editing tools provide various openings in the field of crop improvement hence enabling greatly precise and efficient gene editing up to single nucleotide level. Clustered regularly interspaced short palindromic repeats/CRISPR-associated proteins (CRISPR/Cas9) is the latest genome editing technology. In the past few years, CRISPR/Cas9 has gained so much attention from researchers because of its versatile advantages, including ease in target designing; efficacy of introducing mutations through direct injection of RNAs coding Cas9 protein; and the simplicity of multiplexing, instigating targeted mutations in numerous genes in just one go (Ma et al., 2015; Malzahn et al., 2017). The CRISPR/Cas9 system is a very efficient, robust, and practical methodology used to develop plant varieties resistant to temperature change (Zafar et al., 2020). With the help of multiplex-genome editing, it's possible to develop crops with improved tolerance against many abiotic stresses in just a single transformation event. Klap et al. (2017) with the help of CRISPR/Cas9 facilitated genome editing successfully targeted a sensitivity gene, *slagamous-like 6* (*SLAGL6*) in tomato, which enhanced the fruit setting under high-temperature stress.

14.4.5 Development of Thermotolerance Using Omics Approaches

New advances in science have developed various techniques and tools in this modern world which provide hopes and opportunities to the researchers. Availability of the refined "omics" technologies have offered excellent breaks for the recognition of transcription, translation, and post-translation mechanisms and also signalling pathways that are involved in the regulation of plant response(s) to several abiotic stress including HT (Hasanuzzaman et al., 2013c). Omics approaches have effectively been used for the identification of high-temperature responsive genes; their expression modulates the regulatory network and the related functional proteins and metabolic alterations in plants (Khan et al., 2020). Omics strategies includes genomics, transcriptomics, proteomics, metabolomics, and some other methodologies.

Contemporary as well as conventional breeding methods are not effective enough for the development of heat-tolerant crops until the detailed molecular mechanisms linked with grain yield stability are known.

DNA is the preliminary point of every molecular evidence associated to high-temperature stress response in plants and encompasses numerous HT stress-responsive genes in their genome. With the aid of genome-wide expression studies and genetic screens, researchers have identified an enormous number of genes with potential functions in HT stress response (Yeh et al., 2012). Such genes from the genome further form transcriptomic products (mRNAs). Plants commonly activate defence mechanisms in response to environmental stress to mitigate adverse conditions, making it important to understand the cascades for such responses. Thus, transcriptome cumulation and comparison of various tissue types, as well as developmental stages, is the finest strategy to inspect plant response regulation and to detect genes associated in stress tolerance mechanisms. It would also lead to a better understanding of corresponding phenotypic alteration (Shinde et al., 2015). This transcriptomic approach has been practised to study the changes occurring in gene expression under various stress conditions in several crop species like wheat (Qin et al., 2008), barley (Mangelsen et al., 2011), brassica (Dong et al., 2015), soybean (Gillman et al., 2019), and rice (Sarkar et al., 2014; Mangrauthia et al., 2017).

For the evaluation of transcriptomic data, several methods and tools are accessible to get expression profiling for each gene and broadly for numerous genes at a time. The production and activation of various heat shock factors (HSFs) and heat shock proteins (HSPs) and the rise in ROS scavenging activity play a significant role in plant acclimation and responses to HT stress (Maestri et al., 2002; Comastri et al., 2018). The genes related to ROS are in the core upregulated category under HT stress in many plants species and are linked with basal thermotolerance (Chao et al., 2008; Mittal et al., 2012; Bhattacharjee, 2012). Moreover, both HSFs as well as HSPs are prime to the HT stress response and in acquiring thermotolerance in plants (Scharf et al., 2012; Janni et al., 2020). With the help of whole-genome sequencing, supplementary information has been collected for the HSP gene family (Chen et al., 2018). The HSF gene family has been recognized and characterized in numerous crop species like wheat (Xue et al., 2014), rice (Lavania et al., 2018), soybean (Chung et al., 2013), tomato (Yang et al., 2016), and grape (Liu et al., 2012). Further advancement in gene annotation has led to the progress of a specialized HEATSTER database which can not only identify but also properly annotate novel HSF genes in plant species (Berz et al., 2019). In tomato leaves, gene expression analyses were performed applying massive analysis of cDNA ends (MACE), which indicated that 9.6% of the total genes (approx. 2203 genes) had increased transcript abundance in response to HT stress (Fragkostefanakis et al., 2015).

Conventionally, a majority of the HT stress-responsive genes are embraced in translation, transcription, primary and secondary metabolism, regulation, and responses to processes like phytohormone, calcium, sugar, and lipid signalling, or protein modifications (Takahashi and Shinozaki, 2019).

Proteomes are interconnected in numerous biochemical processes and produce various metabolic products (metabolomes). Plant responses to high-temperature stress are accompanied by alterations in the expression of various proteins, which makes a proteomics approach an efficient aid in identifying and characterizing stress-responsive candidate proteins. A detailed study of these proteins reveals their role in thermotolerance mechanisms (Priya et al., 2019). After exposure to elevated temperatures, the synthesis and accumulation of heat shock proteins is a quick response that is considered a significant adaptive tactic to overcome the harmful effects of HT stress (Keller and Simm, 2018). HSPs are mostly molecular chaperones, well-known for their functions in signal transduction and protein stabilization during HT stress (Arce et al., 2018). They prevent intracellular proteins with anomalous confirmations and preserve their stability. Proteomics strategy permits the investigation of direct gene products, which often varies with gene regulation level. This method had been applied in various crop plants such as chickpea (Parankusam et al., 2017), grape (Jiang et al., 2017), maize (Hu et al., 2015), rice (Mu et al., 2017), and tomato (Lin et al., 2016) to investigate heat responses. Several types of techniques utilized for these studies have involved 2D gel electrophoresis for the separation of proteins, followed by MS:MALDI-TOF or MS or MS/MS TOF/TOF for the isolation of protein, as well as iTRAQ with LCMS/MS for the relative quantification of proteins. Various plant developmental stages and tissues have been investigated by proteomics, from seed, to flower, to fruit and also spikelet. Whole classes of HSPs have been found in rice among the major protein categories, where HT stress has been studied in isolation, although additional protein categories as well as gene ontology classes varied widely with part of the plant studied. For example, the presence of late embryogenesis abundant (LEA) proteins in high numbers was noticed in pollen and anthers (Janni et al., 2020). Plant responses to HT stress also involve numerous post-translational modifications of proteins as an adaptive strategy. Detailed literature is available on the HT stress effects on post-translational modifications like protein phosphorylation in wheat and rice (Chen et al., 2011; Wu et al., 2014).

Metabolomics is an efficient tool with multiple methods for unravelling as well as characterizing stress-responsive plant metabolites. By comparing metabolomics between high temperature and several other kinds of abiotic stresses, researchers have identified metabolites that are usually essential in stress responses or are particular to each stress. In several plants like wheat (Thomason et al., 2018), maize (Qu et al., 2018), soybean (Xu et al., 2016), and barley (Templer et al., 2017), metabolism reprogramming under HT stress has been investigated by untargeted metabolomics. Templer et al. (2017) found that two cultivars of barley were highly affected by quantities of antioxidants such as α-tocopherol and glutathione, and amino acids under both drought and heat stress. In particular, a combination of both stresses led to the accumulation of free amino acids in the leaf. Moreover, heat stress and water stress, either alone or in combination, change the metabolite levels in citrus associated with glycolysis, the Krebs cycle, and the chloroplastic side of the dicarboxylate/glyoxylate cycle, depending on the stress level. Many other examples come from untargeted metabolomics trials, which revealed that ascorbic acid, salicylic acid, phenolic secondary metabolites, sulfur-containing metabolites, and, more broadly, all antioxidants can decrease HT stress symptoms in various plant species (Mobin et al., 2017; Muhlemann et al., 2018; Ihsan et al., 2019). Therefore, significant analysis of the metabolome of both HT-tolerant and HT-susceptible varieties may unravel the important fundamentals of HT stress adaptation.

14.5 CONCLUSION AND FUTURE PERSPECTIVE

Rising temperature as a consequence of climate change in this era is driving us near an unsecure sustainable future. Emission of greenhouse gases from various anthropogenic activities is believed to be responsible for the gradual increase in the world's ambient air temperature (Meehl et al., 2007). HT has become a vital concern for crops as it negatively impacts various plant processes by causing alterations. Numerous tolerance strategies can be implemented to alleviate HT effects such as the use of phytohormones, manipulation of nutrients, osmotic regulation, and the development of thermotolerant lines. There is a need to better understand plant responses and adaptations towards elevated temperature along with mechanisms underlying the improvement of thermotolerance. In recent years, researchers have studied intensively about the plant responses to HT stress, yet a comprehensive understanding of heat-tolerance mechanisms remains elusive. Various endeavours have been made to induce thermotolerance in many plant species with the aid of several different approaches. Exogenous applications of osmoprotectants, nutrients, and phytohormones have shown beneficial outcomes on plant growth, owing to growth-stimulating and antioxidant activities under HT stress. Such applications provide promising results and further research is impending. Engineering plants for the synthesis of these compounds and breeding tolerant cultivars against HT is a sustainable and eco-friendly tactic and epitomizes a potentially significant area of research on HT tolerance. Recent advancements in genome editing tools will provide new breaks for crop improvement and substantially contribute in raising the novel plant varieties. CRISPR/Cas9 is one such promising tool. Being a very efficient and robust approach, CRISPR/Cas9 is well adopted in plant developmental biology for the characterization of genes as well as underpinning the mechanisms of different traits. Moreover, omics approaches have been exploited to improve thermotolerance in various crop species and have shown promising results.

ACKNOWLEDGEMENTS

The authors wish to thank all the contributors in the field of HT-thermo tolerance. In addition, Shreya Chhabra acknowledges the financial support by the DST-Inspire fellowship.

REFERENCES

Aggarwal PK (2008). Global climate change and Indian agriculture: impacts, adaptation and mitigation. Indian *J. Agric. Sci.* 78(11): 911.

Ahammed GJ, Li X, Zhou J, Zhou YH, Yu J-Q (2016). Role of hormones in plant adaptation to heat stress. In: Ahammed GJ, Yu J-Q (eds), *Plant Hormones Under Challenging Environmental Factors*. Springer Science+Business Media, Dordrecht, The Netherlands.

Ahmed FE, Hall AE, DeMason DA (1992). Heat injury during floral development in cowpea (*Vigna unguiculata*). *Am. J. Bot.* 79: 784–791.

Alonso-Ramirez A, Rodriguez D, Reyes D, Jimenez JA, Nicolas G, Lopez-Climent M, Gomez-Cadenas A, Nicolas C (2009). Evidence for a role of gibberellins in salicylic acid-modulated early plant responses to abiotic stress in *Arabidopsis* seeds. *Plant Physiol.* 150: 1335–1344.

Alscher RG, Erturk N, Health LS (2002). Role of superoxide dismutases (SODs) in controlling oxidative stress in plants. *J. Exp. Bot.* 53: 1331–1341.

Andrews M, Hodge S (2010). Climate change a challenge for cool season grain legume crop production. In: Yadav S, Redden R (eds), *Climate Change and Management of Cool Season Grain Legume Crops*, 1–9. Springer, Dordrecht, The Netherlands.

Anza M, Riga P, Garbisu C (2005). Time course of antioxidant responses of *Capsicum annuum* subjected to a progressive magnesium deficiency. *Ann. Appl. Biol.* 146: 123–134.

Arce D, Spetale F, Krsticevic F, Cacchiarelli P, Las Rivas J, Ponce S, Pratta G, Tapia E (2018). Regulatory motifs found in the small heat shock protein (sHSP) gene family in tomato. *BMC Genom.* 19: 860.

Ashraf M, Foolad M (2007). Roles of glycine betaine and proline in improving plant abiotic stress resistance. *Environ. Exp. Bot.* 59: 206–216.

Asseng S, Ewert F, Martre P, Rötter RP, Lobell D, Cammarano D, Kimball B, Ottman M, Wall G, White JW, Reynolds MP (2015). Rising temperatures reduce global wheat production. *Nat. Clim. Change* 5: 143.

Balal RM, Shahid MA, Javid MM, Iqbal Z, Anjum MA, Garcia-Sanchez F, Mattson NS (2016). The role of selenium in amelioration of heat-induced oxidative damage in cucumber under high-temperature stress. *Acta Physiol. Plant.* 38: 158.

Barnabás B, Jager K, Fehér A (2008). The effect of drought and heat stress on reproductive processes in cereals. *Plant Cell Environ.* 31: 11–38.

Berz J, Simm S, Schuster S, Scharf KD, Schleiff E, Ebersberger I (2019). HEATSTER: A database and web server for identification and classification of heat stress transcription factors in plants. *Bioinform. Biol. Insights* 13: 1177932218821365.

Bhandari K, Sharma KD, Rao BH, Siddique KHM, Gaur P, Agarwal SK, Nair RM, Nayyar H (2017). Temperature sensitivity of food legumes: A physiological insight. *Acta. Physiol. Plant* 39: 68.

Bhattacharjee S (2012). An inductive pulse of hydrogen peroxide pretreatment restores redox-homeostasis and oxidative membrane damage under extremes of temperature in two rice cultivars. *Plant Growth Regul.* 68: 395–410.

Bita CE, Gerats T (2013). Plant tolerance to high-temperature in a changing environment: Scientific fundamentals and production of heat stress-tolerant crops. *Front. Plant Sci.* 4: 273.

Blum A (1988). *Plant Breeding for Stress Environments*, 223. CRC Press, Boca Raton, FL.

Bohnert HJ, Jensen RG (1996). Strategies for engineering water-stress tolerance in plants. *Trends in Biotech.* 14: 89–97.

Borrás L, Westgate ME (2006). Predicting maize kernel sink capacity early in development. *Field Crops Res.* 95: 223–233.

Cai H, Yang S, Yan Y, Xiao Z, Cheng J, Wu J, Qiu A, Lai Y, Mou S, Guan D, Huang R, He S (2015). CaWRKY6 transcriptionally activates CaWRKY40, regulates *Ralstonia solanacearum* resistance, and confers high-temperature and high-humidity tolerance in pepper. *J. Exp. Bot.* 66: 3163–3174.

Cakmak I (1994). Activity of ascorbate-dependent H2O2-scavenging enzymes and leaf chlorosis are enhanced in magnesium-and potassium-deficient leaves, but not in phosphorus-deficient leaves. *J. Exp. Bot.* 45: 1259–1266.

Camejo D, Rodíguez P, Morales MA, Dell'amico JM, Torrecillas A, Alarcón JJ (2005). High-temperature effects on photosynthetic activity of tomato cultivars with different heat susceptibility. *J. Plant Physiol.* 162: 281–289.

Cao YY, Zhao H (2008). Protective roles of brassinolide on rice seedlings under high-temperature stress. *Rice Sci.* 15: 63–68.

Chao Y-Y, Hsu YT, Kao CH (2008). Involvement of glutathione in heat shock- and hydrogen peroxide-induced cadmium tolerance of rice (*Oryza sativa* L.) seedlings. *Plant Soil* 318: 37.

Chen B, Feder ME, Kang L. 2018. Evolution of heat-shock protein expression underlying adaptive responses to environmental stress. *Mol. Ecol.* 27: 3040–3054.

Chen P, Yu S, Zhan Y, Kang XL (2006). Effects of jasmonate acid on thermotolerance of grape seedlings. *J. Shihezi. Unvi. (Nat. Sci.)* 24: 87–91.

Chen X, Zhang W, Zhang B, Zhou J, Wang Y, Yang Q, Ke Y, He H (2011). Phosphoproteins regulated by heat stress in rice leaves. *Proteome Sci.* 9: 37.

Chung E, Kim KM, Lee JH (2013). Genome-wide analysis and molecular characterization of heat shock transcription factor family in *Glycine max*. *J Genet. Genomics* 40: 127–135.

Clarke SM, Cristescu SM, Miersch O, Harren FJ, Wasternack C, Mur LA (2009). Jasmonates acts with salicylic acid to confer basal thermotolerance in *Arabidopsis thaliana*. *New Phytologist*. 182: 175–187.

Clausen SK, Frenck G, Linden LG, Mikkelsen TN, Lunde C, Jørgensen RB (2011). Effects of single and multifactor treatments with elevated temperature, CO2 and ozone on oilseed rape and barley. *J. Agronomy Crop Sci.* 197: 442–453.

Comastri A, Janni M, Simmonds J, Uauy C, Pignone D, Nguyen HT, Marmiroli N (2018). Heat in wheat: Exploit reverse genetic techniques to discover new alleles within the *Triticum durum* sHsp26 family. *Front. Plant Sci.* 9: 1337.

Crafts-Brandner SJ, Salvucci ME (2002). Sensitivity of photosynthesis in a C4 plant, maize to heat stress. *Plant Physiol.* 129: 1773–1780.

Das S, Krishnan P, Nayak M, Ramkrishnan B (2014). High-temperature stress effects on pollens of rice (*Oryza sativa* L.) genotypes. *Environ. Exp. Bot.* 101: 36–46.

Demir I, Van de Venter HA (2000). The effect of heat treatment of watermelon seed on germination, hypocotyl emergence and abscisic acid content. *Exp. Agric.* 36: 453–458.

Dhaubhadel S, Browning KS, Gallie DR, Krishna P (2002). Brasssinosteroid functions to protect the translational machinery and heat-shock protein synthesis following thermal stress. *Plant J.* 29: 681–691.

Ding W, Song L, Wang X, Bi Y (2010). Effect of abscisic acid on heat stress tolerance in the calli from two ecotypes of *Phragmites communis*. *Biol. Plant.* 54: 607–613.

Djanaguiraman M, Prasad PV (2010). Ethylene production under high temperature stress causes premature leaf senescence in soybean. *Funct. Plant Biol.* 37(11): 1071–1084.

Djanaguiraman M, Prasad PVV, Seppanen M (2010). Selenium protects sorghum leaves from oxidative damage under high-temperature stress by enhancing antioxidant defense system. *Plant Physiol. Biochem.* 48: 999–1007.

Djanaguiraman M, Sheeba JA, Devi DD, Bangarusamy U (2009). Cotton leaf senescence can be delayed by nitrophenolate spray through enhanced antioxidant defense system. *J. Agron. Crop Sci.* 195: 213–224.

Dong X, Yi H, Lee J, Nou IS, Han CT, Hur Y (2015). Global gene expression analysis to identify differentially expressed genes critical for the heat stress response in *Brassica rapa*. *PLoS One* 10: e0130451.

Drouin P, Prevost D, Antoun H (2000). Physiological adaptation to low temperatures of strains of *Rhizobium leguminosarum* bv. viciae associated with *Lathyrus* spp. *FEMS Microbiol. Ecol.* 32: 111–120.

Dubos C, Stracke R, Grotewold E, Weisshaar B, Martin C, Lepiniec L (2010). MYB transcription factors in Arabidopsis. *Trends Plant Sci.* 15: 573–581.

El-Bassiony AM, Ghoname AA, El-Awadi ME, Fawzy ZF, Gruda N (2012). Ameliorative effects of brassinosteroids on growth and productivity of snap beans grown under high-temperature. *Gesunde Pflanzen* 64: 175–182.

Farhad M, Kumar U, Tomar V, Bhati PK, Krishnan JN, Barek V, Hossain A. (2023). Heat stress in wheat: A global challenge to feed billions in the current era of the changing climate. *Front. Sustain. Food Syst.* 7: 1203721.

Farooq M, Bramley H, Palta JA, Siddique KHM (2011). Heat stress in wheat during reproductive and grain filling phases. *Crit. Rev. Plant Sci.* 30: 33–41.

Farrell TC, Fox KM, Williams RL, Fukai S (2006). Genotypic variation for cold tolerance during reproductive development in rice: Screening with cold air and cold water. *Field Crop Res.* 98: 178–194.

Feder ME (2006). Integrative biology of stress: molecular actors, the ecological theater, and the evolutionary play. *International Symposium on Environmental Factors, Cellular Stress and Evolution*, Varanasi, India, October 13–15.

Feng B, Liu P, Li G, Dong ST, Wang FH, Kong LA, Zhang JW (2014). Effect of heat stress on the photosynthetic characteristics in flag leaves at the grain-filling stage of different heat-resistant winter wheat varieties. *J. Agro. Crop Sci.* 200: 143, 155.

Foyer CH, Kingston-Smith AH, Harbinson J, Arisi ACM, Jouanin L, Noctor G (1998). The use of transformed plants in the assessment of physiological stress responses. In: Kok LJ, Stulen I (eds), *Responses of Plant Metabolism to Air Pollution and Global Change*, 251–261. Backhuys Publishers, The Netherlands.

Fragkostefanakis S, Simm S, Paul P, Bublak D, Scharf KD, Schleiff E (2015). Chaperone network composition in *Solanum lycopersicum* explored by transcriptome profiling and microarray meta-analysis. *Plant Cell Environ.* 38: 693–709.

Gesch RW, Kang IH, Gallo-Meagher M, Vu JCV, Boote KJ, Allen LH, Bower JRG (2003). Rubisco expression in rice leaves is related to genotypic variation of photosynthesis under elevated growth CO2 and temperature. *Plant Cell Environ.* 26: 1941–1950.

Gillman JD, Biever JJ, Ye S, Spollen WG, Givan SA, Lyu Z, Joshi T, Smith JR, Fritschi FB (2019). A seed germination transcriptomic study contrasting two soybean genotypes that differ in terms of their tolerance to the deleterious impacts of elevated temperatures during seed fill. *BMC Res. Notes* 12: 522.

Goswami S, Kumar RR, Sharma SK, Kala YK, Singh K, Gupta R, Rai RD (2015). Calcium triggers protein kinases-induced signal transduction for augmenting the thermotolerance of developing wheat (*Triticum aestivum*) grain under the heat stress. *J. Plant Biochem. Biotechnol.* 24: 441–452.

Guilioni L, Wery J, Tardieu F (1997). Heat stress-induced abortion of buds and flowers in pea: is sensitivity linked to organ age and to relations between organ reproductive organs? *Ann. Bot.* 80: 159–168.

Gupta SC, Sharma A, Mishra M, Mishra R, Chowdhuri DK (2010). Heat shock proteins in toxicology: How close and how far. *Life Sci.* 86: 377–384.

Hare PD, Cress WA, van Staden J (1998). Dissecting the roles of compatible osmolyte accumulation during stress. *Plant Cell Environ.* 21: 535–553.

Hartikainen H, Xue T, Piironen V (2000). Selenium act as anti-oxidant and prooxidant in ryegrass. *Plant Soil.* 225: 193–200.

Hasanuzzaman M, Hossain MA, da Silva JAT, Fujita M (2012). Plant response and tolerance to abiotic oxidative stress: antioxidant defense is a key factor. In: Bandi V, Shanker AK, Shanker C, Mandapakab M (eds), *Crop Stress and its Management: Perspectives and Strategies*, 261–315. Springer, Berlin.

Hasanuzzaman M, Nahar K, Alam MM, Roychowdhary R, Fujita M (2013a). Physiological, biochemical and molecular mechanisms of heat stress in plants. *Int. J. Mol. Sci.* 14: 9643–9684.

Hasanuzzaman M, Nahar K, Fujita M (2013b). Extreme temperatures, oxidative stress and antioxidant defense in plants. In: Vahdati K, Leslie C (eds), *Abiotic Stress-Plant Responses and Applications in Agriculture*, 169–205. InTech, Rijeka, Croatia.

Hasanuzzaman M, Nahar K, Fujita M, Ahmad P, Chandna R, Prasad MNV, Öztürk, M (2013c). Enhancing plant productivity under salt stress – relevance of poly-omics. In: Ahmad P, Azooz MM, Prasad MNV (eds), *Salt Stress in Plants: Omics, Signaling and Responses*, 113–156. Springer, Berlin, Germany.

Horváth E, Szalai G, Janda T (2007). Induction of abiotic stress tolerance by salicylic acid signaling. *J. Plant Growth Regul.* 26: 290–300.

Hu X, Wu L, Zhao F, Zhang D, Li N, Zhu G, Li C, Wang W (2015). Phosphoproteomic analysis of the response of maize leaves to drought, heat and their combination stress. *Front. Plant Sci.* 6: 298.

Huang ZA, Jiang DA, Yang Y, Sun ZW, Jin SH (2004). Effects of nitrogen deficiency on gas exchange, chlorophyll fluorescence, and antioxidant enzymes in leaves of rice plants. *Photosynthetica* 42: 357–364.

Hungria M, Vargas MAT (2000). Environmental factors affecting N2 fixation in grain legumes in the tropics, with an emphasis on Brazil. *Field Crop Res.* 65: 151–164.

Ihsan MZ, Daur I, Alghabari F, Alzamanan S, Rizwan S, Ahmad M, Waqas M, Shafqat W (2019). Heat stress and plant development: Role of sulphur metabolites and management strategies. *Acta Agriculturae Scandinavica*, Section B – *Soil Plant Sci.* 69: 332–342.

Intergovernmental Panel on Climate Change (IPCC) (2014). Mitigation of climate change contribution of working group III. In: Edenhofer O, Pichs-Madruga R, Sokona Y, Farahani E, Kadnar S, Seyboth K, Adler A, Baum I, Brunner S, Eickemeier P, Kriemann B, Savolainen J, Schlömer S, Von Stechow C, Zwickel T, Minx JC (eds), *The Fifth Assessment Report of the Intergovernmental Panel on Climate Change*. Cambridge University Press, Cambridge/New York.

Iqbal M, Hussain I, Liaqat H, Ashraf MA, Rasheed R, Rahman AU (2015). Exogenously applied selenium reduces oxidative stress and induces heat tolerance in spring wheat. *Plant Physiol. Biochem.* 94: 95–103.

Iqbal N, Umar S, Khan NA, Khan MIR (2014). A new perspective of phytohormones in salinity tolerance: regulation of proline metabolism. *Environ. Exp. Bot.* 100: 34–42.

Jagadish SVK, Craufurd P, Shi W, Oane R (2014). A phenotypic marker for quantifying heat stress impact during microsporogenesis in rice (*Oryza sativa* L.). *Funct. Plant Biol.* 41: 48–55.

Janni M, Gulli M, Maestri M, Marmiroli M, Valliyodan M, Nguyen HT, Marmiroli N (2020). Molecular and genetic basis of heat stress responses in crop plants and breeding for increased resilience and productivity. *J Exp. Bot.* 71: 3780–3802.

Jenner CF (1994). Starch synthesis in the kernel of wheat under high-temperature conditions. *Aust. J. Plant Physiol.* 21: 791–806.

Jiang J, Liu X, Liu C, Liu G, Li S, Wang L (2017). Integrating omics and alternative splicing reveals insights into grape response to high-temperature. *Plant Physiol.* 173: 1502–1518.

Johkan M, Oda M, Maruo T, Shinohara Y (2011). Crop production and global warming. In: Casalegno S (ed.), *Global Warming Impacts-Case Studies on the Economy, Human Health, and on Urban and Natural Environments*, 139–152. InTech, Rijeka.

Karagezov TG (2002). Ethylene as a factor in the adaptability of wheat cells to culture. *Proceed Azerb. Nat. Acad. Sci. Biological Sciences Series* 1: 344–354.

Karim MA, Fracheboud Y, Stamp P (1999). Photosynthetic activity of developing leaves of *Zea mays* is less affected by heat stress than that of developed leaves. *Physiol. Plant.* 105: 685–693.

Katiyar-Agarwal S, Agarwal M, Grover A (2003). Heat tolerant basmati rice engineered by over-expression of hsp101. *Plant Mol. Biol.* 51: 677–686.

Kato MC, Hikosaka K, Hirotsu N, Makino A, Hirose T (2003). The excess light energy that is neither utilized in photosynthesis nor dissipated by photoprotective mechanisms determines the rate of photoinactivation in photosystem II. *Plant Cell Physiol.* 44: 318–325.

Kaushal N, Awasthi R, Gupta K, Gaur P, Siddique KH, Nayyar H (2013). Heat-stress-induced reproductive failures in chickpea (*Cicer arietinum*) are associated with impaired sucrose metabolism in leaves and anthers. *Funct. Plant Biol.* 40: 1334–1349.

Kaushal N, Bhandari K, Siddiqui KH, Nayyar H (2016). Food crops face rising temperatures: an overview of responses, adaptive mechanisms, and approaches to improve heat tolerance. *Cogent Food Agric.* 2: 1134380.

Kaushal N, Gupta K, Bhandari K, Kumar S, Thakur P, Nayyar H (2011). Proline induces tolerance in chickpea (*Cicer arietinum* L.) plants by protecting vital enzymes of carbon and antioxidative metabolism. *Physiol. Mol. Biol. Plants* 17: 203–213.

Keller M, Simm S; SPOT-ITN Consortium (2018). The coupling of transcriptome and proteome adaptation during development and heat stress response of tomato pollen. *BMC Genom.* 19: 447.

Khalil U, Ali S, Rimzan S, Rahman KU, Ata-ul-karim ST, Najeeb U, Ahmad MN, Adrees M, Sarwar M, Hussain SM (2018). Role of mineral nutrients in plant growth under extreme temperatures. In: Hasunazzaman M, Fujita M, Oku H, Nahar K, Hawrylak-Nowak B, (eds), *Plant Nutrients and Abiotic Stress Tolerance*, 499–524. Springer Nature, Singapore.

Khan MIR, Asgher M, Khan NA (2013a). Rising temperature in the changing environment: A serious threat to plants. *Climate Change and Environmental Sustainability* 1: 25–36.

Khan MIR, Iqbal N, Masood A, Per TS, Khan NA (2013b). Salicylic acid alleviates adverse effects of heat stress on photosynthesis through changes in proline production and ethylene formation. *Plant Signal. Behav.* 8: 11.

Khan MIR, Fatma M, Per TS, Anjum NA, Khan NA (2015). Salicylic acid-induced abiotic stress tolerance and underlying mechanisms in plants. *Front. Plant Sci.* 6: 462.

Khan MIR, Palakolanu SR, Chopra P, Rajurkar AB, Gupta R, Iqbal N, Maheshwari C (2020). Improving drought tolerance in rice: Ensuring food security through multi-dimensional approaches. *Physiol. Plant.* 1–24.

Khan SA, Singh S (2008). *Abiotic Stress and Plant Responses*. IK International, New Delhi.

Kim K, Portis Jr AR (2005). Temperature dependence of photosynthesis in *Arabidopsis* plants with modifications in Rubisco activase and membrane fluidity. *Plant Cell Physiol.* 46: 522–530.

Klap C, Yeshayahou E, Bolger AM, Arazi T, Gupta SK, Shabtai S, Usadel B, Salts Y, Barg R (2017). Tomato facultative parthenocarpy results from SLAGAMOUS-LIKE 6 loss of function. *Plant Biotechnol. J.* 15: 634–647.

Kumar J, Kant R, Kumar S, Basu PS, Sarker A, Singh NP (2016). Heat tolerance in lentil under field conditions. *Legume Genomics Genet.* 7: 1–11.

Kumar S, Kaur R, Kaur N, Bhandari K, Kaushal N, Gupta K, Nayyar H (2011). Heat stress-induced inhibition in growth and chlorosis in mungbean (*Phaseolus aureus* Roxb.) is partly mitigated by ascorbic acid application and is related to reduction in oxidative stress. *Acta Physiol. Plant* 33: 2091–2101.

Kumar S, Kaushal N, Nayyar H, Gaur P (2012). Abscisic acid induces heat tolerance in chickpea (*Cicer arietinum* L.) seedlings by facilitated accumulation of osmoprotectants. *Acta Physiol. Plant.* 34: 1651–1658.

Kumar S, Thakur P, Kaushal N, Malik JA, Gaur P, Nayyar H (2013). Effect of varying high-temperatures during reproductive growth on reproductive function, oxidative stress and seed yield in chickpea genotypes differing in heat sensitivity. *Arch. Agron. Soil Sci.* 59: 823–843.

Kweku DW, Bismark O, Maxwell A, Desmond KA, Danso KB, Oti-Mensah EA, Adormaa BB. (2018). Greenhouse effect: Greenhouse gases and their impact on global warming. *J. Sci. Res. Rep.* 17(6): 1–9.

Lavania D, Dhingra A, Grover A (2018). Analysis of transactivation potential of rice (*Oryza sativa* L.) heat shock factors. *Planta* 247: 1267–1276.

Lee JH, Hubel A, Schoffl F (1995). Derepression of the activity of genetically engineered heat shock factor causes constitutive synthesis of heat shock proteins and increased thermotolerance in transgenic *Arabidopsis*. *Plant J.* 8: 603–612.

Li S, Zhou X, Chen L, Huang W, Yu D (2010). Functional characterization of *Arabidopsis thaliana* WRKY39 in heat stress. *Mol. Cells* 29: 475–483.

Lin H-H, Lin K-H, Syu J-Y, Tang S-Y, Lo H-F (2016). Physiological and proteomic analysis in two wild tomato lines under waterlogging and high-temperature stress. *J Plant Biochem. Biot*. 25: 87–96.

Liu GT, Wang JF, Cramer G, Dai ZW, Duan W, Xu HG, Wu BH, Fan PG, Wang LJ, Li SH (2012). Transcriptomic analysis of grape (*Vitis vinifera* L.) leaves during and after recovery from heat stress. *BMC Plant Biol*. 12: 174.

Liu H-T, Liu Y-Y, Pan Q-H, Yang H-R, Zhan J-C, Huang W-D (2006). Novel interrelationship between salicylic acid, abscisic acid, and PIP2-specific phospholipase C in heat acclimation-induced thermotolerance in pea leaves. *J. Exp. Bot*. 57: 3337–3347.

Lobell DB, Asner GP (2003). Climate and management contributions to recent trends in US agricultural yields. *Science* 299: 1032–11032.

Lobell DB, Field CB (2007). Global scale climate-Crop yield relationships and the impacts of recent warming. *Environ. Res. Lett*. 2: 014002.

Lobell DB, Schlenker W, Costa-Roberts J (2011). Climate trends and global crop production since 1980. *Science* 333: 616–620.

Ma X, Zhang Q, Zhu Q, Liu W, Chen Y, Qiu R, Wang B, Yang Z, Li H, Lin Y, Xie, Y (2015). A robust CRISPR/Cas9 system for convenient, high efficiency multiplex genome editing in monocot and dicot plants. *Mol. Plant* 8: 1274–1284.

Mackova H, Hronkova M, Dobra J, Tureckova V, Novak O, Lubovska Z, Motyka V, Haisel D, Hajek T, Prasil IT, Gaudinova A, Storchova H, Ge E, Werner T, Schmulling T, Vankova R (2013). Enhanced drought and heat tolerance of tobacco plants with ectopically enhanced cytokinin oxidase/dehydrogenase gene expression. *J. Exp. Bot*. 64: 2805–2815.

Maestri E, Klueva N, Perrotta C, Gulli M, Nguyen HT, Marmiroli N (2002). Molecular genetics of heat tolerance and heat shock proteins in cereals. *Plant Mol. Biol*. 48: 667–681.

Malzahn A, Lowder L, Qi Y (2017). Plant genome editing with TALEN and CRISPR. *Cell Biosci*. 7: 21.

Mangelsen E, Kilian J, Harter K, Jansson C, Wanke D, Sundberg E (2011). Transcriptome analysis of high-temperature stress in develop in barley caryopses: Early stress responses and effects on storage compound biosynthesis. *Mol. Plant*. 4: 97–115.

Mangrauthia SK, Bhogireddy S, Agarwal S, Prasanth VV, Voleti SR, Neelamraju S, Subrahmanyam D (2017). Genome-wide changes in microRNA expression during short and prolonged heat stress and recovery in contrasting rice cultivars. *J. Exp. Bot*. 68: 2399–2412.

Marschner H (1995). *Mineral Nutrition of Higher Plants*, 2nd edn. Academic, New York.

Matsui T, Namuko OS, Ziska LH, Horie T (1997). Effects of high-temperature and CO2 concentration on spikelet sterility in indica rice. *Field Crop Res*. 1: 213–219.

Matsui T, Omasa K, Horie T (2000). High-temperature at flowering inhibits swelling of pollen grains, a driving force for thecae dehiscence in rice (*Oryza Sativa* L.). *Plant Prod. Sci*. 3: 430–434.

Mazorra LM, Nunez M, Echerarria E, Coll F, Sanchez-Blanco MJ (2002). Influence of brassinosteroids and antioxidant enzymes activity of tomato under different temperatures. *Plant Biol*. 45: 593–596.

Meehl GA, Stoker TF, Collins WD, Friedlingstein P, Gaye AT, Gregory JM, Kitoh A, Knutti R, Murphy JM, Noda A, Raper SCB, Watterson IG, Weaver AJ, Zhao ZC (2007). Global climate projections. In: Solomon S, Qin D, Manning M, Chen Z, Marquis M, Averyt KB, Tignor M, Miller HL (eds), *Climate Change*. Cambridge University Press, Cambridge, UK.

Mengel K, Kirkby EA (2001). *Principles of Plant Nutrition*, 5th edn. Kluwer Academic Publishers, Dordrecht, The Netharlands.

Miller PR, McConkey BG, Clayton GW, Brandt SA, Staricka JA, Johnston AM, Lafond GP, Schatz BG, Baltensperger DD, Neill KE (2002). Pulse crop adaptation in the northern Great Plains. *Agron. J*. 94: 261–272.

Mishra N, Saxena P (2009). Effect of salicylic acid on proline metabolism in lentil grown under salinity stress. *Plant Sci*. 177: 181–189.

Mittal D, Madhyastha DA, Grover A (2012). Gene expression analysis in response to low and high-temperature and oxidative stresses in rice: Combination of stresses evokes different transcriptional changes as against stresses applied individually. *Plant Sci*. 197: 102–113.

Mittler R, Finka A, Goloubinoff P (2012). How do plants feel the heat? *Trends Biochem. Sci*. 37: 118–125.

Mizoi J, Shinozaki K, Yamaguchi-Shinozaki K (2012). AP2/ERF transcription factors in plant abiotic stress responses. *Biochim. Biophys. Acta* 1819: 86–96.

Mobin M, Khan MN, Abbas ZK, Ansari HR, Al-Mutairi KA (2017). Significance of sulfur in heat stressed cluster bean (*Cymopsis tetragonoloba* L. Taub) genotypes: Responses of growth, sugar and antioxidative metabolism. *Arch. Agron. Soil Sci*. 63: 288–295.

Morgan PW (1990). Effects of abiotic stresses on plant hormone systems. In: Alscher RG, Cumming JR (eds), *Stress Responses in Plants Adaptation and Acclimation Mechanisms*, 113–146. Wiley-Liss, Inc., New York, NY.

Morita S, Yonermaru J, Takahashi J (2005). Grain growth and endosperm cell size under high night temperature in rice (*Oryza sativa* L.). *Ann. Bot.* 95: 695–701.

Mu Q, Zhang W, Zhang Y, Yan H, Liu K, Matsui T, Tian X, Yang P (2017). iTRAQ-based quantitative proteomics analysis on rice anther responding to high-temperature. *Int. J Mol. Sci.* 18: 1811.

Muhlemann JK, Younts TLB, Muday GK (2018). Flavonols control pollen tube growth and integrity by regulating ROS homeostasis during high-temperature stress. *PNAS, USA* 115, E11188–E11197.

Muller M, Munne-Bosch S (2015). Ethylene response factors: A key regulatory hub in hormone and stress signalling. *Plant Physiol.* 169: 32–41.

Munns R (2005). Genes and salt tolerance: Bringing them together. Tansley review. *New Physiol.* 645–656.

Murakami T, Matsuba S, Funatsuki H, Kawaguchi K, Saruyama H, Tanida M, Sato Y (2004). Overexpression of a small heat shock protein, sHSP17.7, confers both heat tolerance and UV-B resistance to rice plants. *Mol. Breed.* 13: 165–175.

Myllärinen P, Schulman AH, Salovaara H, Poutanen K (1998). The effect of growth temperature on gelatinization properties of barley starch. *Acta Agric. Stand. B.* 48(2): 85–90.

Nakashima K, Ito Y, Yamaguchi-Shinozaki (2009). Transcriptional regulatory networks in response to abiotic stresses in Arabidopsis and grasses. *Plant Physiol.* 149: 188–195.

Navrot N, Rouhier N, Gelhaye E, Jaquot JP (2007). Reactive oxygen species generation and antioxidant systems in plant mitochondria. *Physiol. Plant* 129: 185–195.

Nemeskeri E (2004). Heat tolerance in grain legumes. *Bodenkultur* 55: 3–11.

Ni Z, Li H, Zhao Y, Peng H, Hu Z, Xin M, Sun Q (2017). Genetic improvement of heat tolerance in wheat: Recent progress in understanding the underlying molecular mechanisms. *The Crop J.* 6: 32–41.

Ogweno JO, Song XS, Shi K, Hu WH, Mao WH, Zhou YH, Yu JQ, Nogués S (2002). Brassinosteroids alleviate heat-induced inhibition of photosynthesis by increasing carboxylation efficiency and enhancing antioxidant systems in *Lycopersicon esculentum*. *J. Plant Growth Regul.* 27: 49–57.

Oshino T, Miura S, Kikuchi S, Hamada K, Yamo K, Watanabe M, Higashitani A (2011). Auxin depletion in barley plants under high-temperature conditions represses DNA proliferation in organelles and nuclei via transcriptional alterations. *Plant Cell Environ.* 34: 284–290.

Parankusam S, Bhatnagar-Mathur P, Sharma KK (2017). Heat responsive proteome changes reveal molecular mechanisms underlying heat tolerance in chickpea. *Environ. Exp. Bot.* 141: 132–144.

Peng S, Huang J, Sheehy JE, Laza RC, Visperas RM, Zhong X, Cassman KG (2004). Rice yields decline with higher night temperature from global warming. *Proc. Natl. Acad. Sci. USA* 101: 9971–9975.

Perez-Perez JG, Robles JM, Tovar JC, Botia P (2009). Response to drought and salt stress of lemon 'Fino 49' under field conditions: Water relations, osmotic adjustment and gas exchange. *Sci. Hortic.* 122: 83–90.

Petkova V, Denev ID, Cholakov D, Porjazov I (2007). Field screening for heat tolerant common bean cultivars (*Phaseolus vulgaris* L.) by measuring of chlorophyll florescence induction parameters. *Sci. Hortic.* 111: 101–106.

Piramila BHM, Prabha AL, Nandagopalan V, Stanley AL (2012). Effect of heat treatment on germination, seedling growth and some biochemical parameters of dry seeds of black gram. *Int. J. Pharm. Phytopharmacol. Res.* 1: 194–202.

Popelka JC, Terryn N, Higgins TJV (2004). Gene technology for grain legumes: can it contribute to the food challenge in developing countries? *Plant Sci.* 167: 195–206.

Prasad PVV, Boote KJ, Allen LH, Sheehy JE, Thomas JMG (2006). Species, ecotype and cultivar differences in spikelet fertility and harvest index of rice in response to high-temperature stress. *Field Crops Res.* 95: 398–411.

Priya M, Dhanker OP, Siddique KHM, HanumanthaRao B, Nair RM, Pandey S, Singh S, Varshney RK, Prasad PVV, Nayyar H (2019). Drought and heat stress-related proteins: An update about their functional relevance in imparting stress tolerance in agricultural crops. *Theor. Appl. Genet.* 132: 1607–1638.

Qin D, Wu H, Peng H, Yao Y, Ni Z, Li Z, Zhou C, Sun Q (2008). Heat stress-responsive transcriptome analysis in heat susceptible and tolerant wheat (*Triticum aestivum* L.) by using wheat genome array. *BMC Genomics* 9: 432.

Qu M, Chen G, Bunce JA, Zhu X, Sicher RC (2018). Systematic biology analysis on photosynthetic carbon metabolism of maize leaf following sudden heat shock under elevated CO2. *Sci. Rep.* 8: 7849.

Radwan UAA (2007). Plant water relations, stomatal behaviour, photosynthetic pigments and anatomical characteristics of *Solenostemma agrel* (Del.) hayne under hyper-arid environmental conditions American-Eurasian. *J. Sci. Res.* 2: 80–92.

Rahman MM (2004). Response of wheat genotypes to late seeding heat stress (Doctoral dissertation, MS Thesis). Department of Crop Botany. Bangabandhu Sheikh Mujibur Rahman Agriculture University, Gazipur, Bangladesh.

Rasheed R, Wahid A, Farooq M, Hussain I, Basra SMA (2011). Role of proline and glycinebetaine pretreatments in improving heat tolerance of sprouting sugarcane (*Saccharum* sp.) buds. *Plant Growth Regul.* 65: 35–45.

Reis SP, Lima AM, de Souza CRB (2012). Recent molecular advances in downstream plant responses to abiotic stress. *Int. J. Mol. Sci.* 13: 8628–8647.

Ren C, Bilyeu KD, Beuselinck PR (2009). Composition, vigor, and proteome of mature soybean seeds developed under high-temperature. *Crop Sci.* 49: 1010–1022.

Rio LA del., Sandalio LM, Corpas FJ, Palma JM, Barroso JB (2006). Reactive oxygen species and reactive nitrogen species in peroxisomes. Production, scavenging, and role in cell signaling. *Plant Physiol.* 14: 330–335.

Ristic Z, Bukovnic U, Momčilović I, Fu JM, Prasad PVV (2008). Heat-induced accumulation of chloroplast protein synthesis elongation factor, Ef-Tu, in winter wheat. *Aust. J. Plant Physiol.* 165: 192–202.

Rodríguez M, Canales E, Borrás-Hidalgo O (2005). Molecular aspects of abiotic stress in plants. *Biotechnol. Appl.* 22: 1–10.

Saitoh H (2008). *Ecological and Physiology of Vegetable*. Nousangyoson Bunka Kyoukai, Tokyo.

Sakamoto A, Murata N (2002). The role of glycinebetaine in the protection of plants from stress: Clues from transgenic plants. *Plant Cell Environ.* 25: 163–171.

Sakata T, Oshino T, Miura S, Tomabechi M, Tsunaga Y, Higashitani N, Miyazawa Y, Takahashi H, Watanabe M, Higashitani A (2010). Auxin reserves plant male sterility caused by high-temperatures. *Proc. Natl. Acad. Sci.* 107: 8569–8574.

Salvucci ME, Osteryoung KW, Crafts-Brandner SJ (2001). Thermal denaturation in vitro and in vivo. *Plant Physiol.* 127: 1053–1064.

Sarkar NK, Kim YK, Grover A (2014). Coexpression network analysis associated with call of rice seedlings for encountering heat stress. *Plant Mol. Biol.* 84: 125–143.

Savicka M, Škute N (2010). Effects of high-temperature on malondialdehyde content, superoxide content and growth changes in wheat seedlings (*Triticum aestivum* L.). *Ekologija* 56: 26–33.

Scharf KD, Berberich T, Ebersberger I, Nover L (2012). The plant heat stress transcription factor (Hsf) family: structure, function and evolution. *Biochim. Biophys. Acta* 1819: 104–119.

Sendhil R, Kumara TK, Kandpal A, Kumari B, Mohapatra S (2023). Wheat production, trade, consumption, and stocks: Global trends and prospects. In: *Wheat Science*, 33–55. CRC Press.

Sepú lveda G, Kliewer M (1986). Effect of high-temperature on grapevines (*Vitis vinifera* L.) II: Distribution of soluble sugars. *Am. J. Enol. Viticult.* 37: 20–25.

Shao HB, Chen XY, Chu LY, Zhao XN, Wu GY, Yong B (2006). Investigation on the relationship of proline with wheat anti drought under soil water deficits. *Colloids Surf. B. Bio. Interfaces.* 53: 113–119.

Sharkey TD, Badger MR, Von-Caemmerer S, Andrews TJ (2001). Increased heat sensitivity of photosynthesis in tobacco plants with reduced *Rubisco activase*. *Photosyn. Res.* 67: 147–156.

Shaul O (2002). Magnesium transport and function in plants: The tip of the iceberg. *Biometals* 15: 307–321.

Shinde S, Behpouri A., McElwain, JC, Ng CKY (2015). Genome-wide transcriptomic analysis of the effects of sub-ambient atmospheric oxygen and elevated atmospheric carbon dioxide levels on gametophytes of the moss, *Physcomitrella Patens*. *J. Exp. Bot.* 66: 4001–4012.

Singh DP, Singh BB (2011). Breeding for tolerance to abiotic stresses in mungbean. *J. Food Legumes* 24: 83–90.

Sita K, Sehgal A, HanumanthaRao B, Nair RM, Prasad PVV, Kumar S, Gaur PM, Farooq M, Siddique KHM, Varshney RK, Nayyar H (2017). Food legumes and rising temperatures: effects, adaptive functional mechanisms specific to reproductive growth stage and strategies to improve heat tolerance. *Front. Plant Sci.* 8: 1658.

Srivastava S, Pathak AD, Gupta PS, Shrivastava AK, Srivastava AK (2012). Hydrogen peroxide-scavenging enzymes impart tolerance to high-temperature induced oxidative stress in sugarcane. *J. Environ. Biol.* 33: 657–661.

Stavang JA, Gallego-Bartolomé J, Gómez MD, Yoshida S, Asami T, Olsen JE, García-Martínez JL, Alabadí D, Blázquez MA (2006). Hormonal regulation of temperature-induced growth in *Arabidopsis*. *Plant J.* 60: 589–601.

Stone P (2001). The effects of heat stress on cereal yield and quality. In: *Crop Responses and Adaptations to Temperature Stress*, 243–291. CRC Press.

Stone PJ, Nicolas ME (1995). Effect of timing of heat stress during grain filling on two wheat varieties differing in heat tolerance. I. Grain growth. *Austr. J. Plant Physiol.* 22: 927–934.

Sun T-p, Gubler F (2004). Molecular mechanism of gibberellin signaling in plants. *Annu. Rev. Plant Biol.* 55: 197–223.

Takahashi F, Shinozaki K (2019). Long-distance signaling in plant stress response. *Curr. Opin. Plant Biol.* 47: 106–111.

Takeoka Y, Hiroi K, Kitano H, Wada T (1991). Pistil hyperplasia in rice spikelets as affected by heat stress. *Sex. Plant Reprod.* 4: 39–43.

Tao Z, Chen Y, Li C, Zou J, Yan P, Yuan S, Wu X, Sui P (2016). The causes and impacts for heat stress in spring maize during grain filling in the North China plain – a review. *J. Integr. Agric.* 15: 2677–2687.

Teixeira EI, Fischer G, Velthuizen HV, Walter C, Ewert F (2013). Global hot-spots of heat stress on agricultural crops due to climate change. *Agric. For. Meteorol.* 170: 206–215.

Templer SE, Ammon A, Pscheidt D, Ciobotea O, Schuy C, McCollum C, Sonnewald U, Hanemann A, Förster J, Ordon F, von Korff, M (2017). Metabolite profiling of barley flag leaves under drought and combined heat and drought stress reveals metabolic QTLs for metabolites associated with antioxidant defense. *J. Exp. Bot.* 68: 1697–1713.

Tester RF, South JB, Morrison WR, Ellis RP (1991). The effects of ambient temperature during the grain-filling period on the composition and properties of starch from four barley genotypes. *J. Cereal Sci.* 13: 113–127.

Tewari RK, Kumar P, Sharma PN (2006). Magnesium deficiency induced oxidative stress and antioxidant responses in mulberry plants. *Sci Hortic.* 108: 7–14.

Tewari RK, Kumar P, Tewari N, Srivastava S, Sharma PN (2004). Macronutrient deficiencies and differential antioxidant responses-influence on the activity and expression of superoxide dismutase in maize. *Plant Sci.* 166: 687–694.

Thakur P, Kumar S, Malik JA, Berger JD, Nayyar H (2010). Cold stress effects on reproductive development in grain crops: An overview. *Environ. Exp. Bot.* 67: 429–443.

Thomason K, Babar MA, Erickson JE, Mulvaney M, Beecher C, MacDonald G (2018). Comparative physiological and metabolomics analysis of wheat (*Triticum aestivum* L.) following post-anthesis heat stress. *PLoS One* 13: e0197919.

Toh S, Imamura A, Watanabe A, Nakabayashi K, Okamot M, Jikumaru Y, Hanada A, Aso Y, Ishiyama K, Tamura N, Iuchi S, Kobayashi M, Yamaguchi S, Kamiya Y, Nambara E, Kawakami N (2008). High-temperature-induced abscisic acid biosynthesis and its role in the inhibition of gibberellin action in *Arabidopsis* seeds. *Plant Physiol.* 146: 1368–1385.

Turhan E, Karni L, Aktas H, Deventurero G, Chang DC, Bar-Tal A, Aloni B (2006). Apoplastic anti-oxidants in pepper (*Capsicum annuum* L.) fruit and their relationship to blossom-end rot. *J. Horti. Sci. Biotechnol.* 81: 661–667.

Tzudir L, Bera PS, Chakraborty PK (2014). Impact of temperature on the reproductive development in mungbean (*Vigna radiata*) varieties under different dates of sowing. *Int. J. Bioresour. Stress Manage.* 5: 194–199.

U.S. Environmental Protection Agency (EPA) (2011). A student's guide to global climate change. www.epa.gov

Uchida A, Jagendorf AT, Hibino T, Takabe T, Takabe T (2002). Effects of hydrogen peroxide and nitric oxide on both salt and heat stress tolerance in rice. *Plant Sci.* 163: 515–523.

Vara Prasad PV, Boote KJ, Allen LH, Thomas JM (2002). Effects of elevated temperature and carbon dioxide on seed-set and yield of kidney bean (*Phaseolus vulgaris* L.). *Glob. Change Biol.* 8: 710–721.

Vollenweider P, Gunthardt-Goerg MS (2005). Diagnoses of abiotic and biotic stress factors using the visible symptoms in foliage. *Environ. Pollut.* 137: 455–465.

Wahid A, Close TJ (2007). Expression of de hydrins under heat stress and their relationship with water relations of sugarcane leaves. *Biol. Plant* 51: 104–109.

Wahid A, Gelani S, Ashraf M, Foolad MR (2007). Heat tolerance in plants: An overview. *Environ. Exp. Bot.* 61: 199–223.

Wang LJ, Fan L, Loescher W, Duan W, Liu GJ, Cheng JS, Luo HB, Li SH (2010). Salicylic acid alleviates decreases in photosynthesis under heat stress and accelerates recovery in grapevine leaves. *BMC Plant Biol.* 10: 34.

Wang LJ, Li SH (2006). Salicylic acid-induced heat or cold tolerance in relation to Ca2+ homeostasis and antioxidant systems in young grape plants. *Plant Sci.* 170: 685–694.

Wang W, Vincour B, Shoseyov O, Altman A (2004). Role of plant heat-shock proteins and molecular chaperones in the abiotic stress response. *Trends Plant Sci.* 9: 244–252.

Wang Y, Wang L, Zhou J, Hu S, Chen H, Xiang J, Zhang Y, Zeng Y, Shi Q, Zhu D, Zhang Y (2019). Recent progress on Heat stress of rice at flowering stage. *Rice Sci.* 26: 1–10.

Waraich EA, Ahmad R, Ashraf MY, Saifullah, Ahmad M (2011). Improving agricultural water use efficiency by nutrient management in crop plants. *Acta Agric. Scand. Sect. B Soil Plant Sci.* 61: 291–304.

Waraich EA, Ahmad R, Halim A, Aziz T (2012). Alleviation of temperature stress by nutrient management in crop plants: A review. *J. Soil Sci. Plant Nutr.* 12: 221–244.

Wendehenne DA, Klessig D, Durner J (2001). Nitric oxide: Comparative synthesis and signaling in animal and plant cells. *Trends Plant Sci.* 6: 177–183.

White PJ, Broadley MR (2003). Calcium in plants. *Ann. Bot.* 92: 487–511.

Wise RR, Olson AJ, Schrader SM, Sharkey TD (2004). Electron transport is the functional limitation of photosynthesis in field-grown Pima cotton plants at high-temperature. *Plant Cell Environ.* 27: 717–724.

Wu X, Gong F, Yang L, Hu X, Tai F, Wang W (2014). Proteomic analysis reveals differential accumulation of small heat shock proteins and late embryogenesis abundant proteins between ABA-deficient mutant vp5 seeds and wild-type Vp5 seeds in maize. *Front. Plant Sci.* 5: 801.

Xia XJ, Zhou YH, Shi K, Zhou J, Foyer CH, Yu JQ (2015). Interplay between reactive oxygen species and hormones in the control of plant development and stress tolerance. *J. Exp. Bot.* 66: 2839–2856.

Xing J, Xu Y, Tian J, Gianfagna T, Huang B (2009). Suppression of shade-or heat-induced leaf senescence in creeping bentgrass through transformation with the ipt gene for cytokinin synthesis. *J. Am. Soc. Horti. Sci.* 134: 602–609.

Xu W, Cai SY, Zhang Y, Wang Y, Ahammed GJ, Xia XJ, Shi K, Zhou YH, Yu JQ, Reiter RJ, Zhou J (2016). Melatonin enhances thermotolerance by promoting cellular protein protection in tomato plants. *J. Pineal Res.* 61, 457–469.

Xue GP, Sadat S, Drenth J, McIntyre CL (2014). The heat shock factor family from *Triticum aestivum* in response to heat and other major abiotic stresses and their role in regulation of heat shock protein genes. *J. Exp. Bot.* 65: 539–557.

Xue T, Hartikainen H, Piironen V (2001). Antioxidative and growth-promoting effect of selenium on senescing lettuce. *Plant Soil.* 237: 55–61.

Yadav AS (2008). Resistance to sodium azide and its effect on symbiotic effectivity of rhizobial strains: A review. *Natl. J. Pl. Improv.* 10: 75–79.

Yadav AS, Nehra K (2013). Selection/isolation of high-temperature tolerant strains of Rhizobium for management of high-temperature stress on Rhizobium-legume symbiosis. *Int. J. Microb. Resour. Technol.* 2: 47–57.

Yamamoto Y, Aminaka R, Yoshioka M, Khatoon M, Komayama K, Takenaka D, Yamashita A, Nijo N, Inagawa K, Morita N, Sasaki T (2008). Quality control of photosystem II: Impact of light and heat stresses. *Photosynth. Res.* 98: 589–608.

Yang H, Huang T, Ding M, Lu D, Lu W (2017). High-temperature during grain filling impacts on leaf senescence in Waxy Maize. *Agron. J.* 109: 906–916.

Yang X, Liang Z, Lu C (2005). Genetic engineering of the biosynthesis of glycine betaine enhances photosynthesis against high-temperature stress in transgenic tobacco plants. *Plant Physiol.* 138: 2299–2309.

Yang X, Tang X, Chen B, Tian Z, Zhong H (2013). Impacts of heat stress on wheat yield due to climatic warming in China. *Prog. Geogr.* 32: 1771–1779.

Yang X, Zhu W, Zhang H, Liu N, Tian S (2016). Heat shock factors in tomatoes: Genome-wide identification, phylogenetic analysis and expression profiling under development and heat stress. *PeerJ* 4: e1961.

Yeh CH, Kaplinsky NJ, Hu C, Charng YY (2012). Some like it hot, some like it warm: Phenotyping to explore thermotolerance diversity. *Plant Sci.* 195: 10–23.

Yin X, Kropff MJ, Ellis RH (1996). Rice flowering in response to diurnal temperature ampliture. *Field Crops Res.* 48: 1–9.

Young LW, Wilen RW, Bonham-Smith PC (2004). High-temperature stress of *Brassica napus* during flowering reduces micro- and mega-gametophyte fertility, induces fruit abortion, and disrupts seed production. *J. Exp. Bot.* 55: 485–495.

Yuan ZC, Haudecoeur E, Faure D, Kerr KF, Nester EW (2008). Comparative transcriptome analysis of *Agrobacterium tumefaciens* in response to plant signal salicylic acid, indole-3-acetic acid and gamma-amino butyric acid reveals signalling cross-talk and *Agrobacterium*-plant co-evolution. *Cell Microbiol.* 10: 2339–2354.

Zafar SA, Zaidi SS-A, Gaba Y, Singla-Pareek SL, Dhanker OP, Li X, Mansoor S, Pareek A (2020). Engineering abiotic stress tolerance via CRISPR-cas mediated genome editing. *J. Exp. Bot.* 71: 470–479.

Zandalinas SI, Mittler R, Balfagón D, Arbona V, Gómez-Cadenas A (2018). Plant adaptations to the combination of drought and high-temperatures. *Physiol. Plant* 162: 2–12.

Zhang J, Nguyen HT, Blum A (1999). Genetic analysis of osmotic adjustment in crop plants. *J. Exp. Bot.* 50: 291–302.

Zhang ZS, Li G, Gao HY, Liu P, Yang C, Meng XL, Meng QW (2013). Changes of photochemistry activity during leaf senescence of leaves in stay green and quick-leaf-senescence inbred lines of maize. *Acta Agron. Sin.* 39: 93–100.

Zhao D, Oosterhuis D (2002). Cotton carbon exchange, non-structural carbohydrates, and boron distribution in tissues during development of boron deficiency. *Field Crops Res*. 78: 75–87.

Zhao H-J, Zhao X-J, Ma P-F, Wang Y-X, Hu W-W, Li L-H, Zhao Y-D (2011). Effects of Salicylic acid on protein kinase activity and chloroplast D1 protein degradation in wheat leaves subjected to heat and high light stress. *Acta Ecol. Sin*. 31: 259–263.

Zhu XC, Song FB, Liu SQ (2011). Effects of arbuscular mycorrhizal fungus on photosynthesis and water status of maize under high-temperature stress. *Plant Soil* 346: 189–199.

Zhu Z, Gerendas J, Bendixen R, Schinner K, Tabrizi H, Sattelmacher B, Hansen UP (2000). Difference tolerance to light stress in $N0_3^-$ and NH_4^+- grown *Phaseolus vulgaris* L. *Plant Biol*. 2: 558–570.

Zinn KE, Tunc-Ozdemir M, Harber JF (2010). Temperature stress and plant sexual reproduction: Uncovering the weakest links. *J. Exp. Bot*. 61: 1959–1968.

Index

Note: Page numbers in *italics* indicate a figure and page numbers in **bold** indicate a table on the corresponding page.

D

E

F

G

N

O

P

Q

R

T

U

V

W

X

Z

Kolibri 360 GmbH
Ferdinand-Jühlke-Straße 7
99095 Erfurt, Deutschland
produktsicherheit@kolibri360.de